American Seacoast Forts

A Directory with Period Military Maps 1890-1950

Volume 1

North Atlantic Coast: Portland to New York

PREPARED BY

TERRANCE C. MCGOVERN
MARK A. BERHOW
GLEN M. WILLIFORD

Published by the Coast Defense Study Group Press
2025

PLEASE DIRECT ANY COMMENTS OR CORRECTIONS TO THE PUBLISHER - INFO@CDSG.ORG

IBSN 978-0-9748167-6-0 (Hardcover B&W)
LIBRARY OF CONGRESS CATALOG CARD NUMBER 2025941598

Library of Congress Cataloging-in-Publication data
American Seacoast Forts: A Directory / Terry McGovern, Mark Berhow and Glen Williford
p. cm.
Includes bibliographical references and index.
Library of Congress Control Number: 2025941598
ISBN 978-0-9748167-6-0 (h.c.)
1. Military History, 2. Artillery I. Terry McGovern, Mark Berhow and Glen Williford

First Edition: August 2025
Printed in the USA by Ingram Spark

Cover Photos
Front cover: Fort Strong, Long Island, Boston, Massachusetts (Terry McGovern)

Rear cover (clockwisw from upper left): Map of the Entrances to Portland Harbor ME 1915 (National Archives); Battery J.M.K. Davis on Great Gull Island, NY (Terry McGovern); Fort Hamilton, NY 1924 (National Archives); Fort Duval, MA 1924 (National Archives)

THE COAST DEFENSE STUDY GROUP, INC.
CDSG.ORG

The Coast Defense Study Group, Inc. (CDSG) is a tax-exempt corporation dedicated to the study of seacoast fortifications. The purposes of the CDSG include educational research and documentation, preservation and interpretation of historic sites, and assistance to other organizations dedicated to the preservation and interpretation of coast defense sites. Membership is open to any person or organization interested in the study or history of coast defenses and fortifications. Membership in the CDSG will allow you to attend annual conferences, special tours, and receive quarterly newsletter and journal. To find our more about the CDSG, please visit the CDSG website at **cdsg.org.**

Acknowledgments

This book is dedicated to three key members of the CDSG who began research on this subject in the 1970s and provided much of the initial information for those that followed in this study. Robert Zink (member number 1) and Glen Williford (member number 2) were both dedicated in their studies and unfailing supportive to those that sought more information on American seacoast defenses. The third key CDSG member was Bolling W. Smith who did a large amount of research work in obtaining copies of documents from the National Archives of which many were used in this compilation.

Sighting a 10-inch gun at Fort Standish, Massachusetts
(Leslie Jones, Digital Commonwealth, Massachusetts Online Collection, Boston Public Library)

The CDSG Press

Coast Defense Study Group Press is a division of Coast Defense Study Group (cdsg.org), which publishes books of historical interest, especially concerning seacoast fortifications. The CDSG Press also offers an ever-expanding number of key reprints reports and manuals in electronic PDF format on compact disks. To order these books and other **CDSG Press** publications, please access the **CDSG Press** pages on the **CDSG web site** at **cdsg.org.**

CDSG Press is interested in new titles, especially those dealing with fortifications, please contact Terry McGovern at 703/538-5403 or at tcmcgovern@att. net if you have a title that you are seeking to have published. Visit www.cdsg.org/press.

Under the CDSG Press label, our organization has published:

Notes on Seacoast Fortification Construction by Col. Eben E. Winslow, 1920, 428 pp. 1994 reprint HC with bound drawings
Seacoast Artillery Weapons Technical Manual (TM) 9-210 by U.S. War Dept. 1944, 202 pp. 1995 reprint PB
The Service of Coast Artillery by F. Hines & F. Ward, 1910, 736 pp. 1997 reprint HC
Permanent Fortifications & Sea-Coast Defenses by U.S. Congress, 1862, 544 pp. 1998 reprint HC
American Coast Artillery Material Ordnance Dept. Doc#2042 by U.S. War Dept., 1922, 528 pp., 2001 Reprint HC
American Seacoast Defenses: A Reference Guide (3rd Edition) by Mark A. Berhow, (2015) 732 pp. HC
The Endicott & Taft Board Reports, reprint of original reports of 1886 and 1905 by U.S. Congress, 525 pp. 2007 HC
Artillerists and Engineers: The Beginnings of US Fortifications 1794-1815 by Col. Wade, U.S. Army, 226 pp. 2011 PB
World War II Harbor Defenses of San Diego by Commander (Ret.) Everett, U.S. Navy, 226 pp. 2020 HC

The CDSG Presss offers these Hole in the Head Press Books

Artillery at the Golden Gate by Brian B. Chin (Hole-in-the-Head Press), 176 pp. 2019 PB
Fort Baker Through the Years by Kristin L Baron and John A. Martini (Hole-in-the-Head Press), 99 pp. 2013 PB
Rings of Supersonic Steel (3rd Edition) by Mark Morgan and Mark Berhow (Hole-in-the-Head Press), 358 pp. 2010 PB
The Last Missile Site by Stephen Hailer and John A. Martini (Hole-in-the-Head Press), 158 pp. 2010 PB
To Defend and Deter by John Lonnquest and David Winker (Hole-in-the-Head Press), 432 pp. 2014 PB

The CDSG Press
1700 Oak Lane
McLean, VA 22101-3322 USA

AMERICAN SEACOAST FORTS
A DIRECTORY OF AMERICAN SEACOAST DEFENSES 1890-1950

Table of Contents

Introduction ... 13

Chronology ... 15
• Key Events

Design and Function of American Seacoast Forts ... 16
• Historic Development of Coast Defenses during the Modern Era ... 16
    The Endicott Program (1885-1904) ... 16
    The Taft Program (1905-1916) ... 19
    The World War I Program and the Interwar Period (1917-1939) ... 22
    The 1940 Modernization Program and World War II (1940-1950) ... 23
• Coastal Defense Objective ... 25
• Coast Defense Armaments and Equipment ... 27
• Coast Defense Organization and Fire Control ... 38
• A Typical U.S. Coast Artillery Fort ... 39
• Coast Artillery Garrison - Daily Life ... 41
• Aftermath and Today ... 43

Period Military Maps ... 43
• Confidential Blueprint Series Maps (1915-37) ... 46
• WW II-Era Maps (1933-46) ... 45
• Glossary of Terms used in Maps ... 47
• Symbols and Abbreviations (1921) ... 49
• Symbols and Abbreviations (1940) ... 52

Comments on what is included in this Directory ... 58

U.S. Coast Artillery 1890-1945 Harbor Defense Locations

Directory to American Seacoast Forts 1885-1950 in 4 volumes
    (21 Continental US Harbor Defenses and Overseas Harbor Defenses)
• Volume 1 North Atlantic Coast: New England – Long Island Sound - New York
• Volume 2 Mid-Atlantic, South Atlantic, and Gulf Coasts: Chesapeake Bay to Galveston
• Volume 3 Pacific Coast San Diego to Puget Sound
• Volume 4 Alaska and Overseas Bases: Hawaii - Philippines - Panama - Caribbean -Newfoundland-Bermuda

VOLUME 1: THE NORTH ATLANTIC COAST: PORTLAND TO NEW YORK

Portland & Kennebec River, ME—Fort and Gun Battery Descriptions ... 59

    FORT BALDWIN ... 61
        Batteries   Hardman; Hawley; Cogan
    Jewell Island Military Reservation ... 56
        Batteries   #202; AMTB #267; AMTB #268
    LONG ISLAND Military Reservation ... 68
        Batteries   AMTB #266; AMTB #267
    Bailey's Island Military Reservation ... 70
        Battery   AMTB #270
    Great Chebeague Island Military Reservation ... 70
        Batteries   AMTB #269
    PEAKS ISLAND Military Reservation ... 70
        Batteries   Steele (#102); Cravens (#203); not named;
                AMTB #263; AMTB #264
    FORT LYON ... 74
        Batteries   Bayard; Abbot
    FORT McKINLEY ... 79
        Batteries   Berry; Ingalls; Thompson; Honeycutt; Weymouth
                Acker; Carpenter; Farry; Ramsey
    FORT LEVETT ... 89
        Batteries   Bowdoin; Kendrick; Ferguson; Daniels; Foote;
                AMTB #962
    FORT PREBLE ... 94
        Batteries   Chase-Kearny; Rivardi; Mason; not named
    FORT WILLIAMS ... 99
        Batteries   Sullivan; DeHart; Blair;; Garesche; Hobart; Keyes
                AMTB #961
    Cape Elizabeth Military Reservation ... 105
        Batteries   #101 (planned); #201

    Portland—WWII Program site list ... 109

Portsmouth, NH—Fort and Gun Battery Descriptions ... 110
    FORT FOSTER ... 110
        Batteries   Bohlen; Chapin; #205; AMTB #2
    FORT CONSTITUTION ... 115
        Batteries   Farnsworth; Hackleman
    FORT STARK ... 119
        Batteries   Hunter; Kirk; Lytle; Hays
    Camp Langdon ... 123
    Fort Dearborn ... 123
        Batteries   Seaman (#103); #204

Pulpit Rock Military Reservation ... 126
    Battery   AMTB #1

Portsmouth—WWII Program site list ... 128

Boston, MA—Fort and Gun Battery Descriptions ... 129
  East Point Military Reservation Nahant ... 131
    Batteries   Murphy (#104); #206
  FORT RUCKMAN ... 134
    Battery   Gardner
  FORT BANKS ... 138
    Battery   Lincoln-Kellogg
  FORT HEATH ... 144
    Batteries   Winthrop; AMTB #945
  Fort Dawes/Deer Island ... 148
    Batteries   #105; #207; Taylor; AMTB #944
  Calf Island ... 153
  Outer Brewster Island ... 153
    Battery   Jewell (#209)
  Great Brewster Island ... 155
    Battery   AMTB #944
  FORT STANDISH ... 157
    Batteries   Burbeck-Morris; Terrill; Whipple; Vincent
                Weir; Williams; AMTB #943
  FORT STRONG ... 162
    Batteries   Ward-Hitchcock; Drum; Basinger; Smyth
                Stevens; Taylor
  FORT WARREN ... 169
    Batteries   Jack Adams; Bartlett; Stevenson; Plunkett; Lowell
  Point Allerton ... 175
  FORT REVERE ... 176
    Batteries   Ripley; Field; Sanders-Pope; AMTB #941
  FORT DUVALL ... 181
    Battery   Long
  FORT ANDREWS ... 185
    Batteries   Whitman; Cushing; McCook; Rice; Bumpus
  Strawberry Hill ... 190
  Fourth Cliff Military Reservation ... 190
    Batteries   #106 (planned); #208

Boston—WWII Program site list ... 195

New Bedford, MA—Fort and Gun Battery Descriptions ... 196
  FORT RODMAN ... 196
    Batteries   Barton; Walcott; Cross; Gaston; Craig; Milliken
  Mishaum Point Military Reservation ... 202
    Battery   #210

Barney's Joy Military Reservation ... 202
    Battery   AMTB #931
Cuttyhunk Island Military Reservation ... 202
    Battery   AMTB #932
Nashawena IslandMilitary Reservation ... 202
    Battery   AMTB #933
Butler's PointMilitary Reservation ... 204
    Battery   AMTB #934

New Bedford—WWII Program site list ... 205

Narrangansett Bay, RI—Fort and Gun Battery Descriptions ... 206
    Fort Church ... 206
        Batteries   Gray (#107); Reilly; #212
    FORT ADAMS ... 212
        Batteries   Edgerton-Greene; Reilly; not named; Bankhead;
                    Talbot; Belton
    Brenton Point Military Reservation ... 217
        Battery   AMTB #925
    Mount Prospect   fire control ... 219
    FORT WEATHERILL ... 221
        Batteries   Varnum; Wheaton; Walbach; Dickenson; Zook;
                    Cooke; Crittenden
    FORT GETTY ... 228
        Batteries   Tousard; House; Whiting; Armistead; AMTB #922
    FORT GREBLE ... 233
        Batteries   Sedgwick; Hale; Mitchell; Ogden
    Fort Burnside ... 241
        Batteries   #110 (planned); #213; New Whiting
    FORT KEARNY ... 242
        Batteries   French; Cram; Armistead
    Fort Varnum ... 246
        Batteries   House; Armestead; AMTB #921
    Fort Greene ... 248
        Batteries   Hamilton (#108); #109; #211

Narragansett Bay—WWII Program site list ... 254

Long Island Sound, CT & NY—Fort and Gun Battery Descriptions ... 255
    FORT MANSFIELD ... 255
        Batteries   Wooster; Crawford; Connell
    Goshen Point Military Reservation ... 260
        Battery   AMTB #914
    Pine Island Military Reservation ... 260
        Battery   AMTB #915

FORT H.G. WRIGHT ... 260
    Batteries   Barlow; Butterfield; Clinton; Dynamite; Dutton; Hamilton;
        Marcy; Hoffman; Hoppock; #215; AMTB #913
North Hill, Fishers Island ... 270
    Batteries   Hackleman
Wilderness Point Military Reservation, Fishers Island ... 271
    Batteries   #111 (planned); #214
East Point, Fishers Island ... 272
    Battery   AMTB #913
FORT MICHIE ... 272
    Batteries   Palmer; North; J.M.K. Davis; Benjamin; Maitland
        Pasco; AMTB #912
FORT TERRY ... 282
    Batteries   Steele; Stoneman; Bradford; Dimick; Floyd; Kelly
        Campbell; Eldridge; Greble; Hagner; Dalliba
        #114 (planned); #217
        AMTB #911
FORT TYLER ... 294
    Battery   Smith
Camp Hero ... 299
    Batteries   Dunn (#113); #112; #216

Long Island Sound—WWII Program site list ... 303

New York, East —Fort and Gun Battery Descriptions ... 304
    FORT SLOCUM ... 304
        Batteries   Haskin-Overton; Kinney; Fraser; not named
    FORT SCHUYLER ... 308
        Batteries   Gansevoort; Hazzard; Beecher; Bell
    FORT TOTTEN ... 311
        Batteries   King; Mahan; Graham; Sumner; Stuart; Baker; Burnes

West Point Military Academy practice batteries ... 316

New York, NY & NJ—Fort and Gun Battery Descriptions ... 317
    Jamacia Bay M.R. ... 317
        Battery   #117 (planned)
    FORT TILDEN ... 319
        Batteries   Harris; Ferguson (East); Kessler (West); #220
    Norton Point ... 325
        Batteries   Catlin; AMTB #19
    Rockaway Point ... 326
        Batteries   AMTB #20; AMTB #21
    Swinbourne Island ... 326
        Battery   AMTB #12

FORT HAMILTON   ... 328
    Batteries   Brown; Doubleday; Neary; Piper; Gillmore; Spear;
                           Burke; Johnston; Livingston; Mendenhall; Griffin
FORT WADSWORTH   ... 335
    Batteries   Ayres; Dix; Hudson; Richmond; Barry; Upton; Duane;
                           Mills; Bacon; Catlin; Turnbull
                           #115 (planned); #218; AMTB #14
  Fort Jay   ... 346
FORT HANCOCK   ... 348
    Batteries   Potter; McCook-Reynolds; Alexander; Bloomfield; Richardson;
                           Halleck; Granger; Arrowsmith; Gunnison; Peck; Engle;
                           Morris; Urmiston; Kingman; Mills
                           AMTB #6; AMTB #7; AMTB #8
  Highlands Military Reservation (Navesink)   ... 368
    Batteries   Lewis (#116); #219

  New York—WWII Program site list   ... 375

VOLUME 2: THE MID-ATLANTIC, SOUTH ATLANTIC, AND GULF COASTS: DELAWARE BAY TO GALVESTON

Delaware Bay—Fort and Gun Battery Descriptions
Chesapeake Bay (Baltimore, Potomac River, James River, Hampton Roads)
Baltimore, MD—Fort and Gun Battery Descriptions
Potomac River MD & VA—Fort and Gun Battery Descriptions
Entrance to the Chesapeake Bay, VA—Fort and Gun Battery Descriptions
Cape Fear River, NC—Fort and Gun Battery Descriptions
Charleston, SC—Fort and Gun Battery Descriptions
Port Royal Sound, SC—Fort and Gun Battery Descriptions
Savannah, GA—Fort and Gun Battery Descriptions
St. Mary's River - Fort Clinch—Fort and Gun Battery Descriptions
St. John's River, FL—Fort and Gun Battery Descriptions
Key West, FL—Fort and Gun Battery Descriptions
Tampa Bay, FL—Fort and Gun Battery Descriptions
Pensacola, FL—Fort and Gun Battery Descriptions
Mobile Bay, AL—Fort and Gun Battery Descriptions
Mississippi River, LA—Fort and Gun Battery Descriptions
Galveston, TX—Fort and Gun Battery Descriptions

VOLUME 3: THE PACIFIC COAST
SAN DIEGO TO PUGET SOUND

San Diego, CA—Fort and Gun Battery Descriptions
Los Angeles, CA—Fort and Gun Battery Descriptions
San Francisco, CA—Fort and Gun Battery Descriptions
Columbia River, OR & WA—Fort and Gun Battery Descriptions
Willapa Bay, WA—Fort and Gun Battery Descriptions
Greys Harbor, WA—Fort and Gun Battery Descriptions
Puget Sound, WA—Fort and Gun Battery Descriptions

VOLUME 4: ALASKA AND THE OVERSEAS BASES

The Harbor Defenses of Sitka, Alaska—WWII Program sites
The Harbor Defenses of Seward, Alaska—WWII Program sites
The Harbor Defenses of Kodiak, Alaska—WWII Program sites
The Harbor Defenses of Dutch Harbor, Alaska—WWII Program sites

The Harbor Defenses of Honolulu—Fort and Gun Battery Descriptions
The Harbor Defenses of Pearl Harbor—Fort and Gun Battery Descriptions
The Harbor Defenses of Kaneohe Bay and the North Shore of Oahu—Fort and Gun Battery Descriptions

The Harbor Defenses of Manila Bay—Fort and Gun Battery Descriptions
The Harbor Defenses of Subic Bay—Fort and Gun Battery Descriptions

The Harbor Defeses of Cristobal—Panama Canal Zone Atlantic side—
    Fort and Gun Battery Descriptions
The Harbor Defenses of Balboa—Panama Canal Zone Pacific side—Fort and Gun Battery Descriptions

The Harbor Defenses of Vieques Sound, Puerto Rico, Virgin Islands (Roosevelt Roads)
    —WWII Program sites
The Harbor Defenses of San Juan, Puerto Rico—WWII Program sites
Planned Defenses in Guantanamo Bay, Cuba
Planned Defenses in Jamaica
Planned defenses in Trinidad

Harbor Defenses in Newfoundland, Canada—WWII Program sites

Defenses in Bermuda—WWII Program sites

Fort McKinley on Great Diamond Island, Casco Bay, Maine (Terry McGovern)

The ferry landing on Plum Island (Fort Terry) Long Island Sound, New York (Terry McGovern)

INTRODUCTION

The United States has long focused on defending its seacoasts against overseas enemies due to its geopolitical situation with its long coastlines and generally peaceful borders with Canada and Mexico. Earlier American fortification efforts resulted in the First System, the Second System, and the Third System of coastal defenses. The great brick and stone forts built or remodeled during 1820 to 1860 are well known from their military importance and use during the American Civil War. For many years after that great internal conflict most U.S. fortification efforts languished. After 1885, due to the great advances in military technology and America's increasing worldwide economic presence that the United States embarked on a new round of fortification building to protect its shores. The U.S. Army expended much of its limited manpower and resources to protecting America's coast from 1890 to 1950.

The technical development and tactical objectives of the coast defenses of the modern era (1890 to 1950) is a product of America's earlier policies and experiences. Until the advent of air power and the missile age, the defense of the United States has primarily been one of defending our shores from naval attack. Only during the nation's early years did the threat of land invasion exist. Accordingly, the United States relied on the ships of its Navy to provide the first line of defense. Its U.S. Army was called upon to provide the second line of defense by building forts at key points along its coastline to defend major harbors. This defense policy of denying an enemy fleet access to its major harbors and anchorages developed into the array of former coastal fortifications that remain today.

Based on concerns over external threats and on internal politics, the United States government has built coastal fortification in a series of construction programs. After inheriting the remains of the fortifications from the colonial era and the revolutionary war, the first in a series of national fortification construction programs began in 1794 and these programs continued into the late 1940s. For the ease of use in this book, these fortification periods have been organized into distinct groups and have been named the following: First System (1794-1801), Second System (1802-1815), Third System including the American Civil War (1816-1867), 1870s Period (1868-1879), Endicott Program (1885-1904), Taft Program (1905-1916), World War I & Interwar Period (1917-1939), and the 1940 Modernization Program including World War II (1940-1950).

The legacy of these seacoast defenses is a series of concrete structures scattered along America's coastline, many now in public shoreline parks. The nature of fortifications, the fact that they were designed to withstand the pounding of naval artillery, has allowed these massive structures to withstand the attack of both the natural elements and economic development. That these structures are still standing many years after their effective use ended draws our attention to them. They captivate us regardless of whether it's a large brick and stone multi-story structure surrounded by a dry ditch or an odd shaped, concrete structure covered with thick vines or bright graffiti and surrounded by worn fences sporting weathered warning signs. Visitors to our seashores are curious about the nature of these structures. Some of the questions that they ask are: What are these structures? Why are these structures here? When were these structures built? This directory provides answers to many of these questions. Many of these parks have visitor's centers and gift shops selling range of books and other items, but few have any books which explain the fortifications that once existed at that site. This directory will fill that void.

This directory is a companion work to the CDSG Press's *American Seacoast Defenses – A Reference Guide* by Mark Berhow. This directory is to aid students of American seacoast fortifications to locate and visit the key defenses of the "Modern Era" of coastal defense (1890 to 1950), which is defined by its use of concrete, steel, and breech-loading rifles. The following pages provide a brief review of the function and history of the development of coastal fortifications in the United States. This history reflects the politics of changing external threats to our nation and rapid advancement of military technology. The directory's focus is a guide to the "Modern Era" of fortifications along the Atlantic, Pacific, and Gulf coasts, as well as U.S. overseas bases by providing key maps, plans, photographs, and short description of their history.

The directory is organized into several sections. The first section is brief history of the modern era of American coastal defenses, including background on the U.S. coast artillery material, organization, armament, design of military reservations, and garrison life. The second section is a brief description of the history and its current status (ownership, public access, remaining assets, things to see, etc.) of the major military reservations that had seacoast artillery and a short history of each major concrete gun battery. This history includes each battery is described as to its rational, authorization, construction and transfer dates, engineering cost, naming citation, armament, service history, ultimate disarming, and current status. These battery histories do not include railway and mobile artillery sites, including those with Panama mounts. Also excluded are temporary batteries, especially those using loaned naval guns, and anti-aircraft batteries that did not also serve a seacoast role. The directory is organized by Harbor Defense around the United States clockwise from Portland Maine to the Puget Sound in Washington State, followed by the Alaskan defenses of World War II, and the defenses in Hawaii, the Philippines, Panama, the Caribbean, Newfoundland and Bermuda.

Supporting the directory of defense sites is a compilation of maps for all the harbor defense reservations utilized during the period of 1900 to 1946. The map collection includes general maps of the location of elements (sites) for each defended harbor and the individual location site maps showing buildings, gun emplacements, fire control stations and other elements. Each Harbor Defense section has an overall 1920s-30s period map of the defense sites and a selected set of Confidential Blueprint Maps for each military reservation.

The maps are arranged more or less in order from the south to the north. A series of map symbol and abbreviation keys from 1921 and 1945 are included in the introduction. Some harbors (Baltimore, Potomac River, Cape Fear, Port Royal, Savannah, Tampa, Mobile, and the Mississippi River) did not receive new defenses during World War II.

12-inch Rifle on a barbette mount, Battery Godfrey, Fort Winfield Scott, California,
(Golden Gate National Recreation Area Collection, NPS)

CHRONOLOGY
Key events during the Modern Era of American coast defenses

1875 – Funding for new construction of coast defenses is stopped by the U.S. Congress

1883 – The U.S. Navy begins the first new construction program since the Civil War

1885 – President Cleveland appoints a joint army, navy, and civilian board headed by the Secretary of War, William Endicott, to evaluate the threats and needs for U.S. coastal defenses (Endicott Board)

1886 – The Endicott Board reports on state of the U.S. defenses and recommends a $126 million construction program of breech-loading cannons and mortars, floating batteries, warships, and submarine mines in 29 locations around the nation

1888 – Congress creates the Board of Ordnance and Fortifications to test weapons and implement the Endicott Program
– Dynamite guns were developed to fire high explosive shells using compressed air

1890 – Congress approves funding for the construction of first Endicott Program batteries

1892 – The first Endicott Program battery is completed (Gun Lift Battery Potter)

1893 – First group of controlled mine casemates are completed

1894 – Buffington-Crozier disappearing carriage for 8-inch and 10-inch guns developed

1896 – Development of the Buffington-Crozier disappearing carriage for 12-inch guns

1898 – The Spanish-American War – 150 coast artillery pieces mounted
– U.S. adds the Philippines, Guam, Puerto Rico as colonies; establishes military bases in Cuba; annexes Hawaiian Is.

1899 – 288 heavy coast artillery guns, 154 rapid-fire guns, and 312 mortars have been mounted

1901 – Reorganization of U.S. Army artillery corps to 30 batteries of field artillery and 126 companies of coast artillery

1902 – Work begins on the fortifications of Corregidor, Philippines

1904 – Work begins on the Panama Canal
– First specially built mine planters constructed

1905 – President Roosevelt appointed a joint army, navy, and civilian board headed by the Secretary of War William Taft, to review the Endicott Program and to bring it up to date

1906 – The Taft Board reports on state of the U.S. defenses and recommends improvement in existing defenses by adding searchlights, electrification of defenses, and a modern system of fire control, as well as new defenses for newly acquired overseas bases

1907 – Establishment of the separate U.S. Army Coast Artillery Corps

1914 –World War I begins; Panama Canal opens

1915 – Report of the Board of Review on the coast defenses of the U.S., Panama Canal, and the Insular Possessions

1916 – First Coast Artillery anti-aircraft units formed

1917 – U.S. enters the World War II
– Construction begins on the first long-range barbette batteries using existing 12-inch gun barrels

1918 – End of the World War I

1920 – Construction of the first Panama mount for 155m GPF guns in the Canal Zone

1922 – Washington Naval Treaty limits naval construction and Pacific fortifications
– 16-inch gun and howitzer barbette batteries are constructed

1925 – Ten U.S. Harbor Defenses on active status and 15 are on caretaker status

1937 – Construction begins on first 16-inch casemated gun battery (Battery Davis)

1939 – Outbreak of World War II; U.S. Coast Artillery Corps has 4,200 troops

1940 – Congress approves the 1940 Modernization Program for 19 harbors in the U.S.
– U.S. draft begins, coast artillery units brought up to wartime strength, national guard units federalized

1941 – U.S. enters the World War II
– Establishment of the Harbor Entrance Control Posts (HECP)

1942 – U.S. Coast Artillery Corps has 70,000 troops

1945 – End of the World War II

1948 – All construction efforts cease, the coast defenses are abandoned, and armament salvaged

1950 – Disestablishment of the U.S. Coast Artillery Corps; remaining units reunited with Field Artillery

DESIGN AND FUNCTION OF AMERICAN SEACOAST DEFENSES IN THE MODERN ERA 1890-1950

Historical Development of American Coast Defenses during the Modern Era

The key development that led to the new American coast defense era was the development of new heavy rifled breech-loading guns that had a longer range, were more accurate and delivered a heavier projectile than the muzzle-loading smoothbore cannons of the Civil War. These new guns were made of high-quality steel that were lighter and stronger, which took advantage of new propellants that replaced gunpowder. Equally important was the development of effective breech mechanisms that could withstand the high pressures and temperatures generated by the new guns and allow for the gun to be loaded from the rear instead of the muzzle, which increased the rate of fire and allowed for improved protection of the gun crew. The new guns and mortars could accurately fire projectiles at effective ranges that were two to three times farther than the muzzle-loading smoothbore cannons used during the American Civil War. These developments coincided with the building of the new steel naval vessels that featured these new big guns starting in 1875. However, between 1875 and 1890 the U.S. Congress did not appropriate any funds for the construction of new coastal fortifications.

Seacoast Defenses built after the Endicott Board Report 1886-1904

As U.S. coastal fortifications were allowed to deteriorate in 1870s and early 1880s, new steam powered, ocean navigating iron warships were being built by foreign navies. As the U.S. Navy embarked on its new construction program, it required protected bases for its operations. The military began to lobby to overhaul the obsolete existing defenses. In 1885, a board was created by U.S. Congress to examine and report upon the state of U.S. coastal defenses. The board headed by the U.S. Secretary of War, William C. Endicott, was comprised of four officers from U.S. Army, two officers from the U.S. Navy, and two civilians. This joint board made an extensive study of fortifications, type of armament, and defense that would be needed, by evaluating current European developments. In 1886, the Endicott Board published its recommendations for new coastal fortifications to be built at 29 key harbors, along with floating batteries, torpedo boats, and submarine mines. The board's original plan called for over 1,300 guns and mortars of 8-inch or larger of the newest design to be installed. The costs of board's recommendations were estimated to be $126 million dollars (in 1886-dollar value). While the U.S. Congress took no immediate action on the board's report, the estimates provided in the report would be cited for the next 20 years as a measure of the construction progress of this new generation of U.S. coastal fortifications.

In 1888, Congress established an U.S. Army board for ordnance and fortifications who as charged with testing new weapons and to design new coastal fortifications. In 1890 Congress made the first appropriations for the first new construction of coastal fortifications in 16 years with an initial funding of $1.2 million dollars. This funding was for the first of new defenses: a 12-inch barbette battery at San Francisco; an 8-inch disappearing battery at New York Harbor; a 12-inch gun lift battery at Sandy Hook, New Jersey; and for 12-inch mortar batteries at Sandy Hook and at the Presidio, San Francisco. The design of these new coastal batteries would set the pattern of coastal defenses that would be duplicated at all of America's major harbors, and this was the beginning of what would become known as the "Endicott Program" of American seacoast defenses.

The designs used for the Endicott Program coastal fortifications demonstrated the shift in importance from the large multi-tiered multi-gunned "fortresses" to weapons emplaced in dispersed concrete "batteries" protected by earthen embankments. The "fort" became a defined reservation of land that contained guns of a range of calibers, along with the housing for the men required to man these defenses, and supply and maintenance buildings. The weapons were grouped into batteries containing from one to sixteen guns. The batteries were located along the shoreline to maximize their range and field of fire and were designed to blend into the landscape as not to be seen from the sea. The armament of these batteries ranged from weapons to engage enemy capital ships to small-caliber rapid fire guns to knock out fast moving torpedo boats, as well as to protect the fields of electrically controlled submarine mines from minesweepers. The dominance of the armament during this period is reflected by the dramatic increase in the time and cost in constructing a gun barrel and breech mechanism along with its carriage over that of the weapons of the earlier periods.

The primary weapons of the Endicott Program were 8-inch, 10-inch and 12-inch rifled breech-loading guns, a growth in size that reflected the need to match the increase in the size of opposing naval guns. These guns were mounted on both barbette and disappearing carriages that had a maximum elevation of 15 degrees and range of about seven to eight miles. The relatively unique American "disappearing" carriage allowed the gun to be raised over the parapet by using a counterweight to fire. The energy from the recoil caused the gun to drop back down behind the parapet into the emplacement to be reloaded while being protected from direct fire from attacking warships. These heavy weapons were mounted in large concrete emplacement with thick frontal walls that were in turn protected by many feet of earthen fill. Located below or adjacent to the firing platform were support areas that included the ammunition magazines containing projectiles and powder propellants. About three hundred of these heavy guns were installed around the United States during the Endicott Program in batteries of from one to six guns. It was a less expensive alternative to the armored turret mount favored by several European nations.

10-inch disappearing guns in Battery Hale, Fort Greble, Rhode Island (C.T. Gardner Collection)

The other large caliber weapon installed during the Endicott Program was the short-barreled 12-inch mortar. The mortar was designed to fire a shell in a high arc that descended down onto the lightly armored decks of warships of that era. To increase the opportunity of making a hit, these mortars were emplaced in groups of eight to sixteen mortars in square concrete pits that were protected by earthen hills. The use of these pits would give maximum protection to the mortars and their magazines from the flat trajectory of naval gunfire of the era. About four hundred of these mortars were installed around the United States during the Endicott Program.

The secondary smaller caliber weapons were installed to protect the controlled submarine mine fields from small craft that could sweep paths through the mines for larger warships, and to protect from attack by newly developed fast torpedo boats that could potentially penetrate the harbor and torpedo the

12-inch mortar firing at Battery Alexander, Fort Barry, California (B.W. Smith Collection)

shipping within. These threats called for guns that could be aimed, loaded and fired very rapidly. While not specified by the Endicott Report, several new gun and carriage systems were developed for this role ranging from 3-inch to 6-inch is size. These guns were generally mounted on either disappearing carriages or on pedestal carriages with simple steel shields. The concrete emplacements for these guns had low parapets and magazines below the guns. A rapid-fire battery had between two and six guns per battery. Over five hundred rapid-five weapons were installed during the Endicott Program.

While the use of submarine mines or "torpedoes," as well as channel obstructions or barriers, has a long history in defense of harbors, it was during the Endicott Program that a widespread and a structured use of submarine mines occurred. The U.S. Army developed a system of controlled submarine mines; stationary explosive devices located below the surface of water where ships were likely to pass. The submarine mines used from 1890 to 1930 were the buoyant type (floating but anchored to the sea bottom), though during World War II the buoyant mines were replaced with ground mines (stationed on the sea floor). The mines were only deployed during times of war or for practice, otherwise they were stored disassembled ashore—the mines and their control cables became defective after extensive exposure in the water. The controlled mines were connected to shore by undersea cables and could be exploded by electrical switches from a control board on shore by the soldiers manning the mine defenses when a warship passing over these mines or by direct contact. Controlled mines were usually laid in rows across the key shipping channels to create a group of mines, usually 19, which would cover a space of about 2,000 feet long in water up to 250 feet deep. Several groups of mines were to be deployed to create a field of mines. The U.S. Army Coast Artillery Corps had dedicated units to man the mine planting vessels, fire control stations, mine and cable storage facilities, mine casemates and switchboards, and loading wharves.

The Endicott Program roughly covered a period from 1885 to 1905, and the coast artillery function was a key mission of U.S. Army during this time (and made up a large percentage of total U.S. Army manpower). This also required a more technical trained soldiers to man them which led to the U.S. Army's Artillery branch to be reorganized in 1901 and 1907 to create the U.S. Coast Artillery Corps.

Planting a mine (Stillion Collection NPS, Gulf Shores Natl. Seashore)

Seacoast Defenses built after the Taft Board Report 1906-1916

In 1905, a new National Coast Defense Board headed by U.S. Secretary of War William Howard Taft was organized by President Theodore Roosevelt and charged with reviewing the progress of Endicott Program construction and update it. In the 20 years since the original Endicott Board report was presented, numerous technical and political developments had taken place. The Board, informally known as the "Taft Board" after its chairman, established new cost estimates and its recommendations were primarily concerned with modernization of existing coastal fortifications and adding coastal defenses to the overseas territories gained after the Spanish-American War including Hawaii and the Philippines and other locations.

The modernization of existing defenses included the electrification of lighting, communications, and ammunition handling equipment, both at the batteries and throughout the fort. The early emplacements had loading platforms widened and projectile hoists were installed to improve the rate of fire. The report recommended the use of searchlights for nighttime illumination of harbor entrances. During the Taft Program was the finalized development and implementation of a coordinated system of target information gathering and processing that greatly improved the target accuracy of the major caliber guns and mortars. Up to this time, the aiming of guns at a target had been generally done from each battery with basic sighting instruments and combination of luck and experience. The new system was based on triangulation using two observers with telescopic instruments at separate position finding stations or "base end" stations communicating with the newly developed telephones to a centralized battery plotting room that provided real-time tracking and firing coordinates on a moving target. The battery plotting room personnel would

The 14-inch gun on a disappearing carriage the Taft-era Battary Osgood, Fort MacArthur, California
(Fort MacArthur Museum)

mathematically process this sighting information and other data into aiming instructions that would then be transmitted to each gun emplacement.

While the Taft Board's recommendations on the construction of new fortifications was largely limited to existing defenses at Eastern Long Island Sound, San Diego, Puget Sound, Columbia River, and Chesapeake Bay in the continental United States, major new construction projects were planned for the Philippines, Panama, Hawaii, Cuba, Puerto Rico, Alaska, and Guam. Plans for Cuba, Puerto Rico, Alaska, and Guam were not carried out. New defenses were added for the port of Los Angeles in 1909. These "Taft Program" defenses varied little from the overall designs used during the Endicott Program. Variations from the Endicott Program were the product of advancing naval armaments and the U.S. Army's twenty years of experience of operating coast defenses. To match the increased caliber of naval guns, a new disappearing gun of 14-inch caliber was developed. Another characteristic of the Taft Program batteries was the increased dispersion of batteries. The reduce density of weapons can be seen in the construction of several one-gun, 14-inch batteries and the reduction of mortar batteries from 8 mortars to 4 mortars (4 per pit to 2 per pit).

During the Taft Program, several one of kind of projects were undertaken. The Endicott Board Report called for 16-inch guns but work on the development stalled after the construction of one gun tube in 1895. Two unique 16-inch disappearing batteries were finally built in Panama and the Long Island Sound. In the Philippines army-designed armored turrets were custom built for a very small island in Manila Bay. Four 14-inch guns were mounted in two turrets at Fort Drum, which also became known as the "concrete battleship."

World War I and the Interwar Period (1917-1939)

The march of technical improvements in naval weapons continued through improvements in naval fire control and the ability for naval turrets to elevate their guns. By 1915 the newer battleships had guns that could out range the effective range of the coast artillery emplaced during the Endicott and Taft Programs. The increased angle of fire of the newer battleships also threatened the disappearing carriage batteries which were not protected from the plunging fire of these new battleships. In 1915, a National Board of Review on the coast defenses of the U.S., Panama Canal, and the Insular Possessions recommended the construction

Firing one of the 12-inch guns of the post-WWI Battery Kingman, Fort Hancock, New Jersey (NARA)

of new batteries mounting 12-inch and 16-inch guns on higher elevation, longer range barbette carriages. While efforts to introduce these coast artillery weapons had begun, the demands of World War I placed the modernization of coast defenses on hold. Many Coast Artillery units were transformed into field and heavy artillery units for service in France. As the United States was short on long range field artillery, 12-inch mortars, 10-inch, 8-inch, and 6-inch gun barrels were removed from several coast artillery batteries. These existing gun barrels, ranging from 6-inch to 14-inch in caliber were quickly mounted on railway and tractor-drawn carriages. While the United States involvement in World War I was brief, it resulted in the Coast Artillery Corps mission to be divided into three specialized areas as compared to its single mission before the war. These missions, based on armament type, were fixed coast defense weapons (including controlled mines), mobile seacoast artillery, and anti-aircraft artillery.

The development of the airplane as a ground attack weapon during the World War I added the task of defending both the mobile ground army and the shores of United States from attacks by aircraft to the Coast Artillery Corps' mission. The U.S. Army developed fixed and mobile anti-aircraft weapons, as well as accessory equipment such as aircraft sound locators, rangefinders, searchlights, specialized fuses, and mechanical fire direction calculators. The primary weapon for the defense against aircraft was the 3-inch gun on a fixed carriage (in batteries of three or four guns) located at existing coast defense posts. This weapon was later supplemented with .50 caliber machine guns and mobile 3-inch AA guns. By 1938 larger caliber anti-aircraft guns were introduced including the 90 millimeters (mm), the 105 mm, and 120 mm guns.

The mobile coastal defense mission came about because of the lack of U.S. heavy artillery for the troops in Europe. Existing gun barrels, ranging from 6-inch to 14-inch in caliber were quickly mounted on railway and tractor-drawn carriages. The construction of the new mobile carriages for guns, such as the railway mounts, took months and most of these weapons never reached European theater before the war ended. The availability of this ordnance material, especially considering the economics of using existing weapons and increased desirability of weapon mobility in the interwar period, made mobile coast artillery an attractive alternative to building new fixed coast defenses. The primary railway guns selected for coast artillery use from this large stock of World War I material were 8-inch guns and the 12-inch mortars mounted on new railway carriages. Added later was an improved version of the wartime 14-inch railway gun of which only four were constructed by 1920. The surplus mobile field artillery mounted on carriages designed for road movement included the 155-mm GPF gun (derived from a 1917 French design) of which almost a thousand were available. This powerful gun became the standard tractor drawn weapon for coast defense use against secondary targets.

a 155 mm G.P.F. mobile mount in a field position at Long Point, California (Ruhlen Collection)

While mobile coast artillery had the advantage of being able to respond to coastal areas most threaten when enemy naval forces approached, both railway and tractor-drawn weapons lacked the accuracy and protection of fixed coast artillery. Without solid and steady firing platforms and the precision of pre-calibrated fire control networks, as well as the inability of the carriages of the mobile guns to quickly track horizontally moving targets made mobile artillery much less effective than weapons in fixed emplacements. Prepared locations with circular arcs of track were prepared at a few select locations. For the 155 mm GPF mobile artillery, simple circular concrete bases were designed. These circular bases improved stability during firing and provided for rapid azimuth adjustment for horizontal tracking. One of the most common base designs developed for the 155 mm GPF guns was a central pivot and a curved rail embedded in concrete, which the gun's split carriage would traverse. This design was first constructed in the Panama Canal Zone, so this design became known as the "Panama Mount". Given the limitations of mobile coast artillery, their use was primarily an augmentation of existing defenses or to provide protection during the construction of permanent fixed coast artillery. Due to the low level of military appropriations during the 1920s and 1930s, mobile coast artillery was the only available weapon to defend vital locations until new permanent defenses could be funded and constructed.

Given the low level of overall U.S. military funding during the 1920s and 1930s, the construction and development of new fixed coast defenses were limited. The need for economy and to allow for higher gun elevations led to the abandonment of the disappearing carriage and its complex two-level emplacements. Among the last disappearing carriages built were for two 16-inch single-gun batteries (one in Panama and the other in the Long Island Sound). A newly designed high angle barbette carriage for existing stocks of 12-inch Model 1895 guns allowed effectively doubled the range of the guns over the same 12-inch gun mounted on a disappearing carriage. Construction of fifteen long-range dual gun 12-inch batteries was started in 1917 and completed by the late 1920s. The emplacement design was a departure from those of the Endicott Program. The battery design had two guns located much further apart, each gun in the center of a large ground level concrete pad to allow for an all-around field of fire. Located between the two guns was an earth-covered reenforced concrete structure containing magazines for shells and powder, the power and plotting rooms, and storage rooms. Protection of the guns from naval fire was based on dispersion; the wide separation of the key elements of the battery. Other than camouflage and nearby anti-aircraft guns these batteries had no protection from air attack. The development of a new 16-inch gun and carriage with a range of nearly thirty miles which exceeded the range of all existing naval warships was completed in 1919. The 16-inch in emplacements that were very similar to those used for the long-range 12-inch barbette batteries, with an increased distance between the two guns of the battery and the dispersed location of magazines in simple storehouses connected by a rail system. Only a few of the U.S. Army designed barrels had been constructed when nearly sixty 16-inch barrels became available from the U.S. Navy. This windfall was due to the Washington Naval Treaty of 1922 that resulted in the cancellation of several U.S. battleships and battle cruisers then under construction. The naval 16-inch barrels were to be installed in modified U.S. Army barbette carriages after 1925. Six new twin-gun 16-inch batteries were built between 1922 and 1934. During the Interwar Period, the construction of new batteries including both long range 12-inch and 16-inch guns, amounted to little more than twenty new batteries. The coming of World War II would inject new life into building modern U.S. coast defenses.

The 1940 Modernization Program and World War II (1940 -1950)

During the 1930s the U.S. Army began discussing how to protect new coast artillery batteries from attack by aerial bombardment. The debate centered on the expense of designing and construction of turret mounts for 12-inch and 16-inch guns as compared to developing protective structures made of concrete and steel. It was practical economic and time frame requirements that resulted in the eventual selection of a concrete casemate structure design to protect the current type of barbette mounts.

The prototypes of this new type of major caliber battery were built at the San Francisco defenses, during 1937-1940. These emplacements were designed for two 16-inch guns located about six hundred feet apart with complete overhead cover. Located between the two guns along a service gallery were the ammunition magazines, power generators, and support areas. The 16-inch guns were enclosed in reinforced concrete casemates. The battery's structure was made up of eight to twelve feet of steel reinforced concrete which was topped by up to twenty feet of earth as additional protection. The entire battery structure was designed to withstand a direct hit from a naval projectile or an aerial bomb. When completed the southern San Francisco battery at Fort Funston emplacement looked like a small hill, especially when camouflage and natural ground cover was added to the structure. The only exposed portions of the battery were the casemates where the gun barrels projected out through armor shields and concrete canopies. A second casemated battery on a hilltop north of San Francisco was also undertaken. Four more casemated batteries were begun at Narragansett Bay, the Delaware River, and Chesapeake Bay in 1940-1941.

In 1940 the Harbor Defense Board was charged with developing a master plan to update the harbor defenses of the continental United States. Eighteen coastal areas in United States were selected for modernization due to their military and economic importance - Portland, Portsmouth, Boston, New Bedford, Narraganset Bay, Long Island Sound, New York, Delaware Bay, Chesapeake Bay, Charleston, Key West, Pensacola, Galveston, San Diego, Los Angeles, San Francisco, Columbia River, and Puget Sound. The Harbor Defense Board recommended the adoption existing stocks of 16-inch gun as the primary weapon and 6-inch gun as the secondary weapon for the modernization program. In all the board proposed building twenty-seven new 16-inch casemated batteries; the casemating of 23 existing primary batteries (both long-range 12-inch batteries and older 16-inch batteries; and building fifty new 6-inch two-gun barbette carriage batteries, which would provide long-range fire (15 miles maximum) against secondary warships.

The new 6-inch batteries would be supported by 63 existing secondary batteries, mostly 6-inch and 3-inch barbette guns from the Endicott and Taft Programs, which would be retained. Upon completion of these new defenses 128 existing obsolete coastal batteries would be eliminated. The board estimated that the whole program would require three years to complete and cost about $82 million during 1941-3. Formal approval of this modernization plan, which would become known as the "1940 Harbor Defense Modernization Program" or the "1940 Program," was approved in September 1940.

One of the 16-inch guns of Battery Steele, Peaks Island M.R., Maine (Joel Eastman Collection)

A 6-inch gun of Battery Cravens, Peaks Island M.R., Maine, with a disguised SRC 296A radar behind
(Joel Eastman Collection)

The 1940 Harbor Defense Modernization Program greatly simplified the task of Coast Artillery Corps by reducing the number of types of batteries as well as the overall number of batteries needed to carry out their coast defense mission. This allowed a reduction in personnel and the level of effort to maintain, training and supply the pre-1940 batteries. Some of coast artillery that was declared obsolete was shipped to Allied nations to supplement their defenses, but most were scrapped for the war effort. As the nation moved closer to war, additional coastal defense projects were added to the 1940 Program, especially at newly acquired overseas bases, such as Trinidad, Bermuda, Newfoundland, and in areas where the enemy threats seem greater, such as Alaska, Hawaii, Puerto Rico, and the Canal Zone. It also became apparent that planning, construction and emplacement of the many new batteries called for the 1940 Program was going to take a much longer time then original envisioned, especially as the program was competing with rapid expansion of the whole U.S. Army and U.S. Navy. By the middle of July 1941, only four 16-inch gun batteries were ready for action and construction work had been started on just five others. With pressure from the U.S. Army Air Corps, it was decided to limit active work to those batteries that could be completed by July 1944. As a result, all work on fourteen of the thirty-seven 16-inch batteries planned for the continental U.S. was discontinued. The expansion of overseas bases during 1941 impacted the construction of the new 6-inch gun batteries in the continental U.S. by priority assigned to the completion of twenty 6-inch batteries to guard these overseas bases.

The new batteries constructed under the 1940 Program were much more standardized that those of proceeding periods. The Army developed standardized designs for the 16-inch gun batteries and the 6-inch gun batteries which were used with only minor variations for local topography and soil conditions. Both the 16-inch and 12-inch guns, whether newly installed or retained from the Interwar Period, were emplaced within reinforced concrete casemates that limited their field of fire to about 180 degrees but gave them superior protection over the old open emplacements. The new 6-inch batteries were not casemated. A cast steel shield from four to six inches thick was placed around the gun and carriage. This shield would protect the gun and its crew from all but direct hits by heavy projectiles. Between the two 6-inch guns was an earth covered steel reinforced concrete structure contain the magazine, power generators, communications, air filtering equipment, storage, and plotting room. As these batteries were being built, they were assigned a "Battery Construction Number" for record keeping purposes. As many of the new batteries were never formally named, these construction numbers were the only designation they received. While the Army never referred to the 16-inch series of batteries as whole as the "100" series or the 6-inch series of batteries as whole as the "200" series, these terms are used by modern historians and are referred to as such in this work.

As the range of these new batteries was far greater than earlier batteries it was also necessary to update the fire control networks. The 16-inch batteries received new base end stations as far as twenty-five miles away from the gun's position to allow for gun's maximum range to be effectively used. These stations were built in wide variety of forms: houses, windmills, silos, water tanks, office buildings, or buried into hillsides. Radar was added as an early warning device and as a fire control instrument allowing the operation of coast artillery at maximum range during all weather conditions.

By the start of World War II, the Coast Artillery Corps' mobile coast artillery units had dwindled from the plans of the Interwar Period, especially the railway guns units. Several tractor-drawn 155mm GPF gun regiments were available in the continental U.S., but only part of one 8-inch railway regiment was on hand. The four 14-inch railway guns continued their role in Los Angeles and Canal Zone. The primary use of mobile coast artillery was to fill in for fixed coast artillery weapons until their completion or at secondary locations. The 155mm GPF gun units were reorganized into seventy-two 2-gun batteries along the Atlantic, Gulf and Pacific coasts. Using 12-inch railway mortars and 8-inch railway guns from storage, several CAC units were formed and sent to both domestic and overseas locations to provide temporary harbor defenses until permanent works could be constructed.

As with earlier periods, an integral part of harbor defenses was the use of controlled mines across key ship channels. These mine defenses were supplement by U.S. Navy contact mines and the use of submarine nets and booms. As the primary threat during World War II turned out to be enemy submarines at most of these ports, the U.S. Navy added detection devices in outer harbor approaches and conducted offshore patrols. Because of the need for both the U.S. Army and U.S. Navy to coordinate their coast defense activities, a centralized harbor entrance command was created in 1941. The Harbor Entrance Control Post (HECP) used both army and navy personnel to provide a link between higher command and all subordinate elements of a harbor defense. These centers were responsible for monitoring all movement of shipping in and out of the harbor. To support this effort a secondary gun battery was on duty as commercial shipping traffic was examined upon entering the harbor. One of the concerns at this time was an attack by fast moving torpedo boats combined with the lack of modern rapid-fire guns. To fill this void, the 90mm anti-aircraft gun was selected to replace the existing 3-inch pedestal guns of the Endicott and Taft Programs. In late 1942, special anti-motor torpedo boat (AMTB) batteries were installed along the Pacific and Atlantic coastlines. These batteries usually consisted of two fixed mounted 90mm guns and two mobile mounted 90mm guns, and two mobile 37mm or 40mm anti-aircraft guns. These guns would be protected by earthen revetments with protected magazines. The active harbor defenses received two, three, four, or more of these AMTB batteries beginning in 1943.

Outside the continental United States, where the threat of attack and invasion was greater, new coast defense construction proceeded with greater speed and with the use of armament on hand rather than waiting for weapons sto be provided by 1940 Program. The coast defenses of Hawaii are a good example, as the Japanese attack on Pearl Harbor made new defenses the highest priority due to concerns of an invasion attempt. A series of batteries were constructed, using excess naval guns, ranging from the 14-inch turrets from the battleship USS *Arizona* to 8-inch gun mounts from the aircraft carriers USS *Lexington* and USS *Saratoga*. Throughout the Pacific Islands and Alaska, surplus U.S. Navy guns (5-inch, 6-inch, 7-inch & 8-inch) were mounted on shore to defend U.S. Navy installations.

With the tide of the war shifting toward the Allies after 1942 and the demands to produce war material for the mobile army, the navy, and the army air corps, the 1940 Program was pared back. While the construction of structures could keep pace with the original plan, the manufacture of weapons and their accessories could not. In response to these pressures, the 1940 Program was scaled back even further. By the war's end, the modernization program resulted in the completion of nearly 200 new batteries in the continental United States at a cost of $220 million, or about one-half the number of installations proposed in the 1940 Program, but still the most powerful collection of coastal defenses in America's military history.

The development in military tactics and technology during World War II brought about numerous changes to the concept of coast defense. It was no longer thought necessary to defend one's seacoast using just coast artillery and controlled mines. Air power and naval forces were to replace breech loading rifles and reinforced concrete. Already at the end of World War II, all except a few 90mm AMTB batteries were placed on caretaking status. During the transition years of 1946 to 1948, some new batteries started during the war were completed while many other batteries were being disposed of and guns scrapped. By 1949, the process was completed as the last of guns were scrapped. In 1950, the remaining harbor defense commands were disbanded, and the Coast Artillery Corps was abolished as a separate U.S. Army branch with its remaining units, all anti-aircraft artillery, recombined into the Field Artillery. After 150 years of being one of America's military prime missions, the building and manning of permanent coastal fortifications was over.

The U.S. Coast Defense Objective in the Modern Era

The objective of seacoast defense is to provide protection of the coastline from invasion by an enemy, and specifically the defense of important harbors, which includes securing the anchorages and bases needed for naval operations. Coast defense is not only protective in its strength but protects the nation's ability to carry war beyond its own coastline.

It is impractical to fortify the entire extent of any nation's long coastline in such a way that an enemy in command of the sea could not land upon some portion of it. The cost of such an undertaking would be excessive, as maintenance of these defenses and number of men required would make it prohibitive expensive. An example of this type of defense was the "Atlantic Wall" built by Germany in World War II (which stretched from Norway to Spain), which failed to prevent the Allies from landing in Europe in 1944. It was essential, however, that certain selected points be permanently fortified to make invasion more difficult and to protect key naval shore installations and fleet anchorages and important commercial harbors that support the nation's economy.

The resources to defending the coastline during this era were divided into two kinds of troops. The first was the Coast Artillery troops, made up the regular U.S. Coast Artillery Corps and the U.S. Coast Artillery Reserves. These technical troops manned both the fixed and mobile seacoast artillery and controlled mines defenses. The second resources were the supporting troops of the mobile ground forces of the U.S. Army which protected the both the coast defenses and unfortified coastline from enemy landings. The second would been the local National Guard troops (formally militia), while the U.S. Navy's role in coast defense was through both offensive and defenses operations against enemy warships.

To carry out this mission, seacoast weapons were divided into classifications according to their capabilities against enemy warships. Primary armament were those weapons that could theoretically destroy the primary or capital warships of enemy naval force. Throughout most of the Modern Era primary weapons were defined as seacoast artillery of initially 8-inch and larger caliber. Controlled submarine mines were also considered part of the primary armament. The second group of seacoast weapons was the secondary armament, which were designed to counter secondary or non-capital warships, such as cruisers, destroyers, and torpedo boats.

The selection of the numbers and type of seacoast weapons was determined by such factors as the importance of the coastal area, the hydrograph profile of the approaches, the topography of area, and effectiveness of seacoast weapons in defending the coastal area. The positioning of seacoast artillery was based on the attainment of effective fire and protective factors, such as concealment, other weapons, and local defense against ground or air attacks. Attainment of effective fire refers to a position which offers the widest field of fire and greatest range over navigable water. Also considered was the need to provide coverage to all areas in which an enemy warship may operate and the placement of a suitable concentration of fire on critical areas such as harbor entrances, approaches to mine fields, and narrow portions of the channel. Consistent with these requirements, batteries were sited to provide mutual support and defense against all forms of attack. The considerations for the location of primary armament included the ability to protect friendly naval forces while entering, within or leaving the harbor, and preventing hostile naval forces from approaching within effective range of the defended coastal areas. Submarine mine fields would be placed in the seaward area of the harbor entrance and within effective range of searchlight and rapid-fire secondary armaments. Both controlled and uncontrolled submarine mines are located to prevent entry into or close approach to the harbor of enemy surface warships or submarines at all times, including during night or during conditions of heavy fog or smoke. The secondary armament would be located to provide protection for mine fields, nets, booms, and other obstacles; and the attack of hostile secondary warships engaged in raids, reconnaissance, laying of mines, and torpedo fire. Since targets of the secondary armaments were within range of visual observation and assumed to move at high speed on rapidly changing courses, these batteries were sited in direct fire positions. Protective factors in site selection included protection for the power plant, plotting room, magazines, communications, exposure of the gun crew and ammunition during the service of piece, gas protection for command post and plotting room, distances between emplacements, and concealment.

U.S. Coast Defense Armament and Equipment in the Modern Era

Few weapons of the Modern Era of coastal fortification remain today. This is the result of the advancing technology that quickly made weapons obsolete and given the economic value of high-grade steel the military sold these obsolete weapons to salvage companies. The scrapping of coast artillery material also holds true for most its supporting equipment, machinery, and instruments. As a result, today we mainly only have period images of these armaments or supporting equipment.

The development of new armament and equipment over this era usually went through cycles where the level of perceived external threats to the United States generated appropriations from Congress to allow the funding of new weapon systems. The development process for new weapons required several steps. First, was the design stage which led to the prototype and testing period and then to production and installation phase. Finally, while the weapon was in service it received modifications and improvements until it was declared obsolete. The life cycle of seacoast artillery varied from a few years to as long as fifty years.

The construction of the Endicott and Taft Programs defenses relied on the growth of heavy industry in the United States. Many of items used in coast artillery forts were invented specially for that purpose and represented the cutting edge of that technology. Early defense works relied on steam, coal, and manual energy to make things work. The use of oil and the advances in electricity brought motor driven equipment, telephones, radar, computers, and electric lighting to become key ingredients in U.S. coastal defenses.

It is also important to note that different U.S. Army branches had specialized functions that need to work together to complete a weapon system. A seacoast weapon would be designed and constructed by Ordnance Department while the emplacement was designed and constructed by the Corps of Engineers. These activities were all support by the Quartermaster Corps, Signal Corps, and so forth. The final product was then turned over to the Coast Artillery Corps for use. As you may imagine sometimes the priorities of these various organizations were not always in agreement, so delays or undesired weapons systems did occur.

For coast artillery material, the U.S. Army insured that all items were assigned a "type" and a "model". For seacoast artillery, the type for the gun or barrel refers to the size of bore (diameter) in inches while type for carriage or mount refers to the style of operation. Associated with the type is the model which refers to year of development and any subsequent modifications until 1930 when use of the year was dropped. This nomenclature extends to projectiles, fire control instruments, searchlights, submarine mines, ammunition hoists, power generators, radar, etc.

Carriage or mount types were either fixed or mobile, they allowed the guns to elevate and provide for some horizontal movement while taking up the recoil of the discharge and return the piece to the loading position. The major caliber fixed carriages were classified as Barbette (BC) carriage, which allowed the gun to remain above the parapet for loading and firing; Mortar (MC) carriage, which allowed a short-barrel gun to fire in a high arc; the Barbette long-range (BCLR) which allowed for greater firing elevations and ranges; and the Turret (TM) mount which was a barbette carriage protected by an armored housing with ammunition supplied from below. Guns of 7-inch or lesser caliber were mounted on the Pedestal (PM) mount, which had a fixed cylindrical base on which rotated a yoke that held the gun in a cradle equipped with recoil absorbing cylinders; the Anti-aircraft (AA) mount, a pedestal mount that allowed fire at high attitude. The Fixed retractable carriages included the Gun-Lift (GLC) carriage which was a BC on an elevator platform; the Disappearing (DC) carriage where the gun is raised above the parapet for firing and retracts behind the parapet for loading. The earlier smaller caliber guns had the Balanced Pillar (BPM) mounts and the Masking Parapet (MPM) mounts, which enabled the gun to be lowered below the parapet to protect it from view. Guns on mobile carriages were used in the Interwar Period as the Railway (RY) mount cars and Tractor-drawn (TD) mounts. Other temporary coast defenses made use of available weapons with a range of carriage types, primarily former naval models.

A 10-inch rifle on a disappearing carriage
Battery Benson, Fort Worden, Washington (Puget Sound Coast Artillery Museum Collection)

6-inch rifles on dissapearing carriages
Possibly Battery Tolles, Fort Worden, Washington (Puget Sound Coast Artillery Museum)

A 6-inch gun on a pedestal mount, Battery Carpenter, Fort McKinley, Maine
(Joel Eastman Collection)

3-inch guns on pedestal mounts in Battery O'Rorke, Fort Barry, California (NPS, GGNRA)

12-inch mortars in pit A, Battery Worth, Fort Pickens, Florida (Stillions Collection, NPS)

Controlled mines were anchored to the bottom of a harbor, either sitting on the bottom itself (ground mines) or floating (buoyant mines) at depths which could vary widely, from about 20 to 250 feet. These mines were fired electrically through a vast network of underwater electrical cables at each protected harbor. Mines could be set to explode on contact or be triggered by the operator, based on reports of the position of enemy ships. The networks of cables terminated on shore in concrete bunkers called mine casemates, that were usually partly buried beneath protective coverings of earth. The mine casemate housed electrical generators, batteries, control panels, and troops that were used to test the readiness of the mines and to fire them when needed. Each protected harbor also maintained a small fleet of mine planters and

On the deck of mine planter (Stillions Collection, NPS)

tenders that were used to plant the mines in precise patterns, haul them back up periodically to check their condition (or to remove them back to the shore for maintenance), and then plant them again. Each of these harbors also had onshore facilities to store the mines and the TNT used to fill them, rail systems to load and transport the mines (which often weighed over 750 lbs.) each when loaded), and to test and repair the electrical cables. Fire control structures were also built that were used first to observe the mine-planting process and fix location of each mine and second to track attacking ships, reporting when specific mines should be detonated. The preferred method of using the mines was to set them to detonate a set period of time after they had been touched or tipped, avoiding the need for observers to spot each target ship.

Key to the successful use of coast artillery was fire control and position finding as if the guns, mortars, and controlled mines failed to strike their intended targets their mission was incomplete. Early aiming efforts relied on the skill of the gunner to hit the target, but as the weapon's range increased so did the need for specialized fire control. Using geometry, optical instruments, telephones, timing interval bells, and mechanical devices a system was devised to point weapons successfully at their targets. Key equipment included the Depression Position Finder (DPF), the Azimuth Instrument (AI), Coincidence Range Finder (CRF), Plotting Board, Range Correction Board, Fire Adjustment Board, Deflection Board, Spotting Board, Range Percentage Corrector, Data Transmission Devices, Telephone Sets, and Timing Interval Bells. Many of the devices were replaced or supplemented by the development of radar (for both surveillance and fire control duties) and gun computers (combining many of plotting room devices) during the 1940 Program.

Until the advent of radar, the use of searchlights (plus star shells and airplane flares) was used to illuminate naval targets at night. Both mobile and fixed searchlights were used for both harbor defenses and anti-aircraft defense. At first 36-inch and 60-inch searchlight were used, but the 60-inch became the standard. Searchlights were located as close as possible to the water-edge to maximize their effective range of between 8,000 and 15,000 yards. Fixed searchlights were provided with a shelter to protect the searchlight from elements, to house the electrical generator, and provide concealment of the light when it was not in use. Some positions placed the searchlight on small rail cars that allowed the searchlight to move a short distance to a more exposed operating site. Searchlights were also housed in towers, pits, and even tower that "disappeared" by pivoting. After 1940 all new searchlights assigned to the defenses were the mobile type.

By 1943, two technological advances significantly changed coast artillery fire control. The most striking was the development of radar, which, as noted, could function in any weather or visibility. The

Searchlight and shelter/powerhouse, Fort Flagler, Washington (Puget Sound C.A. Musuem)

use of radar greatly reduced the need for searchlights and for fire control stations as spotting enemy war-ships and aircraft could now be undertaken by radar units. In addition, after decades of experimentation and development, largely stymied by inadequate funding, the coast artillery adopted gun data computers, primarily for the last generation of batteries. These replaced the plotting boards and, coupled with direct-reading observation instruments, substantially automated the fire control process, reducing the human error that had always plagued the system.

A Depression Range Finder (left) and a azimuth scope (right) in a base end station
(Al Schroeder Collection)

U.S. Coast Defense Fire Control Structures in the Modern Era

The development and changes in the optical instrument fire control system from 1900 to 1945 was a long and complicated process that changed equipment, operating procedures, and designations frequently. The reader is encouraged to consult *American Seacoast Defenses: A Reference Guide* and articles in the *Coast Defense Journal* for more detail and references to U.S. Army manuals and reports. The maps included in this guide have an extensive set of symbols indicating the locations of the various fire control structures.

By 1909, each battery was under the immediate command of the officer stationed at the battery commander's station (BC). Each battery may have had one or more additional base end stations (B) with optical spotting instruments. Small caliber batteries usually had a coincidence range finder station (CRF) nearby. Mine commanders manned their posts at the mine primary (M') station. In the defended harbor areas, called the Coast Defense Command, batteries were grouped into Fire Commands, each under the overall command of the fire commander stationed at the fire command primary station (F'). The Fire Commands were then grouped together by geographical areas under the command of the officer in command of that entire sector of the coast defense. This command was initially called the Battle Command but later was changed to the Fort Command. This officer was stationed at the primary fort command station (C'). In 1925, this chain of command was changed slightly. All forts and/or groups were under the Harbor Defense command (H). Forts (F) were also used as tactical commands. Individual gun batteries were assigned to a gun group (G). Later an additional tactical organization, the groupment (C), was added below the Harbor Defense command composed of two or more groups.

In general, batteries in each harbor defense were assigned tactical number designations, generally in numerical sequence from the south (Tactical Battery #1) to the north (Tactical Battery #2, 3, 4, etc.) on the Atlantic coast; from the north to the south along the Pacific coast; and from the east to west on the Gulf coast and along the Puget Sound during the 1940s. Note that by 1940 base end stations (B) and spotting stations (S) were often combined. This is useful in deciphering the symbols for designating the fire control observation stations on the maps: $B^1_1S^1_1$, $B^2_1S^2_1$, etc. The lower number is the tactical battery number to which the station is assigned, the upper number (or "prime" mark) is the station designation number in the series of stations assigned to that tactical battery. The number of base end stations assigned to each battery ranges from a single station to as many as 14 stations. Each station had at least one azimuth scope and/or depression range finder (DPF) scope as well as connected telephone communication equipment. 3-inch small caliber batteries had one base station with a coincidence range finder (CRF) located close to the battery.

During the period 1905 to 1940 the fire control structures were generally located on existing military reservations. The location and identities of these stations can be found on the confidential blueprint maps; in the reports of completed batteries; in the reports of completed works; and in the harbor defense engineer notebooks that are part of the CDSG ePress harbor defense document collection. After 1940 the ranges for the new guns were longer and the fire control stations were more dispersed, which resulted in the acquisition of a number of new small reservations along the coastline of each active harbor defense.

Battery Plotting Room circa 1944 (NARA)

The U.S. Coast Defense Organization Before World War II

The following organization structure of the administration and tactical command of the U.S. Coast Artillery Corps is for the 1930s period. The organization prior to 1924 was on a company basis and after 1942 on a separate battalion basis and is not discussed here.

Earlier organizations had similar purposes but used different terminology. Earlier tactical structures had Artillery Districts that were divided into Battle Commands, Fire Commands, Mine Commands, and Battery Commands, each with their own commanders, while for administrative and training purposes the CAC was divided into companies which in turn were assigned to coast artillery forts or posts.

A Harbor Defense Command is a subdivision of a Defense Command, which would cover an entire region. All elements, including materiel and personnel of a Harbor Defense Command, were located at one or more coast artillery forts. These forts consisted of defined land areas within a harbor defense in which the harbor defense elements were assigned. The forts were organized primarily to provide a centralized

WAR DEPARTMENT   CORPS OF ENGINEERS, U.S. ARMY.

F.C. DIAGRAM OF THE COAST DEFENSES OF COLUMBIA

FORTS	STEVENS				CANBY		COLUMBIA
F & M. COMMANDS	FIRST		THIRD	FIRST MINE	FOURTH		SECOND MINE
BATTERIES	RUSSELL	MISHLER	CLARK	PRATT	GUENTHER NEW BATTERY	ALLEN	MURPHY
ARMAMENT	2-10"	2-10"	4-12"M	2-6"	4-12"M	2-6"	2-6"
CARRIAGE	D.C.-L.F.	D.C.-A.R.F.	M.C.-A.R.F.	D.C.-L.F.	M.C.-A.R.F.	D.C.-L.F.	D.C.-L.F.

Plotting Room At Battery
Battery Commander At Battery
Plotting Room Remote from Battery
Battery Commander Remote from Battery
Battery Primary
Battery Secondary
Auxiliary Station
Auxiliary Station
Auxiliary Station
FIRE COMMANDER Primary
Emergency Station
Fire Commander Secondary
Mine Commander Primary
Plotting Room Mine Commander
Mine Commander Secondary
Mining Casemate
Mine Loading Room
FORT COMMANDER
Signal Station
Meteorological Station
Tide Station
Radio Station
Post Tel. Switchboard
Switchboard Room
Controller Booth
Searchlights

LEGEND

TELEPHONE
TIME INT. BELLS
SPEAKING TUBE
SEARCHLIGHT CONTROLLER

Drawn by P.A.W.
Traced by C.F.G.
Checked by A.H.

Submitted: 3-9-21.
Arthur Healey
Assistant Engineer.
Approval Recommended
Major, Corps of Engineers.

DMCR. 803

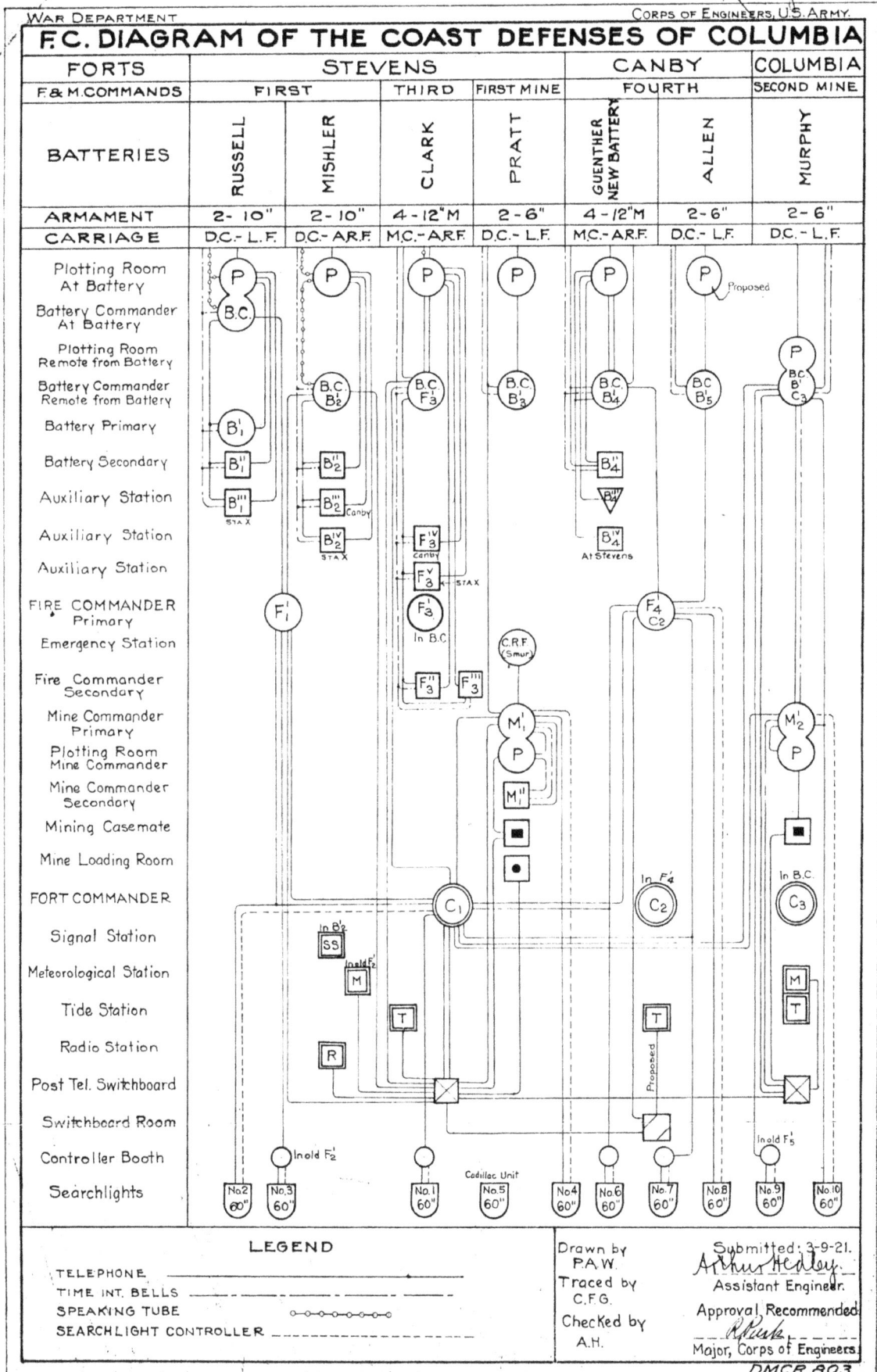

A fire control diagram showing the communications lines between the various stations.

control over administrative and technical components of the harbor defense. The materiel provided for a harbor defense may have included various types of seacoast artillery guns, anti-aircraft guns, searchlights, controlled submarine mines, underwater listening posts, radar, observation and fire control systems, and harbor patrol boats. Harbor defenses were designated by the name the harbor or coastal area which they were defending, or by the name of the largest city in their immediate area. Examples are "The Harbor Defenses of San Francisco" or the "Harbor Defenses of Chesapeake Bay."

A senior U.S. Coast Artillery Corps officer was usually designated the harbor defense commander responsible for both the administration and tactical commands. He was supported by a harbor defense headquarters staff and service units from the Quartermaster, Ordnance, Medical, Signal, Engineers, and Military Police organizations. The service units usually staff the administrative headquarters and the Coast Artillery Corps the tactical headquarters. Each fort was organized with its own headquarters and fort commander, who was responsible for the administration of the post. While the fort commander was not included in the tactical chain of command, he was responsible for the training and supervision of damage control to all the fort's structures and the activities of the service units.

The basic units of the coast defense tactical command were the battery, battalion, and group. The battery was the basic combat unit of the harbor defense and contained enough men required to man one primary battery. Batteries were classified by the type according to the material with which they were equipped. The gun battery consisted of one or more fixed or mobile guns of the same caliber and characteristics to be employed against a single target and of being commanded by a single individual. It included all structures, equipment, and personnel necessary for emplacement (or mobile weapons), the conduct of fire, and the performance of service. The strength and organization of a battery depended upon the type, number, and caliber of the guns of the battery. It was divided into a battery headquarters section, a range section (containing a battery commander's detail, an observing detail, and a range detail), a maintenance section, and a gun section for each gun or mortar. Special gun batteries were the anti-motor torpedo boat (AMTB) battery and the fixed anti-aircraft battery. The mine battery consisted of the personnel, structures, and equipment other than mine planters necessary for the installation, operation, and maintenance of all or part of the controlled mine fields. It was divided into a battery headquarters section, an operations section (containing a command post detail and range detail), a casemate section, a loading and property section (consisting of loading, cable, explosive, and maintenance details), a planting section (consisting of mine planter, distribution box boat, and small boat (yawls) details) and a maintenance section. The searchlight battery consisted of the personnel, material, equipment, and structures necessary for the operation and maintenance of seacoast and anti-aircraft searchlights.

These batteries were normally administerial combined into battalions with each battery commander reporting to the battalion commander. The battalion was organized to provide administrative, training, and tactical functions. Gun battalions were composed of from two to five-gun batteries, while a mine battalion consisted of the personnel, submarine mine material, structures and vessels necessary to plant, operate, and maintain part or all the controlled mine fields. The primary purpose of the coast defense battalion was providing effective fire direction through the coordination of various types of batteries. When a harbor defense command was large, battalions will be organized into groups. A group was a tactical command containing from two to five battalions or independent batteries. As with battalions, the primary mission for the group and the group commander was to provide effective fire direction. The use of groups occurred when the number of units is greater than can be controlled by the harbor defense commander. The basis for battalions or groups was to organize batteries that covered same field of fire or water area. When large number of batteries covered the same water area then the organization was based on target selection, such as primary and secondary armaments.

For administrative and training purposes, battalions were organized into regiments up to 1942. The garrison of a harbor defense consisted of part or all of one or more regiments, and the organization of different regiments varied to conform to the special requirements of the different harbor defenses. Generally, a coast artillery regiment assigned to fixed armament consisted of a headquarters battery, a searchlight battery, a band, and three battalions. The forts were assigned to Coast Artillery Districts. The district commander commands all coast artillery troops stationed within the territorial limits of the district, including the coast artillery units of the Organized Reserves and those of the National Guard when in the service of the U.S. At the start of World War II, the headquarters for Coast Artillery Districts were in Boston, MA (1st CAC District), New York, NY (2nd CAC District), Fort Monroe, VA (3rd CAC District), Fort MacPherson, GA (4th CAC District), and Presidio, CA (9th CAC District). Overseas coast artillery units were assigned to local U.S. Army Departments, such as the Hawaii Department, etc.

A Typical U.S. Coast Artillery Fort in the Modern Era

While each coast artillery fort has its own unique design, it is possible to provide a general blueprint of the type and purpose of structures that you would find at a U.S. coast artillery fort built during the Modern Era. It is important to remember that each fort was like small self-contained city. All the services that were required to support the daily needs of its garrison and to operate the fort's weapon systems were included within the military reservation.

View of Fort Flagler, Washington (D. Kirchner Collection)

The reservation was typically surrounded by a fence. There was a main entrance gate with a guard house. While very few coast artillery forts had any land defenses, the use of security fencing was widespread. Recognizing this fencing is usually the first indication of a former U.S. military reservation. The main cantonment area contained a variety of buildings spread over a large area, not much different in appearance of a rural college campus. This support area was subdivided into functional sections surrounding by a large parade ground area. While the overall fort was under the Coast Artillery Corps, each of the support services (Quartermaster, Engineer, Medical, etc.) had their own buildings or reservations within the fort.

The main parade ground is the focal point of the post. The fort headquarters, officer's quarters, non-commission officer housing, service clubs, and enlisted barracks usually surround it. Most of the non-tactical structures at the forts constructed during the Endicott-Taft Programs were designed to be permanent structures. These wood-frame buildings were built on stone foundations with slate roofs, sided with local brick, clapboard, or stucco. The Quartermaster Corps architect's office created standard plans for all types of buildings. Those designed at the turn-of-the-century—when most Coast Artillery forts were constructed—were of Colonial Revival style with elements of Queen Anne style in the officers' quarters. As the century progressed, new styles were adopted, such as Italianate and Spanish Revival, and these styles were used when additional buildings were constructed. Store houses and pumping plants used more practical industrial or utilitarian styles.

Parade ground and officer's quarters, Fort Casewell (BW Smith Collection)

Officer's quarters varied in size and elaborateness depending upon the rank of officer for whom the building was intended. The Commanding Officer's Quarters was usually the largest and most elaborate of the officer's quarters, and it was placed, if possible, on the highest and most prominent location on the parade ground. Other senior officers were assigned single quarters, while many of the quarters were double quarters for two families. Large forts had a Bachelor Officer's Quarters with its own mess. Non-Commissioned Officer's quarters were usually double sets.

The interiors of buildings were finished with wood floors, plaster walls with wood trim, and pressed metal ceilings. All structures where officers and men lived or worked had electricity, running water and flush toilets. Each barracks was designed to house a company or battery of 100 men and was self-contained with its own kitchen, dining room, day room, barber shop, and tailor shop. Sleeping quarters were on the second floor, while the lavatory and latrine were located in the basement in northern climates. In the south, separate lavatory and latrine buildings were sometimes built. Large forts had double barracks–two 100-man barracks-built end-to-end–which functioned as two separate barracks. Forts which served as the headquarters post for a harbor defense usually had a band barracks.

Although the parade ground was used as a general athletic field, tennis and handball courts, and baseball fields were also built in open areas of the fort. A system of permanent roads served the entire fort, and the streets were usually named. Railroads and tramways were built during the construction of the forts, and these lines often continued to be used. These forts eventually had their own water, sewer, telephone, and electrical systems. If municipal water and commercial power services were available, the army used them, but at many sites the engineers built their own water and electrical plants and distribution systems. Sewer pipes ran into the ocean. Ice houses, and in northern areas, ice ponds, were also built to provide refrigeration for food in the years before electrical cooling became available. Systems for the disposal of garbage and rubbish were also created. Garbage and combustible waste were burned in crematoria, while non-combustible materials were disposed of in landfills or dumped into the ocean. The major fuel at forts was coal, and a system of unloading, transporting, and storing the fuel was developed, usually relying on mule-drawn wagons.

A large portion of fort's reservation would be devoted to the Quartermaster Corps. The Quartermaster was tasked with providing housing, supplies, and transportation for all the troops assigned to the fort. The Quartermaster oversaw the construction of most of fort's support buildings, as well as the installation of its own quartermaster wharf and tramway to transport supplies within the fort. Storehouses, commissary, workshops, and stables were usually centered near the quartermaster wharf.

Fort Terry buildings (BW Smith Collection)

The Corps of Engineers were responsible for construction the actual fortifications known as the tactical structures (emplacements, fire control stations, casemates, power houses, etc.); the Ordnance Department provided the weapons, machinery, and instruments that went into these structures; and the Signal Corps provided the technical equipment as new technology was developed. Near the shoreline were located the fire control stations along with protected telephone exchanges, command posts, meteorological stations, seacoast searchlights positions, and reserve magazines that support the fort's weapons systems.

The fort's two main coast defense weapon systems were controlled mine fields and seacoast artillery. Controlled mine fields required an extensive infrastructure within the fort. Principal structures for the mine defense included the mining casemates from which the mines were operated; the conduits connecting the casemates with the shore; the cable terminals on the shore; the cable tanks in which the mine cables were stored when not in use; the mine storehouses in which were kept the mine cases; the loading rooms in which the mines were loaded; the magazines in which the dynamite was stored; the range stations, plotting rooms, and dormitories, the mine wharves at which the mine planter used to land and receive the loaded mines; and the tramway connecting the wharves with the cable tanks, storehouses and loading rooms.

Closer to the shoreline are the emplacements of the fort's other main weapon system the large caliber gun batteries. These gun batteries consisted of both of large caliber breach loading rifles mounted on disappearing or barbette carriages and smaller rapid-fire guns on pedestal mounts during the Endicott and Taft Programs. The purpose of these gun emplacements was to provide a stable base for these guns and carriage and a convenient platform for the personnel serving the gun. The emplacement also designed to provide the armament and the personnel the maximum protection as possible, as well as providing a safe storage place of the ammunition. These thick concrete structures were covered with earthen fill on the seaward side, while they were partially buried, these batteries are easily accessible from the rear due their open back design.

Fort Mott 1936 (NARA)

The gun emplacements from the Interwar Period and 1940 Program are quite different from these earlier designs as American coast artillery responded to the progression of larger caliber naval weapons with longer firing ranges and the advent of military aviation with aerial bombs. These emplacements are usually one-story high but usually completely buried. The gun position consists of a gun well surrounded by a circular concrete pavement. Later many of these emplacements were completely rebuilt with thick reinforced concrete casemates to protect their weapons from aerial attack and naval bombardment.

Another primary seacoast weapon of the Endicott and Taft Programs was the seacoast mortar, actually a short barrel breech loading rifle. These batteries by definition did not require direct fire, so they were often located away from the shoreline. They were located within or behind the fort's cantonment area. A typical mortar battery had a high reinforced concrete parapet with traverses that formed a series of pits. These pits were usually open to the rear, but early designs were completely surrounded with access through a tunnel. A battery had one to four pits with two or four mortars in each. Between the pits or around their sidewall were ammunition magazines, power generator rooms, shot truck areas, storerooms, and a plotting room.

The secondary gun batteries mounted rapid fire guns for the defense of the controlled mine fields from minesweepers and to repulse fast moving naval vessels and were installed after the first round of primary gun batteries. These are simple emplacements that basically provided a stable firing platform for the weapons and a protected magazine for their ammunition. Also located around some forts were groups of three or four concrete gun blocks for anti-aircraft guns that were added in later years.

A key feature of all coast artillery forts are the fire control stations which provided the target information for the mine and gun defenses. These stations come in all shapes and size. They range from a single below-grade room with observing slots to large multi-level, multi-room towers. Constructed of both wood and concrete, these stations have been disguised as non-military structures ranging from summer cottages to grain silos. Associated with World War II fire control stations were radar stations that by 1944 replaced their function. These radar stations had antennas which were mainly located on steel towers but could be mount on other structures. These antennas sent their signals to operating rooms where measurements provide location data to plotting rooms. Support these stations were power rooms and dorms for troops manning the stations.

Several to many of the structures at most of remaining U.S. coast artillery forts. However, you may only view piles of rubble and mounds of dirt as their status and condition are constantly changing. Nearly all the seacoast armament and equipment were scrapped after World War II which accounts for the lack of actual coast artillery at the forts.

Garrison Life at a U.S. Coast Artillery Fort in the Modern Era

The soldiers assigned to the defenses experienced a great change in quality of life during the years from 1890 to 1950. The early years were certainly the roughest. In general military service in the U.S. armed forces was not well compensated or widely respected in some quarters. As the permanent posts were being established, physical living conditions were sometimes poor, and relationship with the local, civilian community at times strained. Officers could afford higher standards of living for themselves and their families as well as greater social involvement with the local community.

63rd Coast Artillery Company on parade gound at Fort Worden, Washington in 1908
(Puget Sound C.A. Museum)

By the end of the early modernization programs in the 1910s, the living and work conditions had greatly improved. In particular the Coast Artillery was an elite assignment, with considerable prestige. The Coast Artillery Corps was relatively well funded and equipped, had a strong technical and professional dedicated career officer contingent, and was based on teamwork activity that encouraged close camaraderie. Opportunities for duty at oversea bases in exotic tropical locations like Hawaii, the Philippines, and the Panama Canal had its advantages, especially as many of the tropical diseases had been conquered. Training was emphasized, but in all the workload was reasonable. Pay was not extravagant for the enlisted man– but decent food, recreation, and athletic events were provided on post. Soldiers tended to stay in this branch of service, often re-enlisting, and were quite good at what they were taught and with what equipment they practiced on.

The daily schedule of the Coast Artillery troops focused on drill, inspections, maintenance, meals, and recreation. The center of activity for enlisted men was their barracks which was designed to house one or two companies or batteries of 100 men each with its own kitchen, lavatories, dining room, day (recreation) room, barber shop, and tailor shop. The barrack along usually arrayed around a parade ground. The

Soldiers in barracks (Stillions Collection, NPS GSNS)

officer quarters were usually located on the opposite side of the parade ground. The day would begin with meal, roll call, and assignment of duties. This usually was training/drill in the mornings with maintenance tasks or recreation events in the afternoons. Recreation was considered important by the U.S. Army after 1900 as it was believed that it not only maintained physical fitness but promoted competitiveness which made the men more effective in combat. Most large forts were provided with a gymnasium and bowling alley, as well as athletic fields, handball and tennis courts. Other recreation activities included visiting the post theater, service clubs, libraries, chapel, and the Post Exchange, as well as leave to visit the local cities and towns. Another aspect of garrison life were weekly inspections and parades, and soldiers who failed these inspections would end up spending their weekend cleaning barracks and latrines, rather than having a weekend pass to visit the local communities.

Mess hall set for Christmas Dinner 1911, 126th Coast Artillery Company, Fort Worden Washington
(Puget Sound CA Museum)

MODERN ERA SEACOAST FORTS TODAY

In 1950, the remaining harbor defense commands were disbanded, and the U.S. Coast Artillery Corps was abolished as a separate branch with its remaining units, all anti-aircraft artillery, moved into the Field Artillery. Meanwhile, the responsibility for limited harbor defense, primarily underwater defenses, was transferred to the U.S. Navy. The U.S. Army retained several of the old coast artillery forts for other missions, while the Navy acquired several reservations for thier use including for its new role in harbor defense. Other federal agencies had an opportunity to claim all or portions of the former coast artillery sites. Those not transferred were turned over for disposition to the U.S. General Services Administration (GSA), who offered them to state, county, and local governments, and finally to private citizens. Many of the smaller, independent plots of land which had been leased or purchased for fire control and searchlight positions were returned to original owners or sold to private owners, before selling or transferring these former forts, the U.S. Army either returned to its depots all usable equipment or auctioned items in lots to the public.

Several coast defense sites had been abandoned by the U.S. Army as active defenses by 1928 including those at the Mississippi River, Mobile Bay, Tampa Bay, Savannah, Port Royal Sound, Cape Fear River, Baltimore, and the Potomac River, and the smaller inner harbor defenses at East New York, San Francisco, and Puget Sound. Many of these reservations were reclaimed for use during World War II. The next large-scale transfer of harbor defense properties from the U.S. Army began in 1947 and continued through the mid-1950s. In the early 1970s a general series of military base closures occurred throughout the U.S. Department of Defense to reduce basing costs. Several large former harbor defense sites, including military reservations around San Francisco, New York, and Pensacola, were included. Given the large size and value of these properties, Congress passed several laws that directed the ownership of these former forts to be transferred to the U.S. National Park Service (NPS). Base closure commissions in the 1990s, 2000s, and 2010s recommended the closure and transfer of other former harbor defense sites, which included the Presidio of San Francisco, Fort Wadsworth on Staten Island, Fort Monroe in Hampton, and Fort Trumbull in New London. Only a handful of old coast defense reservations remain in military hands in 2025 — Fort Story, VA, Fort Hamilton, NY, a large part of Fort Rosecrans, CA, Fort Kamehameha, HI, Fort Hase, HI and a few other sites. Other national agencies, state agencies, and local governments acquired numerous coast artillery sites for parks and recreation areas, since they inevitably had scenic river or ocean views. Depending on how diligently the GSA protected the sites, and the length of time it took to dispose of them, some sites and structures survived in excellent condition, while others suffered at the hands of salvagers and vandals.

While many of the Modern Era forts and batteries are now located within parks, they have not been accorded the same level of protection or care as the remaining brick and stone forts. Most of the old coast defenses structures are considered to be, at the worst, a legal liability or at best, an eyesore to the park. Remaining structures have been built on, fenced in, buried, or destroyed. They have been removed as interfering with the park's primary mission of providing recreation space. Vandalism has caused considerable damage over the years. Abandoned and neglected coast defense structures have suffered from freeze-thaw cycles cracking and spalling the concrete and brick, rusting metal rebar and materials has hastened deterioration. Unchecked vegetation growth has caused some structures to collapse. And rising sea levels and increasingly violent storm surges are eroding away shoreline and destroying major structures. While most gun emplacements have been constructed in such a way to resist these attacks, many other tactical structures have collapsed, and even brick structures have been damaged or destroyed by vandals and neglect. Non-tactical structures, particularly officers' quarters, have survived at many parks and government-owned sites through adaptive reuse, but at some former posts such structures have been completely removed.

However, public interest in the history of American coast defenses has grown since the publication of *Seacoast Fortifications of the United States: An Introductory History*, by E.R. Lewis in 1970. The book publication was a pivotal event, giving the public and park personnel a well-documented interpretive history of American coast defenses. A group of coast defense history enthusiasts gathered at a meeting in 1978 and organized the Coast Defense Study Group (CDSG) in 1985. The CDSG's annual conferences, Journal, Newsletter, web site, and reprints of key coast defense books have played important roles in fostering interest in the history of American coast defense and assisting both the public and park staffs in understanding the fascinating history of these defenses and to interpret their surviving elements. These massive seacoast batteries have been able to withstand both the natural climate and economic development longer than other military features from the same periods. These structures incorporated the leading edge of technology of their time and that draws interest in studying them and interpreting purpose and history. Hopefully this will translate into efforts to preserve and restore these sites for current and future generations.

Battery Winchester, Fort Armistead, Baltimore, Maryland (Terry McGovern)

PERIOD MILITARY MAPS

This book contains a compilation of maps for all the harbor defense reservations utilized during the period 1900 to 1950. The harbor defense projects show a general map of the location of elements (sites) for each harbor and the individual site maps showing the fire control elements. A series of map symbol and abbreviation keys from 1921 and 1945 are included.

The directory is organized by Harbor Defense around the United States clockwise from Portland Maine, down the Atlantic coast to Key West Florida, across to Gulf coast to Galveston Texas, then up the Pacific Coast from San Diego California to the Puget Sound in Washington, then up to the Alaskan defenses of WWII, and followed by the defenses in Hawaii, the Philippines, Panama, the Caribbean, Bermuda and Newfoundland, Canada. While the status information is fairly comprehensive of the larger fort and military reservations, the status of many of smaller WWII-era fire control stations is not. The authors would appreciate receiving any updated information to correct or add to what has been presented here.

Notes on Coast Defense Maps

Site maps; site plans; exhibits from project plans, supplements and annexes; confidential blueprints; D-series maps—these are all terms that have been used to describe various maps which depict sites used by the U.S. Army, at one time or another, in connection with harbor defense fortifications and fire control. These maps have been keys to ferreting out the identification of the various remaining structures during site visits, yet there is some confusion over where these maps come from, what their cryptic symbols mean, and even what they are called.

Most maps of harbor defense installations are located in the Cartographic Branch of the National Archives. Many of the more frequently seen maps have come from a variety of National Archives holdings. The two concrete-era (1890-1950) map formats most frequently seen are the Confidential Blueprint map series (1900-1935 and 1940-1948) and the exhibits from the annexes/supplements to the harbor defense projects (1940s), which cover the 1940 Modernization Program (WWII-era) construction.

Confidential Blueprint Series Maps (1915-37)

As new construction finished, maps were created, revised, and updated by the Corps of Engineers. A series of maps was reproduced as negatives from a master positive in blueprint style, which meant maps were composed of white lines on a blue or dark background. As they were classified "confidential" by the War Department, they became known as "confidential blueprints."

A number of these confidential blueprints have been found in various cartographic and textual Corps of Engineers records in the National Archives. The confidential blueprint series of maps have general maps of each defended harbor, and general maps of each of the forts and military reservations in the harbor defense. If it was warranted, larger scale maps of parts of some forts were also included. These were labeled "D" for "detail" and followed in series, D-1, D-2, D-3, etc., as required. These maps show the location of batteries, various components of the fire control and communication system, mine facilities, and all the post buildings. Identification of each structure was shown by name, symbol, abbreviation, or number.

After 1900 an optical system for fire control based on trigonometric principles was developed for more precisely aiming coast artillery guns. The structures that were built to house the optical and communication elements of this system were often numerous and small in relation to the other major buildings on a military reservation, and many required a detailed description making it complicated to label them on a map, so a set of map symbols was developed to indicate the fire control structures. As these fire control structures were built in the years following 1905, they were incorporated into the maps on which the

Corps of Engineers recorded the location of all the structural elements of the fortifications in the seacoast defenses.

Keys to the fire control map symbols began appearing in coast artillery manuals, such as drill regulations, training regulations, and later field manuals. A complete update of these maps was performed during the years 1920-1922, just after the major construction projects of the Endicott and Taft programs were completed and before some of the smaller harbor defense areas were eliminated. These maps were kept as part of the records of the various Corps of Engineer district offices around the country. Copies were turned over to qualified parties in the army, such as the Coast Artillery Corps, the Quartermaster Corps, etc. On July 12, 1922, the Coast Artillery Board at Fort Monroe requested a complete list and set of these maps for their records, which were provided in August 1922. The 1922 collection contains about 290 maps of 29 harbor areas. Other versions of these maps were found in the notebooks kept by the engineer assigned to each harbor defense. In due course, the records of the Corps of Engineers and the other branches of the army have been turned over to National Archives. The map collections have been scanned and digitally "cleaned up" to remove extraneous lines and smirches from the scanning process.

WW II-Era Harbor Defense Project Maps

The 1940 "Modernization Program" brought a new set of harbor defenses, some on existing reservations, some on entirely new reservations. The fire control system was much more widespread and frequently located on newly obtained smaller reservations located around the harbor defense shoreline. Maps for these works in this guide come from the 1944-46 supplements to the various harbor defense projects published by the army.

A Harbor Defense Project was a written document which described all existing and projected harbor defense elements, including structures, first prepared in 1932-33. Supplements to the Harbor Defense Projects were prepared 1943-44 and updated during 1945-46. The supplements detailed the progress on the construction of the new 1940s modernization program defenses with descriptions and a set of maps that showed where these new structures were located, the field of fire of the guns, radar coverage, etc. The supplements provide extensive detailed information on all tactical and physical aspects of the harbor defenses on the date of the annex, both existing and proposed, and a number of exhibits detailing the locations of elements. The supplements are generally composed of 7 annexes:

A-   Armament
B-   Fire Control (including optical instruments and radar installations)
C-   Seacoast Searchlights
D-   Underwater Defenses (mines)
E-   Antiaircraft Artillery
F-   Gas Defense
G-   Equipment (usually detailing what was on hand and what was needed)
H-   Real Estate Requirements (usually detailing sites not yet obtained

These supplements and other forms of the Harbor Defense Projects have been scanned from the National Archives and are available from the CDSG ePress as electronic PDFs. These supplements contain a very comprehensive listings and exhibits of everything that was to be in place at the completion of new rearmament program and are the key references to consult for information on the final state of the American seacoast fortifications in 1945.

A few comments on the items that appear on the confidential blueprint maps and the Harbor Defense Project maps—

A **Harbor Defense** (called a "Coast Defense" before 1925) consisted of a series of land reserves (named as "Forts" and in some cases "Camps" and "Military Reservations") on which the various components of the seacoast defense fortifications were built to guard a major commercial and naval seaport. When the harbor defenses of the United States were modernized in 1890-1910, a new system of defensive works were created. The modern forts consisted of tactical and non-tactical structures spread over hundreds of acres of land. The U.S. Army Corps of Engineers selected the locations, purchased additional land, sited, designed, and constructed the tactical structures—gun batteries, mine facilities, observation stations, plotting rooms, power plants, switchboard rooms, and searchlight shelters.

Gun Batteries: The modern seacoast artillery consisted of guns, mortars and antiaircraft weapons mounted in concrete support structures varying from the simple to the quite complex. Guns were mounted on barbette, pedestal and "disappearing" carriages. Mortars were emplaced in protected pits. Antiaircraft weapons, usually the 3-inch guns, were mounted in simple concrete platforms. The term "battery" was used to describe a set of guns under a single commander together with the entire structure erected for the emplacement, protection, and service of those guns.

Fire Control Structures: The target range and azimuth for seacoast artillery guns were determined using command and equipment systems collectively referred to as fire control and position finding. The standard systems of position finding used by seacoast artillery were based on trigonometry. Components of the system included widely spaced base end stations, command stations, plotting rooms, tide stations, meteorological stations, and cable linked telephone communication systems with protected switchboards. Radar installations were deployed for the major gun batteries and as general surveillance after 1942. The radar installations included power/control buildings and antenna towers.

Searchlights: Most searchlights installed during the period 1901-1920 were fixed, located in a structure for concealment and protection during the day, with their electrical power generator. Over the years after WWI, the mobile searchlights became more reliable, durable, and rugged. By the late 1930s, the Coast Artillery switched to using mobile searchlights and replaced fixed searchlights where at all possible so after 1940 the US seacoast defenses used mostly mobile searchlights.

Controlled Mine Facilities: Throughout the modern or "concrete" era of American harbor defenses (1890-1950), mines were considered to be one of the primary harbor defense weapons. Mines were only deployed during times of war or during limited training expertises. The mines and cables were stored ashore between use. The mine shore facilities included torpedo storerooms, loading rooms, mine wharfs, explosives storage, tramway systems, cable tanks, mine casemates, and cable vaults.

Electrical Generator Power Plants: By the turn of the century, electricity had become a vital necessity for the Coast Artillery. It was used to traverse and elevate some of the large guns, to light emplacements, to operate ammunition hoists, to power searchlights, to control submarine mines, and for communications, in addition to standard garrison uses. Most large forts had a central power plant with electrical generators. The requirement that coast defenses be self-contained resulted in power rooms being included in most batteries and mining casemates, and separate searchlight powerhouses were constructed.

Protected Switchboard Rooms: As seacoast defense artillery covered increasing distances, a need for remote accurate and instant communications was required. Telephones connected by phone lines were integrated into the fire control system utilizing protected switchboard rooms after 1906. As radio communication developed in the 1930s, fixed radio sets were often integrated with the telephone communication system in their protected switchboard rooms or housed in separate protected structures.

Garrison Buildings: These are shown in the Confidential Blueprint series maps but not on Supplement series maps.

The system of numbering for buildings was the same for all Confidential Blueprint maps in the period 1915 to 1937. All buildings of the same "type" were given the same number on all the maps. For example all barracks buildings were numbered "7."

1.	Administration Building
2.	Commanding Officer's Quarters
3.	Officer's Quarters
4.	Hospital
5.	Hospital Steward's Quarters
6.	Non-commissioned Officer's Quarters
7.	Barracks
8.	Guard House
9.	Post Exchange

10 to 19 and 100 to 199	Post Buildings
20 to 29 and 200 to 299	Quartermaster Buildings
30 to 39 and 300 to 399	Ordnance Buildings
40 to 49 and 400 to 499	Engineer Department Buildings
50 to 59 and 500 to 599	Signal Corps Buildings
60 to 69 and 600 to 699	Reserved for future requirements
70 to 79 and 700 to 799	Religious and Social Buildings
80 to 89 and 800 to 899	Government Buildings not under War Dept. Control
90 to 99 and 900 to 999	All Private Buildings (Private dwellings, stores, contractor's buildings and buildings purchased with the land but not assigned to public use.)

Fort Columbia, Washington 1913 (NARA)
From left to right is the Post Exchange, a Company Barracks, the Administration Building,
a Double Officer's Quarters and the Commanding Officer's Quarters.
Just visible behind the front row of buldings is the Quartermaster's Storehouse and the Post Hospital

Symbols and Abbreviations—1921 Confidential Blueprints

Name	Abbr.	symbol	Sta. w/o roof
Fort Commander's Station	C		
Primary Station, Fire Command	F'		
Secondary Station, Fire Command	F''		
Supplementary Station, Fire Command	F'''		
Primary Station of a Battery	B'		
Secondary Station of Battery	B''		
Supplementary Station of a Battery	B'''		
Battery Commander's Station	BC		
Primary Station, Mine Command	M'		
Secondary Station, Mine Command	M''		
Supplementary Station, Mine Command	M'''		
Double Primary Station, Mine Command	M'-M'		
Double Secondary Station Station, Mine Command	M''-M''		
Separate Plotting Room	P		
Separate Observing Room	O		
Self-contained Horizontal Base	C.R.F.		
Emergency Station	E		
Spotting Station	Sp		
Meteorological Station	Met		
Tide Station	T		
Searchlight (30, 60, etc., relates to the size of the lights)	S		
Controller Booth	C.B.		
Watchers Booth	W		
Signal Station	S.S.		
Radio Station	R		
Cable Terminal	C.Ter.		
Post Telephone Switchboard	P.S.B.		
Mining Casemate	M.C.		

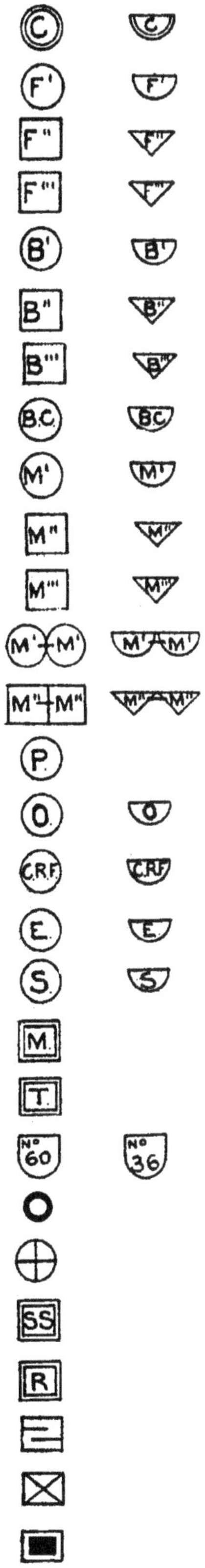

Name	Abbr.	symbol
Loading Room	L.R.	◉
Switchboard Room	S.W.B.	 ◩
Central Powerhouse	C.P.H.	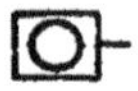 ◎⊢
Powerhouse (and Searchlight Powerhouse)	P.H.	□⊢
Combined Stations, in same room		 Ⓑ/Ⓑᴄ
Combined Stations, in communicating rooms		  Ⓕ'Ⓑ Ⓑᴄ/Ⓟ
Combined C and F' Station in same room		 Ⓕ'
Differentiation of auxiliary plants		 ⓐ⊢ ⓑ⊢ ⓒ⊢ etc.

Abbreviations used on maps

Cable Gallery	C.Gal.
Cable Tank	C.T.
Cable Hut (commercial cable)	C.H.
Coast Guard Station	C.G.S.
Engineer Wharf	Engr. Whf.
Gasoline Tank	G.Tk.
Guard House	G.H.
Latrine	L.
Lighthouse	L.H.
Lighthouse Wharf	L.H.Whf.
Magazine	Mg.
Mining Boathouse	M.B.H.
Mining Derrick	M.D.
Mining Tramway	M.T.
Ordnance Machine Shop	O.M.S.
Mine Wharf	M.Whf.
Private Wharf	Pvt.Whf.
Radio (commercial station)	Rad.
Railway Wharf	Ry.Whf.
Saluting Battery	Sl.B.
Searchlight Shelter	S.Sh.
Service Dynamite Room	S.D.R.
Steamship Wharf	S.S.Whf.
Sunset Gun	S.G.
Tide Gauge (not a Tide Station)	T.G.
Torpedo Storehouse	T.S.
Tower	Tw.
Water Tank	W.Tk.
Weather Bureau	W.B.

Additional Symbols and Abbreviations
Name   Abbr.   symbol

Name	Abbr.	symbol
Pumping Plant	P.P.	
Radio Powerhouse	R.P.H.	
Searchlight Powerhouse	S.P.H.	
60 inch Searchlight No. 7	$S._7^{60}$	
Coincidence Rangefinder	C.R.F.	
Quartermaster Wharf	Q.M.Whf.	

Subscripts for use in both Legend and on Face of Plat are—
Imp.   Improvised.   B″ imp.   B″imp.
            (for temporary fire control structures only.)
p.   Portable. S_{p2}^{36}
            (Principally used for portable searchlights etc.)
s.   Superseded.   24s.
            (for abandoned buildings, etc.)
t.   Temporary.   19t.
            (For all uses except fire control structures.)

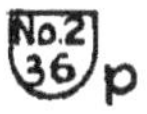

Datum Point—location indicated by intersection of lines
or by dot at end of arrow.

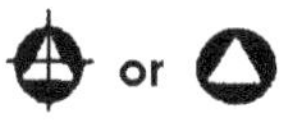

Triangulation Station.   TIMM   TIMM

Intersection Point.   ○ BLACK BEACON

Benchmark.   B.M.
                                                                    ✕
                                                                   1232

Lighthouse.   ✸ L.H.

Such other topographic signs as were necessary were taken from the *Engineer Field Manual* (Professional Papers, Corps of Engineers, No. 29) pages 74 to 97.

Note: Maneuver buildings were classed as post buildings.

SYMBOLS and ABBREVIATIONS 1940
FM 4-155, Reference Data (Seacoast Artillery and Antiaircraft Artillery) 1940
TABLE C.-Symbols for seacoast artillery fire-control maps, diagrams, and structures

Part 1.—Basic symbols

Name	Abbreviation	Symbol
Harbor defense command post	H D C P	 (H)
Groupment command post	Gpmt C P	 (C)
Fort command post	Ft C P	 (F)
Gun group command post	G C P	 (G)
Mine group command post	M C P	 (M)
Seacoast battery command post	B C P	 (BC)
Harbor defense observation station	H D O P	 △H
Groupment observation station	Gpmt O P	 △C
Fort observation station	Ft O P	 △F
Gun group observation station	G O P	 △G
Mine group observation station	M O P	 △M
Battery observation station	B O P	 △B
Emergency observation station	E O P	 △E
Antiaircraft observation post	A A O P	 AA △
Battery spotting station	S O P	 △S
Separate observation station	O P	 △O

Name	Abbreviation	Symbol
Operations and plotting room	O P R	
Plotting room	P	
Self-contained base range-finder station	R F	
Magazine	Mg	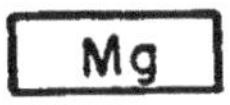
Shellroom	S Rm	
Temporary or improvised fire-control structures	Imp	
Mine casemate	M C	
Mine loading room	L R	
Searchlight, 60-inch seacoast	S L	
Searchlight, seacoast, other than 60-inch	S L	
Antiaircraft searchlight	A A S L	
Searchlight shelter	S Sh	
Searchlight powerhouse	S P H	
Searchlight controller booth	C B	
Data booth	Data B	
Watchers booth	W Bth	
Meteorological station	M E T	

Name	Abbreviation	Symbol
Tide station	Td	 [T]
Signal station	S S	 [SS]
Fire Control switchboard room	F S B	 ◩
Post telephone switchboard room	P S B	 ◹
Combined fire-control & post telephone S B room	F S B P S B	 ⊠
Cable terminal	C Ter	 ⊟
Powerhouse	P H	 ⊟
Radio powerhouse	R P H	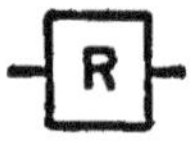 [R]
Central powerhouse	C P H	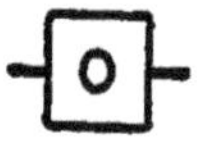 [O]
Pumping plant	P P	 [P]
Datum point		 ● OR
Triangulation station		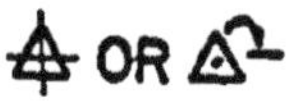 △ OR △
Intersection point		 ○ Black Beacon
Benchmark	B M	 BM ✕ 1232
Lighthouse	L H	 ★

Other abbreviations used in this guide

BS - base end station & spotting station
HECP - harbor entrance command post
HDOP - harbor defense command observation post
HDCP- harbor defense command post
SBR -telephone system switchboard or radio room
AMTB- Anti-motor torpedo boat BC station

BC - battery commander's station
C - fort commanders station
G- group command station
M- mine station
SCR - signal corps radar
SL - searchlight

Part 2.-Numbers for harbor defense installations.—a. In harbor defense, seacoast artillery installations of each type are numbered consecutively from right to left, facing the center of the field of fire of the harbor defense. Antiaircraft installations pertaining to the harbor defense may be numbered in any convenient sequence.

b. Groupments, gun groups, mine groups, batteries, and all installations functioning directly under the harbor defense commander, such as harbor defense observation stations, searchlights, and underwater listening posts, are numbered consecutively, each type in a separate series, beginning with number 1. These numbers normally are shown as subscripts to the letter included in the appropriate symbol. Exceptions are included among the examples that follow.

Name	Abbreviation	Symbol
Harbor defense observation station	$HDOP_3$	
Fort observation station	$FtOP_3$	
Antiaircraft observation post	$AAOP2$	
Magazine or shell room	$Mg2$ or $SRm2$	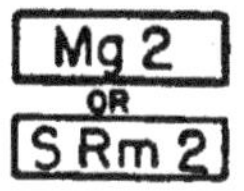

c. Groupment, group, and battery observation and spotting stations assigned to a unit are numbered consecutively within the unit, each type in a separate series, beginning with number 1. These numbers are shown as superscripts to the letter included in the appropriate symbol, the unit number remaining as the subscript.

Name	Abbreviation	Symbol
Groupment observation station	$Gpmt_2 OP_2$	
Gun group observation station	$G_2 OP_1$	
Mine group observation station	$M_2 OP_1$	
Battery observation station	$B^1_1 OP$	
Spotting station	$S^1_3 OP$	
Emergency observation station	$E_2^1 OP$	
Temporary or improvised fire control structures	$B_3^2 Imp.$	

d. In certain cases it is desirable to show additional information regarding an installation, such as its size and whether fixed, portable, or mobile. Such information is placed either in the symbol or to the right thereof.

Name	Abbreviation	Symbol
60-inch seacoast searchlight; fixed, portable or mobile.	SL 2F (P or M)	2F(P or M)
Seacoast searchlight other than 60-inch	SL^{36}_{3P}	36'
Antiaircraft gun battery or composite battery, fixed or mobile.	A A No. 2 (F or M)	AA 2 (F or M)

e. Where two stations are combined in one room, the symbols are superimposed one upon the other, and the letters representing each station are inclosed in the combined symbol.

Name	Abbreviation	Symbol
Combined groupment command post and fort command post.	Gpmt Ft Cp	CF
Combined battery observation and spotting station.	$B^2_1 S^2_1$ O P	$B^2_1 S^2_1$
Combined group command post and battery command post.	$G_1 B_2$ C P	G_1 / BC_2
Combined battery command post and battery observation station.	B_2C P $B_2{}^2$O P	B^2_2 / BC_2

f. Where stations are adjacent in the same structure, the symbols are tangent to each other and are arranged to show the relative location, as:

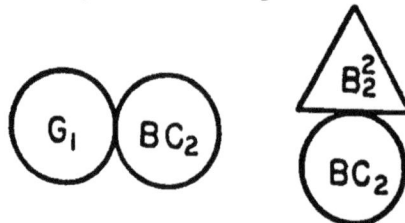

g. Where communication may be had by voice through a passage, door, window, or voice tube, the symbols are left open at the point of contact, as:

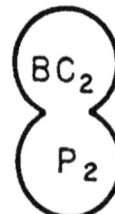

Part 3.—Communications symbols for use on harbor defense fire-control charts and diagrams.

Telephone cable (numerals indicate number of pairs and gage)	 26-19
Speaking tube	
Mechanical data transmission line	
Electrical data transmission line	
Searchlight controller line	
Zone signal and magazine telephone line	
Firing signal line	
Time interval bell line	
Submarine cable (numerals indicate number of pairs and gage)	 50-19

Part 4.-Abbreviations

Cable gallery	C Gal
Cable tank	C T
Cable hut (commercial cable)	C H
Coast Guard station	C G S
Engineer wharf	Engr Whf
Gasoline tank	G Tk
Guardhouse	G H
Latrine	L
Lighthouse wharf	L H Whf
Mine boathouse	M B H
Mine derrick	M Drk
Mine tramway	M Tmy
Mine wharf	M Whf
Ordnance machine shop	O M S
Private wharf	Pvt Whf
Radio (commercial station)	Rad
Railway' wharf	Ry Whf
Saluting battery	Sl B
Service dynamite room	S D R
Steamship wharf	S S Whf
Quartermaster wharf	Q M Whf
Superseded (for abandoned buildings, etc.)	24 s
Temporary (for all uses except fire-control structures)	19 t
Sunset gun	S G
Tide gage	T G
Torpedo storehouse	T S
Tower	Tw
Water tank	W Tk
Weather bureau	W B

A DIRECTORY OF AMERICAN SEACOAST DEFENSES 1890-1950

This directory is a comprehensive guide to all the major locations and sites used for harbor defense, with maps showing what was at each site and comments on the current status of each site (extant, in ruins, destroyed, privately owned, current U.S. military use, federal, state, county, city parks, etc.) as far as information is known to the authors. While the status information is fairly comprehensive for the larger forts and military reservations, the status of many of smaller World War II-era fire control sites is not. The authors would appreciate receiving any updated information to correct or add to what has been presented here. Terms used in this reference work to describe the various periods of construction such as "Endicott-Era," "Taft-Era," "Post-World War I-era," "World War II-Era," the "100-Series" and "200-Series" batteries, etc. are terms used by modern historians and were not used by the Army to describe these programs in progress. Note that several of the planned batteries in the 1940 program were cancelled before any work was done as denoted by their *battery # in italics* and as (planned).

This directory does not cover the following artillery used for seacoast defense at various times between 1898 and 1945—the Rodman guns emplaced or re-emplaced during the Spanish American War; the Navy guns and mounts installed during the World War II years, mostly in the Pacific theater; Hawaii's World War II temporary and provisional defenses; the fixed antiaircraft gun batteries emplaced in the defenses from 1920; mobile artillery which had prepared positions including those for 12-inch railway mounted mortars and 8-inch railway guns; the Panama mount positions for the tractor drawn 155 GPF guns; and the positions on Oahu for the 240 mm howitzers.

The directory is organized by Harbor Defense around the United States clockwise from Portland Maine to the Puget Sound in Washington State, followed by the Alaskan defenses of World War II, and the defenses in Hawaii, the Philippines, Panama, the Caribbean, Newfoundland, and Bermuda.

This directory includes detailed brief histories of modern-era American coast artillery concrete gun batteries. Glen Williford created this as a personal reference guide over many years of research and study of U.S. Coast Artillery history. This battery listing includes the histories of all modern (post-1886) "fixed" or permanent concrete seacoast gun batteries emplaced by the U.S. in the country and outlying territories. The emplacements were mostly built by the Corps of Engineers, and manned by the Coast Artillery. Each battery description includes the following information where possible. The battery name in capital letters if an officially conferred name, in lower case if just an informal, local, or construction designation. The description then briefly covers the purpose for construction and the general location on the reservation, particularly in relation to other elements. In most cases the act or source of original funding (which does not include the cost of coast artillery) and date of plan submission follows. Major design features or significant variations from Mimeograph Type plans are discussed. The general dates of construction, transfer date to the Coast Artillery and engineering costs may also be included. This is followed by a description of the armament, including gun and carriage models and specific serial numbers and date of mounting if known. In general, the manufacturer of the guns and carriages are only designated only when there are multiple producers and thus duplicate serial number runs. The general order and date, that names each battery is included with a brief description of person honored. Subsequent service events, including any major alterations, accidents, armament, or name changes follows. The date of gun dismounting or at least the date of authorization for deletion is covered. A brief statement on whether the battery still exists, or when destroyed, and park or status if on public property concludes the description.

The major sources consulted were: Reports of Completed Batteries, and Reports of Completed Works, Engineer Letters of Submission, surviving Fort Record Books, Seacoast Gun Record cards and earlier Ordnance Department Seacoast Gun Ledger Books, Annexes to Seacoast Projects, General Orders naming citations, supplemented by various records in archive primary engineer correspondence files, annual reports of the Chief of Engineers, private Williford studies on emplacement accidents, temporary defenses, defenses of the Spanish American War.

THE HARBOR DEFENSES OF PORTLAND — MAINE

Portland Maine is the northern-most major US harbor. Protected by a series of forts from the colonial times through World War II, the forts and locations spread out along the southern coastline from Portland and on several Islands. Casco Bay was one of the original deep seaports used by British and French explorers beginning in the late 15th century. A British colony was established in 1623, and fortifications were established over the years for defense against native Americans, the French, and later against the British as well. The area around Casco Bay and the river bays to the north have seven fortifications built under the second system (1794-1812) and third system (1816-1867) of fortifications. New defenses were constructed but not completed during the 1870s. During the Endicott and Taft Period 28 modern concrete batteries were built at 6 locations, with facilities for controlled submarine mines and four major garrison facilities. Upgraded with a new 12-inch battery in 1920s, and four new major gun batteries and 10 AMTB batteries during World War II, the Portland defenses contain examples of all major American seacoast artillery-based weapon systems. No longer needed after seacoast artillery was discontinued in 1948, the reservations were turned over to a variety of civil and private interests. Portland still retains an outstanding collection of remaining seacoast fortifications from unique second and third system forts, largely intact early modern era forts, and World War II era batteries. In all there is a significant collection of seacoast defense structures that are rarely seen elsewhere.

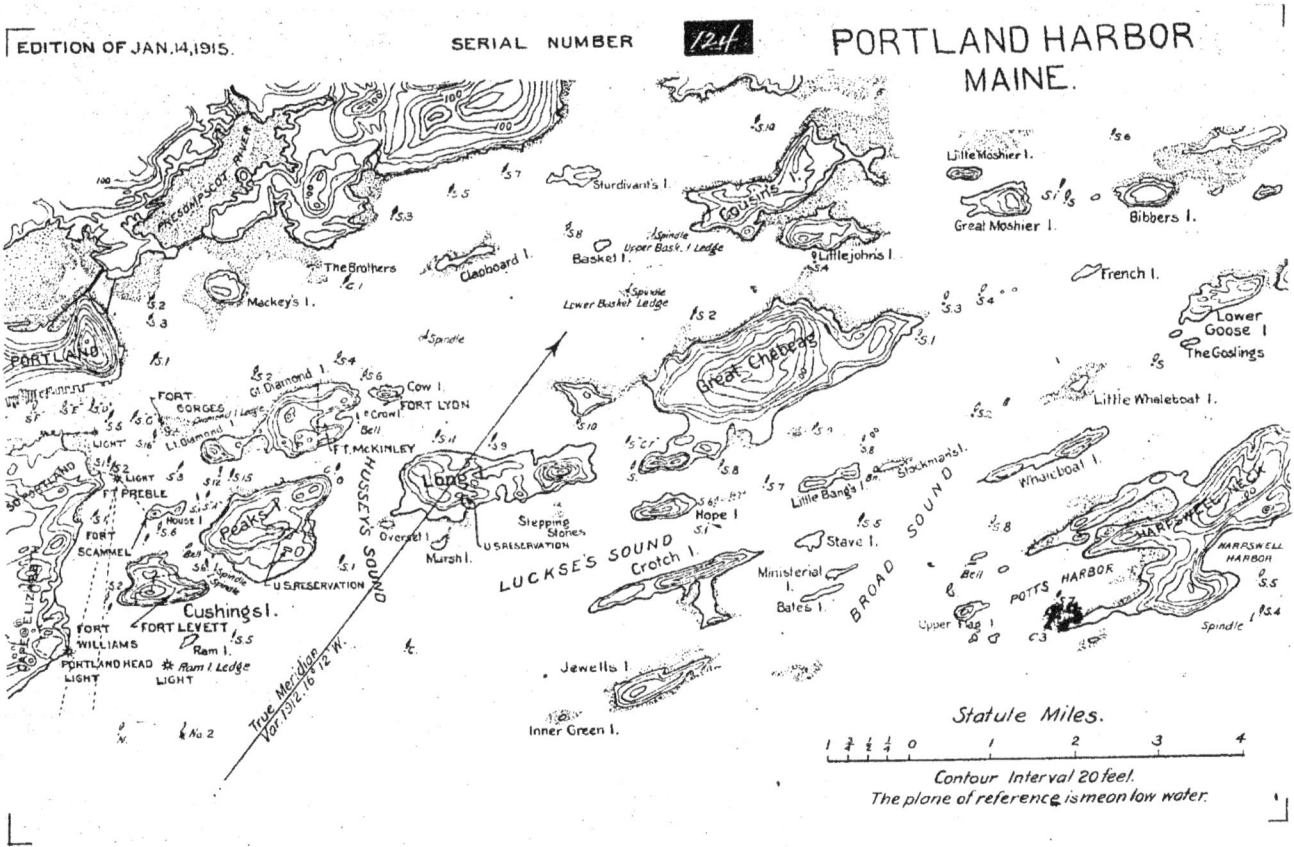

BATTERY BERRY, 12" D.C.
MAGAZINE TO ROAD - 2810 FT.
TO DWELLING - 3000 FT.

BATTERY ABBOTT, 3" G.
MAGAZINE TO ROAD -
TO DWELLING -

BATTERY BAYARD, 15 PDR.
(ABD), MAGAZINE TO ROAD -5550 FT.
TO DWELLING -5700 FT.
TNT STORAGE.

BATTERY HONEYCUTT, 8" D.C.
MAGAZINE TO ROAD - 2400 FT.
TO DWELLING - 2700 FT.

BATTERY RAMSEY, 15 PDR. (ABD)
MAGAZINE TO ROAD - 2400 FT.
TO DWELLING - 2700 FT.

BATTERY WEYMOUTH, 8" D.C.
MAGAZINE TO ROAD - 2850 FT.
TO DWELLING - 2320 FT.

BATTERY THOMPSON, 8" G.
MAGAZINE TO ROAD - 2400 FT.
TO DWELLING - 2330 FT.

BATTERY ACKER, 6" R.
MAGAZINE TO ROAD -2460 FT.
TO DWELLING - 2610 FT.

Fort McKinley

GREAT DIAMOND I.

LITTLE DIAMOND I.

BATTERY CARPENTER, 6" R.
MAGAZINE TO ROAD - 2910 FT.
TO DWELLING - 2580 FT.

BATTERY INGALLS, 12" M.
MAGAZINE TO ROAD - 1800 FT.
TO DWELLING - 1980 FT.

1. MINE STORE HOUSE
TO ROAD -
Ft. Lyons
Cook

L O N G

MARINE

PEAKS IS.

BATTERY FOOTE, 12" B.C.
MAGAZINE TO ROAD - 2100 FT.
TO DWELLING - 1980 FT.

FORT GORGES

SPRING PT. LT. HO.

BATTERY CHASE, 12" M.
MAGAZINE TO ROAD 480 FT.
TO DWELLING 110 FT.

4. BATTERY KEARNEY, 12" M.
MAGAZINE TO ROAD 720 FT.

Ft. Preble

BATTERY MASON, 3" GUN.
MAGAZINE TO ROAD-1300 FT.
TO DWELLING - 1140 FT.

BATTERY RIVARDI, 6" R.
MAGAZINE TO ROAD-1500 FT.
TO DWELLING - 1140 FT.

BATTERY KENDRICK, 10" D.C.
MAGAZINE TO ROAD - 620 FT.
TO DWELLING - 1100 FT.

CUSHING IS.

BATTERY BOWDOIN, 12" D.C.
MAGAZINE TO ROAD - 600 FT.
TO DWELLING - 1825 FT.

BATTERY DANIELS, 15 PDR. (ABD.)
MAGAZINE TO ROAD - 690 FT.
TO DWELLING - 740 FT.

BATTERY FERGUSON, 6" R.
MAGAZINE TO ROAD - 740 FT.

Ft. Levett

SOUTH PORTLAND

1. BATTERY KEYES, 3" G.
MAGAZINE TO ROAD -950 FT.
TO DWELLING - 900 FT.

2. BATTERY SULLIVAN, 10" G.
MAGAZINE TO ROAD - 1710 FT.
TO DWELLING -1500 FT.

3. BATTERY DEHART, 10" D.C.
MAGAZINE TO ROAD - 1200 FT.
TO DWELLING - 1110 FT.

FT. WILLIAMS

BATTERY BLAIR, 12" D.C.
MAGAZINE TO ROAD - 720 FT.
TO DWELLING - 720 FT.

BATTERY GARESCHE, 6" R.
MAGAZINE TO ROAD-18 FT. TO DWELLING-300 FT.

4. MINE STORE HOUSE
TO ROAD 1020 FT.   TO DWELLING-1080 FT.

PORTLAND HEAD LT.

HARBOR DEFENSES OF PORTLAND

EXHIBIT NO. 2-A

PREPARED IN THE OFFICE OF THE ARTY. ENG., Ft. Preble, Me.

Approved:

O. H. Schrader
Lieut. Col. C.A.C., Commanding

Yards

RF
20 000

JAN. 7, 1938

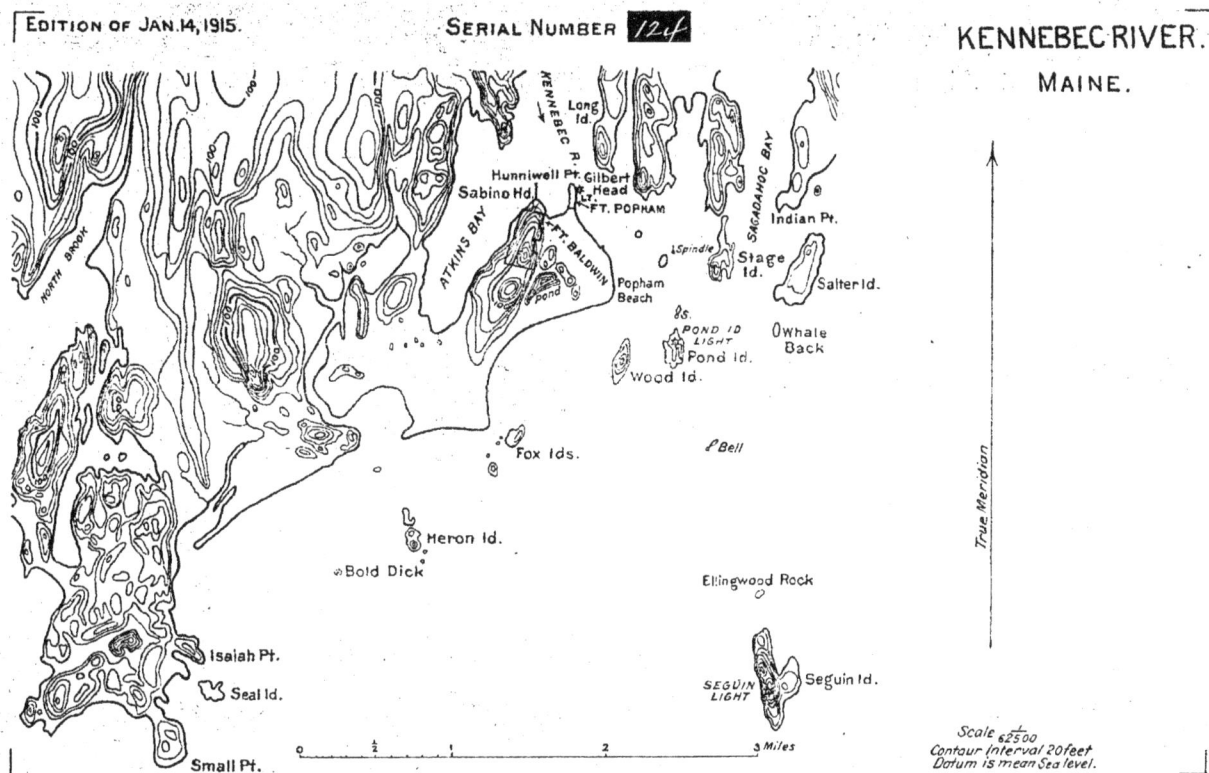

EDITION OF JAN. 14, 1915.　SERIAL NUMBER 124

KENNEBEC RIVER.
MAINE.

Fort Baldwin (1905-1928) is located on Sabino Head, near the Fort Popham (a Third System fort). Located off State Highway 209, Fort Baldwin was established in 1905 as part of the Endicott Program and completed three years later to guard the entrance to the Kennebec River. It was named in General Order 20, Jan. 29, 1906, after Colonel Jeduthan Baldwin, an engineer for the Continental Army during the American Revolution. The 45-acre military reservation was deactivated in 1924 and was acquired by the State of Maine as a state park. In 1942, four Panama mounts for mobile 155mm GPF seacoast artillery were constructed within decommissioned Battery Hawley. A fire control tower for a 16-inch battery at Portland was also built on this site. With only three batteries and controlled mine facilities, Fort Baldwin was a small coast artillery fort. Today, the former fort continues to be a state historic site with only its primary concrete structures remaining. The day use park is open during daylight hours.

Fort Baldwin Gun Batteries

- During the Spanish American War an emergency emplacement for a single modern 8-inch gun on a strengthened 15-inch Rodman carriage was authorized for old Fort Popham on the Kennebec River. It was built just to the south of the uncompleted old work in a series of three old Rodman emplacements. It consisted of just the concrete platform and the adjacent magazine. Work was done in 1898-1899. The gun was shipped there on August 13, 1898 (Watervliet Model 1888 #15) and mounted on the carriage in early 1899. The emplacement was not named. Armament returns for 1907 and 1910 show it was still mounted, but it was probably removed right after the latter date. The emplacement still exists on the Fort Popham State Historic Site. The battery is open to public.

- **HARDMAN:** An emplacement for a single 6-inch disappearing gun emplaced near the center of the reservation. Actual construction was done in 1902-04, for transfer on November 13, 1908 at a cost of $33,000. It was named on General Orders No. 20 on January 25, 1906 for 2nd Maryland Infantry Regiment Captain John Hardman who died in 1780 in British captivity. The battery after

KENNEBEC RIVER, ME.

FORT BALDWIN

SABINO HEAD.

SERIAL   NUMBER

EDITION OF JAN.14,1915.
REVISIONS: DEC.7,1915; APRIL 10,1920.

LEGEND.

3. OFFICERS QRS.
4t. HOSPITAL.
6. N.C.OFFICERS QRS.

8t. GUARD HOUSE.

12 BOAT HOUSE.
13. STABLE.
14. WOOD SHED.
15t. STOREHOUSE.

31. ORDNANCE ST.HO.
4t. HOISTER HOUSE AND CEMENT SHED.
80. LIGHT-KEEPER'S HO.
81 WOOD SHED.
82. OIL HOUSE.

BATTERIES.

HAWLEY ___ 2-6"P.
HARDMAN ___ 1-6"D.
COGAN ___ 2-3"P.

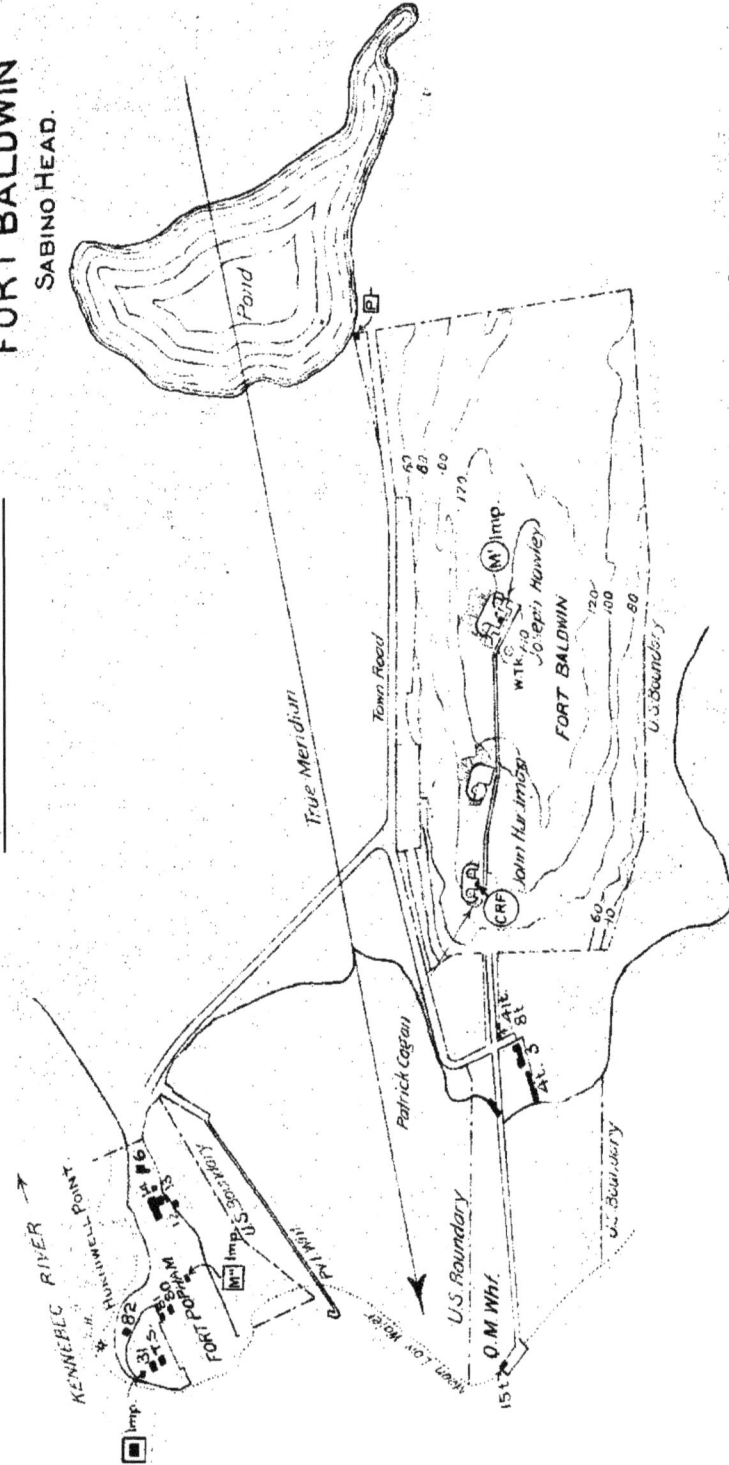

Scale of feet.
100' 0 100 200 300' 400' 500' 600' 700' 800' 900' 1000' 1100'

The plane of reference is mean low water.   Contour Interval 20 feet.

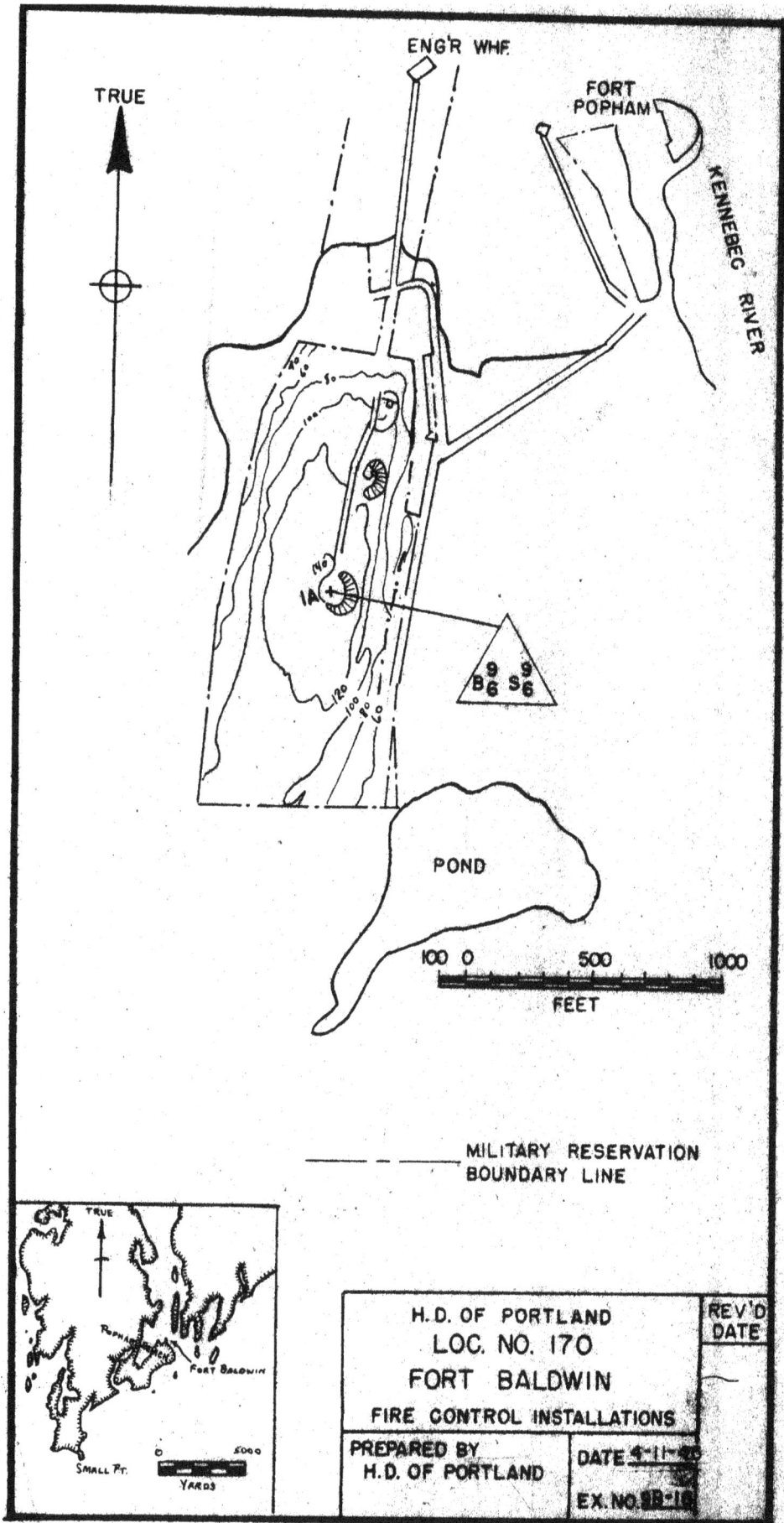

ENG'R WHF.

TRUE

FORT POPHAM

KENNEBEC RIVER

IA

No.

9 9
B6 S6

POND

100 0 500 1000

FEET

MILITARY RESERVATION
BOUNDARY LINE

TRUE

FORT BALDWIN

SMALL PT.

0 5000

YARDS

H.D. OF PORTLAND
LOC. NO. 170
FORT BALDWIN
FIRE CONTROL INSTALLATIONS

REV'D
DATE

PREPARED BY
H.D. OF PORTLAND

DATE 4-11-

EX. NO. SB-18

Fort Baldwin 1920s (NARA)

Fort Baldwin 1936 (NARA)

Fort Popham and the Spanish American War emergency battery (Glen Williford)

Battery Cogan Fort Baldwin State Park (Glen Williford)

Battery Hawley and fire control tower Fort Baldwin State Park (Glen Williford)

some delay was armed with one 6-inch Model 1905 gun tube on a Model 1903 disappearing carriage (#10/#71) The carriage was received and mounted in July 1908, and the tube in mid-1910. This gun fired only five shots during its service at the post, being held in an almost caretaker status for the life of the battery. The armament was dismounted on November 26, 1917 and shipped away to Watervliet on December 8, 1917. The carriage was scrapped in place in May 1920. The emplacement was not used thereafter for armament but still exists on the Fort Baldwin State Historic Site. The battery is open to the public.

- **HAWLEY:** A battery for two 6-inch pedestal guns erected to the north of Hardman also in the 1903-1904 timeframe. The work was a little delayed due to the 1903 maneuvers held in nearby Casco Bay. Transfer was finally made on November 13, 1908 for a construction cost of $45,000. It was named on General Orders No. 20 of January 25, 1906 for Brevet Major General Joseph B. Hawley of Civil War service. The battery was armed with two 6-inch Model 1900 Watervliet gun tubes on Model 1900 pedestal mounts (#47/#31 and #48/45). There had been a considerable delay in mounting the armament, one of the tubes was not received until 1912. The guns were removed in 1918 for use on wheeled guns for Europe, the battery was finally abandoned with the rest of the Kennebec defenses in 1924. In 1941-42 the emplacement was modified to hold Panama mount rings for 155mm guns—actually having the racer rings superimposed on the old gun platforms. So modified, the emplacement still exists on the Fort Baldwin State Historic Site. The battery is open to the public.

- **COGAN:** A battery for two 3-inch pedestal guns, emplaced at the northern end of the sequence of gun batteries, to the north of Battery Hawley. There were siting problems due to the rock ledge at the selected location. Initially plans called for placing the No. 2 gun (left flank) lower than the No. 1 gun and at a greater distance than normal. However, in May 1906 this plan was rejected, and the battery was built to standard type plan designs. Work was done in 1906-1907, and by May 1908 the guns were reported on hand, but not yet mounted. Transfer finally came on November 13, 1908 at a cost of $20,500. It was named on General Orders No. 20 of January 25, 1906 for 2nd Lieutenant Patrick Cogan, who died during Revolutionary War in 1778. It was armed with two 3-inch Model 1903 guns and pedestals (#19/#22 and #21/#23). It was disarmed about 1943 with the authorized abandonment of these defenses. The emplacement still exists on the Fort Baldwin State Historic Site. The battery is open to public.

Jewell Island Military Reservation (1942-1948) is located on Jewell Island in Casco Bay. Reserved for military use in 1940, construction on the island included 6-inch battery (Construction Number #202), two AMTB batteries and two large fire control towers. A base-end station for Battery Foote (Fort Levett) was built here in 1934. After the war the site was disarmed and returned to state control as a state nature preserve. Accessible only by boat, the trails on 221-acre, uninhabited Jewell Island enable visitors to explore one of Casco Bay's outermost islands and the remains of the military installation.

Jewell Island M.R. Gun Batteries

- **Battery #202:** A 1940 Program dual 6-inch barbette battery emplaced at the southwestern end of Jewell Island, Casco Bay. Its field of fire was to the southeast. The construction was originally assigned national priority #25, and its construction was funded under the FY-1942 budget. Work was done in 1942-1944 but was never completed. Structural work was finished on the battery, but the armament and other equipment was never installed. The design was typical of the 200-series 1940 Program emplacements. It was intended to be armed with two 6-inch M1(T2) guns on Model M3

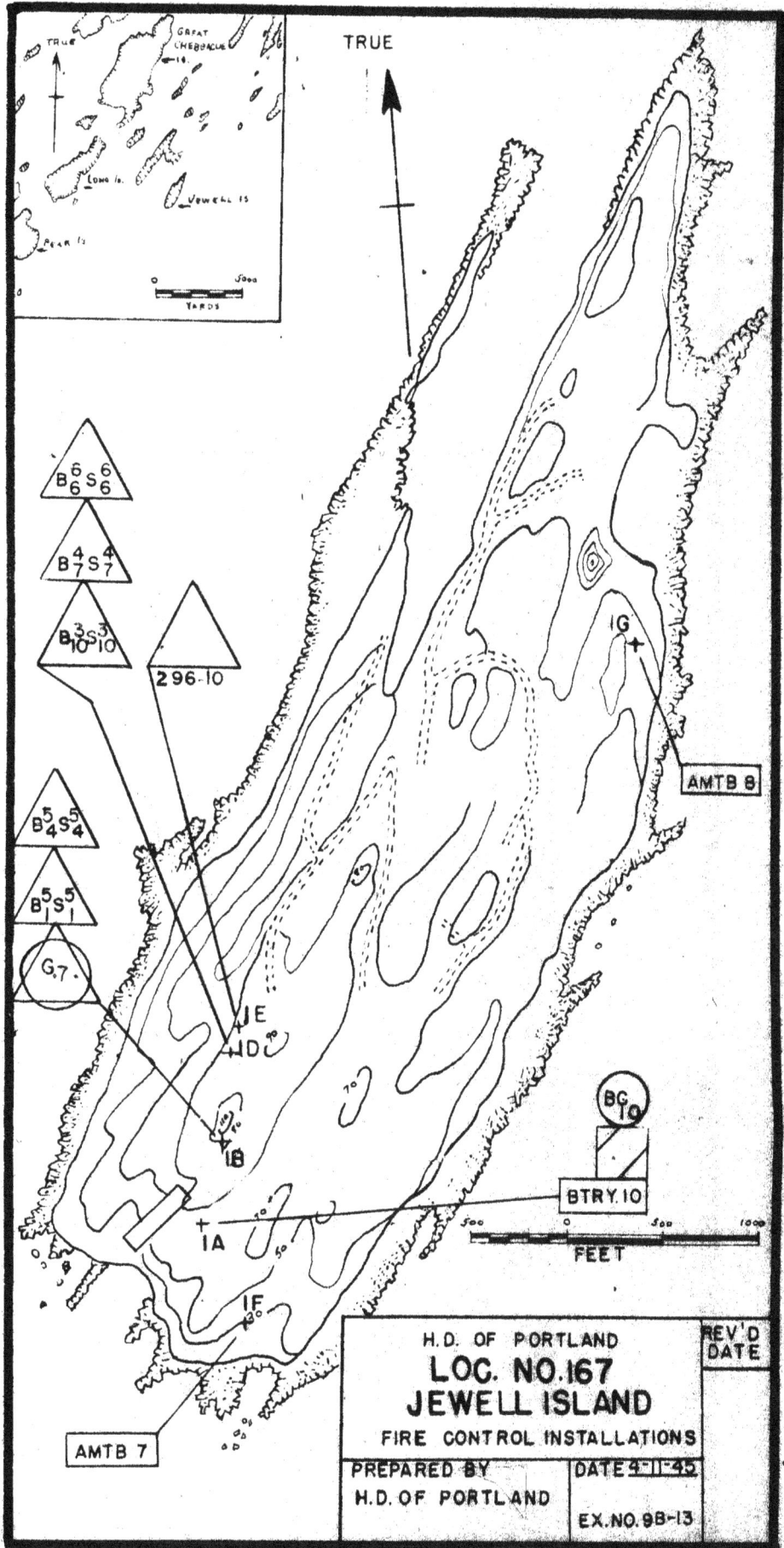

TRUE

GREAT CHEBEAGUE

Long Is.

Jewell Is

PEAK Is

YARDS

TRUE

$B_6^6 S_6^6$

$B_7^4 S_7^4$

$B_{10}^3 S_{10}^3$

296-10

IG

AMTB 8

$B_4^5 S_4^5$

$B_1^5 S_1^5$

G.7

IE

ID

BG_{10}

IB

BTRY.10

IA

500   0   500   1000

FEET

IF

AMTB 7

H.D. OF PORTLAND

LOC. NO. 167
JEWELL ISLAND

FIRE CONTROL INSTALLATIONS

REV'D DATE

PREPARED BY
H.D. OF PORTLAND

DATE 4-11-45

EX.NO. 9B-13

barbettes (slated for #20/#8 and #21/#9). The 1946 Program recommended the battery's completion, but the site was soon abandoned. The emplacement still exists on Jewell Island, which is administered by the Maine Bureau of Parks. The battery is open to the public.

- **AMTB-967**: A 1943 Program AMTB battery consisted of two 90mm fixed and two 90mm mobile guns. It was emplaced on the southern end of the island, 700-feet southeast of Battery #202. Work was done in 1943, and it served locally as Tac-7. After the war the guns were removed. The emplacement still exists on Jewell Island, which is administered by the Maine Bureau of Parks. The battery is open to the public.

- **AMTB-968**: A 1943 Program AMTB battery consisted of two 90mm fixed and two 90mm mobile guns. It was emplaced on the northeast side of the island. Work was done in 1943, and it served locally as Tac-8. After the war the guns were removed. The emplacement still exists on Jewell Island, which is administered by the Maine Bureau of Parks. The battery is open to the public.

Long Island Military Reservation (1942-1948) is located on Long Island in Casco Bay. During World War II, Casco Bay became United States Navy base "Sail" for destroyers escorting HX, SC, and ON convoys of the Battle of the Atlantic. Facilities constructed on the island included the Torpedo Control Officers School of the Portland Naval Training Center, a navy supply pier with a naval fuel annex, and the Naval Auxiliary Air Facility Casco Bay seaplane base operated as part of Naval Air Station Brunswick from 14 May 1943 to 15 December 1946. The United States Army Coast Artillery Corps built two batteries of 90 mm dual-purpose guns on the island as the Long Island Military Reservation, part of the Harbor Defenses of Portland. A concrete fire-control station with an SCR-296A radar generator house still exists between the two AMTB positions. A 1917 double searchlight station is also located nearby. A U.S. Navy fuel depot and seaplane base were located on the opposite side of the island during World War II, now town property (Long Island Conservation Area - about 200 acres). There is direct ferry service from Portland. Most of the former military sites are on private property.

Long Island M.R. Gun Batteries

- Long Island received an interim, temporary AMTB battery consisting of three 3-inch Model 1902 guns in 1942. The guns and pedestals (#15/#15, #12/#12, and #3/#3) were in local storage, having been removed from various Portland batteries in the recent past. These were probably removed in 1944.

- **AMTB-965**: A 1943 Program AMTB battery for two 90mm fixed and two 90mm mobile guns. Serving as Tactical Battery No. 5, it was on a site on the southern end of the island. It apparently served from 1943-1945 and was disarmed after the end of the war. The blocks still exist. The battery is closed to the public.

- **AMTB-966**: A 1943 Program AMTB battery for two 90mm fixed and two 90mm mobile guns. Serving as Tactical Battery No. 6, it was on a site on the northeastern tip of the island. It apparently served from 1943-1945 and was disarmed after the end of the war. The blocks still exist. The battery is closed to the public.

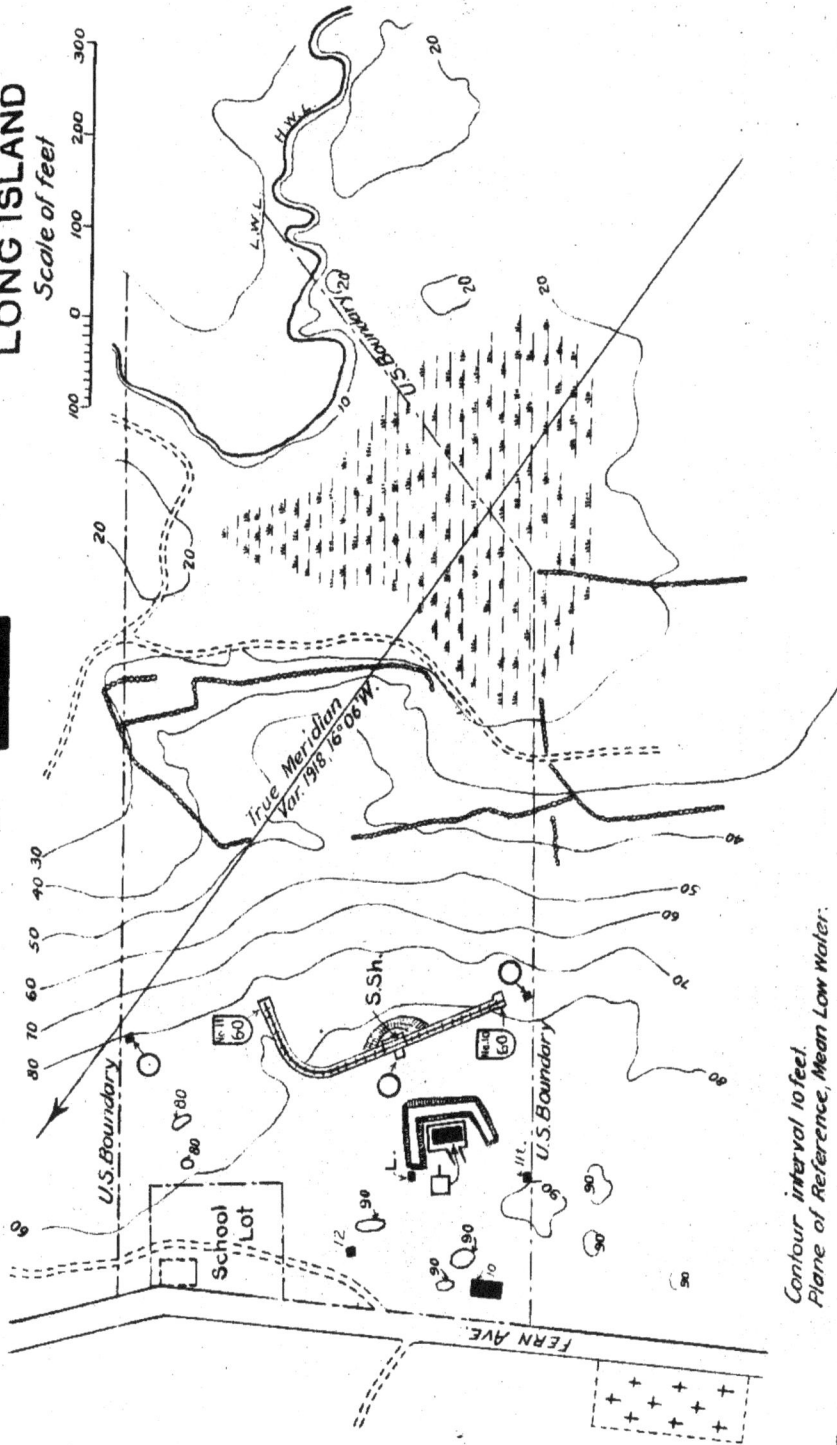

PORTLAND HARBOR, MAINE.
LONG ISLAND
Scale of feet

300 200 100 0 100

SERIAL NUMBER

EDITION OF OCTOBER 20, 1919.

LEGEND

10 DORMITORY.
11 STABLE.
12 WELL HOUSE.

Contour interval 10 feet.
Plane of Reference, Mean Low Water.

U.S. Boundary

True Meridian
Var. 1918 16°06′W.

School Lot

FERN AVE.

Bailey Island Military Reservation (1943-1946) is located on Bailey Island in Casco Bay. Anti Motor Torpedo Boat Battery 970 is here now partly covered. Two fire-control stations are still located behind the AMTB location. An SCR-296A radar was also here. All on Private property. Bailey Island is one of three large islands, including Sebascodegan (Great Island), and Orr's Island, connected by bridges to themselves and the mainland. Some experience the island as part of a scenic cruise to Eagle Island or Mackerel Cove. Others may come by car, taking the 15-mile drive south from Brunswick, or the 50-mile drive from Portland.

Bailey Island M.R. Gun Batteries

- **AMTB-970:** A 1943 Program AMTB battery consisted of two 90mm fixed and two 90mm mobile guns. It was emplaced on the southern tip of Bailey Island and fired to the southwest. It served between 1943 and removal probably early postwar in 1946. The two fixed blocks still exist, though heavily sodded over. The battery is closed to the public.

Great Chebeague Island Military Reservation (1942-1946) was located on Great Chebeague Island in Casco Bay. There were three batteries, two at the same site at the north tip, and one on the west shore. Chebeague Island can be accessed by ferry from Portland or Yarmouth.

Great Chebeague Island M.R. Gun Batteries.

- A detached 37mm AMTB battery (Battery 367) was located at Division Point on the west shore. Private property.

- At the north tip of the island, a temporary, unnamed AMTB battery (1942-1944) was emplaced. It consisted of simple gun blocks for two 3-inch Model 1902 guns (#14 and a second of unknown serial number). Apparently, it was removed in 1944 with the completion of the more modern 90mm battery. The two guns blocks still exist. The battery is closed to the public.

- **AMTB-969:** A 1943 Program AMTB battery consisted of two 90mm fixed and two 90mm mobile guns. The blocks for the fixed emplaced on the northern tip of Great Chebeague Island to cover the open approach into Casco Bay. from the 1943 program was located at the same site as the unnamed battery. The fixed guns covered the northern approaches to the inner Casco Bay through Broad Sound. The guns were removed early postwar. The gun blocks still exist on private property.

Peaks Island Military Reservation (1942-1948) is located on Peaks Island in Casco Bay. As part of the 1940 Modernization Program, a 16-inch casemated battery (Construction Number #102) and 6-inch battery (Construction Number #203), two AMTB batteries, a mine casemate, and several fire control structures were built on the island. Declared surplus after the war, the property reverted to private control. Today Battery Cravens (#203) is in private hands with a residence built on top of it. The Peaks Island Land Preserve began in 1995 as a grass-roots campaign to save Battery Steele (#102) from development. The island community raised $70,000 to buy the 14-acre property and preserved it forever as a recreational and historical site, it was listed on the National Register of Historic places, but little has been done to restore it. Peaks Island is accessible by ferry or boat. Daily visitors are welcome on the island. Access to many of the former coast defense sites is limited by private ownership.

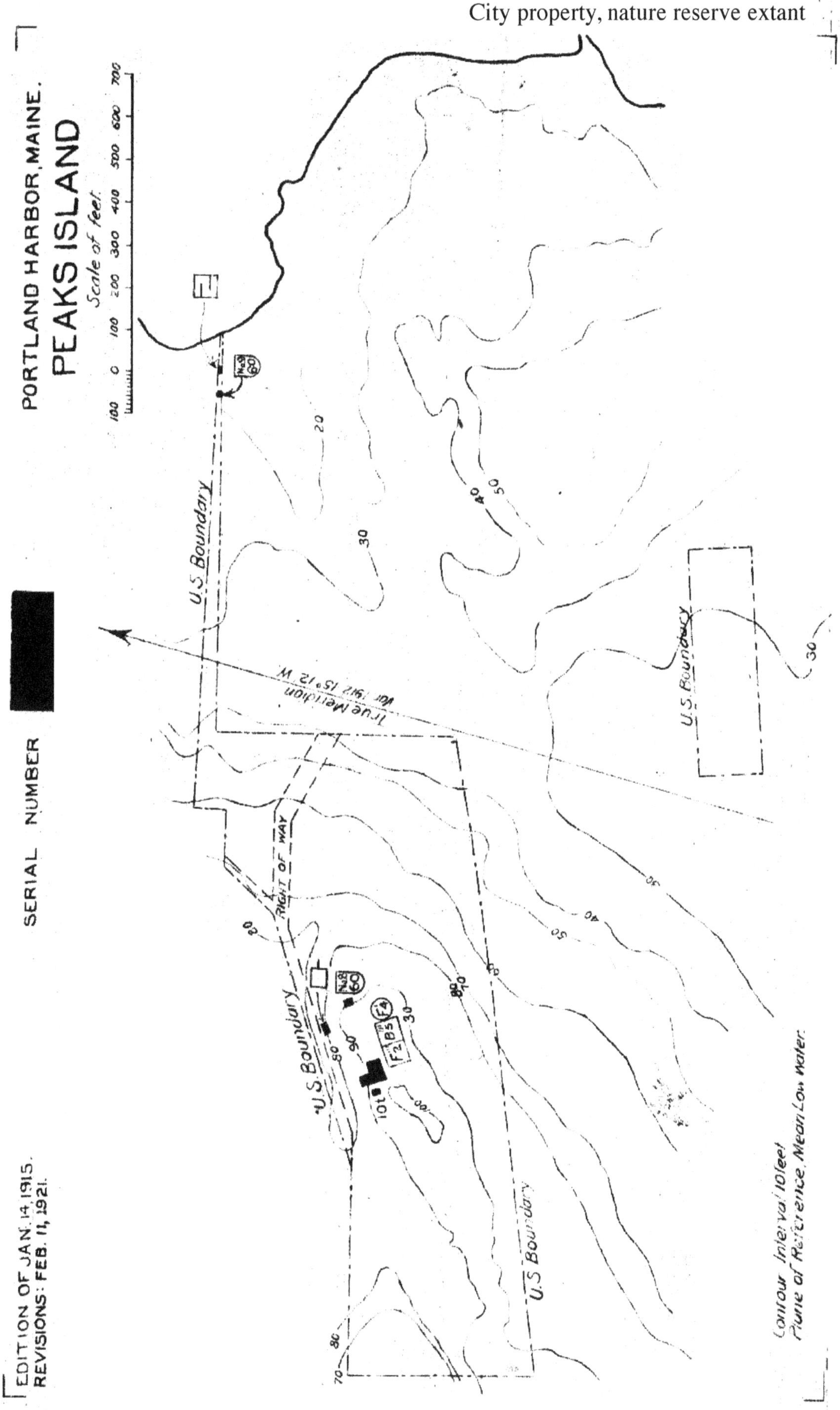

City property, nature reserve extant

PORTLAND HARBOR, MAINE.
PEAKS ISLAND
Scale of feet.

SERIAL NUMBER

EDITION OF JAN. 14, 1915.
REVISIONS: FEB. 11, 1921.

U.S Boundary

True Meridian Var. 1917 15° 12' W.

RIGHT OF WAY

Contour Interval 10 feet
Plane of Reference, Mean Low Water.

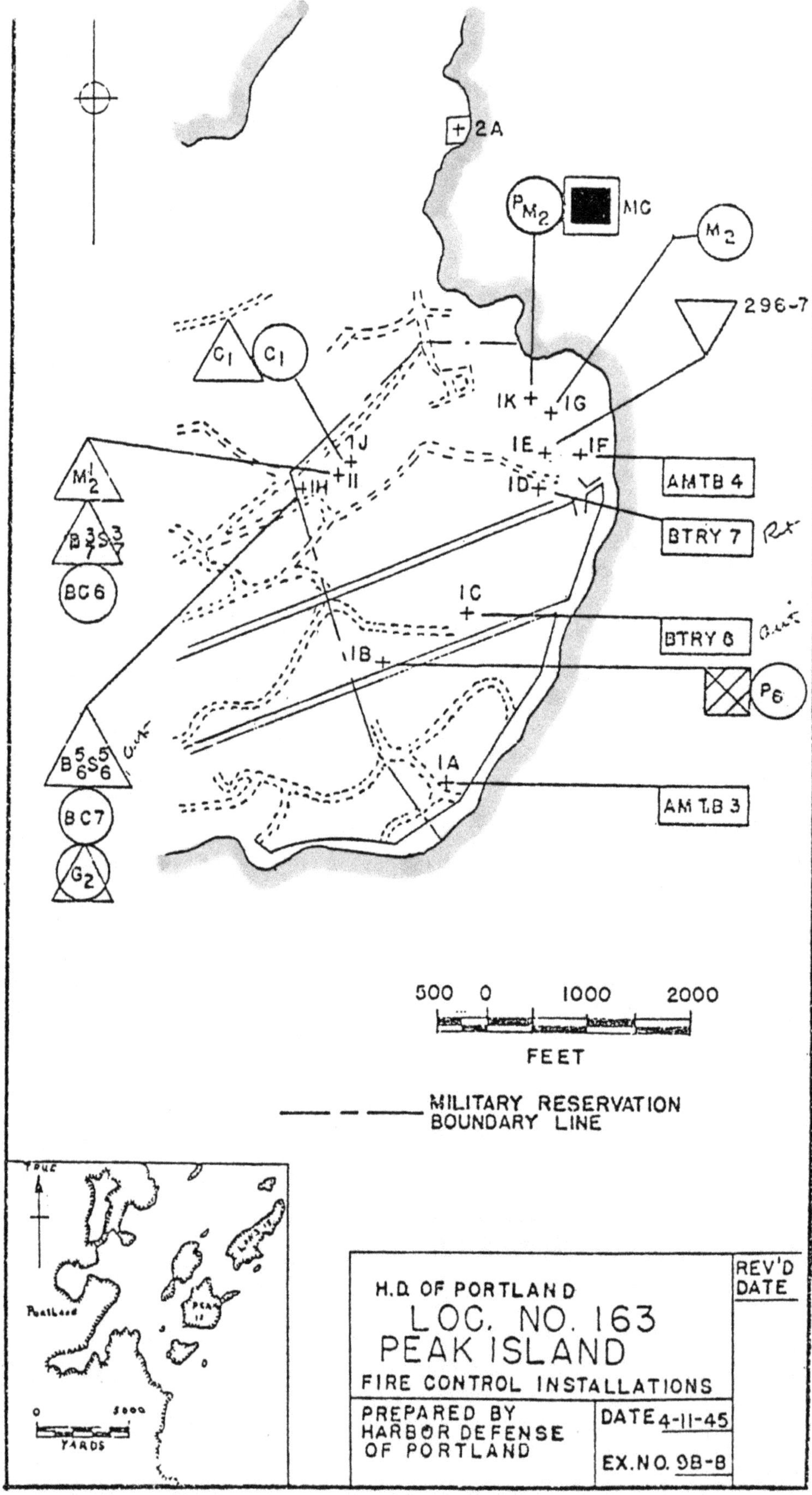

+ 2A

P M₂   ▪ MC   M₂

296-7

C₁ C₁

1K + + 1G

1E + + 1F

1J + + 1I

M¹₂   + 1H

AMTB 4

1D +

B³₇S³   BTRY 7   Rt

BC6

1C +

BTRY 6   Out

1B +

P6

B⁵₆S⁵₆   Cut

1A +

BC7   AMTB 3

G₂

500   0   1000   2000

FEET

MILITARY RESERVATION
BOUNDARY LINE

True

Portland

PEAK IS.

0   5000
YARDS

H.D. OF PORTLAND
LOC. NO. 163
PEAK ISLAND
FIRE CONTROL INSTALLATIONS

REV'D
DATE

PREPARED BY
HARBOR DEFENSE
OF PORTLAND

DATE 4-11-45

EX. NO. 9B-B

Battery Steele Peaks Island (Terry McGovern)
The distinctive house in the distance sits on top of Battery Cravens

Battery #202 Jewell Island (Mark Berhow)

Peaks Island M.R. Gun Batteries

- **STEELE:** A 1940 Program dual 16-inch barbette battery, originally started as Battery Construction No. 102. It was emplaced on the Peaks Island reservation, on the eastern side firing to the east. It was assigned national priority #31 on September 11, 1940. That priority classification was increased to #15 on the August 11, 1941 list. This priority increase was primarily due to the switching of priority with Battery Construction No. 101 planned for Cape Elizabeth. Actual work was done from January 6, 1942 to April 30, 1943 for completion by October 1945. The battery was of conventional design, with a separate PSR room and the usual network of distant base end stations. It was armed with two 16-inch MkIIM1 guns on Model 1919M4 or M5 barbette carriages (#53/#34 and #77/#46). It was named on AGO Memorandum of August 20, 1942 in honor of Major General Harry Lee Steele. The battery was retained as completed under the 1946 review but deleted shortly thereafter. The emplacement still exists at the Peaks Island Land Preserve. The battery is open to the public in abandoned condition.

- **CRAVENS:** A 1940 Program dual 6-inch battery originally started as Battery Construction No. 203. It was built at a location about 1000-feet to the north of Battery Steele on Peaks Island. It was originally assigned national priority #10 and funded under the FY-1942 budget. Construction was done in 1943-1943 and apparently completed in late 1944. It was of typical design for the standard 200-series batteries. It was armed with two 6-inch Model 1905A2 guns on M1 shielded barbettes (#26/#53 and #23/#52). It was named on AGO Memorandum of August 20, 1942 for Col. Richard K. Cravens. The battery was deleted after the war and probably disarmed in 1946-1948. The emplacement was not subsequently used by the military. It is now on private property with a private residence built on top of it. The battery is closed to the public.

- A temporary AMTB battery of two 3-inch Model 1902 guns emplaced on simple concrete gun blocks in 1942 at Peaks Island. The guns used with 3-inch M1902 guns #49/#49 and carraiges #50/#50 which had been in storage locally in the defenses. It was located on the eastern side of the reservation. A concrete battery commander's station was also part of the battery. Apparently, it served between 1943 and 1944 when its function was superseded by the 90mm batteries. The station and part of the blocks still exist on private property at Peaks Island. The battery is closed to the public.

- **AMTB-963:** A 1943 AMTB battery of two 90mm fixed and two 90mm mobile guns. The blocks for the fixed guns were emplaced at the southern end of the Peaks Island Military Reservation, overlooking Casco Bay harbor. It apparently served between completion in 1943 and 1946-1948. The battery commander's station and covered gun blocks still exist on private property. The battery is closed to the public.

- **AMTB-964:** A 1943 AMTB battery of two 90mm fixed and two 90mm mobile guns. It was emplaced northeast of Battery Cravens, near the shore and the 1942 3-inch battery. In fact, in addition to the new gun blocks, it used the battery commander's station of that older work. Apparently, it served between completion in 1943 to shortly postwar, when the fixed armament was removed. The station and covered gun blocks still exist on private property. The battery is closed to the public.

Fort Lyon (1896-1946) is located on Cow Island in Casco Bay. Built as part of the Endicott Program to defend the Hussey Sound entrance to Portland Harbor, its two batteries of secondary weapons covered the same area as the batteries of Fort McKinley. AA Battery #3 (1921-1945) was also located here. Concrete emplacements still exist for the AA height finder, gun director, and BC station. The double Mine Command station (1909) is in ruins, located behind Battery Bayard. It was named in General Orders No. 194 on Dec. 27, 1904 for Brig. Gen. Nathaniel Lyon U.S. Volunteers who was killed in action at Wilsons Creek, Missouri August 10, 1861. This very small fort was a sub-post of Fort McKinley. A detached 37mm AMTB battery (Battery 365) was located here in World War II. Declared surplus after World War II, the fort was turned over to the City of Portland and later sold to private interests. Today Cow Island is managed by Rippleffect, a youth advancement program which uses the island to host events, training, youth outreach, camping and water sports, is only accessible by boat with island owner's permission.

Fort Lyon Gun Batteries

- **BAYARD:** A battery for three 6-inch disappearing guns erected on Fort Lyon at Cow Island. Design submission was made on November 3, 1903. Its location was moved from an original site on a ridge at the northeast end of the island to a low-lying flat space near the sea on the eastern side. It was to cover the channel in the dead space created by Cow Island itself from the heavy guns at Fort McKinley. Construction was done in 1903-1904, and transfer made on December 27, 1907 for a cost of $122,921.93. It was generally of recommended type plan but had a full magazine set (for two guns) in each of the two traverses. The battery was armed with three 6-inch Model 1903 guns on Model 1903 disappearing carriages (#8/#64, #12/#65, and #16/#66). It was named on General Orders No. 194 of December 27, 1904 for Brigadier General George D. Bayard, mortally wounded at Fredericksburg in 1862. The armament was removed in 1917 during World War I for use on field carriages destined for Europe. They were never replaced. The carriages were scrapped in 1921 and the emplacement not subsequently used for armament storage. It still exists on private property. The battery is closed to the public.

- **ABBOT:** A battery of unique design for three 3-inch pedestal guns situated on the southern rise or hogback of Cow Island. Design submission was made using funds from the Fortification Act of March 1, 1901. As the area to cover was a wide arc on either side of Cow Island, one was place on either side of the ridge crest with a third gun emplaced in front and below these two. They were connected with a longitudinal passage connecting the rear gun guns in a T-shaped magazine to the forward gun. Work was done in 1901-1902; but the armament was considerably delayed in being mounted. Transfer was not made until January 28, 1909 for a cost of $18,675. It was armed with three 3-inch Model 1903 guns and pedestals (#40/#26, #51/#27, and #20/#28). The battery was named on General Orders No. 78 of May 25, 1903 for 1st Lieutenant Edward S. Abbot, 17th U.S. Infantry, who died of wounds received at Gettysburg in 1863. The battery existed for a considerable period of time and it guns were not removed when Bayard was taken out of service in 1917. There is some evidence that the gun tubes may have been switched on their carriages in the 1920s. Battery Abbot was retained in the 1932 Review and in the 1940 Program. It was not eliminated until May 16, 1946. The emplacement still exists on private property. The battery is closed to the public.

PORTLAND HARBOR, ME.

FORT LYON

Cow Island

SERIAL NUMBER

EDITION OF JAN. 14, 1915.
REVISIONS: DEC. 7, 1915, MAR. 27, 1916.
APRIL 10, 1920; FEB. 11, 1921.

LEGEND.

6t. N.C.O. QUARTERS.
7t. TEMP. BARRACKS.
10t TEMP. STABLE.

BATTERIES.

BAYARD......3-6"Dis
ABBOT......3-3" P.
A-3"Anti-Air Craft Guns

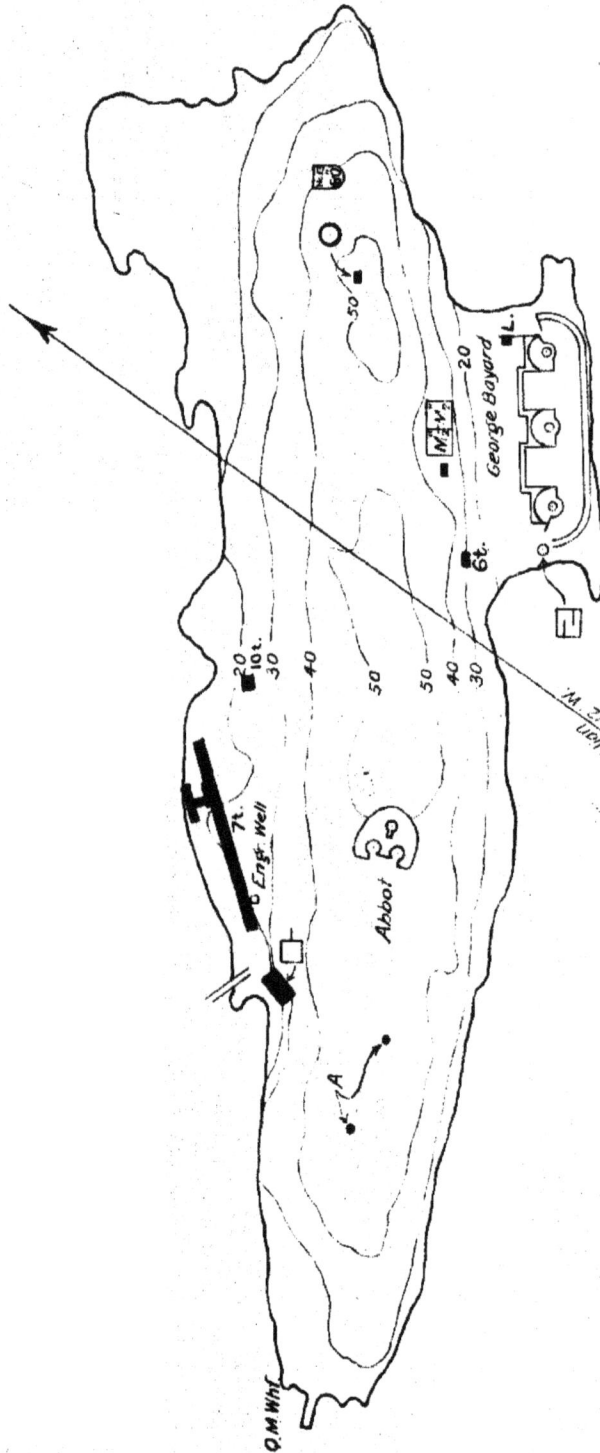

Scale of feet

Contour interval 10 feet.
The plane of reference is M.L.W.

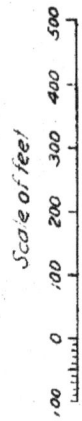

True Meridian
Var. 1912, 15°12'W.

George Bayard

M.L.W.

Abbot

Engr. Well

Q.M. Whf.

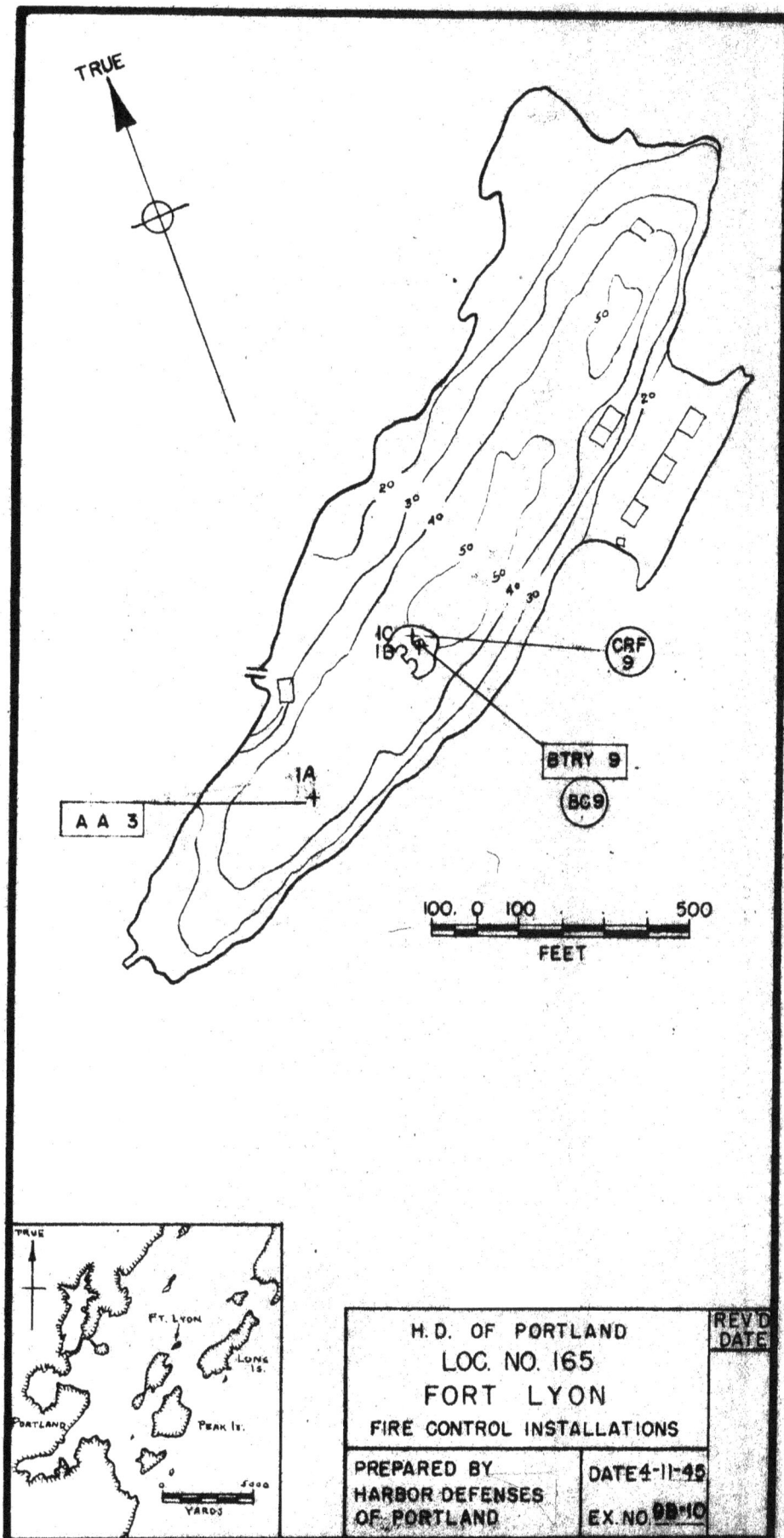

TRUE

50

2°

2°
3°
4°

5°

5°
4° 3°

10
1B 1A

CRF 9

1A
AA 3

BTRY 9

BC 9

100. 0 100 500
FEET

TRUE

FT. LYON

LONG IS.

PORTLAND

PEAK IS.

5000
YARDS

H.D. OF PORTLAND
LOC. NO. 165
FORT LYON
FIRE CONTROL INSTALLATIONS

REV'D
DATE

PREPARED BY
HARBOR DEFENSES
OF PORTLAND

DATE 4-11-45

EX. NO 99-10

Fort Lyon 1920s (NARA)

Fort McKinley 1936 (NARA)

Battery Abbot Fort Lyon (Terry McGovern)

Diamond Cove and Fort McKinley, Great Diamond Island (Terry McGovern)

Fort McKinley (1900-1946) is located on Great Diamond Island in Casco Bay. Construction started in 1893 as part of the Endicott Program to defend the Hussey Sound entrance to Portland Harbor, the fort occupied 192 acres on the eastern half of the island. It was named in General Orders No. 16 on Feb. 14, 1902 for William McKinley, a Civil War veteran and the twenty-fifth President of the United States. At its peak, the fort had nine batteries mounting a total of 21 guns and an extensive controlled submarine mine complex. The fort was turned over the City of Portland after World War II and later sold to private interests as a future site of a petroleum refinery. After many years of local opposition to the refinery project, the former fort was sold for residential development. Currently the Fort McKinley reservation is a privately owned residential community known as the Diamond Cove Homeowners Association. Many of the garrison buildings have been renovated as residences and condominiums. The Inn at Diamond Cove is situated within the former double barracks of Fort McKinley. The on site restaurant serves breakfast, lunch, and dinner. Guests travel around by foot, golf cart, and bicycle. Open May through September. Access to the island is by Ferry or boat. Fort McKinley remains one of the best-preserved coast artillery forts from the Endicott Program.

Fort McKinley Gun Batteries

- **BERRY**: A heavy battery of 12-inch disappearing guns erected on the north fork of Great Diamond Island at Fort McKinley. The first plans were submitted to Washington on November 19, 1896. The plan was fairly conventional and adhered to the approved mimeographs for design features. Most concrete work was done between 1897-98. Transfer to service troops was made on April 16, 1901, for a construction cost of $111,056.92. Berry carried two Model 1888 Bethlehem gun tubes on Model 1896 LF disappearing carriages (#2/#5 and #1/#7). It was named on General Orders No. 43 of April 4, 1900, for Major General Hiram G. Berry, killed in action at Chancellorsville in 1863. It served as a primary battery for Casco Bay for many years. In August 1912 work was done extending the battery's field of fire by cutting away the concrete on the flank of the No. 2 gun (better coverage to the north). Then in 1915-16 the 12-inch guns were given increased elevation (from 10 to 15-degrees) by modifying their disappearing carriages. It was retained in the 1932 Review Program. Finally, it was deleted under authority of August 15, 1943, the armament being removed shortly thereafter. The emplacement still exists on private property in Diamond Cove, but is closed to the public.

- **INGALLS**: A battery for eight 12-inch mortars planned for the north fork of Fort McKinley. The site was on the east shore of the fork, to the southwest of Battery Berry. The battery was funded by the Fortification Act of May 7, 1898. On June 25, 1898, engineer Major Hoxie submitted the battery plan for approval. The submitted plans required no modification from standardized plans. It had two pits of four mortars each and magazines were in the traverses in the center and flanks of the pits. It had the later wide pits and was likely one of the last continental U.S. Endicott mortar batteries built. Work was done in 1901-1903. Transfer came on January 18, 1904, for a cost of $139,000. Ingalls carried eight 12-inch Model 1890M1 mortars (all tubes Watervliet Arsenal:) on Model 1896 carriages (#70/#206, #129/#228, #83/#204, #121/#205, #120/#165, #79/#207, #71/#229, and #116/#164). The battery was named on General Orders No. 78 of May 25, 1903, for Brevet Major General Rufus Ingalls of Mexican and Civil War service. Four guns were removed in 1918 for service as railway artillery (#79, #116, #121, and #129). The final four guns served until deleted under authority of November 6, 1942. The emplacement still exists on the common property of Diamond Cove. The battery is open to the public, but is being used as a refuse transfer station.

PORTLAND HARBOR, MAINE.

FORT McKINLEY

GREAT DIAMOND ISLAND.

EDITION OF OCTOBER 20, 1919.
REVISIONS: APRIL 10, 1920; FEB. 11, 1921;
JAN. 29, 1925; MAY 6, 1929; FEB. 12, 1935

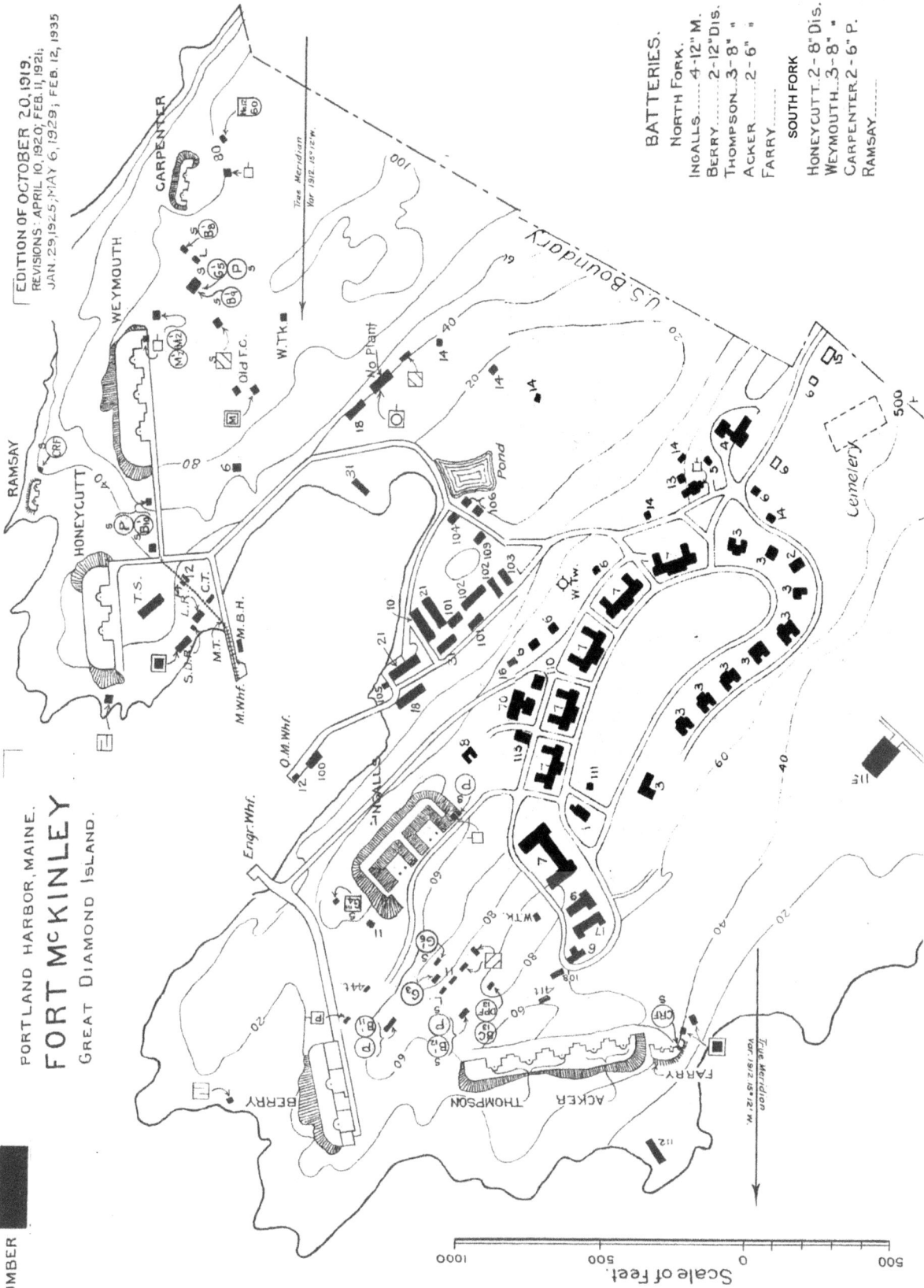

BATTERIES.
NORTH FORK.
INGALLS......4-12" M.
BERRY.......2-12" Dis.
THOMPSON...3-8" "
ACKER.......2-6" "
FARRY......

SOUTH FORK
HONEYCUTT..2-8" Dis.
WEYMOUTH..3-8" "
CARPENTER.2-6" P.
RAMSAY......

NUMBER

CARPENTER

WEYMOUTH

RAMSAY

HONEYCUTT

U.S. Boundary

Pond

Cemetery

THOMPSON

ACKER

FARRY

BERRY

INGALLS

Scale of feet.

1000   500   0   500

PORTLAND HARBOR, MAINE.

FORT McKINLEY D-I.

GREAT DIAMOND ISLAND.

SERIAL NUMBER

Scale of Feet.

EDITION OF OCTOBER 20, 1919.
REVISIONS: APRIL 10, 1920.

LEGEND

1 ADMINISTRATION BLDG.
3 OFFICERS QUARTERS.
6 N. C. O. QRS.
7 BARRACKS
7a TEMP. BARRACKS
8 GUARD HOUSE.
9 POST EXCHANGE
17 BOWLING ALLEY
100 BOAT HOUSE.
105 SAW MILL.
111 BAND STAND.
112 TARGET PITS.
40 ENGINEERS OFFICE.
41t TEMP. CEMENT SHED.
42t TEMP. ENGR. ST. HO.
43t TEMP. ENGR. TOOL HO.
44t TEMP. BLKSM'TH. SHOP.
90 LAUNDRY.

BATTERIES.

NORTH FORK.

INGALLS...... 8-12" M.
BERRY....... 2-12" Dis
THOMPSON..... 3-8" "
ACKER....... 2-6" "
FARRY....... 2-3" P.

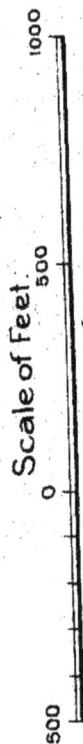

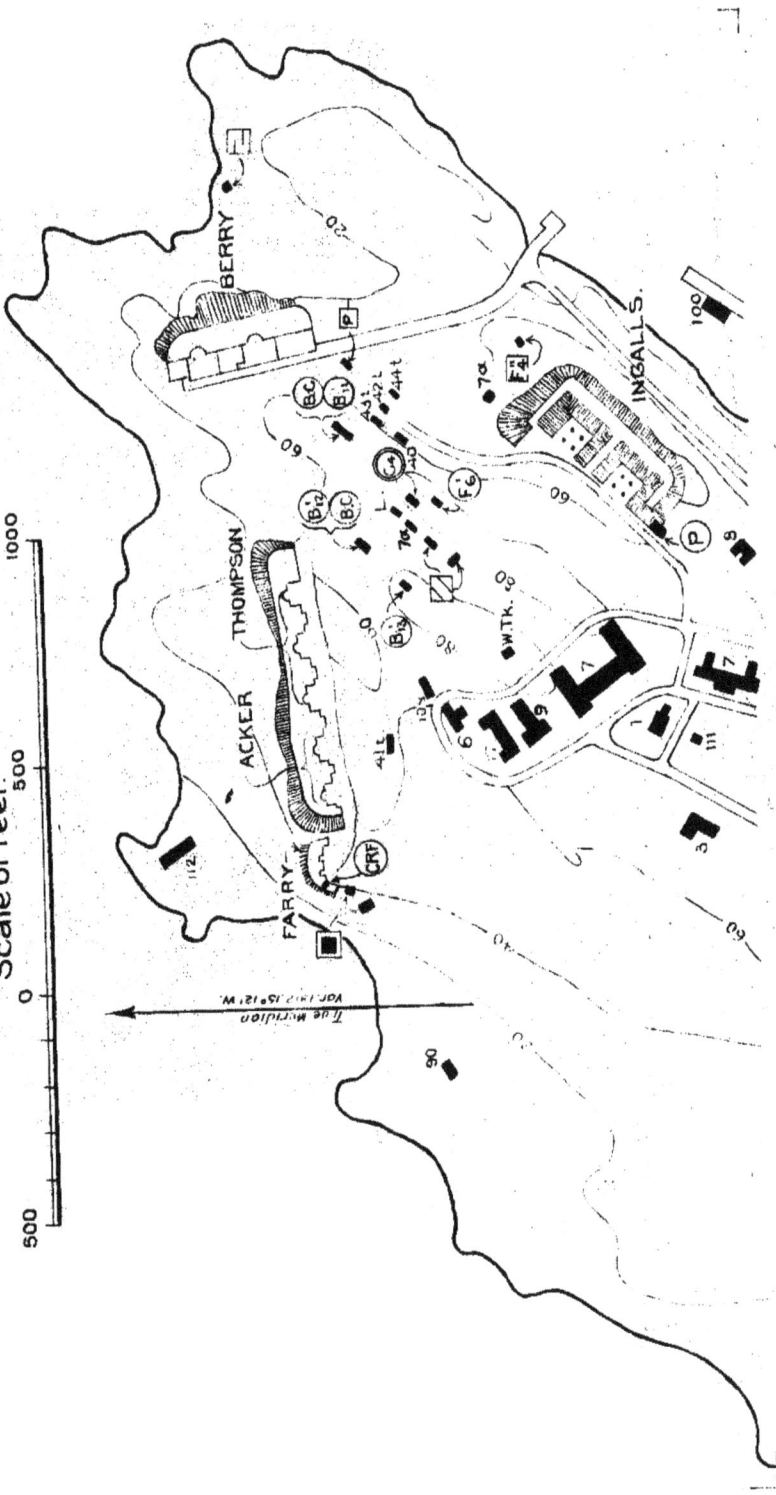

PORTLAND HARBOR, MAINE.

FORT McKINLEY D-2.

GREAT DIAMOND ISLAND.

SERIAL NUMBER

EDITION OF OCTOBER 20, 1919.
REVISIONS: APRIL 10, 1920.

True Meridian
Var. 1912. 15°12' W

Scale of Feet.

U.S. Boundary

Pond

Cemetery

LEGEND

1 ADMINISTRATION BLDG.
2 COMMANDING OFF. QRS.
3 OFFICERS QUARTERS.
4 HOSPITAL.
5 HOSPITAL STWD'S QRS.
6 N.C.O. QUARTERS.
7 BARRACKS.
9 GUARD HOUSE.
10 ARTILLERY ENGR.
13 RESERVOIR.
14 WELL & SHELTER.
15 FIREMANS QRS.
16 CIV. EMPLOYEE'S QRS.
101 WORK SHOP.
102 WAGON SHED.
103 STABLE.
104 OIL HOUSE.
106 ICE HOUSE.
109 TEAMSTER'S QRS.
110 FIRE STATION.
111 BAND STAND.
113 BAKERY.
21 Q.M. STOREHOUSE.
31 ORDNANCE ST. HO.
70 Y.M.C.A.

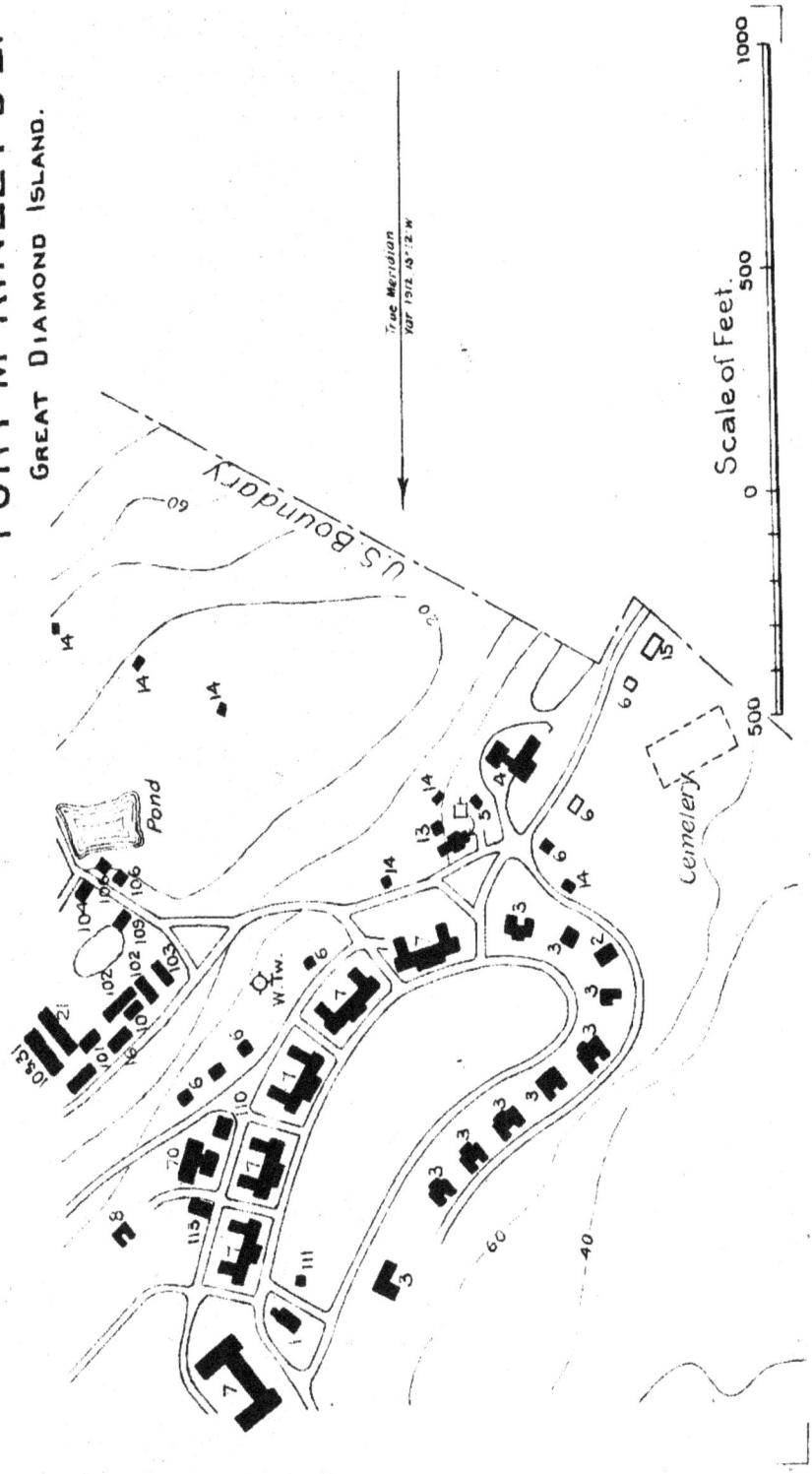

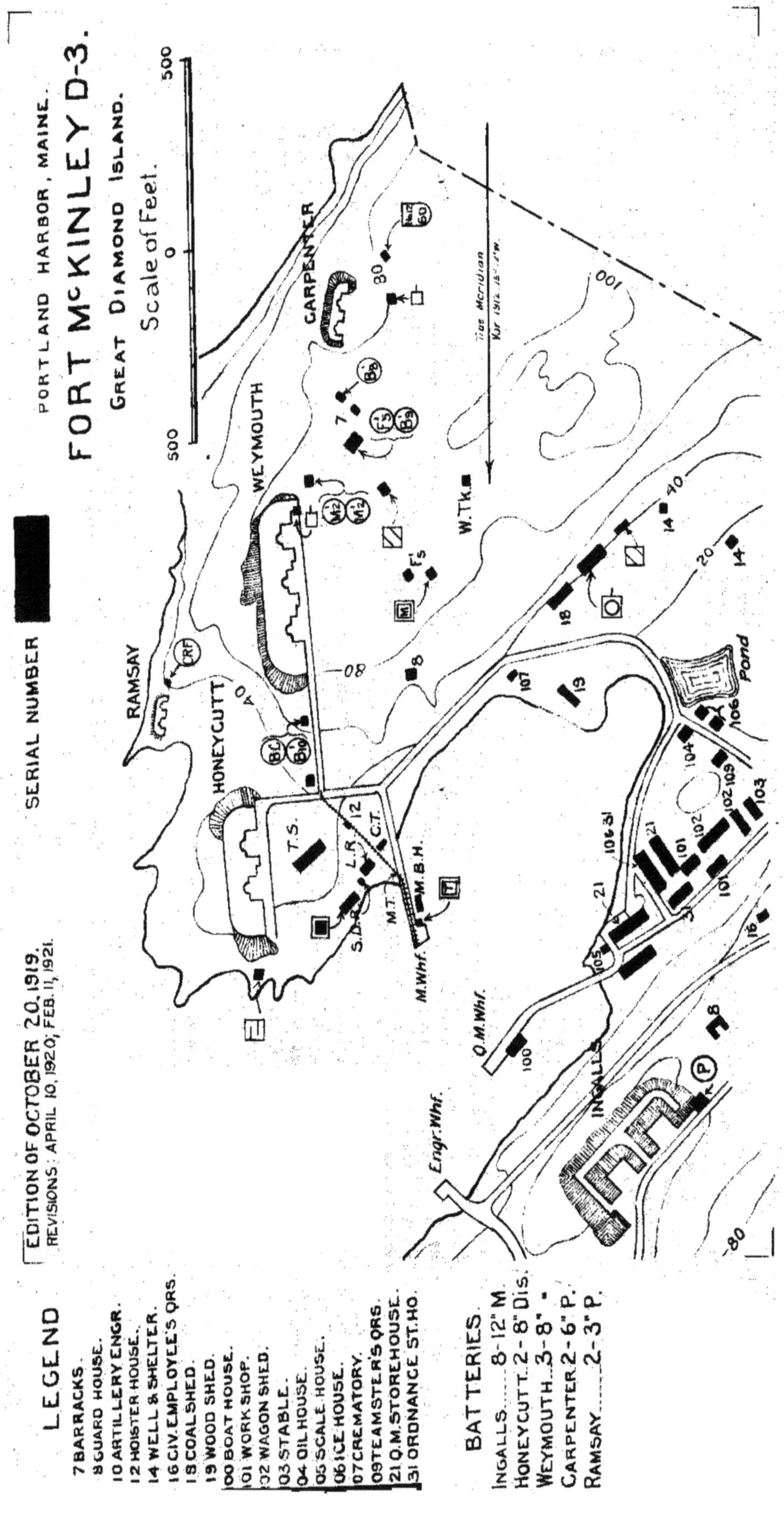

PORTLAND HARBOR, MAINE.
FORT McKINLEY D-3.
GREAT DIAMOND ISLAND.
Scale of Feet.
500 0 500
SERIAL NUMBER
EDITION OF OCTOBER 20, 1919.
REVISIONS: APRIL 10, 1920; FEB. 11, 1921.
RAMSAY
HONEYCUTT
WEYMOUTH
CARPENTER
INGALLS
Engr. Whf.
Q.M. Whf.
M. Whf.
Pond
LEGEND
7 BARRACKS.
8 GUARD HOUSE.
10 ARTILLERY ENGR.
12 HOISTER HOUSE.
14 WELL & SHELTER.
16 CIV. EMPLOYEE'S QRS.
18 COAL SHED.
19 WOOD SHED.
100 BOAT HOUSE.
101 WORK SHOP.
02 WAGON SHED.
03 STABLE.
04 OIL HOUSE.
05 SCALE HOUSE.
06 ICE HOUSE.
07 CREMATORY.
09 TEAMSTER'S QRS.
21 Q.M. STOREHOUSE.
31 ORDNANCE ST. HO.
BATTERIES.
INGALLS ...... 8-12" M.
HONEYCUTT 2-8" Dis.
WEYMOUTH 3-8" "
CARPENTER 2-6" P.
RAMSAY ...... 2-3" P.

TRUE

1C

20

40

60

20

20

T

80

1B ⊕ BC8

1A

80

BTRY 8

500 0 500 1000 1500
FEET

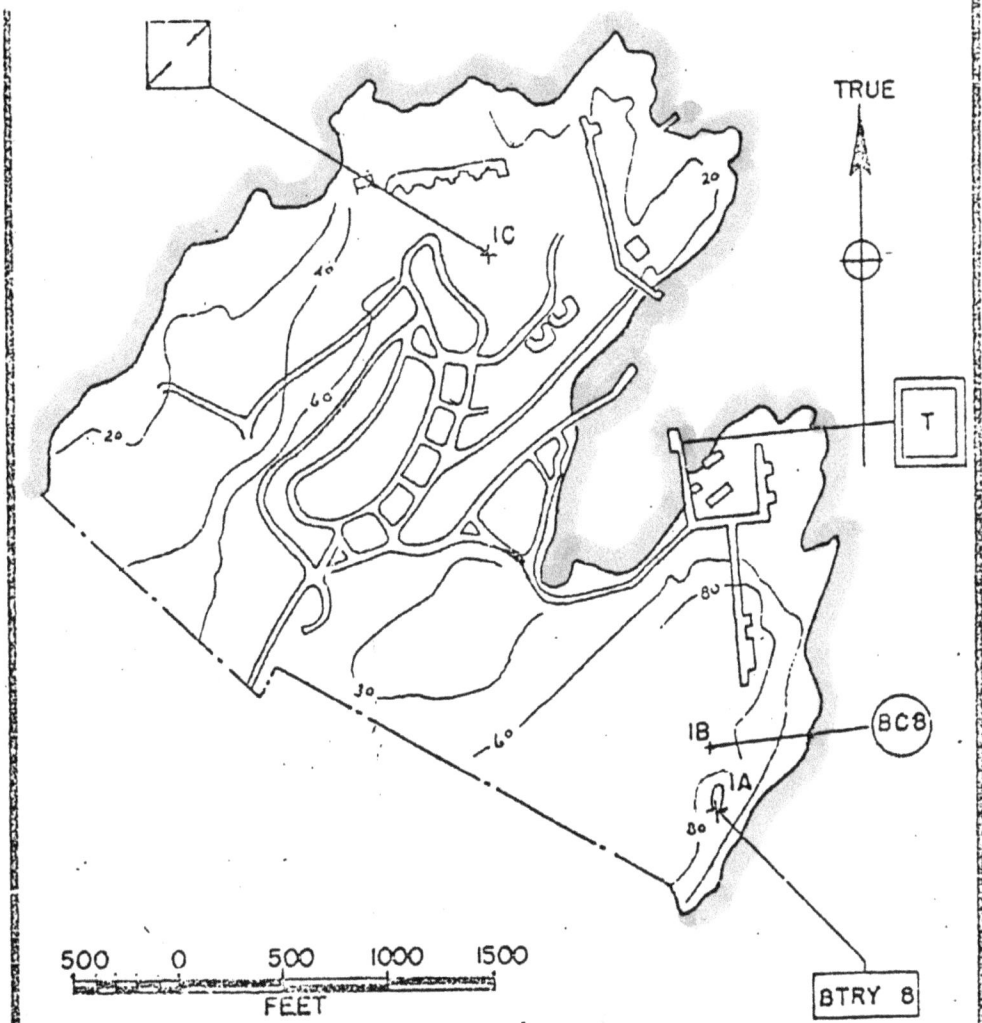

MILITARY RESERVATION
BOUNDARY LINE

TRUE

Ft McKinley

PORTLAND

PEAK IS

YARDS

H.D. OF PORTLAND
LOC. NO. 164
FORT McKINLEY
FIRE CONTROL INSTALLATIONS

REV'D
DATE

PREPARED BY
HARBOR DEFENSES
OF PORTLAND

DATE 4-11-45

EX. NO. 9B-9

Upper and lower photos are of Fort McKinley on Great Diamond Island in the 1920s (NARA)

- **THOMPSON:** A battery for 8-inch disappearing guns erected on the Fort McKinley north fork on a ledge to the west of Battery Berry, firing to the north. The first emplacement was built from 1898-99. Funds were made available on July 26, 1900, for the final two emplacements, which were built in 1900-1901. It was decided, even after one emplacement was already underway, to shift the battery to the right in order to allow space for the two-gun 6-inch Battery Acker on the left flank. Also, the last two emplacements were raised by 3.6-feet in order to minimize the amount of excavation of the underlying ledge, which was proving problematical. The completed battery for three guns was transferred to service troops on December 11, 1902, for a cost of $147,848.27. It was named on General Orders No. 43 on April 4, 1900, for Lt. Colonel Samuel Thompson of the Massachusetts Militia in the Revolutionary War. Thompson carried three 8-inch Model 1888MII Bethlehem Steel tubes on Model M1896 disappearing carriages (#25/#33, #5/#27, and #6/#29). While it was recommended to delete this armament under the 1932 Program, this was in fact not accomplished until authority granted on December 15, 1942. The armament was removed by mid-1943. The emplacement still exists on the common property of Diamond Cove. The battery is open to the public.

- **HONEYCUTT:** On March 19, 1898, the Engineer Department authorized the construction of a series of five 8-inch guns on disappearing carriages for the South Fork of Fort McKinley. Soon organized as two batteries—a three-gun battery and a two-gun battery—construction plans were submitted on June 22, 1898. Plans for Honeycutt (the two gun unit) were of standardized type. It was located on the north end of the fort, with a field of fire to the east. Work was done in 1898-1899. Transfer was made on January 22, 1901, for a cost of $96,385.61. It was named on General Orders No. 43 on April 4, 1900, for Captain John T. Honeycutt, 6th Artillery, who died in 1898. The battery was armed with two 8-inch guns Model 1888MII on Model 1896 LF disappearing carriages (Watervliet #19/#21 and Bethlehem #21/#24). This armament was retained up until World War II. Though recommended for deletion in the 1932 Review, it was not actually authorized for removal until December 15, 1942. The emplacement still exists on the common property of Diamond Cove. The battery is open to the public.

- **WEYMOUTH:** The second three gun 8-inch battery started with the appropriations of March 19, 1898. Like Battery Honeycutt it went to the fort's South Fork, being located south of Honeycutt roughly in line between that battery to the north and eventually Battery Carpenter to the south. Plans were submitted on June 22, 1898, generally following the approved emplacement type plans. Work was done in 1898-1899, and transfer made on January 22, 1901 for a construction cost of$156,247.50, Weymouth carried three 8-inch Model 1888MII Bethlehem Steel tube guns on Model LF 1896 disappearing carriages (#14/#17, #17/#16, and #22/#23). It was named on General Orders No. 43 of April 4, 1900 for Captain George Weymouth, an early local explorer. The eight inch batteries at Fort McKinley were too important to be removed in 1917, unlike most of the other 8-inch batteries during that conflict. The battery was recommended for deletion in the 1932 Review, but that was not actually done until authorization of December 15, 1942. The emplacement still exists on the common property of Diamond Cove. The battery is open to the public.

- **ACKER:** A battery for two 6-inch guns on disappearing carriages. It was a planned RF battery for the North Fork of Fort McKinley and required to cover the minefield here. Intended to be placed along the same ledge line as 8-inch Battery Thompson to the west of Battery Berry. At first it was planned to build it in two separate, single emplacements on either side of the 8-inch battery; though eventually Thompson was shifted, and the two-gun 6-inch battery was built on its left flank. Work was done from 1899-1900. The emplacement closely followed the recommended mimeograph design plans. Transfer was made on December 11, 1902 for a construction cost of $56,100. It carried two 6-inch Model 1897M1 guns on Model 1898 disappearing carriages (#5/#20 and #19/#21). It was named on General Orders No. 78 of May 25, 1903 for Captain William H. Acker, who died at the Battle of Shiloh in 1862. The armament was retained for a considerable period of time, into the 1930s and 1940s. It was finally authorized for removal on August 15, 1943. The emplacement still exists on the common property of Diamond Cove. The battery is open to the public.

- **CARPENTER:** An emplacement for two 6-inch guns for Fort McKinley's south fork needed to defend the southern submarine mine field. A site was fixed in April 1899 on the southern end of the reservation, south of Battery Weymouth. In June 1901 a plan was submitted by local engineer Captain S. Roessler for a two-gun battery, closely following the type plans which at the time stipulated that both guns share the same platform with magazines below, in the flanks. Work began in 1901, but a significant re-design occurred in 1903. Following plans to move the guns to individual platforms (two guns on the same platform interfered with each other too much), the battery was modified. The right-hand, No. 1 gun remained on the emplacement, but the No. 2 gun was moved further left onto a new platform of its own. An additional $10,000 was needed to make this change. It was completed by the time of the battery transfer on May 25, 1906 for a cost of $45,400. It was named on General Orders No. 78 of May 25, 1903 for Major Stephen Carpenter, who died at the Battle of Stone River in 1862. It was armed with two 6-inch Model 1900 guns and pedestals (#9/#3 and #10/#4). The guns were removed in 1917 for use on field mounts, but they were returned and replaced on their pedestals in 1919. It then served through World War II, until final deletion and scrapping of the armament in 1947. The emplacement still exists on private property of Diamond Cove. The battery site is closed to the public.

- **FARRY:** An emplacement for two 3-inch masking parapet guns for the Fort McKinley North Fork. Like Battery Acker it was required to cover the northern mine field. The battery was sited to the immediate western flank of that other battery, and quite close to the fork's mine casemate. Work was done in 1899-1900. Transfer was made on December 11, 1902 for a cost of $10,000. It was named on General Orders No. 78 of May 25, 1903 for 1st Lieutenant Joseph F. Farry, killed in 1847 at the Battle of Molino del Ray. The battery carried two 3-inch, 15-pounder Model 1898 and pillars (Driggs-Seabury #49/#49 and #51/#51). These were removed in common with other such Model 1898s in 1920. The emplacement still exists on the property of Diamond Cove. The battery is open to the public.

- **RAMSAY:** The 3-inch gun emplacement on the south fork of Fort McKinley for defense of the southern minefield. After some debate about location, it was situated on a constricted rock point about equal distance between 8-inch batteries Honeycutt and Weymouth. The narrow confines of the ledge prompted a constricted design with narrower dimensions between gun platforms than normally preferred. Plans for construction were submitted on June 25, 1900 with an estimate of a $13,000 cost. Work was done from 1900-1901. Transfer was delayed until May 26, 1906 for a final cost of $9,103.10. It was named on General Orders No. 78 of May 25, 1903 for 1st Lt. Douglas Ramsay, killed in action at the Battle of the First Bull Run in 1861. It carried two 3-inch, 15-pounder Model 1898 guns and masking parapet mounts (Driggs-Seabury #113/#113 and #114/#114). The battery was disarmed in common with all such armament in 1920. The emplacement still exists on the property of Diamond Cove. The battery site is closed to the public.

Fort McKinley on Great Diamond Island with Cow Island (Fort Lyon) in the distance (Terry McGovern)

Battery Kendrick and Battery Ferguson, Fort Levett on Cushing Island (Terry McGovern)

Fort Levett (1903-1947) is located on Cushing Island in Casco Bay. Constructed as part of the Endicott Program, its four batteries covered the southern entrance to Portland Harbor. It was named in General Orders No. 43 of April 4, 1900 for Christopher Levett who explored and settled Portland harbor in 1623. A new long-range 12-inch battery was built and completed in 1924. This battery was casemated during World War II and an anti-motor torpedo boat battery (AMTB), and several fire control towers were also added at this time. The remaining batteries were disarmed by 1948. In 1957, the 125-acre reservation was sold to a private group, and the entire island is now a private association of summer homes. Due to the protected nature of a private island, the gun emplacements retain many original artifacts. The island is accessible by ferry and boat, and only homeowners and guests are allowed on the island, so arrangements must be made to visit the island. The remaining garrison buildings are now private residences, and the fortification area is now a privately held nature preserve.

Fort Levett Gun Batteries

- **BOWDOIN:** A battery for three 12-inch disappearing guns emplaced on Cushing Island. On July 7, 1898 the emplacement plan was submitted (although initially for just two guns) by local engineer Major Hoxie and approved. It was at the north end of a sequence of batteries with both Battery Kendrick and Ferguson further to the south. The center of its field of fire was to the southeast. It was of conventional two-story design plan, using ammunition hoists for projectile delivery. Construction took place between 1898-1899 for transfer on April 23, 1903 for a cost of $190,290.56. The first two guns were initially built together, with emplacement No. 1 being added slightly later. It carried three 12-inch Watervliet Arsenal gun tubes Model 1895 on Model 1897 LF disappearing carriages (#4/#23, #28/#18, and #13/#13). It was named on General Orders 43 of April 4, 1900 for James Bowdoin, Governor of Massachusetts from 1785-1786. The guns were given increased range with a change in their elevation from 10 to 15-degrees in 1916. The armament remained unchanged in the interwar years. In 1932 it was classified as a retained unit, and authority to abandon the emplacement did not come until August 15, 1942. The guns and carriages were scrapped soon after. The emplacement still exists on private property on Cushing Island. The battery is not open to the public.

- **KENDRICK:** A battery for two 10-inch guns on disappearing carriages built about the same time as Battery Bowdoin, in line to the south of that emplacement, and also firing to the southeast. Plan by Major Hoxie was submitted on July 20, 1898. Work was done from 1900-1902. It was of conventional design, closely following department mimeographs. Transfer was made on April 21, 1903 for a cost of $118,578.20 The battery was named on General Orders No. 43 of April 4, 1900 for Professor Henry Lane Kendrick of the U.S. Military Academy. It was armed with two 10-inch Model 1895 gun tubes on Model 1896 LF disappearing carriages (#19/#67 and #17/#68). The gun tubes were temporarily removed in 1917-19 for potential use on railroad carriages destined for Europe. However, they were returned to the battery in the early 1920s. The emplacement was placed in deletion status in 1932 but was retained as reserve material on a specially reduced maintenance status. Final authority for removal came on November 6, 1942. The emplacement was then abandoned but still exists on private property on Cushing Island. The battery is not open to the public.

- **FERGUSON:** A dual 6-inch pedestal battery intended for minefield coverage at Fort Levett. The site selected was on the southern flank of Battery Kendrick, on the southwestern end of the reservation. It also fired to the southeast. Plans were submitted on May 21, 1901 for a battery of four 6-inch Model 1900 guns in this position (replacing earlier plans for two additional 10-inch guns here).

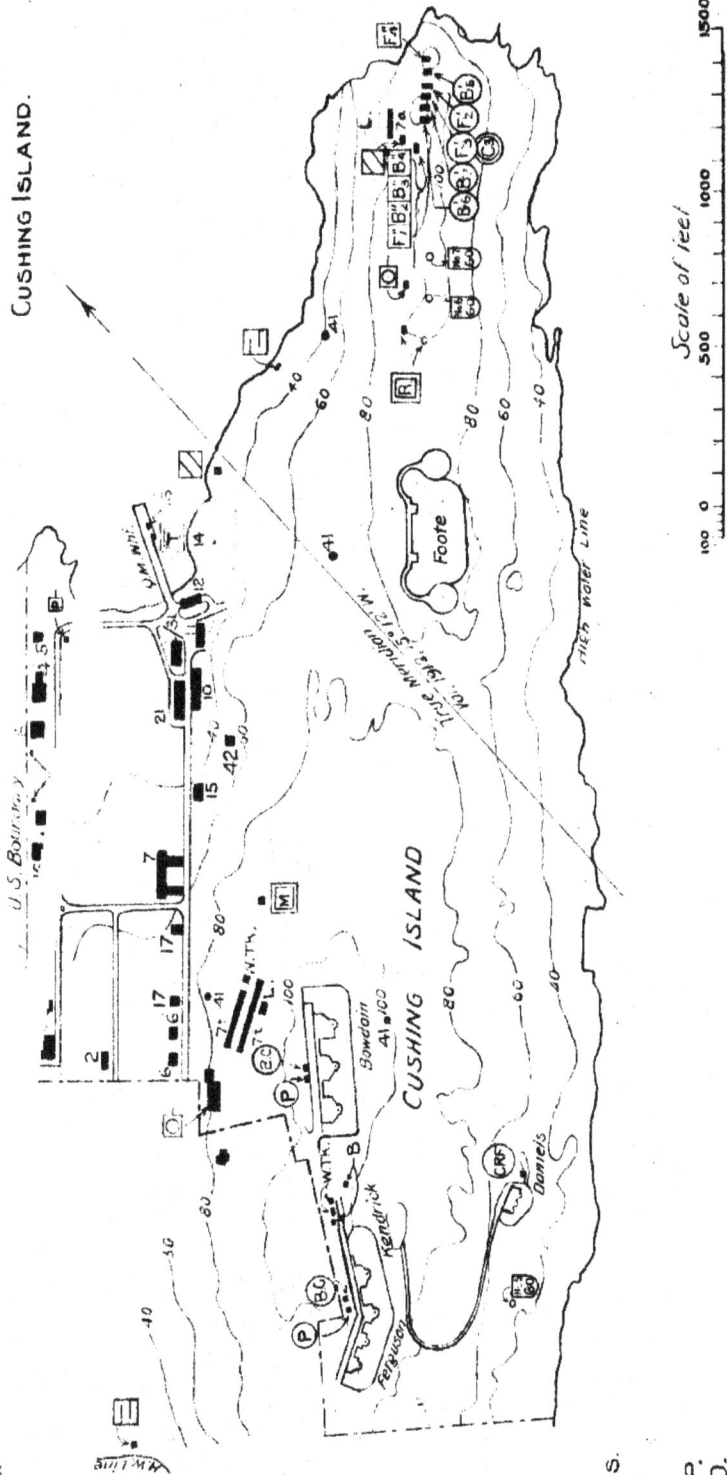

PORTLAND HARBOR, ME.

FORT LEVETT

CUSHING ISLAND.

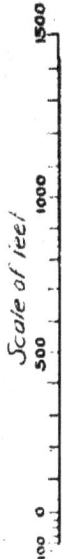

SERIAL NUMBER

EDITION OF JAN. 14, 1915.
REVISIONS: DEC. 7, 1915; MAR. 27, 1916.
APRIL 10, 1920; FEB. 11, 1921.

LEGEND.

1 ADMINISTRATION BLDG.
2 COMMANDING OFFICER'S QRS.
3 OFFICERS QUARTERS.
4 HOSPITAL.
5 HOSPITAL STW'D. QRS.
6 N.C.O. QUARTERS.
7 BARRACKS.
7t TEMP. BARRACKS.
7a DORMITORY.
8 GUARD HOUSE.
9
10 COAL SHED.
11 WAGON SHED.
12 STABLE.
13 SHED.
14 OIL HOUSE.
15 BAKERY.
16 RECREATION BLDG.
21 Q.M. STORE HOUSE.
22 Q.M. CISTERN.
31 ORDNANCE ST. HO.
41 ENGINEER WELLS.
42 ENGINEER OFFICE.
17 FIRE STATION.

BATTERIES.

BOWDOIN ... 3 - 12" DIS.
KENDRICK. 2 - 10" "
FERGUSON 2 - 6" P.
DANIELS ... 3 - 3" P.
FOOTE ...... 2 - 12 N.D.
B - Anti-Air Craft Guns

Scale of feet

Contour interval 20 feet.
The plane of reference is mean low water.

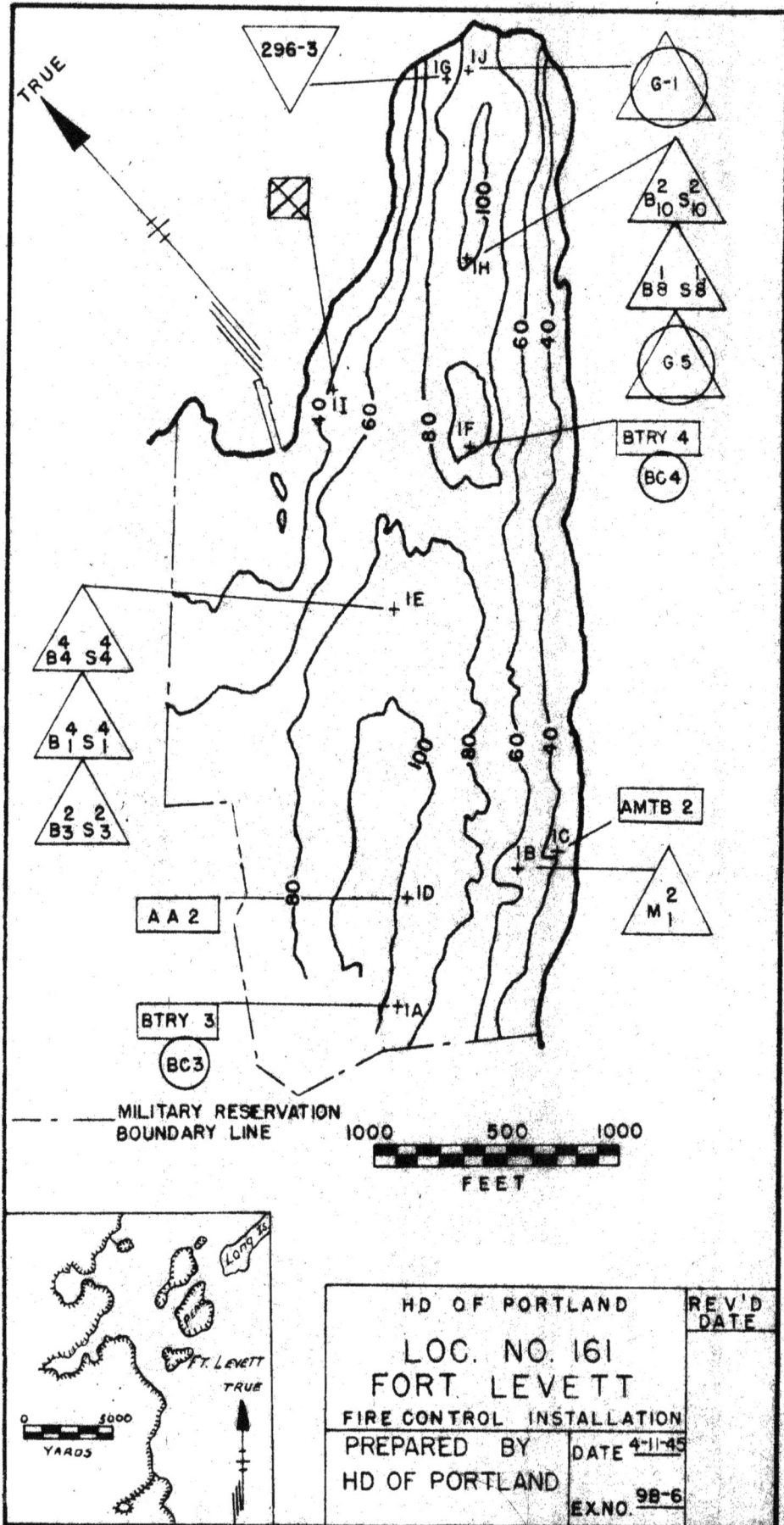

TRUE

296-3

1G   1J

G-1

100

B^2_{10} S^2_{10}

1H

B^1_8 S^1_8

G 5

1I

60

80   1F

BTRY 4

BC 4

1E

B^4_4 S^4_4

B^4_1 S^4_1

100   80   60   40

B^2_3 S^2_3

AMTB 2

B   1C

AA 2

80   1D

M^2_1

BTRY 3

BC3

1A

MILITARY RESERVATION
BOUNDARY LINE

1000   500   1000

FEET

Long Is.

FT. LEVETT

TRUE

0   5000

YARDS

HD OF PORTLAND

LOC. NO. 161
FORT LEVETT

FIRE CONTROL INSTALLATION

PREPARED   BY
HD OF PORTLAND

REV'D
DATE

DATE 4-11-45

EX.NO. 98-6

Upper and lower photos are of Fort Levett on Cushing Island in the1920s (NARA)

First construction plans submitted in January 1902 called for a mimeograph plan type with two platforms for two guns each, however the recommendations for the type of plan soon changed to requiring just a single gun per platform. Consequently, a new plan submitted by the local engineers on February 17, 1903 modified the plan by reducing it to just a single gun per platform, so only a two-gun battery with guns 67-feet apart was built. The left gun on each platform was deleted (they weren't actually mounted yet). Construction work was done from 1902-1904. Transfer was made on June 22, 1904 for a cost of $58,500. It was armed with two 6-inch Model 1900 guns and pedestals (#7/#5 and #16/#6). It was named on General Orders No. 78 of May 25, 1903 for Major William Ferguson, killed in 1791 during St. Clair's Defeat in the Ohio Indian Wars. Unlike other similar batteries, this armament was not removed in during World War I. Between the wars it received several modifications to its loading platforms to allow higher gun elevation and being equipped with ready ammunition boxes. It was finally disarmed in 1947 and subsequently abandoned. The emplacement still exists on Cushing Island private property. The battery is not open to the public.

- **DANIELS:** A battery for three 3-inch, 15-pounder masking parapet guns emplaced at Fort Levett to cover the minefield. Space was difficult to locate for the battery, eventually it was placed in front of and below the main battery line—not the most favorable location but the best available. Work was done from 1901-1902. Transfer was made on April 23, 1903 for a cost of $14,994.18. It was named on General Orders No. 78 on May 25, 1903 for Napoleon H. Daniels, who died in 1866 in the Indian Wars in the Dakota Territory. It was armed with three 3-inch Model 1898 guns and masking parapet mounts (Driggs-Seabury #46/#46, #47/#47, and #48/#48). The armament was removed, and the battery abandoned in 1921. The emplacement still exists, though at times somewhat overgrown, on Cushing Island private property. The battery is not open to the public.

- **FOOTE:** A 1915 Program battery for two 12-inch long-range barbette guns. In common with several other harbors, Portland received a battery of these weapons on the new, long-range barbette carriage to supplement existing armament. Local engineers were directed to prepare plans under letter of December 2, 1915, adhering to type plans with the exception of omitting the crow's nests as the battery was behind a ridge with no clear ocean view. It was one of the higher priority works, being completed in 1919. In most aspects it followed prescribed type plans. The site selected was to the northeast of the original Endicott battery fort. It was named on General Orders No. 129 of December 1, 1919 for Colonel Stephen M. Foote, a Coast Artilleryman who died in 1919. The battery was armed with two 12-in guns Model 1895A4 on Model 1917 carriages (M1895 #3/#12 and M1895M1 #70/#13). This armament was retained throughout the battery's service life. The battery was provided with heavy overhead protection and casemates from March 29, 1942 to December 3, 1942; and had gun shields added in August 1943. It was finally deleted and disarmed under recommendation in 1946. The emplacement still exists on Cushing Island private property. The battery is not open to the public.

- **AMTB-962**: A 1943 Program AMTB battery consisted of two 90mm fixed and two 90mm mobile dual-purpose guns. The gun blocks for the fixed mounts were emplaced near the shoreline on the southern end of the island, firing to the southeast. They were built in mid-1943 and soon armed. Disarmed after the war's end, the blocks still exist. The battery is not open to the public.

Fort Preble (1808-1947) is located on Spring Point in South Portland. Fort Preble was a Second System fort which received a Third System reconstruction, which was only partially completed. Four batteries were constructed on the site during the Endicott Program as well as a set of garrison buildings. It was officially named in General Orders No. 8 of Sept. 29, 1937 for Commodore Edward Preble, U.S.N. of Revolutionary War service. AA Battery #1 (1921-1945) was emplaced on Battery Rivardi. A detached 37mm AMTB battery (Battery 361) was also located on post during World War II. Declared surplus in 1950, the fort is now the Southern Maine Community College-South Portland Campus. The mortar batteries were buried to make room for additional classrooms and open space. Several of the fort's garrison buildings are now used by the college. The remains of the 2nd-3rd system fort and the two remaining batteries (6-inch and 3-inch guns) are open to the public.

Fort Preble Gun Batteries

- **CHASE – KEARNY:** A mortar battery for sixteen 12-inch mortars emplaced as the first new Endicott battery at Fort Preble. Initially there were problems getting the approved plan to fit the available property. Eventually it was decided to fit two double pit batteries offset one in front of the other, somewhat reminiscent of the Abbot quadrangular design. It was located not far from the walls of the old, incomplete third system work. The letter of plan submission was dated October 6, 1896. Construction work started almost immediately, and at one point was scheduled for completion on July 1, 1897. However, in common with many of Portland's sites, problems excavating the rocky surface rock were encountered. It does appear that the armament was installed (maybe somewhat in advance of full completion) in time for the 1898 Spanish American War. Transfer was finally made on March 8, 1901 for a construction cost of $226,205.37. The entire emplacement was named initially on General Orders No. 78 on May 25, 1903 for Brevet Major General Stephen W. Kearny of Mexican War service. However a couple of years later the battery was split tactically: the two western pits were renamed on General Orders No. 20 of January 25, 1906 for Lt. Colonel Constantine Chase, an artillery officer from the Civil War. The battery was armed with sixteen 12-inch Model 1890M1 mortars on Model 1896 carriages (Kearny: Niles #10/#32, Watervliet #24/#43, Builders #38/#14, Bethlehem Steel #34/#30, #36/#33, #44/#38, #45/#49, and #47/#31. Chase: Builders #17/#37, #19/#77, #20/#73, Watervliet #26/#40, #30/#36, #33/#69, #36/#72, and #37/#78). In 1910 two guns and carriages were moved from Battery Kearny and sent to serve as training pieces at the West Point Military Academy (#24/#43 and #44/#38). Then in 1918 six more mortars and carriages were removed, in common with most other domestic mortar batteries. Remaining in Chase were guns #20, #26, #19 and #37 and at Kearny #34, #38, #47, and #45. These then continued to serve for many years, not being authorized for removal until November 6, 1942. The emplacement itself was buried/destroyed in the early 1960s to make way for Southern Maine Community College use. Nothing remains today of the emplacement.

- **RIVARDI:** A battery for two 6-inch disappearing guns emplaced within the parade salient of the old third system work, at the northeastern tip of the reservation. It was found to have an excellent field of fire to the main channel just to the east. This battery and adjacent Battery Mason were the light armament approved on May 4, 1903 for the fort. The battery, conforming to standard plans, was built in 1903-04. It was transferred to service troops on May 16, 1906 for a cost of $44,136.70. It was armed with two 6-inch Model 1903 guns on Model 1903 disappearing carriages (#8/#39 and #40/#40). It was named on General Orders No. 194 of December 27, 1904 for Major John J, U. Rivardi of the Artillerists and Engineers who served during the years 1795-1802 and died in 1808. The battery was disarmed in 1918 to provide gun tubes for mobile mounts. The carriages

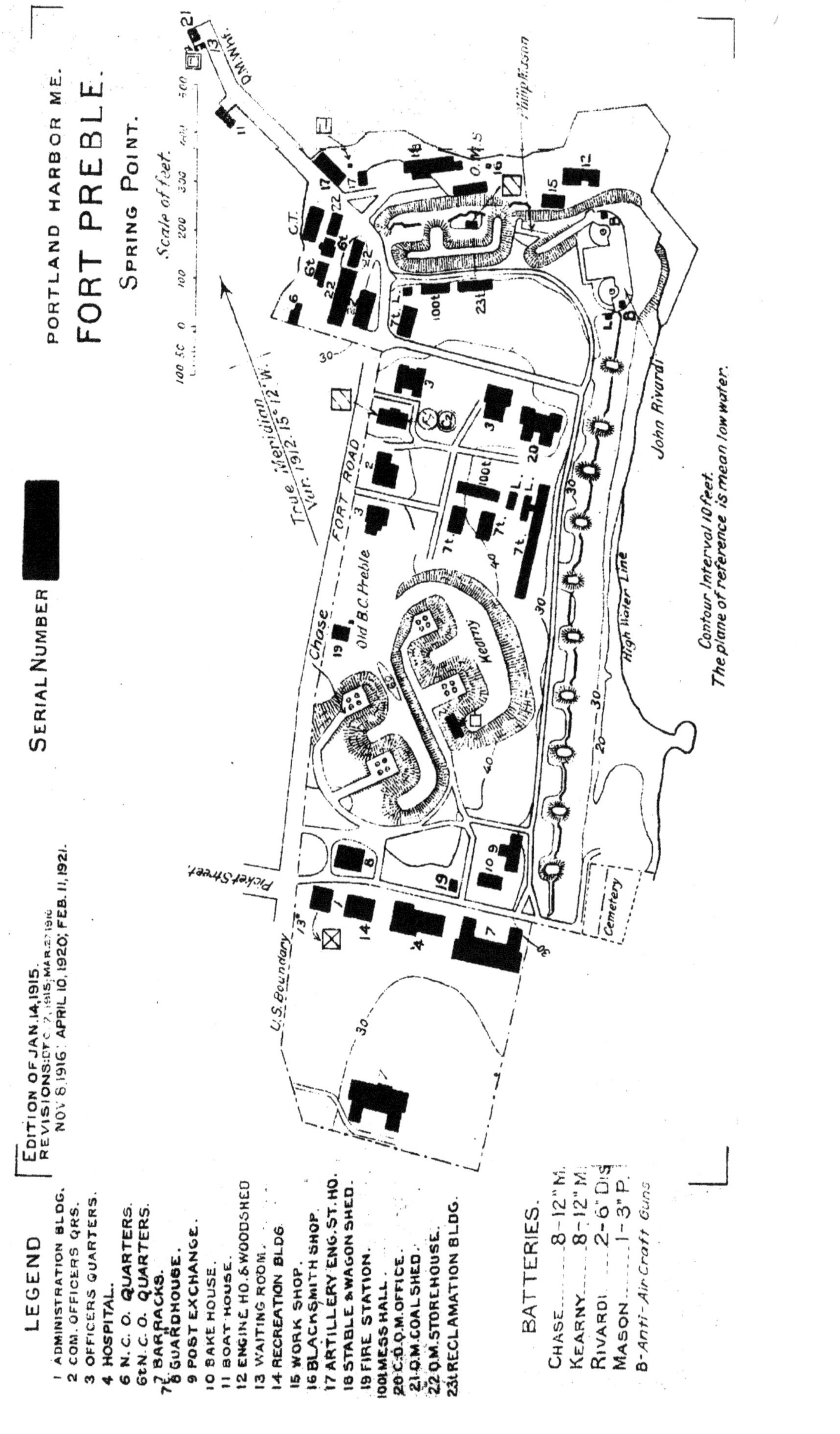

PORTLAND HARBOR ME.

FORT PREBLE.

Spring Point.

Scale of feet.

100 50 0 100 200 300 400 500

EDITION OF JAN.14,1915.
REVISIONS:OCT 7, 1915; MAR.2,1916;
NOV 8,1916; APRIL 10, 1920; FEB.11,1921.

SERIAL NUMBER

True Meridian
Var. 1912-15 12° W.

Contour Interval 10 feet.
The plane of reference is mean low water.

John Rivardi

High Water Line

Cemetery

Old B.C. Preble

Kearny

Chase

FORT ROAD

Picket Street

U.S. Boundary

O.M.M.H.

Philip R. Jean

LEGEND

1 ADMINISTRATION BLDG.
2 COM. OFFICERS QRS.
3 OFFICERS QUARTERS.
4 HOSPITAL.
6 N.C.O. QUARTERS.
6t N.C.O. QUARTERS.
7 BARRACKS.
7t "
8 GUARDHOUSE.
9 POST EXCHANGE.
10 BAKE HOUSE.
11 BOAT HOUSE.
12 ENGINE HO.& WOODSHED
13 WAITING ROOM.
14 RECREATION BLDG.
15 WORK SHOP.
16 BLACKSMITH SHOP.
17 ARTILLERY ENG.ST. HO.
18 STABLE & WAGON SHED.
19 FIRE STATION.
100 MESS HALL.
20 C.Q.M.OFFICE.
21 Q.M.COAL SHED.
22 Q.M.STOREHOUSE.
23t RECLAMATION BLDG.

BATTERIES.

CHASE	8-12" M.	
KEARNY	8-12" M.	
RIVARDI	2-6" DIS	
MASON	1-3" P.	
B-Anti-Air-Craft Guns		

TRUE

BTRY 5

CRF 5

1C

1A

AA 1

B

HIGH WATER LINE

PICKET ST.

100 0 100 200 300 400 500
FEET

MILITARY RESERVATION
BOUNDARY LINE

TRUE

PORTLAND

0 5000
YARDS

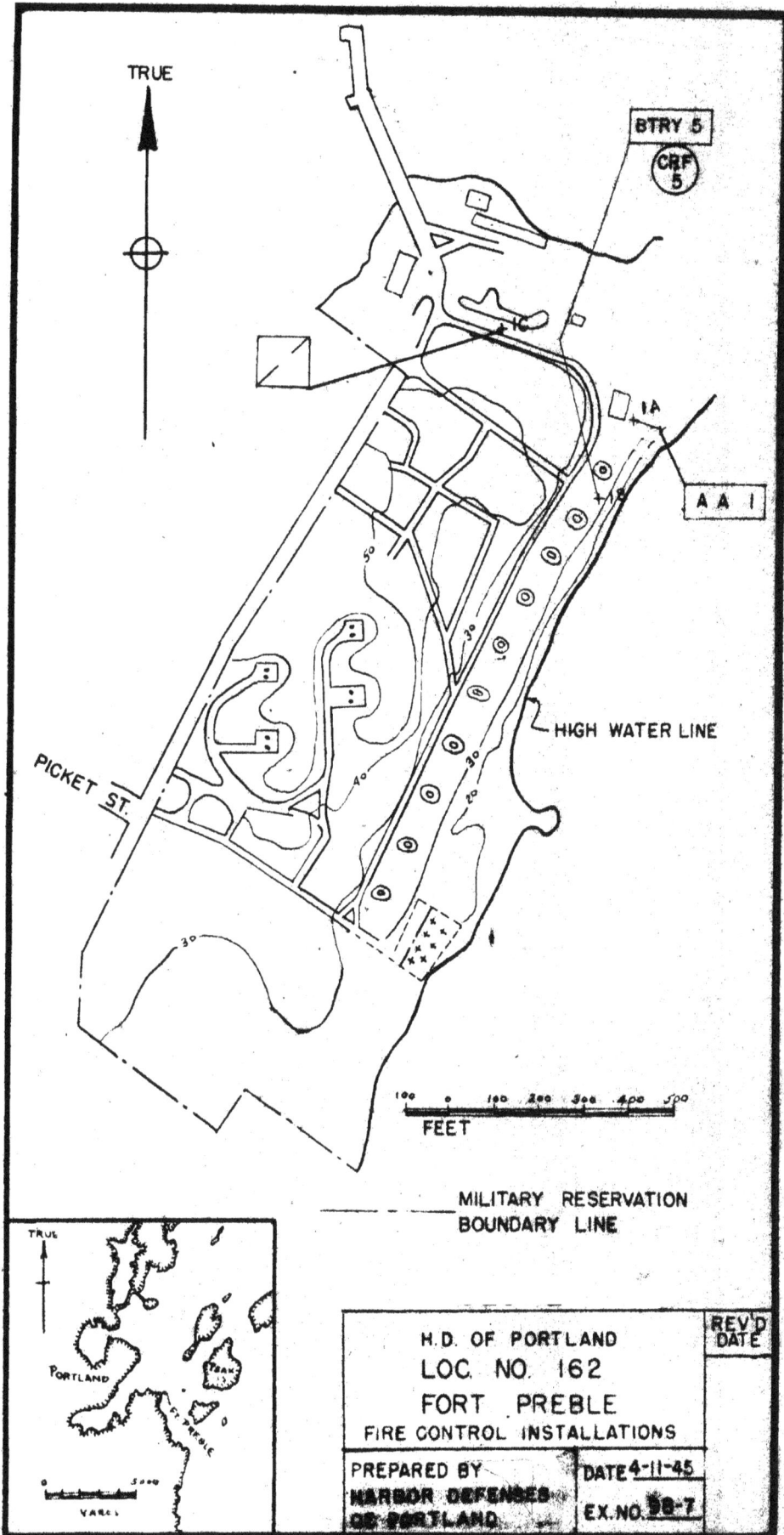

H.D. OF PORTLAND
LOC. NO. 162
FORT PREBLE
FIRE CONTROL INSTALLATIONS

REV'D
DATE

PREPARED BY
HARBOR DEFENSES
OF PORTLAND

DATE 4-11-45

EX. NO. 98-7

Fort Preble 1920s (NARA)

Southern Maine Community College (Fort Preble) (Terry McGovern)

were scrapped in place in 1921. The emplacement still exists on the college campus. The battery site is open to the public, but the interior is closed to the public.

- **MASON**: A battery for a single 3-inch pedestal gun emplaced to the west of Battery Rivardi among the remains of the 1870s Rodman cannon battery at Fort Preble. It fired to the north and covered White Head passage between Cushing and Peaks Islands. The small emplacement had just a single gun platform and adjacent magazine. It was built in 1903-04 and transferred with Rivardi on May 16, 1906 for a cost of $8042.67. It was armed with 3-inch Model 1902M1 Bethlehem gun and pedestal #2. It was named on General Orders No. 194 of December 27, 1904 for 1st Lieutenant Philip D. Mason mortally wounded at the Battle of Trevillian Station, VA in 1864. It retained its armament after World War I, but the gun and pedestal were relocated to a new position in 1942. The emplacement itself still exists on the college campus. The battery site is open to the public, but the interior is closed.

- The 3-inch gun and pedestal from Battery Mason was relocated to a new gun block in 1942. This site was on the eastern side of the fort between two of the old 1870s traverse magazines (one of which was used to serve the ammunition storage for this new position). The block was a simple new cylindrical block with foundation bolts. It served until removed under authority of May 16, 1946. The block was destroyed in the mid-1980s.

Fort Williams (1899-1961) is located on Portland Head in Cape Cottage, which is part of the town of Cape Elizabeth. Constructed between 1898 and 1906 as part of the Endicott Program, its six concrete batteries and a controlled mine casemate covered the southern entrance to Portland Harbor. Named in honor of Bvt. Maj. Gen. Seth Williams, Assistant Adjutant General, U.S. Army (General Orders 71, Apr. 13, 1899). It served as the HQ post for the harbor defenses around Portland and was its largest garrison post. Field emplacements for a four-gun 155mm gun battery (1934-1943) were located in front of Battery Sullivan. An AMTB battery was installed during World War II. By the end of World War II, all of the fort's coast artillery had been scrapped, but the fort continued to serve as an infantry post. Declared surplus in 1961, the fort was transferred to the town of Cape Elizabeth. In the late 1970's, all but 6 of the reaming post garrison buildings were removed and the large gun emplacements of Battery Blair, Sullivan and DeHart and were filled with earth to the top of the parapets. The central powerhouse (1907) and protected post switchboard (1920) (buried) still exist near the main gate. The mine casemate still exists below Battery Hobart, near the ruins of the mine tramway and wharf. Today the reservation is a large city park, with its historic focal point being the Portland Head lighthouse. One of the gun pits of Battery Blair was uncovered in the early 2000s and has some interpretive signage. The three smaller caliber batteries are open to the public. The museum and shop at the lighthouse have information on the history of the fort. The Cape Elizabeth Historical Society has a museum in the BOQ. The park is open to the public.

Fort Williams Gun Batteries

- **SULLIVAN**: The first set of Endicott guns for Casco Bay. In August 1892 plans were submitted by local engineers for two guns. These were in separate emplacements, and designed for the Model 1894 front pintle disappearing carriages. Work progressed relatively quickly, despite the usual problems with excavating the rock ledge. The two emplacements were essentially completed by September 1894. On September 22, 1894 the design of the third emplacement was submitted, connected on the north to the previous northern M1894 type, making this into a two-gun emplacement, albeit eventually with different model carriages. The combined set of three guns was transferred on June 18, 1898 for a cost of $107,063.49. the guns were test fired in their emplacements on August 26,

1898. It was named on General Orders No. 78 of May 25, 1903 for General John Sullivan, of Revolutionary War service. Sullivan carried three 10-inch Model 1888 guns on two Model 1894 and one Model 1896 carriages. The Model 1894 carriage emplacements had Watervliet tubes #22/on carriage #16 and #40/on carraige #10; while the Model 1896 emplacement had Watervliet #50/ on carriage #11. It was located in the center of the reservation, firing to the east. In the eventual sequence of seven heavy guns, Sullivan was on the northern end, emplacements No. 5-7. These served until the battery was dismounted on orders of May 25, 1918—and the tubes taken for possible railroad carriage service during World War I. However, in just a year, in 1919 different tubes were sent to the battery and mounted on the carriages which had been retained (thus having #4/#16, #5/#10 and #11/#11). These served until authorized for removal under the 1932 Program, the guns were gone by 1940. In 1942 parts of the emplacement were converted to become the HDCP for Portland, as which it served until the early postwar years. In the 1960s the emplacement was buried and exists as such at the local Fort Williams Park. The top of the battery is open to the public.

- **DEHART**: battery for two 10-inch disappearing guns emplaced to the southeast of Sullivan in the late 1890s. Plans for the battery were submitted by the local engineer on August 13, 1895. Work was done in 1895-1897. It was completed by early 1898 but not turned over to service troops until June 8, 1898 for a construction cost of $106,757.08. The guns were test fired on August 26, 1898. It was armed with two 10-inch Model 1888MII guns on Model 1896 10-inch disappearing carriages (Watervliet tubes/carriages #32/#9 and #47/#18). The battery was named on General Orders No. 78 of May 25, 1903 for Captain Henry V. DeHart mortally wounded at the Battle of Gaines Mill, in 1862. The armament was carried until authorized for deletion and scrapping under the 1932 review. Final authority for physical scrapping came on November 6, 1942. In postwar years the battery emplacement was buried for public safety concerns and exists in this condition at the local Fort Williams Park. The top of the battery is open to the public.

- **BLAIR**: An emplacement for two 12-inch guns on disappearing carriages built as part of the heavy gun sequence at Fort Williams. They were the last of these heavy guns and placed at the southern end of the sequence Sullivan-DeHart-Blair. Specific plans were submitted for approval on August 7, 1900. The emplacement was to follow typical plans, but with no exposure to fire from the rear, had reduced protection for that side of the ammunition servicing area. Work was done in 1900-1901 and transfer made on July 31, 1903 for sum of $127,845.44. It was named on General Orders No. 78 of May 25, 1903 for Major General Frank P. Blair of Mexican and Civil War service. Blair carried two 12-inch Model 1895 Watervliet Arsenal gun tubes on Model 1897 LF disappearing carriages (#14/#33 and #11/#32). The battery apparently retained this armament throughout its service life. About 1916 the carriages were modified to allow an increase in elevation (to allow more range). The battery was retained under the 1932 Review but then recommended for deletion under the 1940 Program pending completion of more modern defenses. Authorization for removal came in July 1943. The emplacement was subsequently abandoned and partially buried in the 1960s for public safety reasons. One of the gun pits of Battery Blair was uncovered in the early 2000s and has some interpretive signage. It still exists in this state at the Fort Williams Park. The battery's loading platform is open to the public.

- **GARESCHE**: A battery for a pair of 6-inch disappearing guns built at the southern extreme of the Fort Williams reservation between Battery Blair and the reservation fence. Plans were submitted for a rapid-fire, pedestal battery of three guns in July 1901, but they were not immediately approved. Finally on July 27, 1904 it was determined, because of the restricted space possible, to build a battery for just two guns on disappearing carriages. It conformed to recommended design plans and

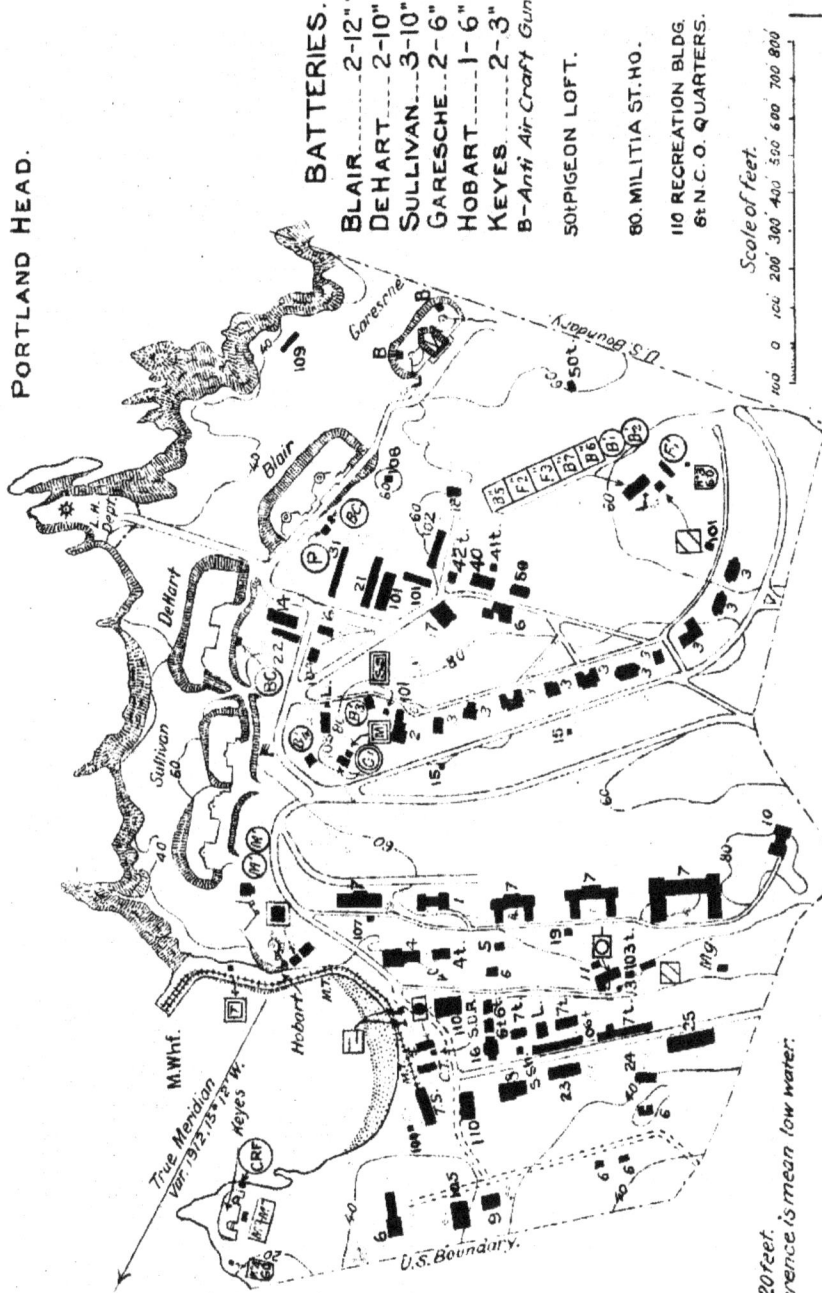

PORTLAND HARBOR ME.

FORT WILLIAMS

PORTLAND HEAD.

SERIAL NUMBER

BATTERIES.

BLAIR	2-12"
DEHART	2-10"
SULLIVAN	3-10"
GARESCHE	2-6"
HOBART	1-6"
KEYES	2-3"
B-Anti Air-Craft Gun	

50† PIGEON LOFT.

80. MILITIA ST. HQ.

110 RECREATION BLDG.
6† N.C.O. QUARTERS.

Scale of Feet.
100   0   100   200   300   400   500   600   700   800

Contour Interval 20 feet.
The plane of reference is mean low water.

LEGEND

EDITION OF JAN.14, 1915.
REVISIONS: DEC.7, 1915; MAR.2, 1916;
APRIL 10, 1920; FEB. 11, 1921.

1 ADMINISTRATION BLDG.
2 COM'D'G. OFF. QUARTERS.
3 OFFICERS QUARTERS.
3†
4 HOSPITAL.
4† TEMPORARY HOSPITAL.
5 HOSPITAL STWD'S. QRS.
6 N.C. OFFICERS QRS.
7 BARRACKS.
7† TEMPORARY BARRACKS.
8 GUARD HOUSE.
9 POST EXCHANGE.
10 BAND QUARTERS.
11 ARTILLERY ENGR. ST. HO.
12
13 TRANSFORMER HOUSE.
14 WORKSHOP.
15 BAND STAND.
16 BAKERY.
17 STABLE.
18 WOOD SHED.
19 FIRE STATION
100 OIL HOUSE.
101 GARAGE.
102 WAGON SHED.
103 DORMITORY.
103† TEMP. DORMITORY
104 GREEN HOUSE.
105 GYMNASIUM.
106† MESS HALL.
107 HOSE SHED.
108 STONE CRUSHER.
109 TARGET PIT.
21 Q.M. ST. HO. & GRANARY.
22 Q.M. WOOD SHED & ST. HO.
23 Q.M. STORE HOUSE.
24 Q.M. COAL SHED.
25 GARAGE.
31 ORDNANCE ST. HO.
40 ENGR. OFFICE & ST. HO.
41† ENGR. BLACKSMITH SHOP.
42† MOTOR TRUCK HOUSE.

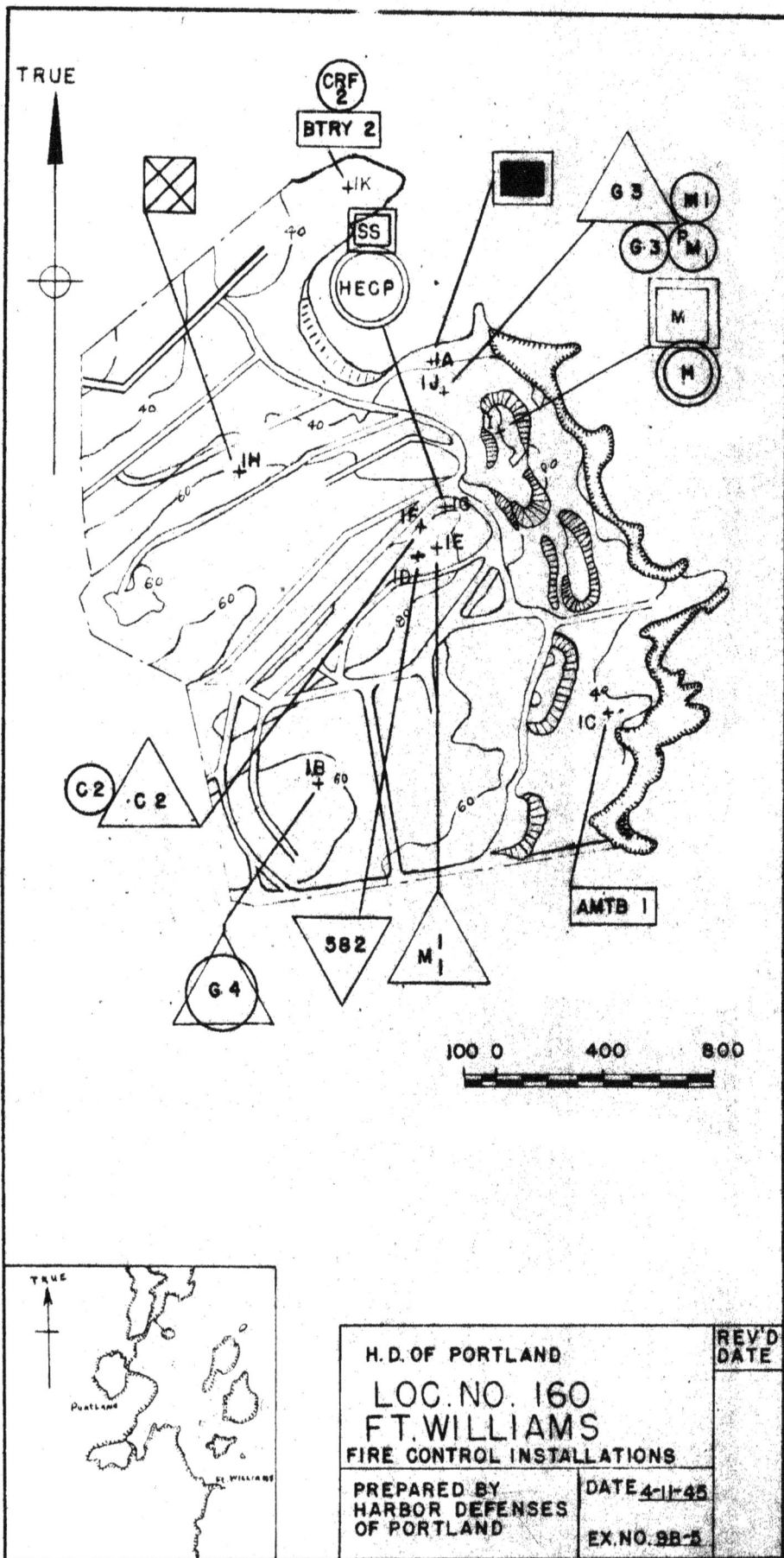

TRUE

CRF
2

BTRY 2

+IK

70

SS

HECP

40

40

40

+IH

60

60

60

G 3

M I

G 3
P
M

M

M

+IA
IJ+

IF+ +IG
+IE

ID+

80

C 2　C 2

IB+　60

60

60

IC +

AMTB I

G 4

582

M I
I

100 0　　400　　800

TRUE

Portland

Ft. Williams

H. D. OF PORTLAND

LOC. NO. 160
FT. WILLIAMS

FIRE CONTROL INSTALLATIONS

PREPARED BY
HARBOR DEFENSES
OF PORTLAND

REV'D
DATE

DATE 4-11-45

EX. NO. 9B-5

Upper and lower photos are of Fort Williams at Portland Head in the 1920s (NARA)

was identical to Battery Rivardi being built at Fort Preble. Work was done from 1904-1905 and transfer made on October 16, 1906 for $48,990.89. The battery was named on General Orders No. 78 of May 25, 1903 (in fact even before work was begun) for Lt. Colonel Julius P. Garesche killed in action at Stone River, TN, in 1862. While the emplacement was of conventional design, the actual assigned armament was not. The battery received the only two 6-inch Model 1900 gun tubes specially fitted with trunnions to allow them to fit on the standard Model 1903 disappearing carriages. One gun and carriage came from proving ground testing, the second set had been on display at the St. Louis Louisiana Purchase Exposition of 1904-05, arriving on March 5, 1905. As armed, the battery carried Model 1900 6-inch guns on Model 1903 disappearing carriages (#45/#11 and #46/#12). They were never fired from the battery due to concerns over damage to neighboring civilian cottages. The armament was removed on November 24, 1917. Carriages were scrapped in place in June 1920. In later years the old emplacement served as a magazine for the nearby 3-inch anti-aircraft battery. The emplacement still exists at the Fort Williams Park. The battery site is open to the public, but the interior is closed.

- **HOBART**: A battery for a single Vickers 6-inch rapid-fire gun acquired during the Spanish American War. Local engineers submitted a plan in May of 1898 to emplace one gun on a modified old 1870s Program gun platform and adjacent magazine to the west of Battery Sullivan near the cove. Construction was done in mid-1898, and the gun was mounted and proof fired on August 26, 1898. Transfer was delayed until January 6, 1900 for a construction cost of $6545.33. It was named for Lt. Henry A. Hobart who was killed in action during the War of 1812. The battery carried a single 6-inch Armstrong-type gun made by Vickers and pedestal (#12133/#11159). The emplacement served for several years, at least through 1910. It was dismounted and shipped to Benicia Arsenal in California on August 25, 1913 (eventually it was sent to Hawaii). The emplacement was not subsequently used for armament but still exists with exposed anchoring bolts at the Fort Williams Park. The battery site is open to the public.

- **KEYES**: A battery for two 3-inch rapid-fire pedestal guns emplaced the point to the north of the fort reservation. It was built in 1903-04 and was of typical mimeograph plan type. Transfer was made on April 27, 1906 for a cost of $16,893.66. The battery was named for Major General Erasmus D. Keyes of Civil War service. It carried two 3-inch Model 1902 guns and pedestal mounts (#4/#4 and #36/#36). This same armament remained throughout the battery's service life. Keyes was kept in the World War I years, interwar period, and under the 1940 Program. In the early 1920s it received a large new BC and coincidence rangefinder station. As virtually the only guns covering the minefield after Battery Hobart's elimination, it was an important part of the defenses. Keyes was finally eliminated under authority of March 16, 1946. The emplacement still exists at the Fort Williams Park. The battery is open to the public.

- **AMTB-961**: A 1943 Program AMTB battery of two 90mm fixed and two 90mm mobile guns emplaced at Fort Williams. It was located east of the major battery line close to the shoreline. It consisted of just two simple concrete gun blocks holding the foundation bolts for the two M3 shielded 90mm mounts. It served until removed postwar. One gun block still remains to be seen at the Fort Williams Park. The battery site is open to the public.

Cape Elizabeth Military Reservation (1942-1948) is located at Cape Elizabeth, south of Portland Head. A site nearby was slated for a 16-inch casemated battery (#101) that was never built. A 6-inch Battery (#201) was built but not armed in 1943. A round concrete fire control tower was built in 1943. Declared surplus after World War II, the military reservation was acquired by the State of Maine. Nearby outside of the state park, the west tower of the Cape Elizabeth Two Lights (built 1874) was discontinued as an active light in 1924 and was converted in 1942 into a fire-control station (private property since 1959). Today, former military reservation is a state park, and the exterior of the 6-inch battery and concrete FC tower can be visited.

Cape Elizabeth M.R. Gun Batteries

- *Battery #101* (planned): A 1940 Program dual 16-inch barbette battery planned for Casco Bay. Originally it was given the highest priority for this type of battery in the harbor, #16 being assigned on September 11, 1940 schedule. Subsequently however, priority was switched with the unit for Peaks Island and Construction No. 101 was reduced national priority #33 by August 11, 1941. No work was ever actually done, the project being deleted on November 13, 1942.

- **Battery #201**: A 1940 Program dual 6-inch barbette battery emplaced on the southern part of the Cape Elizabeth reservation. Its field of fire was to the southeast. It was funded under the FY-1943 Budget. Most of the physical construction was accomplished in 1943-44. While most of the structural concrete work was finished, installation of armament and equipment was never completed. It was to have received two Model M1(T2) 6-inch guns on Model M4 barbette carriage (which were actually delivered to the site, #63 and #64). By 1945 the battery was still incomplete. The 1946 Program advocated the completion of the work, but this was not done. It was soon abandoned. The concrete emplacement still exists at Two Lights State Park, Maine. The battery site is open to the public, but the interior is closed.

Battery #201 Cape Elizabeth State Park (Terry McGovern)

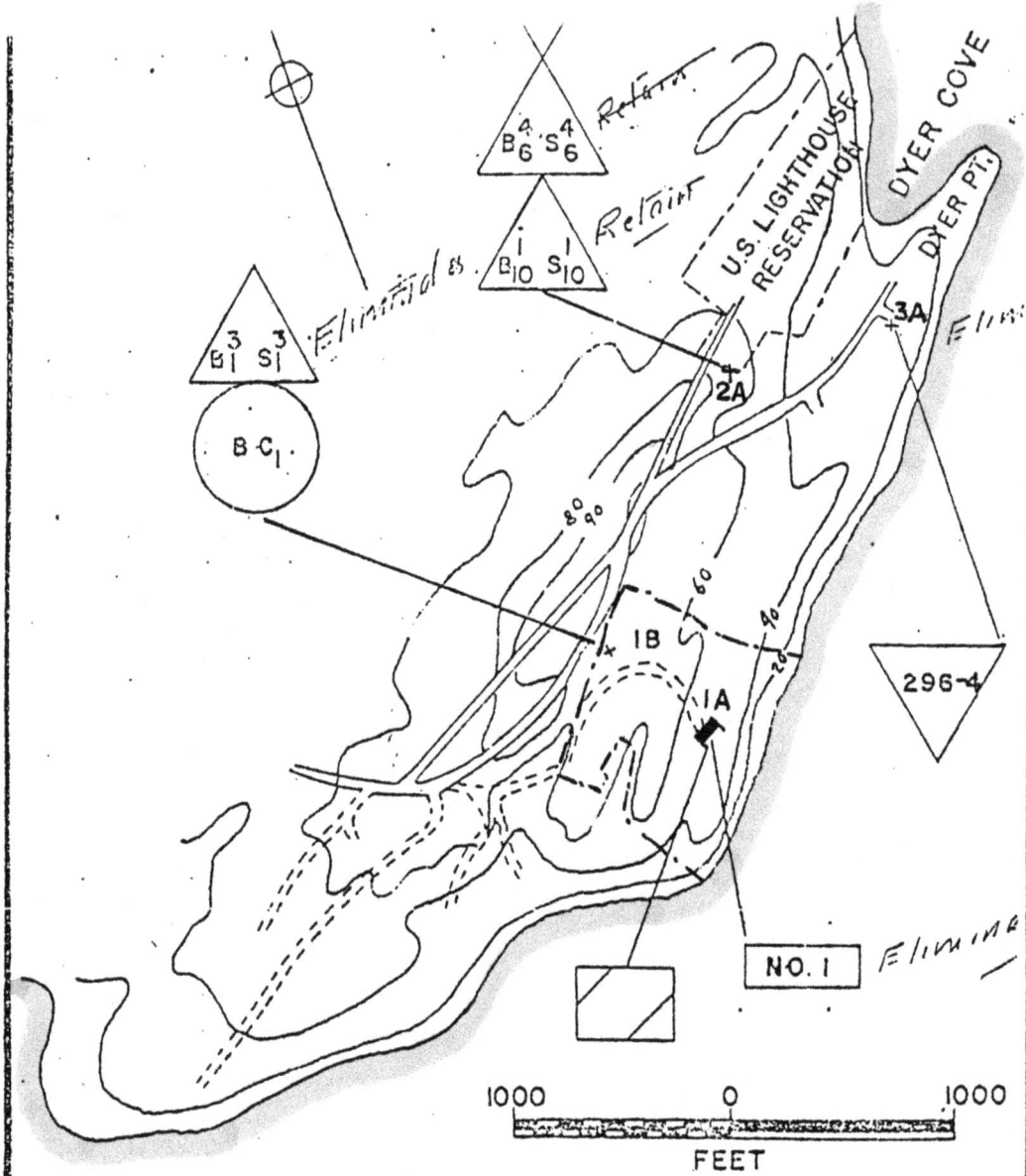

MILITARY RESERVATION
BOUNDARY LINE

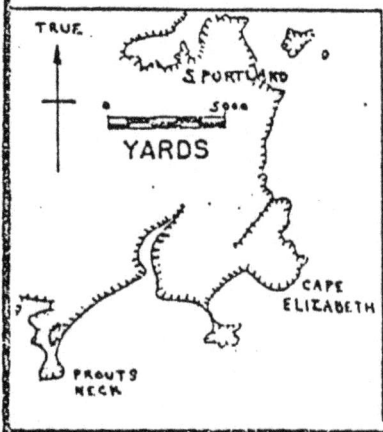

TRUE

S PORTLAND

YARDS

CAPE
ELIZABETH

PROUTS
NECK

H.D. OF PORTLAND
LOC. NO. 158
CAPE ELIZABETH
FIRE CONTROL INSTALLATIONS

REV'D
DATE

PREPARED BY
HARBOR DEFENSES
OF PORTLAND

DATE: 4-11-45

EX.NO. 9B-3

Battery Blair Fort Williams Park

Guardhouse of Fort Preble, now the welcome center for Southern Maine Community College

Double Barracks, Fort McKinley, now the Inn at Diamond Cove

DRINKWATER PT.
37 MM OF
AMTB #10

FORT WILLIAMS
LOC. 160
HDCP-HDOP
HECP
C2- CP AND OP
G3- CP AND OP
M1BC
BTRY #2
BC BTRY #2
G4- CP
M1 — PM1
SCR-582
AMTB #1
LESS 37 MM SECTION
SL #7
SL #8
FCSBR　PSB
SIG. STA.
MET. STA.
CRF BTRY #2
MINE CASEMATE #1

FORT LYON
LOC. 165
AA #3
BTRY #9
BC BTRY #9
37 MM SECTION
FROM AMTB #5
CRF BTRY #9
SL #14

LONG ISLAND
LOC. 166
AAIS　SCR-296
M5　　　BTRY-8
B5 S8
AMTB #5 LESS 37MM
AMTB #6 W/37 MM
G6 CP OP
SL #15　SL #17
SL #16　SL #18

CHEBEAGUE ISLAND
LOC. 166 A
AMTB #9
W/37 MM SECTION
37 MM SECTION
OF AMTB #7
SL #22
SL #23

BAILEY ISLAND
LOC. 168
AMTB #10 LESS
37 MM SECTION
SCR-296 BTRY #6
B6 S8
B10 S10
B6 S9
B6 S7
AAIS
SL #24

FORT BALDWIN
LOC. 170
AAIS
B6 S6
SL #26
SL #27

FORT PREBLE
LOC. 162
BTRY #5
BC BTRY #5
AA #1
37 MM SECTION
FROM AMTB #1
PSB
CRF BTRY #5

FORT McKINLEY
LOC. 164
BTRY #8
BC BTRY #8
PSB
TIDE STA.

PEAK ISLAND
LOC. 163
BTRY #6
BTRY #7
BC BTRY #6
BC BTRY #7
AMTB #3 W/37MM
AMTB #4 W/37MM
SCR-296 BTRY #7
M2 — PM2
G5 CP-OP
B7 S7
C1 - CP-OP
M2- BC
SL #12
SL #13
MINE CASEMATE #2
FCSBR　PSB

JEWELL ISLAND
LOC. 167
AAIS
BTRY #10
BC BTRY #10
G7- CP-OP
SCR-296 BTRY #10
AMTB #7 LESS
37 MM SECTION
AMTB #8 W/37 MM
B6 S6
B6 S9
B6 S4
B7 S7
SL #19
SL #20
SL #21
FCSBR

SMALL POINT
LOC. 169
AAIS
B6 S6
B10 S10
SL #25
FCSBR

TRUNDY POINT
LOC. 159
B6 S9
B6 S4
B6 S5
B7 S7

PROUTS NECK
LOC. 157
B6 S8
B6 S5
B6 S4
B6 S5
SL #3

FORT LEVETT
LOC. 161
BTRY #3
BTRY #4
AA #2
AMTB #2
W/37MM SECTION
BC BTRY #3
BC BTRY #4
B6 S4
B6 S5
B6 S5
AAIS
G1- CP-OP
B7 S10
B6 S5
G5 - CP-OP
M1
SCR-296 BTRY #3
SL'S #9, #10, & #11
FCSBR　PSB

CAPE ELIZABETH
LOC. 158
BTRY #1
BC BTRY #1
AAIS
B7 S3
B6 S6
B10 S10
SCR-296 BTRY #1
SCR-296 BTRY #4
SL #4
SL #5
SL #6
FCSBR

FLETCHERS NECK
LOC. 156
AAIS
B6 S6
B6 S4
B6 S1
SL #1
SL #2

CAPE PORPOISE
LOC. 155
AAIS
B6 S6
FCSBR

H.D. OF PORTLAND

LOCATION OF ELEMENTS

PREPARED BY	DATE 4-11-46
HARBOR DEFENSES	
OF PORTLAND	EX.NO. 4-A

REVISED DATE

TRUE

10　5　0　　　10　　　20
SCALE-THOUSANDS OF YARDS

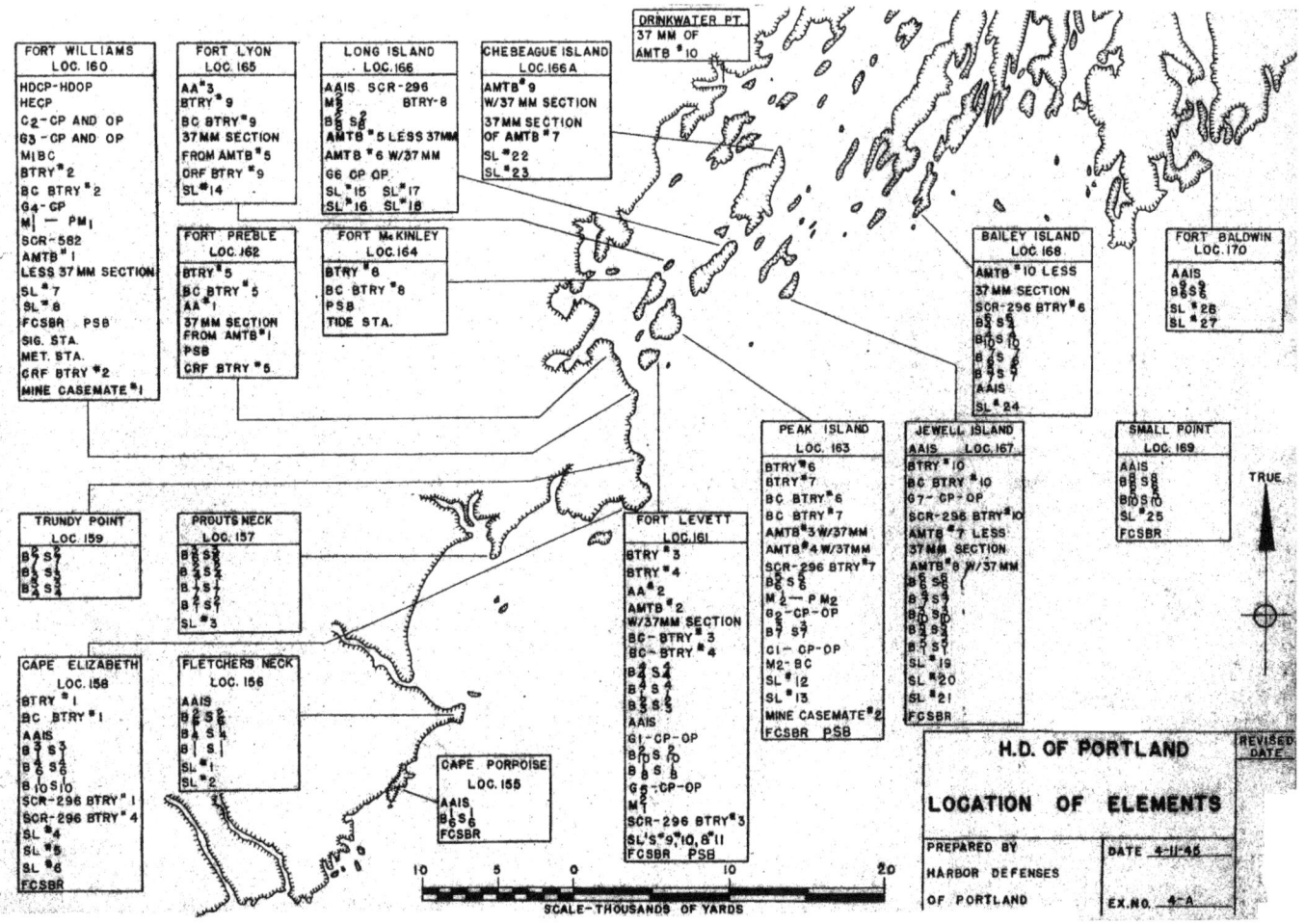

Battery Bowdoin Fort Levett Cushing Island (Mark Berhow)

Portland World War II-era Site Locations. Stations housed in a single structure are connected by dashes (-)

location	Loc#	Purpose
Cape Porpoise	155	BS1/Steele, SBR
Fletcher Neck	156	BS2/Steele-BS1/Foote,-BS1/201, SL 1,2
Prouts Neck	157	BS3/Steele-BS2/Foote-BS1/Cravens-BS2/201, SL 3
Cape Elizabeth	158	Tact. Batt. #1 BCN 201, BC-BS3/201, BS2/Foote-BS1/202, SCR296-201, SCR296-Foote, SBR, SL 4,5,6
Trundy Point	159	BS2/Cravens-BS1/Ferguson, BS3/Foote
Fort Williams	160	Tact. Batt. #2 Keyes, HECP-HDCP-HDOP, C2, G3, M1BC, BC/Keyes, G4, M1/1, SCAR582, AMTB 1, Met, MC, Mines, SL 7,8
Fort Scammel	160A	not used
Fort Levett	161	Tact. Batt. #3 Ferguson, Tact. Batt. #4 Foote, BC/Ferguson, BC/Foote, BS2/Ferguson, BS4/201, BS4/Foote, BS2/202-BS1/Carpenter-G5, G1, AMTB 2, M2/1, SCR296-Ferguson, SL 9,10,11
Fort Preble	162	Batt. Tact. #5 Mason, BC/Mason
Peaks Island	163	Batt. Tact. #6 Steele, Batt, Tact. #7 Cravens, M1/2-BS3/Cravens-BC/Steele, BS5/Steele, BC/Cravens-G2, C1, M2, MC, AMTB 7, AMTB 8, SL 20,21, PSR/Steele, SCR-Foote, SBR
Fort McKinley	164	Batt. Tact. #8 Carpenter, BC/Carpenter, T, Met SBR
Fort Lyon	165	Batt. Tact. #9 Abbott, BC/Abbot, SL 14
Crow Island	165A	not used
Long Island	166	SCR296-Carpenter-M2/2-B2/Carpenter-G6, AMTB 5, AMTB 6, G6, SL15,16,17,18
Great Chebeague Is.	166A	AMTB 9, SL 23,24
Jewell Island	167	Batt. Tact. #10 BCN 202, BS3/202-BS6/Steele-BS4/Cravens, BS5/Foote-BS5/201-G7, BC/202, SCR296-202, AMTB 7, AMTB 8, SL 19,20,21
Bailey Island	168	BS6/Foote-BS4/202-SCR296/Steele, BS7/Steele-BS5/Cravens, AMTB 10, SL 24
Drinkwater Point	168A	37 mm battery
Small Point	169	BS8/Steele-BS5/202, SL 25
Fort Baldwin	170	BS9/Steele, SL 26,27

Thompson, Kenneth E. *Portland Head Light and Fort Williams: An Illustrated History with a Walking Guide Map.* The Thompson Group, Cape Elizabeth, ME 1998.

Gaines, William C. "The Seacoast Defenses of Portland, Maine, 1605-1946, *Coast Defense Journal* Vol. 25, No. 1, Feb. 2011, p. 37; Vol. 25, No. 2, May 2011, p. 15; Vol. 25, No. 3, Aug. 2011, p. 4; Vol. 25, No. 4, Nov. 2011, p. 42.

THE HARBOR DEFENSES OF PORTSMOUTH – NEW HAMPSHIRE

Portsmouth is a deep-water port on the border between the states of New Hampshire and Maine. Settled in 1630 by British colonists, the first fort was built on Castle Island in 1632 to protect the harbor and its citizens from attack. Fought over during the Revolutionary War, the fort was eventually rebuilt during the Second System. Portsmouth was the location of one of the first U.S. Navy shipyards, which is still in operation today. It was to be rebuilt, along with the other Second System fort across the channel, during the Third System but neither was completed. Additional batteries were started after 1870 but halted short of completion. The harbor received eight new concrete batteries at 3 locations during the Endicott Program. There was not much in the way of garrison structures and the defenses were generally staffed by details from the Portland defenses through the 1930s. A 16-inch casemated battery, two new 6-inch batteries and 2 new 90mm AMTBs were built as part of the 1940 Program. Garrison facilities at the seacoast defense sites was limited, so an additional reservation was established on New Castle Island for quarters and additional barracks which became Camp Langdon. The defenses were all inactivated by 1948.

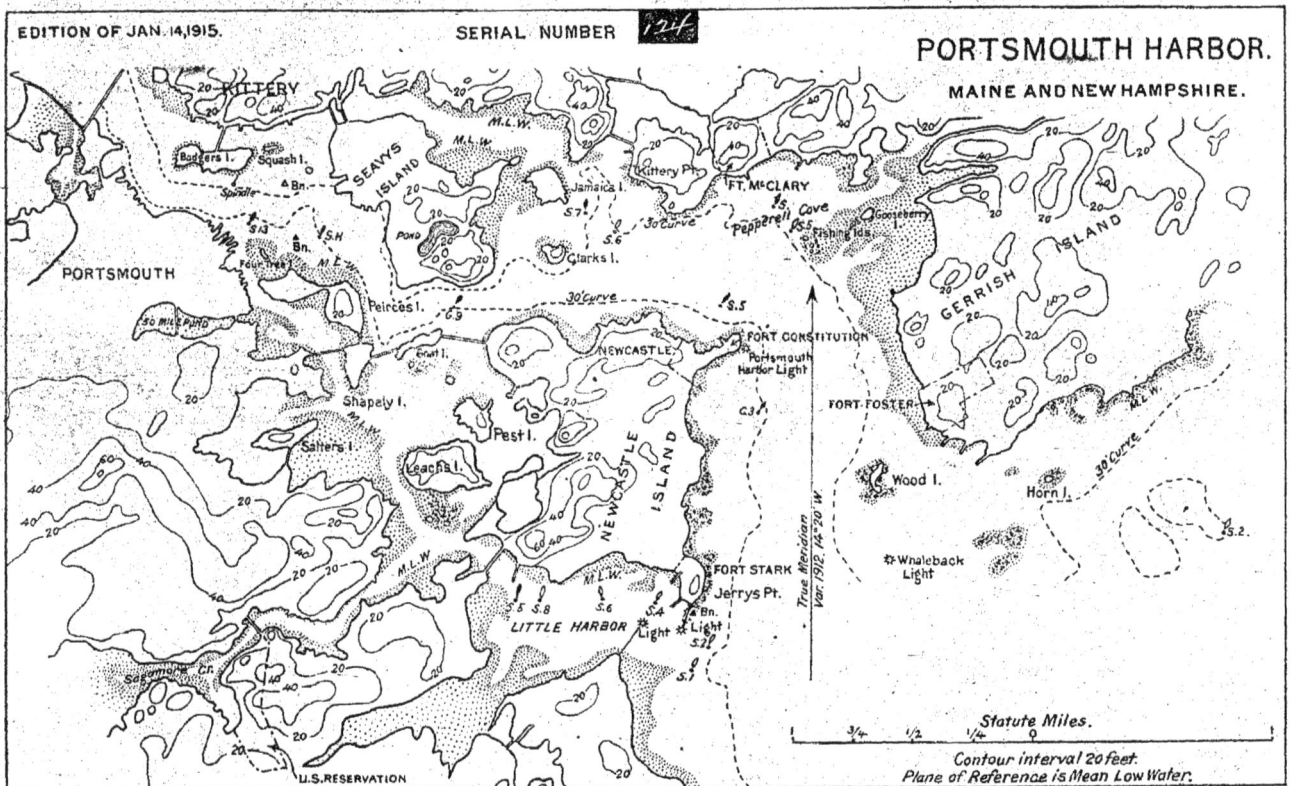

Fort Foster (1900-1946) is located on Gerrish Island near Kittery, Maine, off State Highway 103. The fort was established in 1900 as part of the Endicott Program on 40 acres and two concrete gun batteries were built. It was named in General Orders No. 43 on Apr. 4, 1900 for Maj. Gen. John G. Foster Corps of Engineers, who served during the Mexican and Civil Wars. A 6-inch battery (#205) was built during 1940 Program along with a seven-story concrete fire control tower, controlled submarine mine casemate, and a 90mm AMTB battery. Fort Foster was declared surplus in 1949 and is now a Kittery city park. The lower level of 10-inch DC Battery Bohlen was filled up to the gun platform, and two other batteries are accessible in the park. Vehicle entrance to the park is restricted during the winter months.

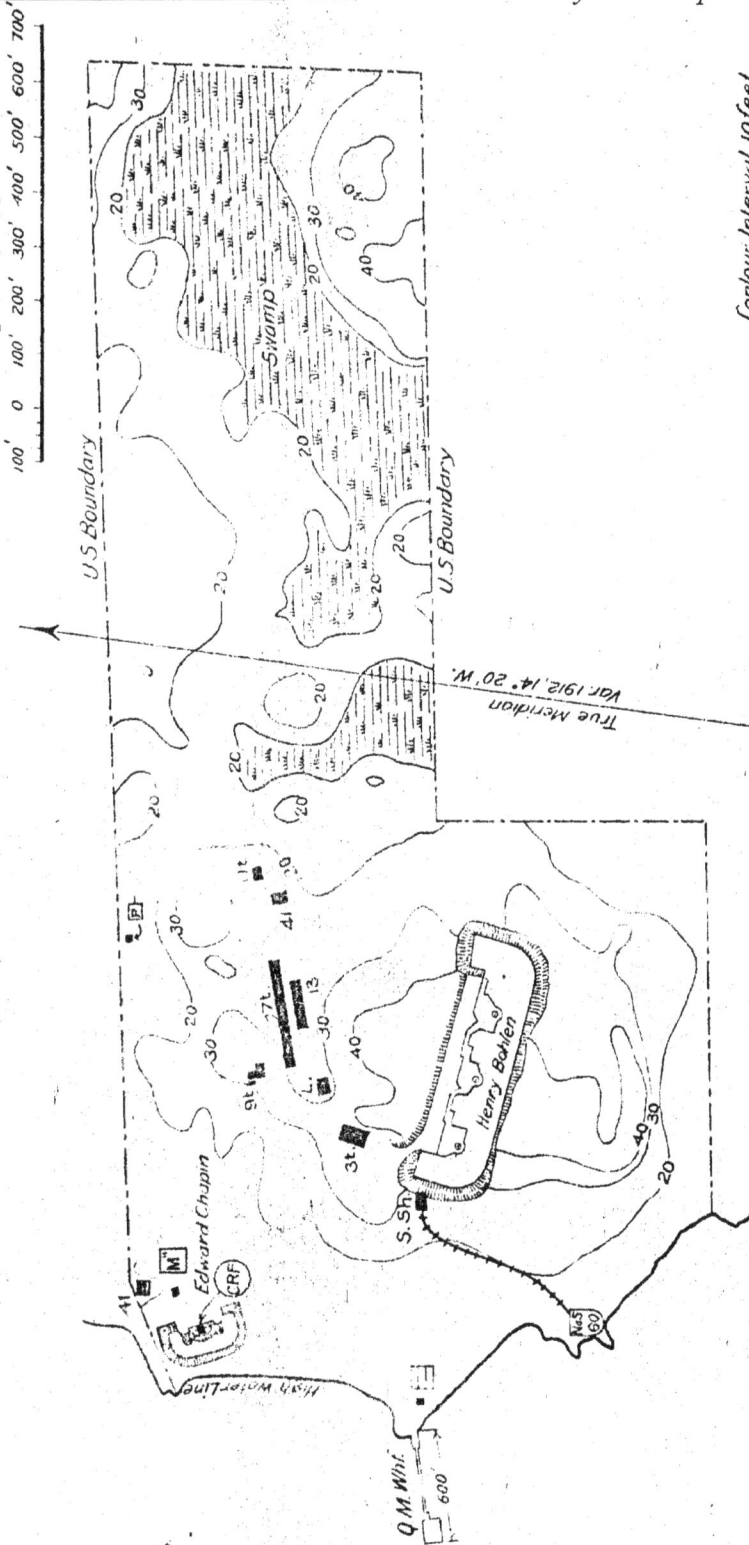

PORTSMOUTH HARBOR, N.H.

FORT FOSTER

GERRISH ISLAND.

Scale of feet.

100' 0 100' 200' 300' 400' 500' 600' 700'

SERIAL NUMBER

124

Contour Interval 10 feet.
Plane of reference is M.L.W.

EDITION OF JAN. 1915.
REVISIONS, DEC. 1915; MAR. 27, 1916.
APRIL 10, 1920; FEB. 11, 1921.

LEGEND.
3t. TEMP. OFFICERS QRS.
7t. TEMP. BARRACKS.
9t. TEMP. POST EXCHANGE.
11t. TEMP. STABLE.

13 GARAGE
41 ENGINEER ST. HO.

BATTERIES.
BOHLEN ... 3-10" DIS.
CHAPIN ... 2-3" P.

U.S Boundary

U.S Boundary

Swamp

True Meridian Var. 1912, 14°. 20'. W.

Henry Bohlen

Edward Chapin

High Water Line

Q.M. Whf.

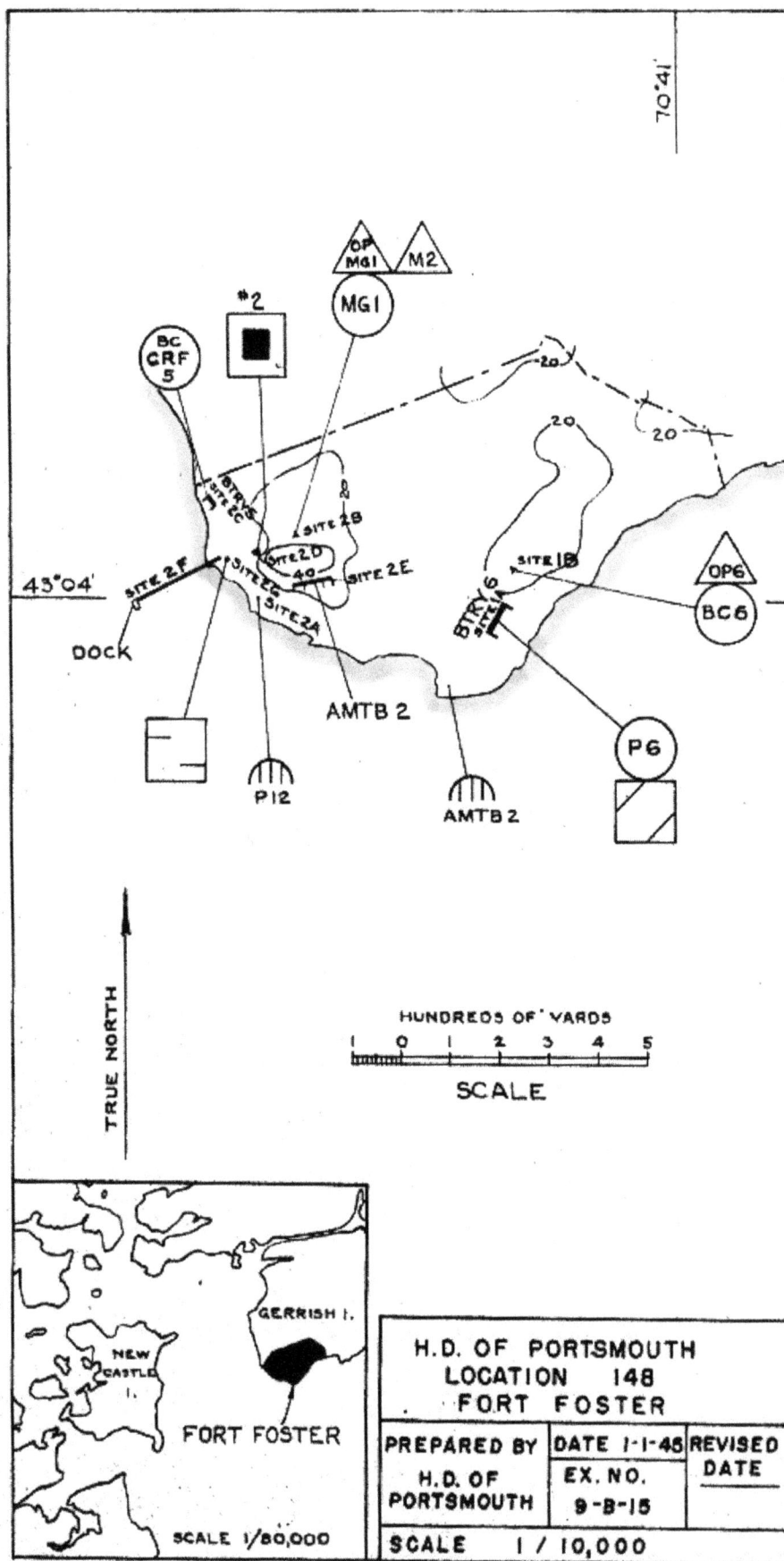

70°41'

OP MG1   M2

MG1

#2

BC CRF 5

20

20   20

BTRY S
SITE 2C
SITE 2B
SITE 2D
40
SITE 2E
SITE 1B
OP6

43°04'   SITE 2F   SITE 2G   BC6
SITE 2A

DOCK   BTRY 16
SITE 16

AMTB 2

P12

AMTB 2

P6

TRUE NORTH

HUNDREDS OF YARDS
1 0 1 2 3 4 5
SCALE

GERRISH I.

NEW CASTLE I.

FORT FOSTER

SCALE 1/80,000

H.D. OF PORTSMOUTH
LOCATION   148
FORT FOSTER

PREPARED BY | DATE 1-1-45 | REVISED
H.D. OF | EX.NO. | DATE
PORTSMOUTH | 9-B-15

SCALE   1 / 10,000

Fort Foster Gun Batteries

- **BOHLEN**: An emplacement for three 10-inch disappearing guns emplaced centrally at the Fort Foster reservation on Gerrish Island. Detailed plans and estimates were submitted for approval on August 31, 1898 for the guns on Model 1896 disappearing carriages. Work was approved and done in 1898-1900. The battery design and gun spacings were conventional and followed design mimeograph specifications. It was reported as almost completed on November 21, 1900. Transfer was made on December 16, 1901 for a cost of $158,000. It was armed with three 10-inch guns Model 1895 Watervliet gun tubes on Model 1896 LF disappearing carriages (#1/#69, #13/#70 and #14/#65). The battery was named on General Orders No. 194 of December 27, 1904 for Brigadier General Henry Bohlen who was killed in action at Second Bull Run in 1862. Modifications to the platforms, hoists and a new battery commander's station were added before World War I. The gun tubes were removed in 1918 for use on railway artillery, but they did not actually leave the reservation but were stored there. In 1920 they were remounted and then served until removed in World War II. Authority for disarmament and abandonment of the emplacement was granted on December 15, 1942. In later years the emplacement was filled-in to the loading platform level, and so modified it still exists at the Fort Foster Park in Kittery, Maine. The battery site is open to the public, but the interior is closed.

- **CHAPIN**: A battery for two 3-inch rapid-fire guns on pedestals erected at Fort Foster. Plans were submitted on May 9, 1903. The design followed the usual recommended type plans closely. Construction work was done from 1903-1904. Transfer was made on December 31, 1904 for a cost of $15,213.09. The battery was armed with two 3-inch Model 1902M1 guns and pedestal mounts (#37/#37 and #38/#38). The battery was named on General Orders No. 194 of December 27, 1904 for Brevet Brigadier General Edward P. Chapin who was killed in action at Port Hudson, LA, in 1863. The battery served for a long time, not being disarmed until postwar in October of 1945. The emplacement still exists at the Fort Foster Park in Kittery, Maine. The battery is open to the public.

- **Battery #205:** A 1940 Program dual 6-inch barbette battery for Fort Foster. It was built on the east side of the reservation, with a field of fire to the east and southeast. Initially it was assigned a relatively low national priority and not funded until the FY-1943 budget. Work began only in early 1944 and was never completed. In 1946, still incomplete, it was recommended for abandonment. The battery was to be of conventional 200-series type design, with guns separated by the usual earth-covered, concrete traverse magazine. No guns were ever delivered or allocated, but the M4 barbette carriages (serial numbers #9 and #10) were delivered but not mounted. No name was ever assigned. The emplacement still exists at the Fort Foster Park in Kittery, Maine. The battery site is open to the public, but interior is closed.

- **AMTB-2:** A 1943 Program AMTB battery for two 90mm fixed and two 90mm mobile guns. The fixed blocks were emplaced adjacent to Battery Bohlen, which was used as a magazine for the battery. Work was authorized on April 12, 1943 and completed later that year. Disarmed with the end of the war, the two-gun blocks still exist at the Fort Foster Park in Kittery, Maine. The battery site is open to the public.

Battery Bohlen Fort Foster Park (Terry McGovern)

Battery #205 Fort Foster Park (Terry McGovern)

Fort Constitution (1791-1949) is located on Fort Point on New Castle Island. The first fort on this point was erected during colonial times with succeeding forts from the First, Second, and Third Systems. Two batteries were built on the point during the early Endicott Program along with the controlled submarine mine complex (with two mine casemate – 1905 & 1921). Fort Constitution was the harbor defense headquarters in World War I. It was designated officially in General Orders 6, on Aug. 4, 1937. After the U.S. Army left after World War II and the mine complex was transferred to the US Navy and then to the Coast Guard in 1965 and the 3-inch battery was destroyed for a Coast Guard helicopter landing pad. Today the 2nd/3rd System fort has been closed to the public due the disintegration of fort's walls. The Endicott Program 8-inch battery and Interwar Period mine casemate are closed to the public as a University of New Hampshire property. A brick mine torpedo storehouse (used by the Coast Guard) also remains. The old fort and the adjacent lighthouse are a state park but are currently closed to the public.

Fort Constitution Gun Batteries

- **FARNSWORTH:** A battery for two 8-inch disappearing guns located to the southwest of the fort reservation, near the old Walbach tower. Field of fire was to the east. Plans were submitted on November 17, 1896. It followed older type plans with small loading platforms, disconnected split-level lower magazines and early dumbwaiter-type ammunition lifts. Work was done from 1897-1898. It was transferred on July 23, 1898 for a construction cost of $65,000. It was armed with two 8-inch Model 1888MII Watervliet tube guns on Model 1894 disappearing carriages (#38/#22 and #43/#23). It was named on General Orders No. 194 of December 27, 1904 for Brigadier General Elon J. Farnsworth, U.S. Cavalry, who died at Fredericksburg in 1862. As early as November 1899 the battery was reported as being extremely wet. The local porous rock provided constant dampness, and despite several attempts to seal the rooms, improve the drains, and cut ventilation conduits nothing seemed to alleviate the problem. The condition was so severe that it essentially ended the service of the battery. In 1905 it was recommended that the battery be entirely rebuilt. While such a drastic step was not taken, the battery was essentially reduced to caretaker, reserve status. By 1910 it was still armed, but out of service. Consequently, the battery did not receive many of the later modifications to major disappearing gun emplacements. The loading platforms were never widened, modern chain hoists were never installed, a new BC station was not constructed, and it was not electrified nor received a latrine. In 1918 the guns were removed for use on railway batteries, the carriages scrapped in place in 1920. In the early 1920s a new post power station was built adjacent to the central traverse of the battery. In 1942 a modern mine battery station was erected in emplacement No. 1. In postwar years the right half of rthe battery was in danger of collapse, and stabilizing steps were taken by the University of New Hampshire. The emplacement still exists but fenced and is not usually available for public visits.

- **HACKLEMAN:** A dual 3-inch battery planned for mine field coverage at Fort Constitution. The plan was submitted on May 9, 1903. It followed the type plans then current. It was located near the eastern side of the reservation, just off the southeast bastion of the old masonry work. It required relocation of the lighthouse keeper's quarters. Work was done in 1903-1904, being transferred on December 31, 1904 for a cost of $16,654.70. It was named on General Orders No. 194 of December 27, 1904 for Brigadier General Pleasant Hackleman, U.S. Volunteers, who died at the Battle of Corinth in 1862. It was not armed until 1908. Eventually it received two 3-inch Model 1903 guns and pedestals (#88/#63 and #89/#64). The battery was retained in the World War I period, and under the 1932 Review. However, in 1942 the armament was transferred to Long Island Sound's Fort H.G. Wright for use at the North Hill position. At some point mid-war replacement guns were

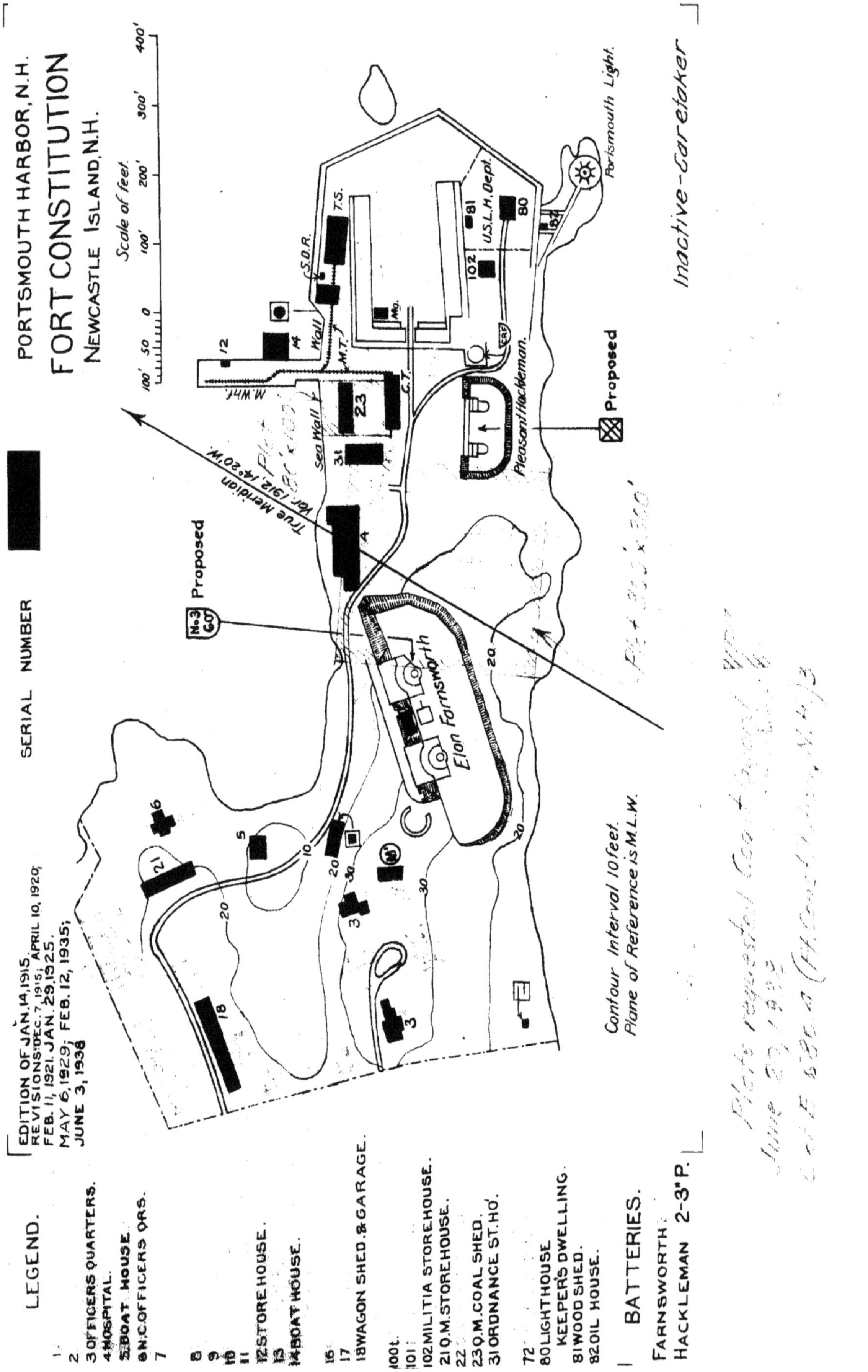
PORTSMOUTH HARBOR, N.H.
FORT CONSTITUTION
NEWCASTLE ISLAND, N.H.
Scale of feet
100' 50 0 100' 200' 300' 400'
SERIAL NUMBER [redacted]
EDITION OF JAN. 14, 1915.
REVISIONS: DEC. 7, 1915; APRIL 10, 1920; FEB. 11, 1921. JAN. 29, 1925. MAY 6, 1929; FEB. 12, 1935; JUNE 3, 1938
LEGEND.
1
2
3 OFFICERS QUARTERS.
4 HOSPITAL.
5 BOAT HOUSE.
6 N.C.OFFICERS QRS.
7
8
9
10
11
12 STORE HOUSE.
13
14 BOAT HOUSE.
15
17
18 WAGON SHED & GARAGE.
100
101
102 MILITIA STOREHOUSE.
21 Q.M.STOREHOUSE.
22
23 Q.M.COAL SHED.
31 ORDNANCE ST.HO.
72
80 LIGHTHOUSE
KEEPER'S DWELLING.
81 WOOD SHED.
82 OIL HOUSE.
BATTERIES.
FARNSWORTH.
HACKLEMAN 2-3"P.
Proposed
Proposed
Elon Farnsworth
Pleasant Hackleman.
True Meridian
Var. 1912, 14°20'W.
Sea Wall
M.Whf.
Wall
M.T.
C.T.
T.S.
S.D.R.
U.S.L.H. Dept.
Portsmouth Light.
Contour Interval 10 feet.
Plane of Reference is M.L.W.
Inactive-Caretaker

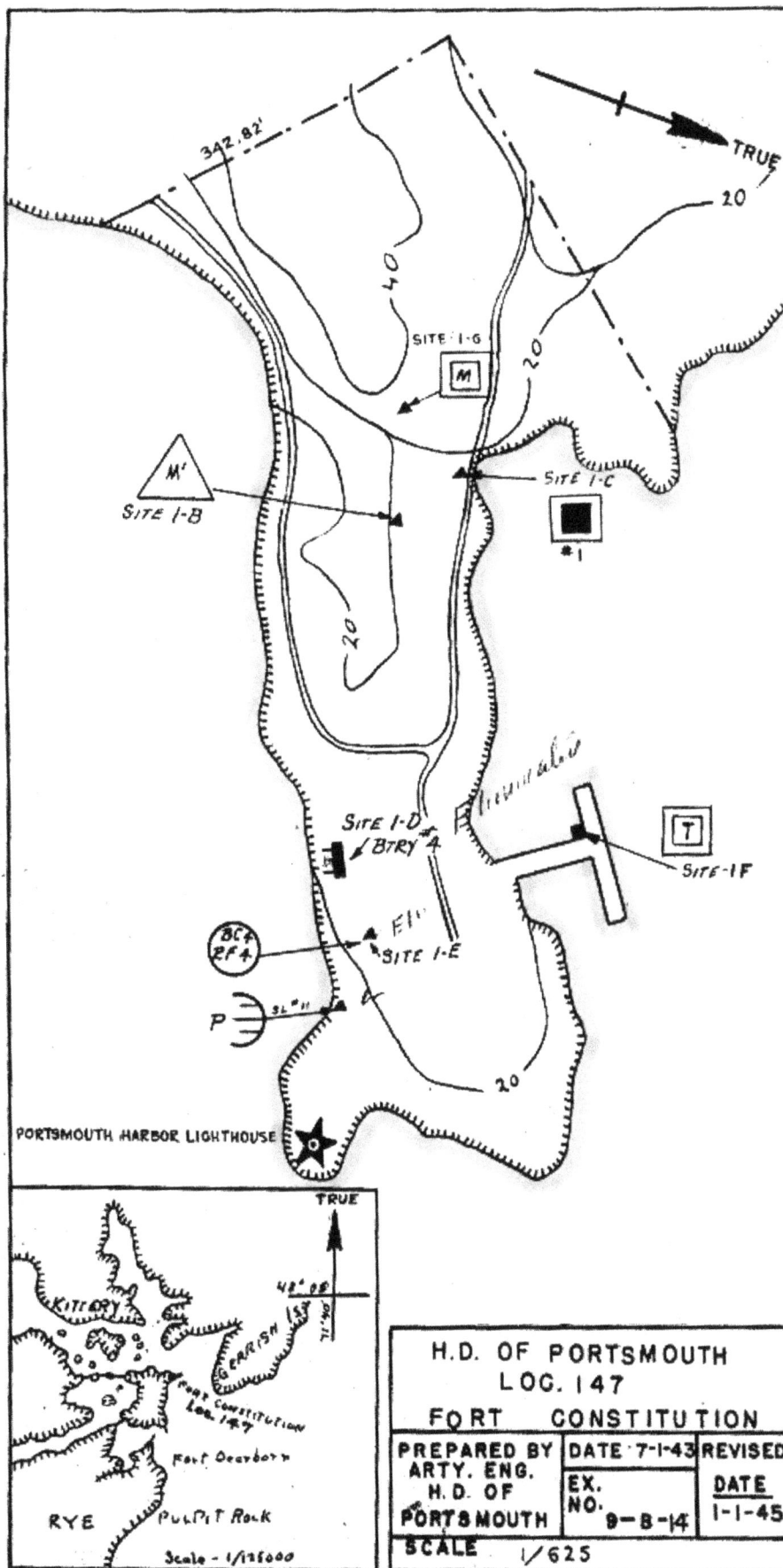

Battery Farnsworth Fort Constitution (Terry McGovern)

Fort Stark Historical Park (Terry McGovern)

sent to Battery Hackleman. It received from Battery Lytle at nearby Fort Stark two 3-inch Model 1902M1 guns and pedestals (#39/#39 and #40/#40). These then served until shortly after the war, probably being removed in 1946. The emplacement was destroyed in the early 1960s to make additional room available at the Coast Guard station. Nothing remains today of the emplacement, except its coincidence rangefinder station on the bastion of the neighboring masonry Fort Constitution.

Fort Stark (1873-1949) is located on Jerry's Point on New Castle Island. Construction of the concrete batteries was begun in 1902 as part of the Endicott Program. The fort occupied 10 acres overlooking the entrance to the Piscataqua River. It was named in General Orders No. 43 on Apr. 4, 1900 for Brig. Gen. John Stark who served during the Revolutionary War. The fort had four batteries mounting a total of 8 guns. During World War II, a large Harbor Entrance Control Post (HECP) was built on top of the fort's 6-inch disappearing battery. Fort Stark was declared surplus in 1949 and transferred to the U.S. Navy who continued to use the post for harbor defense purposes until 1983. The fort eventually became a state historic site. The park has a parking area and is open during daylight hours. A small museum and visitors center is located in the former ordnance machine shop (1902) but is currently closed. Some of the batteries have limited access.

Fort Stark Gun Batteries

- A battery for two modern, breechloading 8-inch guns on reinforced 15-inch Rodman carriages placed at Jerrys Point for the Spanish American War emergency. Plans began on April 27, 1898 to use two of the 21 available 8-inch guns at Portsmouth. The site selected was at two 1870s emplacements for 15-inch Rodmans at the earthen battery of Jerrys Point (emplacements #4 and #7). Very little work was required besides the mounting of the guns. By January 12, 1899 they were reported complete and ready to transfer to troops. Then in October of that same year removal of the guns was authorized. The gun tubes were to go to any defense in need of these Model 1888 guns—in this case that was Fort McKinley in Casco Bay. As it was planned to build modern Endicott batteries at Stark, the temporary battery had to be removed. Soon the guns and carriages were gone and within a year the emplacements destroyed or buried. No sign remains of this battery at the Fort Stark Historic Site.

- HUNTER: A battery for two 12-inch guns on disappearing carriages emplaced at the Jerrys Point reservation of Fort Stark. The emplacement for Model 1897 disappearing carriages was funded by the Fortification Act of March 1, 1901. The emplacement was submitted on May 7, 1901. Generally, the plan used followed mimeograph specifications. However, the frontal protection was reduced somewhat due to the presence of the extensive earthen embankment left by the old 1870s earthen battery in front of it. Work was done from 1901-1902, transfer being made on December 31, 1904 for a construction cost of $109,000. It carried two 12-inch Model 1895 guns on Model LF 1897 disappearing carriages (Bethlehem tube #5/on carriage #34 and Watervliet tube #39/on carriage #35). The battery was named on General Orders No. 194 on December 27, 1904 for Major General David Hunter, U.S. Volunteers, of Civil War service. The emplacement received the usual modifications during its early service. The two carriages were altered for increased elevation in 1916-1917. The armament stayed in place during World War I. It was finally authorized for deletion pending completion of more modern defenses under the 1940 Program. Deletion was authorized on December 21, 1944. The emplacement still exists at the Fort Stark Historic Site. The battery site is open, but interior is closed to the public.

PORTSMOUTH HARBOR, N.H.

FORT STARK

JERRYS POINT

Scale of feet

100' 50' 0 100' 200' 300'

SERIAL NUMBER **124**

L.H.

BREAKWATER.

S.Sh.

Engr. Whf.

William Lytle

CRF

Bs. Hunter.

True Meridian

Edward Kirk

Alexander Hays

CRF

M'

O.M.S.

U.S. Boundary.

Contour Interval 10 feet.
The plane of reference is M.L.W.

EDITION OF JAN.14,1915.
REVISIONS: DEC. 7, 1915; APRIL 10, 1920;
FEB. 11, 1921.

LEGEND.

7 BARRACKS.
7t. TEMP. BARRACKS.
9t. TEMP. POST EXCHANGE
10 COAL SHED & ST. HO.

12 RECREATION BLDG.
30 BRICK OIL HOUSE
(U.S. L.H. DEPT.)
13. GARAGE

BATTERIES.

HUNTER	2-12" D.G.	
KIRK	2-6"	"
HAYS	2-3" P.	
LYTLE	2-3" P.	

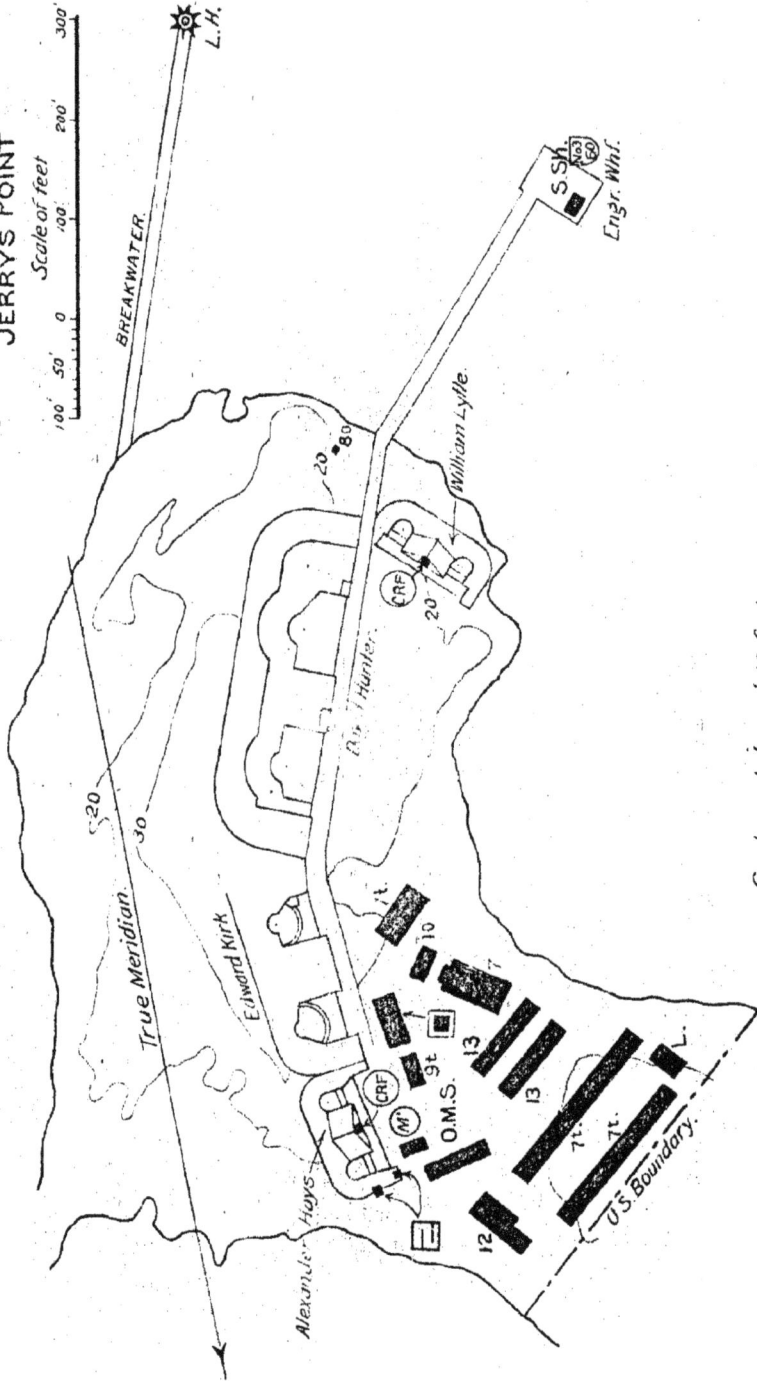

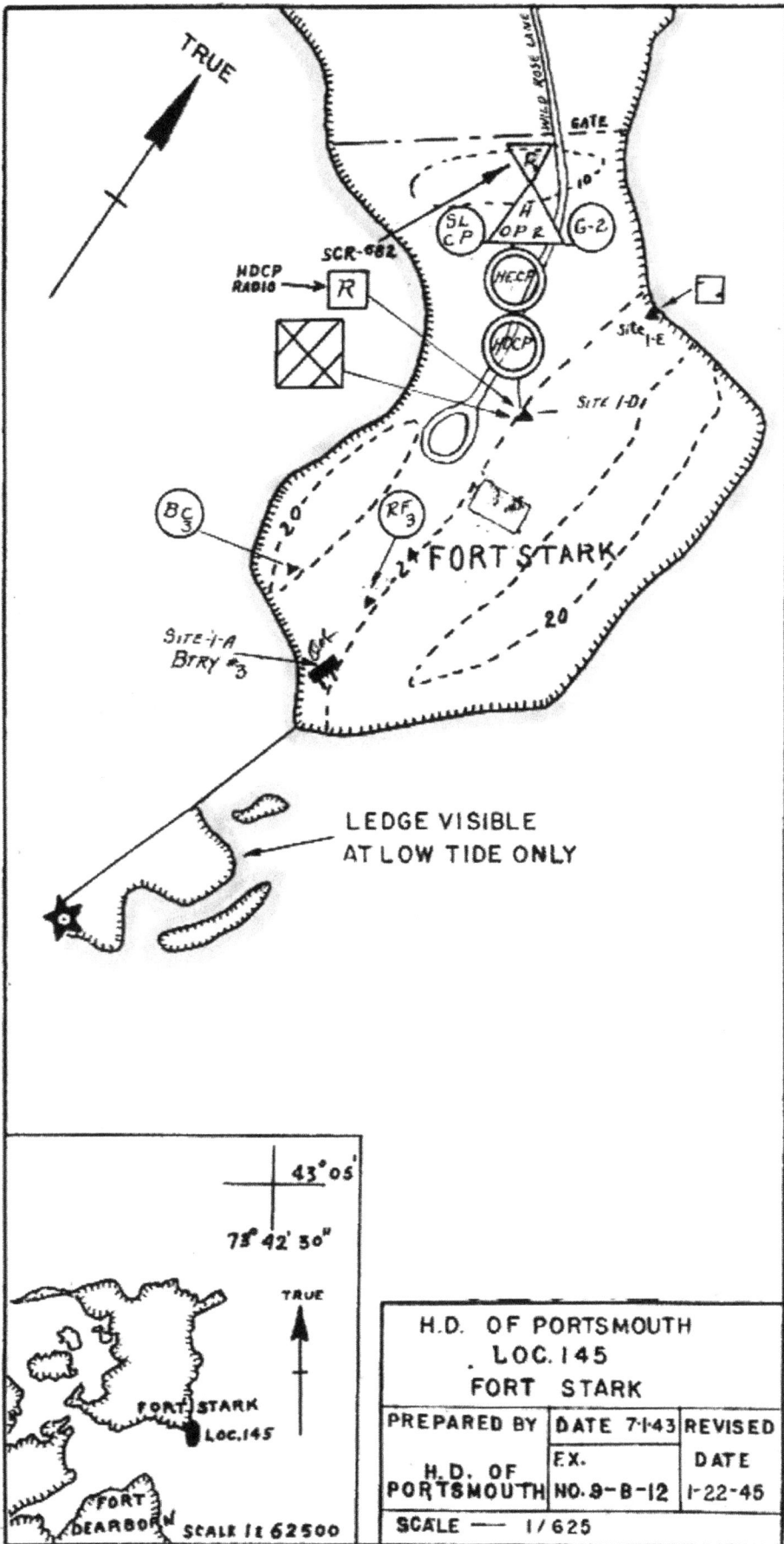

TRUE

INNER ROSE LANE

GATE

10'

SL
C.P

A
OP R

G-2

SCR-682

HDCP
RADIO → R

HECP

HDCP

Site I-E

SITE I-D

B C
3

RF
3

FORT STARK

20

SITE I-A
Btry #3

20

LEDGE VISIBLE
AT LOW TIDE ONLY

43° 05'

75° 42' 30"

TRUE

FORT STARK
LOC.145

FORT
DEARBORN   SCALE 1: 62500

H.D. OF PORTSMOUTH LOC. 145 FORT STARK		
PREPARED BY	DATE 7-1-43	REVISED
H.D. OF PORTSMOUTH	F.X. NO. 9-B-12	DATE 1-22-45
SCALE —— 1/625		

- **KIRK:** A battery for two 6-inch guns on disappearing carriages placed on the left flank of Battery Hunter at Fort Stark. Submission of specific plans was made on March 20, 1903. It followed mimeograph type plans, except it had an additional 3-foot gun center spacing and had several new sets of steps added for convenience for ammunition handling. Work was done in 1903-1904, for transfer on December 31, 1904 for a construction cost of $40,000. It was armed with two 6-inch Model 1903 guns on Model 1903 carriages (#7/#8 and #30/#7). The battery was named on General Orders No. 194 of December 27, 1904 for Brigadier General Edward N. Kirk who died of wounds received at Stones River, TN in 1862. The gun tubes were authorized for removal in 1917 for use on field carriages. They were never returned, and the disappearing carriages were scrapped in place in 1920. During World War II, the Portsmouth Harbor Entrance Command Post was built atop otherwise abandoned Battery Kirk. Thus modified the emplacement still exists at the Fort Stark Historic Site. The battery site is closed to the public.

- **LYTLE:** A battery of two 3-inch pedestal guns emplaced on the southwest corner of the Fort Stark reservation to protect Little Harbor. The plan for the battery was submitted on January 27, 1904. Except for a large oblique retaining wall on the east side along the road cut between the battery and Battery Hunter, it was of standard design. Work was done in 1904, it was reported complete and ready for transfer on March 22, 1905. Transfer was made on April 3, 1905 along with Battery Hays for a total cost of $20,955.20. It was named on General Orders No. 194 of December 27, 1904 for Brigadier General William H. Lytle, U.S. Volunteers who was mortally wounded at Chickamauga in 1863. The battery was armed with two 3-inch Model 1902 guns and pedestals (#39/#39 and #40/#40). It was retained throughout the 1920s and 1930s. However, in 1942 the armament was removed and transferred to Battery Hackleman at Fort Constitution. The emplacement still exists at the Fort Stark Historic Site. The battery site is open to the public.

- **HAYS:** A battery for two 3-inch pedestal guns emplaced on the left flank of Battery Kirk at the northern end of the reservation at Fort Stark. It was an excellent location to cover the minefield and shoreline extending to the north. The plan was submitted on January 27, 1904 and was of standard design. Work was done in 1904, it was reported complete and ready for transfer on March 22, 1905. Transfer was made on April 3, 1905 along with Battery Lytle for a total cost of $20,955.20. It was named on General Orders No. 194 of December 27, 1904 for Brevet Major General Alexander Hays who died at the Battle of the Wilderness in 1864. It was armed with two 3-inch Model 1902 guns and pedestals (#41/#41 and #42/#42). These served until 1942 when they were moved to new gun blocks built on the eastern end of the reservation. Battery Hays was not subsequently used for armament. The emplacement still exists at the Fort Stark Historic Site. The battery site is open to the public.

- In 1942 it was decided to relocate the two 3-inch pedestals gun of Battery Hays to a new position. New cylindrical concrete blocks were built in front of Battery Hunter near the breakwater on the southeastern tip of Jerrys Point. The emplacement was built from June 30, 1942 to August 21, 1942. It transferred on November 18, 1942 for a cost of $2271. The two guns and pedestals (#41 and #42) from Hays served at the battery through the end of the war. At some point two naval 3-inch/50 caliber AA guns were placed on the blocks as saluting guns. Not until the late 1970s were these removed, one is now on display at the fort's local museum, while the other on display at the Portsmouth Naval Shipyard. The blocks have begun to tilt and give way from the shoreline erosion. Even though damaged, they can still be located at the Fort Stark Historic Site. The battery site is open to the public.

Camp Langdon (1903-1963) was a military reservation on Newcastle Island which was used for the military garrison in Portsmouth.

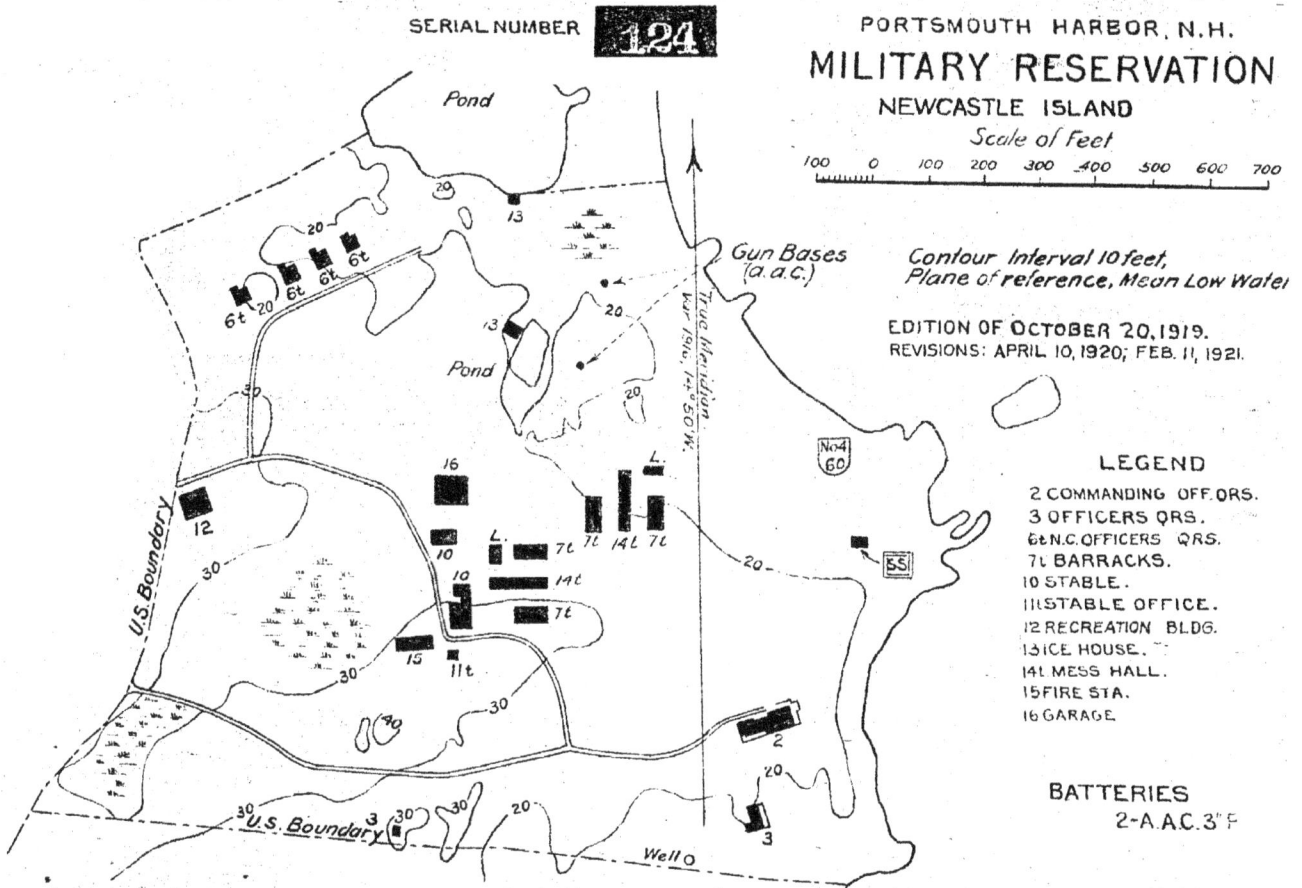

Fort Dearborn (1941-1948) is located on Odiorne Point to the south of Portsmouth Harbor. A set of Panama mounts for four 155mm guns were built there in 1942, followed by the construction of 16-inch casemated battery (#103), a 6-inch battery (#204), and a 90mm AMTB located on nearby Pulpit Rock. It was named in General Orders No. 25 on May 21, 1943 for Henry Dearborn, Secretary of War 1801-09. Transferred from U.S. Army to the US Air Force in 1949 as radar site (Rye Air Force Station) until declared surplus in 1959. The military reservation was transferred to the state in 1961 and opened as a state park in 1972. Today, the reservation is now part of Odiorne Point State Park. The Seacoast Science Center located in the park has exhibits relating to the natural and human history of Odiorne and the seacoast area. Battery Seaman's emplacements are open, but the interior is closed off. The park uses Battery #204 for storage, but the exterior can be visited as well as the 155mm Panama mounts.

Fort Dearborn Gun Batteries

- **SEAMAN**: A 1940 Program dual 16-inch barbette battery built as the heavy new armament for Portsmouth. It was placed on the new Odiorne Point reservation near Frost Point. Its field of fire was to be to the east/southeast. During planning and construction, it was called Battery Construction No. 103. Funding (initially $625,000) was contained in the FY-1942 Budget. It was given national priority for 16-inch batteries of #17 on September 11, 1940 and #16 on August 11, 1941. Most of the structural work was done between April 1942 and August 1943. Completion and transfer were made on September 8, 1944 for a cost of $1,473,506. The battery was of conventional type plan,

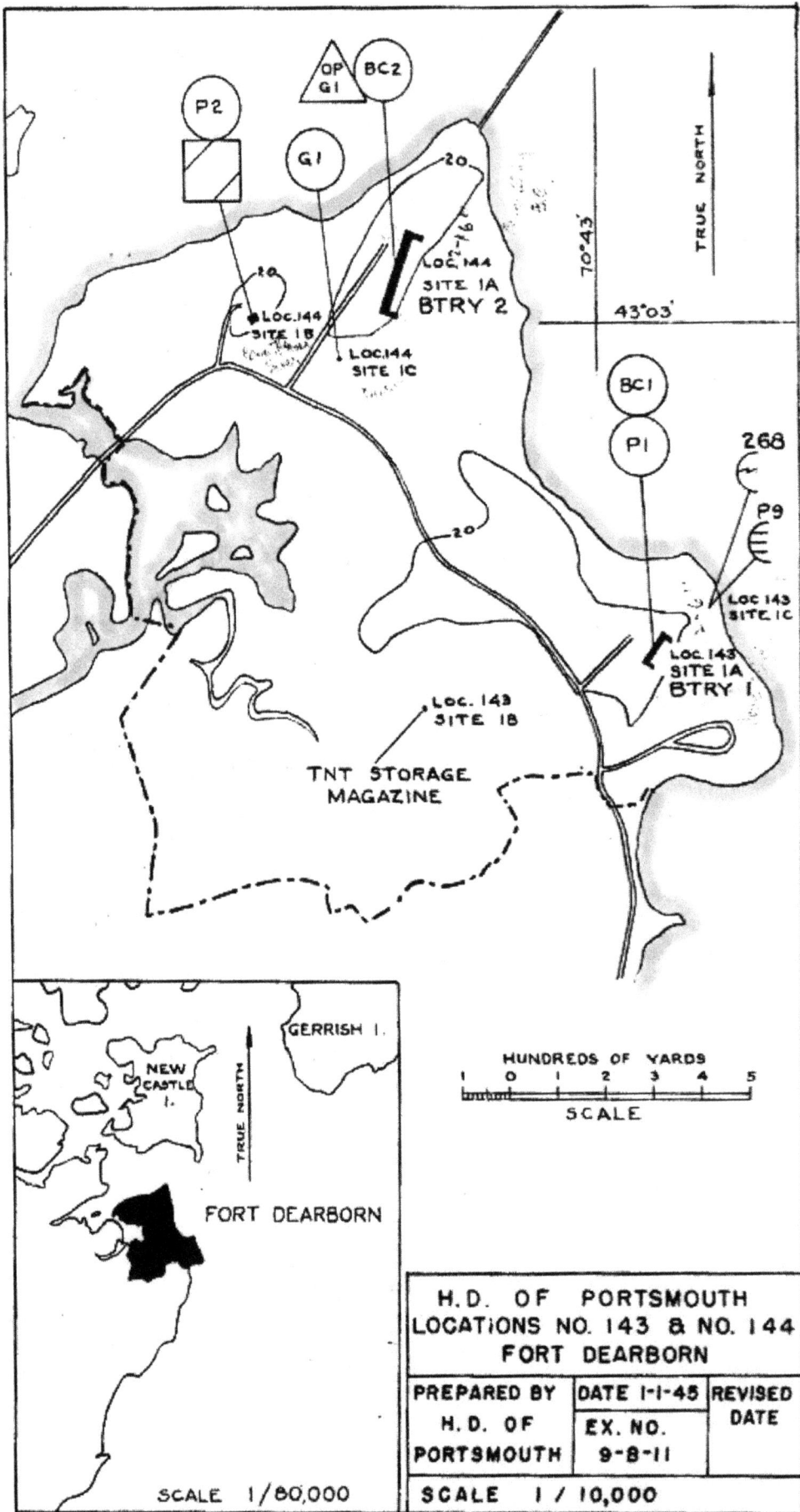

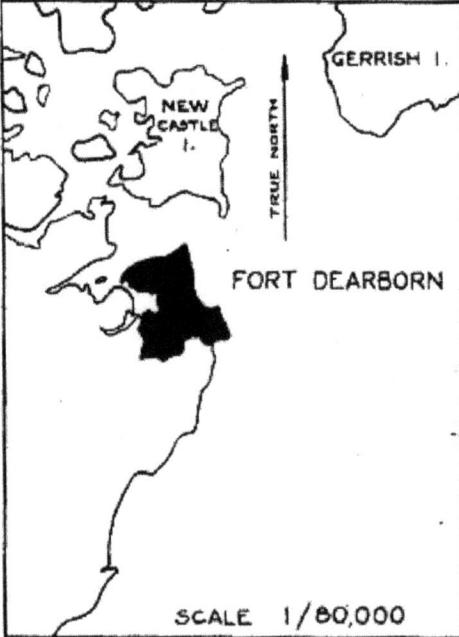

Battery #204 in Odiorne Point State Park (Terry McGovern)

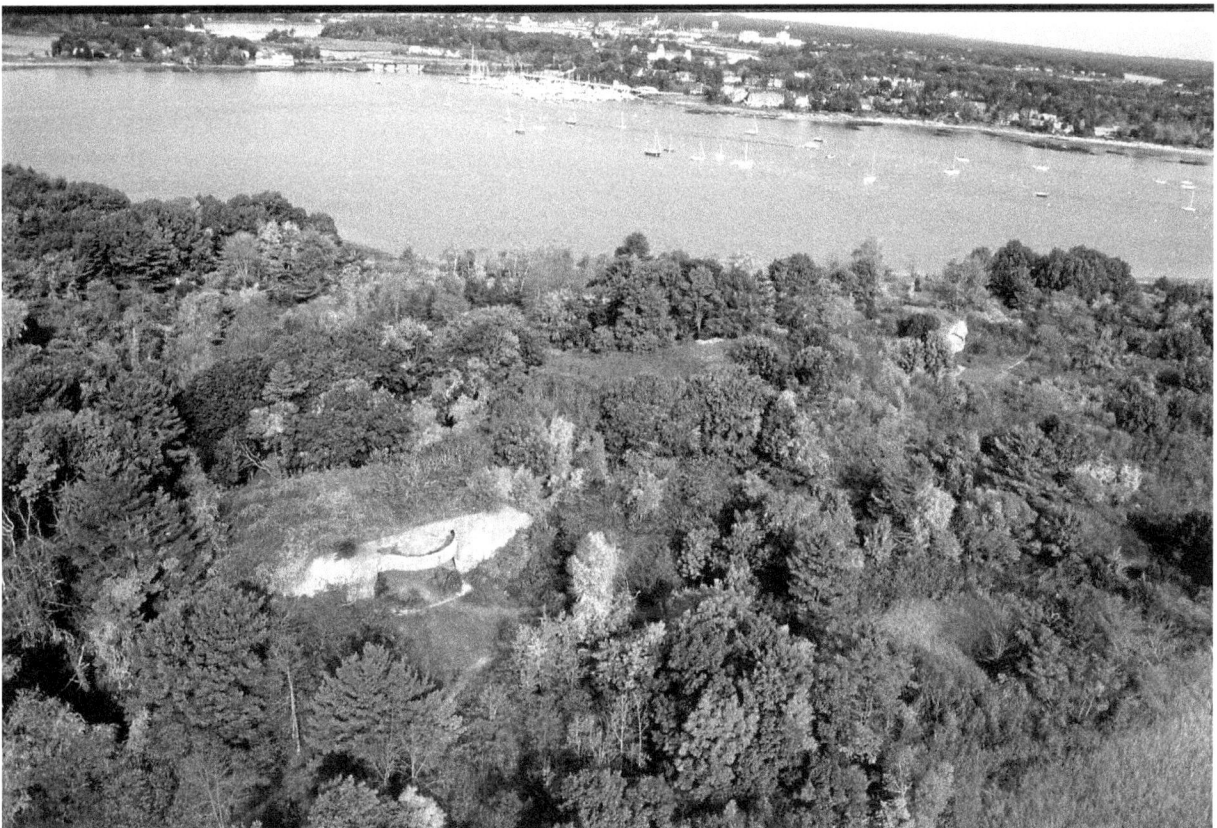

Battery Seaman in Odiorne Point State Park (Terry McGovern)

with the guns slightly fanned out to give a greater field of fire (by 8 ½-degrees). The usual network of battery commander's station and base-end stations was built, along with a separate combined PSR room. It was named on AGO Memorandum of August 20, 1942 for Colonel Claudius A. Seaman, Coast Artillery Corps. The battery was armed with two 16-inch MkIIM1 guns on Model 1919M5 barbette carriages (#99/#56 and #104/#57). The battery served out the war, until it was eliminated and the armament removed in 1948. The emplacement still exists at the Odiorne Point State Park. The battery site is open, but the interior is closed to the public.

- **Battery #204**: A 1940 Program dual 6-inch barbette battery built on the southern end of the Fort Dearborn reservation near Odiorne Point. Its field of fire was to the southeast. Funding came with the FY-1943 Budget, and the initial national priority list of September 1941 assigned it priority #26. Work was done in 1942-1943, the battery reported completing on July 20, 1944. It was of mostly conventional type design, except it has a battery commander's station on the roof connected by ladder to the interior magazine passageway. It was armed with two 6-inch M1(T2) guns on M3 barbette carriages (#22/#14 and #23/#15). The battery was never officially named. Though retained in the 1946 Review, it was deleted shortly thereafter (probably in 1948) and disarmed. The emplacement still exists at the Odiorne Point State Park. The battery site is open, but the interior is closed to the public.

Pulpit Rock Military Reservation (1943-1971) is located in northern Rye, New Hampshire, on a small parcel of land between Neptune Drive and Pulpit Rock Road. One Anti-Motor Torpedo Boat battery and the Pulpit Rock Tower, also known as Pulpit Rock Base-End Station (N. 142), were located at this reservation and was built in 1943 as part of the Harbor Defenses of Portsmouth. Now owned by the state, it is periodically open to the public and maintained by a local non-profit organization.

Pulpit Rock M.R. Gun Batteries

- **AMTB-1**: A 1943 Program AMTB battery for two 90mm fixed and two 90mm mobile guns emplaced south of Fort Dearborn at Pulpit Rock. It was authorized on April 17, 1943 and work was accomplished on the two concrete gun blocks later that same year. It was armed, at least with the two fixed guns shortly after and served until removed in 1945. For years the blocks survived and gradually were eroded and covered over by adjacent road construction. Only partial remains still exist. The battery site is open to the public.

90 mm AMTB gun mount at Fort Levett, Portland ME (Mark Berhow)

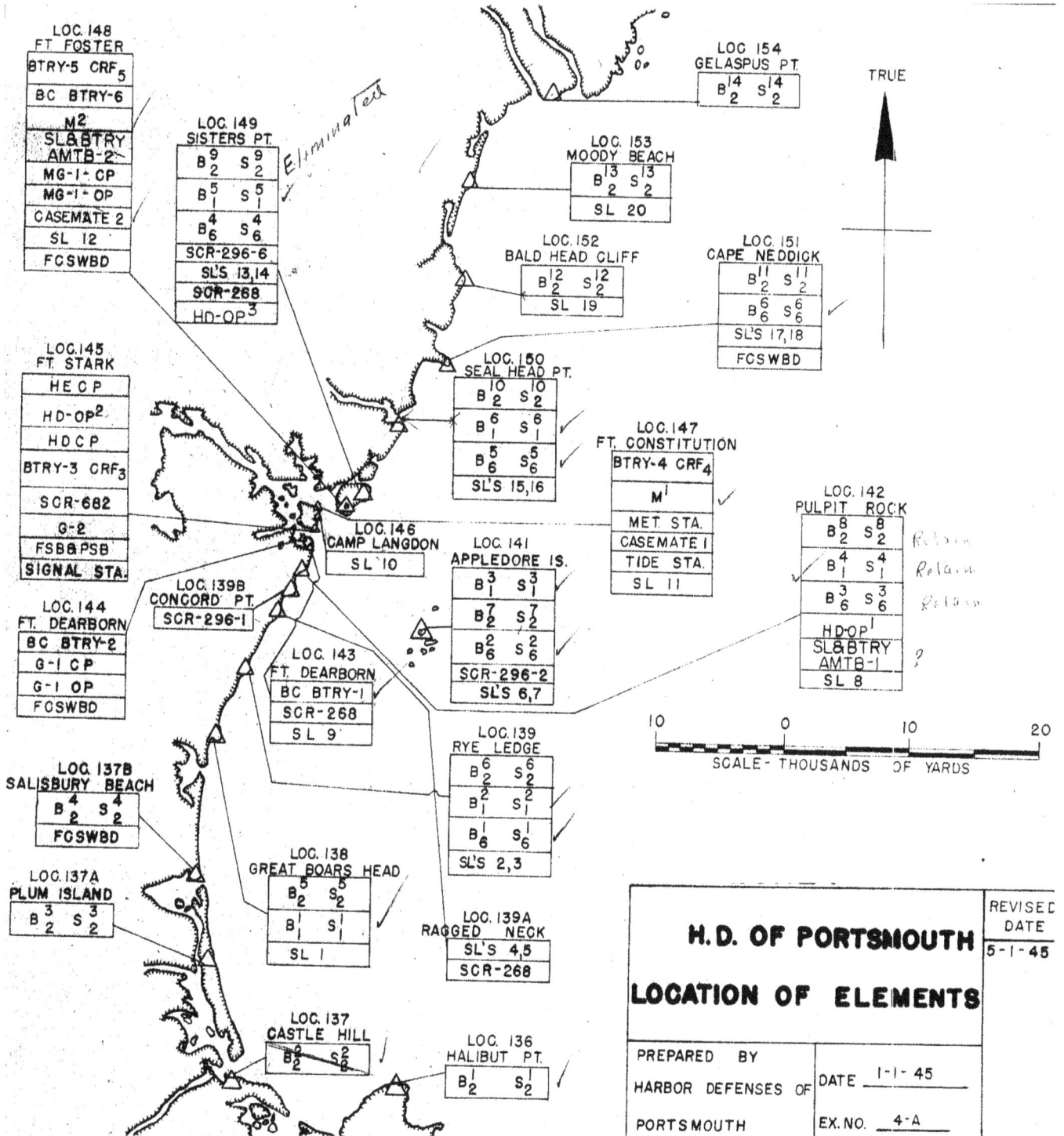

LOC. 148
FT. FOSTER

| BTRY-5 CRF$_5$ |
| BC BTRY-6 |
| M^2 |
| SL&BTRY AMTB-2 |
| MG-1-CP |
| MG-1-OP |
| CASEMATE 2 |
| SL 12 |
| FCSWBD |

LOC. 149
SISTERS PT.

B$_2^9$	S$_2^9$
B$_6^5$	S$_6^5$
B$_6^4$	S$_6^4$
SCR-296-6	
SL'S 13,14	
SCR-268	
HD-OP^3	

Eliminated

LOC. 154
GELASPUS PT.

| B$_2^{14}$ | S$_2^{14}$ |

TRUE

LOC. 153
MOODY BEACH

| B$_2^{13}$ | S$_2^{13}$ |
| SL 20 | |

LOC. 152
BALD HEAD CLIFF

| B$_2^{12}$ | S$_2^{12}$ |
| SL 19 | |

LOC. 151
CAPE NEDDICK

B$_2^{11}$	S$_2^{11}$
B$_6^6$	S$_6^6$
SL'S 17,18	
FCSWBD	

LOC. 145
FT. STARK

| HECP |
| HD-OP^2 |
| HDCP |
| BTRY-3 CRF$_3$ |
| SCR-682 |
| G-2 |
| FSB&PSB |
| SIGNAL STA. |

LOC. 150
SEAL HEAD PT.

B$_2^{10}$	S$_2^{10}$
B$_1^6$	S$_1^6$
B$_6^5$	S$_6^5$
SL'S 15,16	

LOC. 147
FT. CONSTITUTION

| BTRY-4 CRF$_4$ |
| M^1 |
| MET. STA. |
| CASEMATE 1 |
| TIDE STA. |
| SL 11 |

LOC. 142
PULPIT ROCK

B$_2^8$	S$_2^8$	*Retain*
B$_1^4$	S$_1^4$	*Retain*
B$_6^3$	S$_6^3$	*Retain*
HD-OP^1		
SL&BTRY AMTB-1		?
SL 8		

LOC. 146
CAMP LANGDON

| SL 10 |

LOC. 141
APPLEDORE IS.

B$_1^3$	S$_1^3$
B$_2^7$	S$_2^7$
B$_6^2$	S$_6^2$
SCR-296-2	
SL'S 6,7	

LOC. 144
FT. DEARBORN

| BC BTRY-2 |
| G-1 CP |
| G-1 OP |
| FCSWBD |

LOC. 139B
CONCORD PT.

| SCR-296-1 |

LOC. 143
FT. DEARBORN

| BC BTRY-1 |
| SCR-268 |
| SL 9 |

LOC. 139
RYE LEDGE

B$_1^6$	S$_1^6$
B$_1^2$	S$_1^2$
B$_6^1$	S$_6^1$
SL'S 2,3	

10   0   10   20
SCALE - THOUSANDS OF YARDS

LOC. 137B
SALISBURY BEACH

| B$_2^4$ | S$_2^4$ |
| FCSWBD | |

LOC. 138
GREAT BOARS HEAD

B$_2^2$	S$_2^2$
B$_1^1$	S$_1^1$
SL 1	

LOC. 139A
RAGGED NECK

| SL'S 4,5 |
| SCR-268 |

LOC. 137A
PLUM ISLAND

| B$_2^3$ | S$_2^3$ |

LOC. 137
CASTLE HILL

| B$_2^2$ | S$_6^2$ |

LOC. 136
HALIBUT PT.

| B$_2^1$ | S$_2^1$ |

H.D. OF PORTSMOUTH

LOCATION OF ELEMENTS

| REVISED DATE |
| 5-1-45 |

PREPARED BY	
HARBOR DEFENSES OF	DATE 1-1-45
PORTSMOUTH	EX. NO. 4-A

Portsmouth World War II-era Site Locations. Stations housed in a single structure are connected by dashes (-)

location	Loc#	Purpose
Halibut Point	136	BS1/Seaman (+2 Boston stations)
Castle Hill	137	BS2/Seaman
Plum Island	137A	BS3/Seaman
Salisbury Beach	137B	BS4/Seaman, SBR
Great Boars Head	138	BS5/Seaman-BS1/204, SL 1
Rye Ledge	139	BS6/Seaman-BS2/204-BS1/205, SL 2,3
Ragged Neck	139A	SCR268, SL 4,5
Concord Point	139B	SCR296-204
Star Island	140	not used
Appledore Island	141	SCR296-2-BS3/204-BS7/Seaman-BS2/205, SL 6,7
Pulpit Rock	142	BS8/Seaman-BS4/204, HDOP1-BS3/205, AMTB, SL 8
Odiorne Point/ Fort Dearborn	143	Batt. Tact. #1 BCN 204-BC/204, SCR268, SL 9
Frost Point/ Fort Dearborn	144	Batt. Tact. #2 Seaman, BC/Seaman, G1, SBR
Fort Stark	145	Batt. Tact. #3 Lytle, HECP-HDCP-HDOP2-G2-SBR, BC/Lytle, SCR 682,
Camp Langdon	146	SL 10
Fort Constitution	147	Batt. Tact. #4 Hackleman, BC/Hackleman, M2, Met, MC, Mines, SL 8
Fort Foster	148	Batt. Tact. #5 Chapin, Batt. Tact. #6 BCN 205, BC/Chapin, M2-M2OP, AMTB MC, SL 12, BC/205, SBR
Sisters Point	149	BS9/Seaman-BS5/204-BS4/205-HDOP3, SCR268, SCR296/205, SL 13,14
Seal Head Point/ Western Point	150	BS5/205-BS6/204-BS10/Seaman, SL 15,16
Cape Neddick	151	BS6/205-BS11/Seaman, SBR, SL 17,18
Bald Head Cliff	152	BS12/Seaman, SL 19
Moody Beach	153	BS13/Seaman, SL 20
Gelaspus Point	154	BS14/Seaman

Battery Hunter, Fort Stark State Historic Park (Mark Berhow)

THE HARBOR DEFENSES OF BOSTON — MASSACHUSETTS

The large bay area covered by several islands outside of Boston Harbor made this an attractive harbor area that was first settled by British Puritan colonists in 1629. By the early 1700s Boston was a lively shipping and fishing port and an important town in the Thirteen Colonies. Many crucial events leading to the outbreak of the American Revolution occurring in and around Boston and the result of the early days of fighting left Boston in the revolutionaries' hands after mid-1775. The United States built several fortifications to defend the main entrances to Boston Harbor during the First and Second System programs. Two large Third System forts were built on Castle Island near the city and the strategically located Georges Island overseeing the main entrances to the harbor. An extensive set of brick and earthwork batteries were begun in the 1870s but not completed. New modern defenses built under the Endicott construction program included nearly 30 new concrete gun batteries and mine defenses spread out over 8 headlands and islands around the harbor entrances. The defenses were augmented by a 12-inch battery at Nahant and a 16-inch battery near Hull during the 1920s. During the 1940 Program, two new 16-inch casemated batteries were built (with a third planned), along with four new 6-inch batteries and 5 new 90mm AMTB batteries. Defended through the Cold War years by the U.S. Army, the defenses were eventually transferred to a variety of state and local agencies. Today, the Boston Harbor Islands National Recreation Area is managed corroboratively as a partnership park. Eleven different agencies make up the Boston Harbor Islands Partnership, which serves as the federally legislated body that oversees the park management.

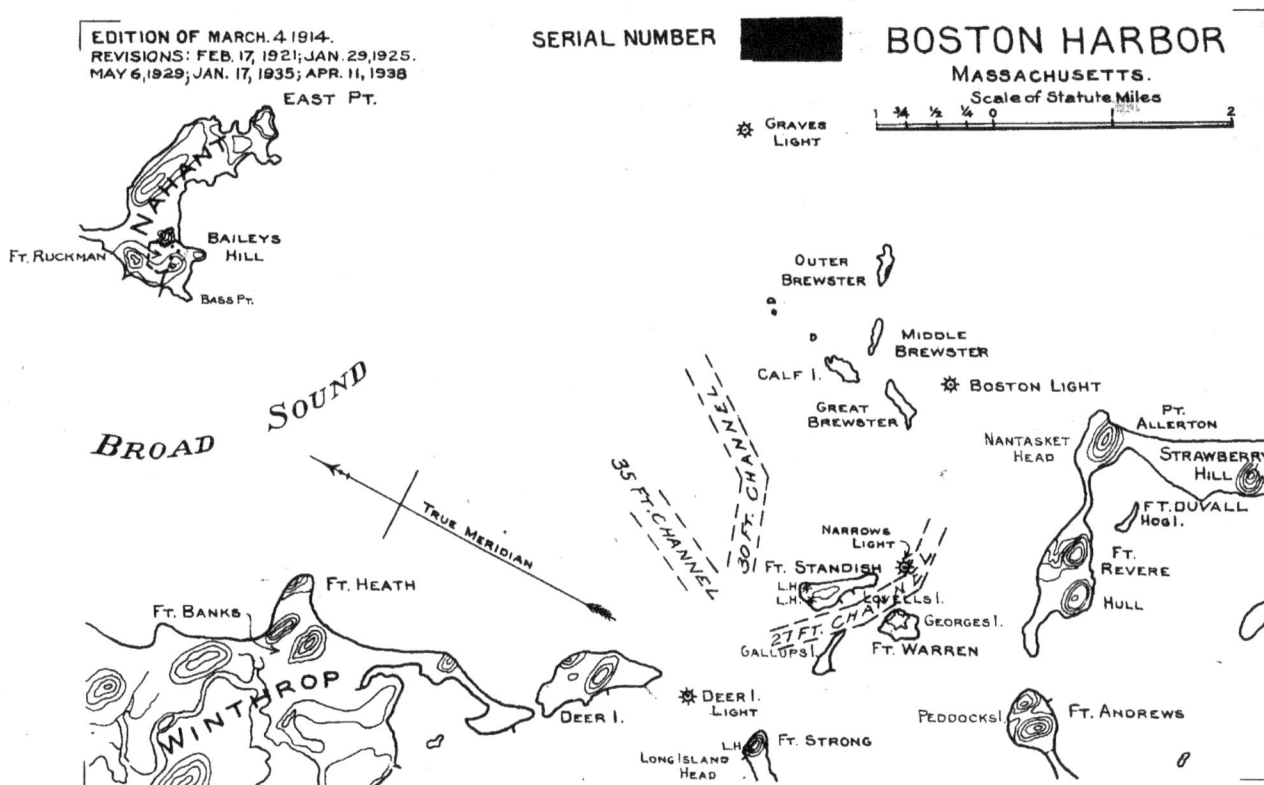

Butler, Gerald W. *Military Annals of Nahant, Massachusetts.* Nahant Historical Society, Nahant, MA 1996.

Butler, Gerald W. *The Guns of Boston.* First Books, 1999.

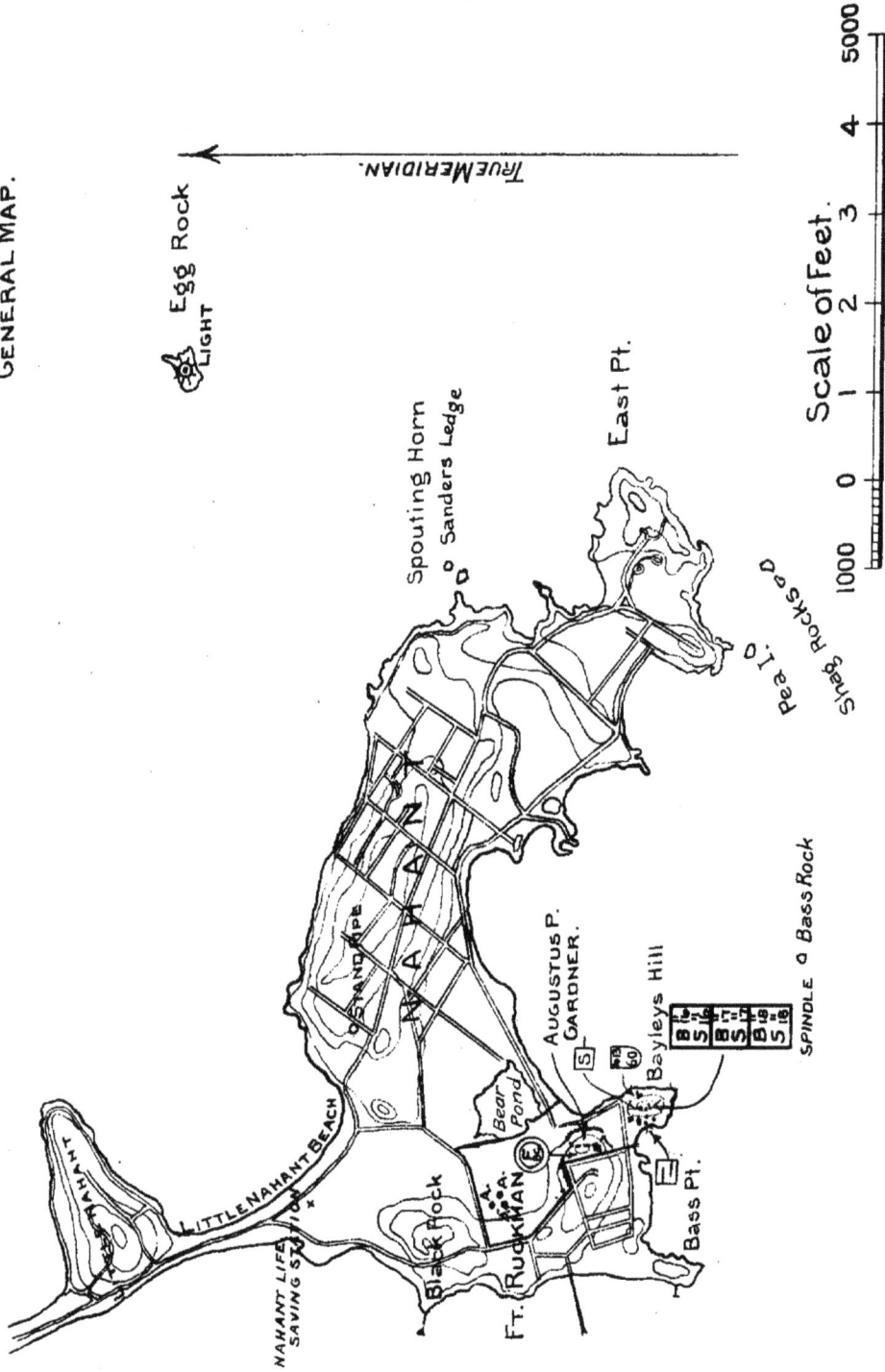

BOSTON HARBOR, MASS.

NAHANT -
GENERAL MAP.

EDITION OF OCTOBER 20, 1919.
REVISIONS: APR. 8, 1920; FEB. 17, 1921.
JAN. 29, 1925; MAY 6, 1929; JAN. 17, 1935;
APR. 11, 1938

BATTERIES
GARDNER 2-12"N.D

A.A. BATTERY No. 5
3-3" A.A. Guns, Mobile

A. – Gun Blocks for
3" Fixed A.A. Guns

SERIAL NUMBER

True Meridian.

Egg Rock
LIGHT

Spouting Horn
o Sanders Ledge

East Pt.

Pea I. o
Shag Rocks o

Scale of Feet
1000 0 1 2 3 4 5000

On Maintenance Status

Augustus P.
GARDNER.

Bayleys Hill

SPINDLE o Bass Rock

Bass Pt.

Ft. RUCKMAN

Black Rock

Bear Pond

LITTLE NAHANT BEACH

NAHANT LIFE
SAVING STATION

NAHANT

GRAND APE

NAHANT

East Point Military Reservation (1940-1967) is located on East Point on the Nahant Peninsula. Built as part of the 1940 Program to cover the northern approaches to Boston Harbor, its initial temporary defenses were two 155mm guns on Panama mounts. Its permanent defenses were Battery Murphy (#104) with two casemated 16-inch guns and a Series #200 battery (#206) with two shielded barbette 6-inch guns. Several large fire control towers were built nearby. During the Cold War a Nike launch area was located out on the point. Declared surplus in 1959, the military reservation was transferred to the state. The site became home to a Marine Science Center, a part of Northeastern University. The Center uses Battery Murphy for a variety of experimental and classroom uses. There is no public parking near East Point, and access to the Center is restricted except for some public events and tours.

East Point M.R. Gun Batteries

- **MURPHY**: A 1940 Program battery for two 16-inch casemated barbette guns built at the East Point Military Reservation. The battery was planned as Battery Construction No. 104. It was given the highest priority of the new program batteries for Boston and ranked nationally as priority #13 on September 11, 1940 and #10 on August 11, 1941. The new reservation was acquired from the private property of the Henry Cabot Lodge estate. The battery was sited on the eastern point of the reservation, firing to the east. Funding was provided in the FY-1942 Budget. Work was done from January 23, 1943 to December 19, 1943 for transfer on June 19, 1944 at a cost of $1,655,582. It was named for John B. Murphy by Adjutant General Memorandum of August 20, 1942. It was armed with two 16-inch MkIIM1 guns on Model 1919M4 barbette carriages (#76/#33 and #49/#40). It was one of the few completed 16-inch batteries to actually have installed the intended armored shield and surrounding embrasure closure plate. Ten separate base-end stations were built and assigned, and the usual separate, protected PSR room was completed near, but behind the gun battery. It was retained in the 1946 Review but finally deleted in 1948. The armament was scrapped shortly thereafter. The emplacement still exists and is currently used as a laboratory at Northeastern University. The battery site is closed to the public.

- **Battery #206**: A 1940 Program battery for two 6-inch barbette guns also emplaced at East Point, to the south of Battery Murphy. It was funded in the FY-1942 Budget and given national priority #20 for initial construction in September 1942. Work was done by early fall of 1943; the guns being mounted and test-fired in January 1944. It held two 6-inch Model 1903A2 on Model M1 barbette carriages (#3/#108 and #35/#107) in a standard 200-series design plan. The battery was never named. It served out the balance of the war, finally being deleted in 1947 or 1948. The emplacement still exists, though with some erosion under the gun platforms. It is on the property of Northeastern University. The battery site is closed to the public.

Northeastern University Marine Science Center, East Point (Terry McGovern)

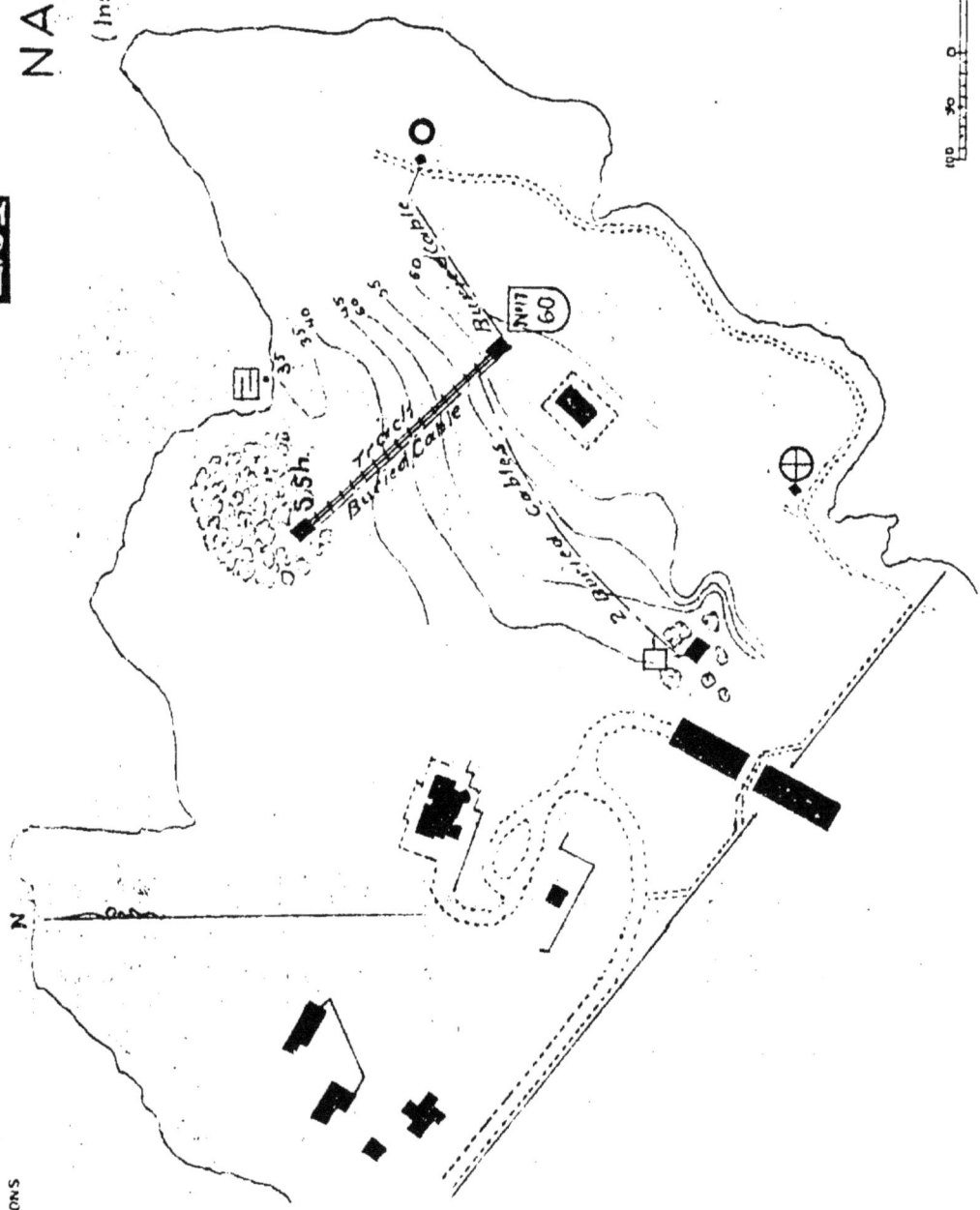

BOSTON HARBOR, MASS.
NAHANT D-2.
EAST POINT
(Installation dismantled.)

SERIAL NUMBER 124

EDITION OF APR. 8, 1920.
REVISIONS

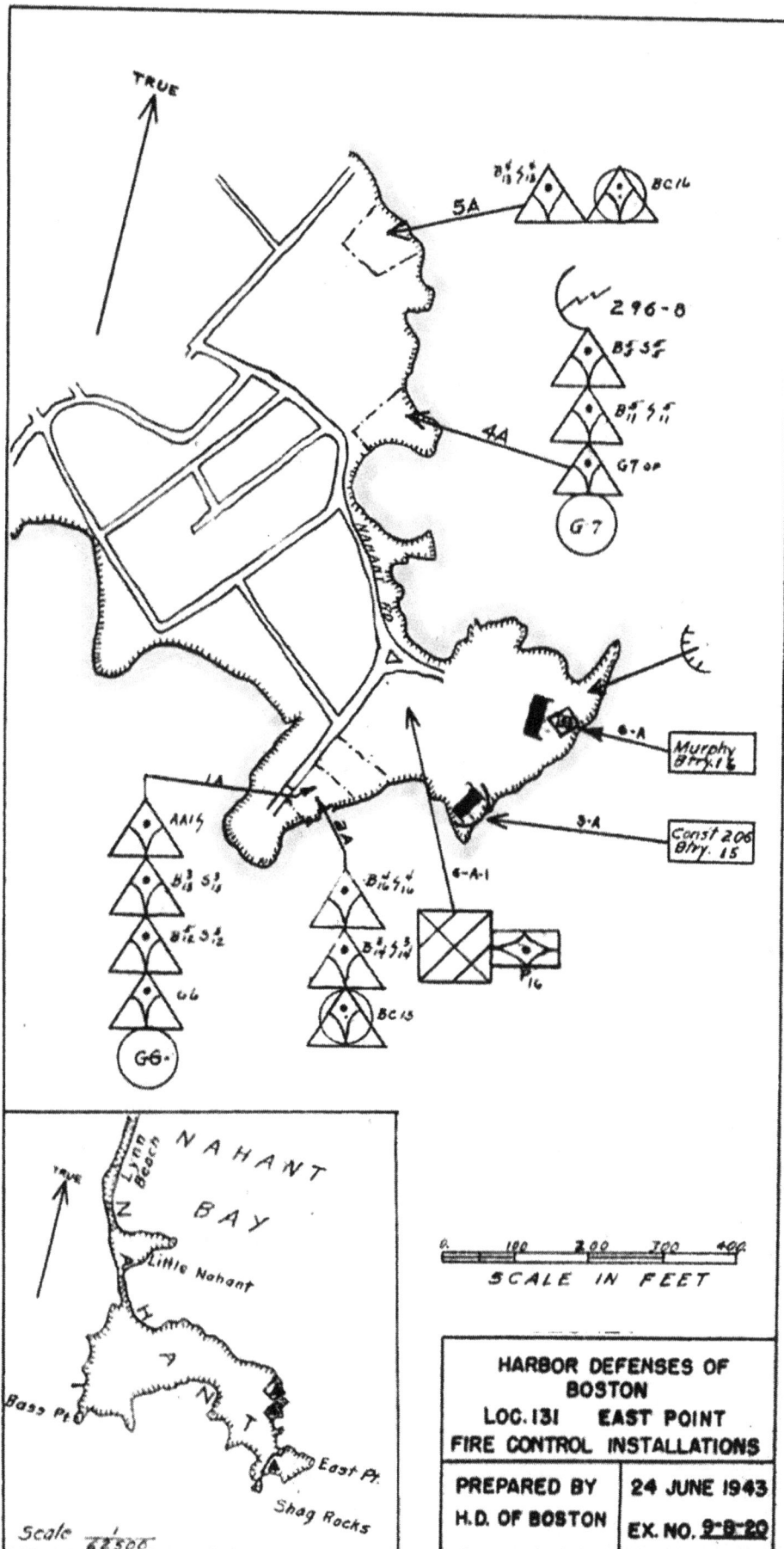

TRUE

5A

$B^d_{13} S^d_{13}$ BC 16

296-8

$B^5_9 S^5_9$

$B^5_{11} S^5_{11}$

4A   G 7 OP

G 7

6-A   Murphy Btry. 16

3-A   Const 206 Btry. 15

1A

AA 15

$B^3_{13} S^3_{13}$

$B^5_{12} S^5_{12}$

G 6

G 6

$B^d_{16} S^d_{16}$

6-A-1

$B^3_{14} S^3_{14}$

BC 15

16

LYNN BEACH   NAHANT BAY

TRUE

N   A   H   A   N   T

Little Nahant

Bass Pt.

East Pt.

Shag Rocks

Scale $\frac{1}{62500}$

SCALE IN FEET

0   100   200   300   400

HARBOR DEFENSES OF BOSTON

LOC. 131   EAST POINT

FIRE CONTROL INSTALLATIONS

PREPARED BY	24 JUNE 1943
H.D. OF BOSTON	EX. NO. 9-8-20

REVISED DATE   31 JAN 45

Fort Ruckman (1921-1957) is located on the south tip of the Nahant Peninsula. It was named in General Orders No. 13 on Mar. 27, 1922 for Brig. Gen. John W. Ruckman, U.S. Army. Battery Gardner was completed in 1924 for two 12-inch guns on long-range barbette carriages cover a larger area of the waters of Massachusetts Bay. The Boston Area Command-2nd (North) Group (C2) command post was also here. AA Battery Five (three guns) was here from the 1920s to 1945, on the post parade ground (relocated in 1942). This battery was casemated during World War II. After the war the adjacent Bailey's Hill was used for a Nike fire control site. In 1957, the 44-acre reservation was transferred to the Town of Nahant. Battery Gardner's gun casemates have been filled with earth; the city uses the interior for storage. Nothing remains of the Nike fire control site. The park is open during daylight hours.

Fort Ruckman Gun Batteries

- **GARDNER**: A 1915 Program dual gun, long-range, 12-inch barbette battery emplaced on the north side of Boston Harbor. On December 20, 1915 the site was selected at a new reservation in Nahant for a standard program 12-inch gun battery. Plans were submitted on August 24, 1917. They specified a standard battery of two barbette guns in open, all-round firing positions separated by 420-feet from each other. A large, earth-covered traverse magazine between the guns had ammunition and support structures. Because it was considered an isolated location, it was a closed-back battery with the rear gallery containing crew quarters and a winter kitchen. Work was approved and started quickly in late 1917 and was essentially completed by early 1919. It was transferred to troops on December 28, 1921. The battery was named on General Orders No. 9 of February 11, 1920 for Major Augustus P. Gardner, 121st Infantry, who was killed in France during World War I in 1918. It was armed with two 12-inch Model 1895A4 guns on new Model 1917 barbette carriages (#25/#28 and #42/#29). It remained in active status though the interwar years. The battery was modified to receive new, heavy casemated gun houses over the open gun positions. That work was carried out between April 23, 1942 and January 23, 1943. It was finally recommended for deletion in 1946 and disarmed within a short time afterwards. The emplacement still exists on town property. The battery site is open, but the interior is closed to the public.

Battery Gardner Bass Point Park Nahant (Terry McGovern)

On Maintenance Status.

FORT RUCKMAN
BOSTON HARBOR, MASS.
NAHANT D-1.

BAYLEYS HILL

BASS POINT

SERIAL NUMBER

SCALE OF FEET.
1000   500   0   500

TRUE MERIDIAN.

EDITION OF OCTOBER 20, 1919.
REVISIONS: APR. 8, 1920, FEB. 17, 1921,
JAN. 29, 1925; MAY 6, 1929; JAN. 17, 1935;
APR. 11, 1938

U.S. GOVERNMENT RESERVATION

BEAR POND

AUGUSTUS P. GARDNER

TRIMOUNTAIN ROAD

CASTLE

HIGH WATER LINE

FLASH ROAD

N.&L.S. RY.

LEGEND

FENCE.

N.C.O. QUARTERS.

BATTERIES.
GARDNER......2-12" N.D

A.A. BATTERY No. 5
3-3" A.A. Guns, Mobile

A.- Gun Blocks for
3" Fixed A.A. Guns

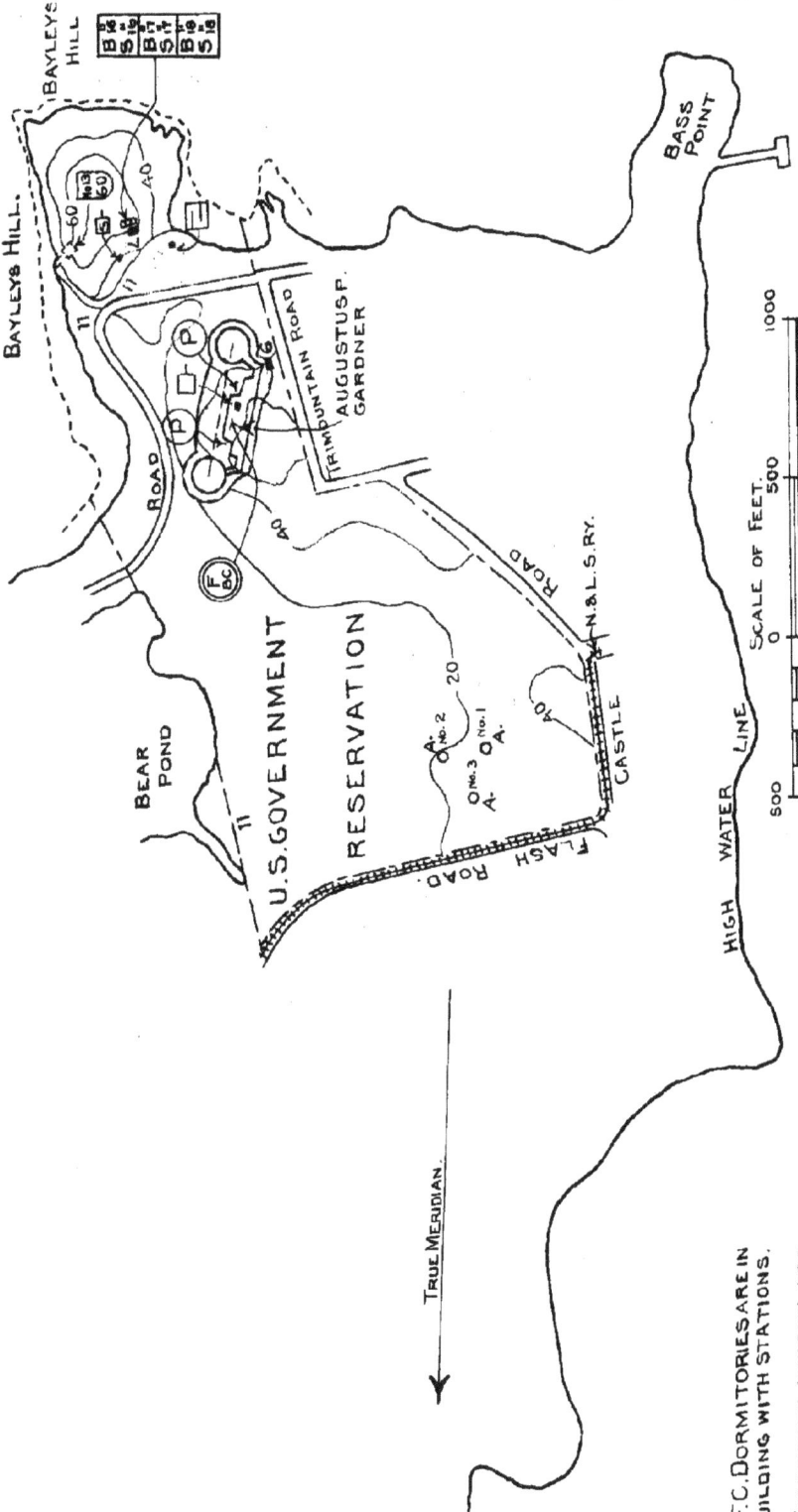

F.C. DORMITORIES ARE IN
BUILDING WITH STATIONS.

CONTOURS APPROX. ARE
REFERRED TO M.L.W.

TRUE

Bc

AA4

2A

40

AA

Gardner
Btry /4

20

40

3A

BC 14

B⁴ 5 4
4 4 4

B⁴ 5 4
10 10

60

1A

40

C 2

BASS
PT

Scale of Feet

500   0   1000

Scale = 1/62,500

NAHANT
BAY

TRUE

LYNN

HARBOR

N A H A N T

Ft. RUCKMAN

Bass Pt

East

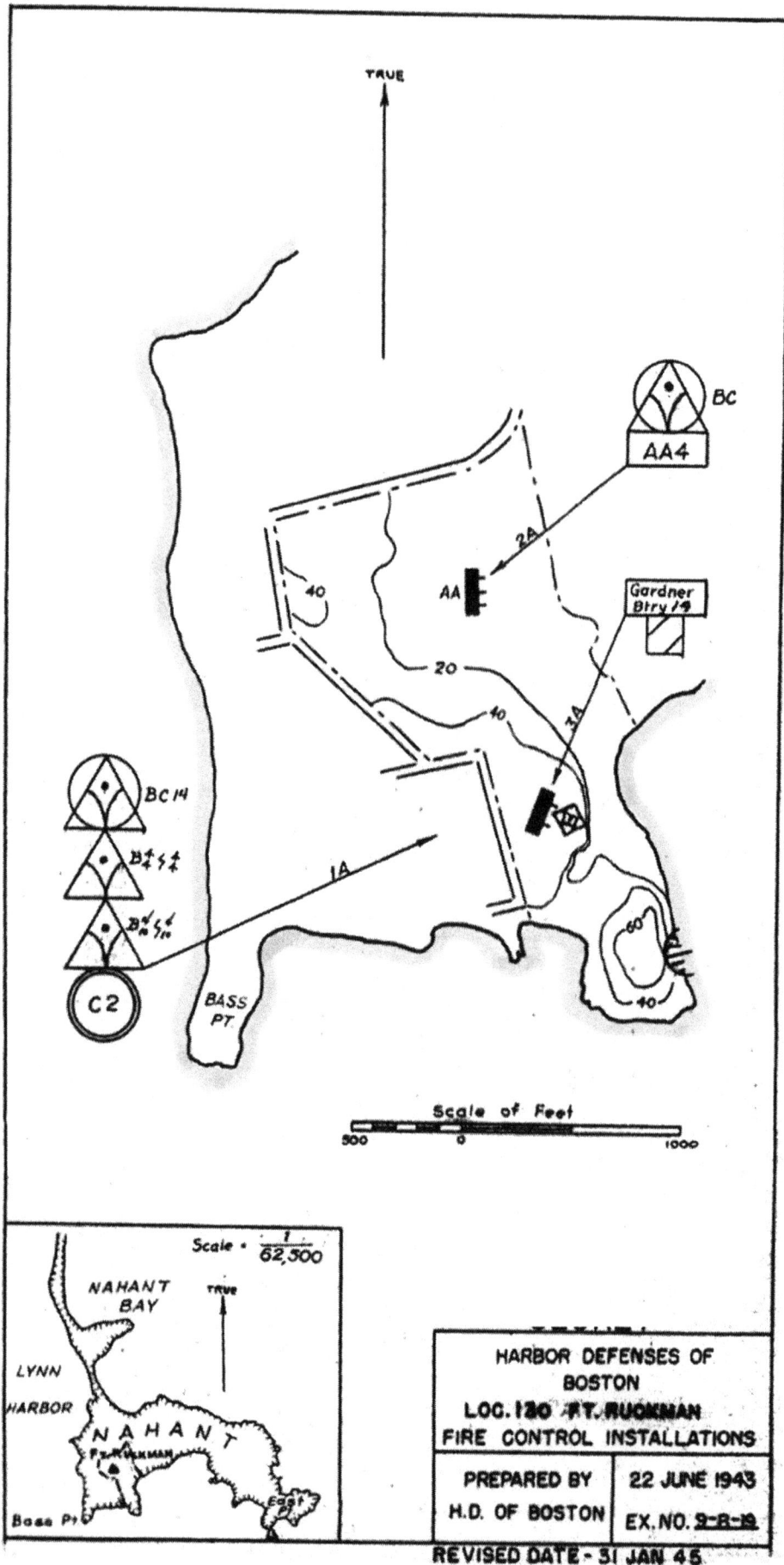

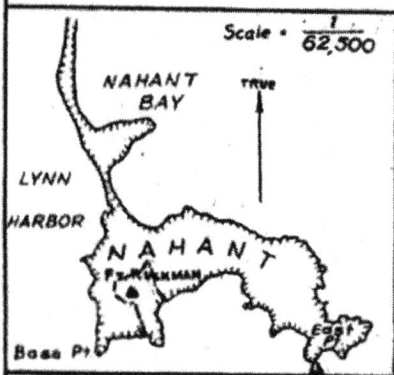

HARBOR DEFENSES OF
BOSTON
LOC. 130   FT. RUCKMAN
FIRE CONTROL INSTALLATIONS

PREPARED BY   22 JUNE 1943
H.D. OF BOSTON   EX. NO. 9-8-19

REVISED DATE - 31 JAN 45

(VI5I-89IC-IOI)(3-7-41- 2:15)(8.25-IO 000)

Nahant 1941 (NARA)

Fort Ruckman, Nahant 1920s (NARA)

Fort Banks (1889-1966) is located in the city of Winthrop. One of the earliest 12-inch M1886 mortar batteries (Abbot Quad) and a garrison area were built at the site beginning in 1891 and completed in 1896 as part of the Endicott Program and the garrison supported nearby Fort Health. It was named in General Orders No. 134 on July 22, 1899 for Maj. Gen. Nathaniel P. Banks, USV for his Civil War service. The mortar batteries were rebuilt in 1910-1914 due to inferior concrete that proved to be unstable. This reconstruction allowed for the batteries' design to be modernized, and newer mortars installed. During World War I, each pit had two 12-inch mortars removed. During World War II, the mortars were declared surplus in 1942, and the emplacement was used as Boston's HDCP. Fort Banks was used by the U.S. Army for Nike defense of Boston from 1955-1963. The fort was declared surplus in 1966 and transferred to the Town of Winthrop. Over the years all the garrison structures have been demolished. The garrison area is now an elementary school, a housing development, and an industrial park. The town has a facility building next to the mortar battery and uses the interior for storage and as a work site. The one of the mortar pits has been built over while two others have been turned into a car park. Public access is limited to the exterior of the mortar battery.

Fort Banks Gun Battery

- **KELLOGG – LINCOLN:** One of four early Endicott initial mortar batteries, using the first 1887, Abbot-type, quadrangular design for four pits. Originally Fort Banks was allotted two separate 16-mortar batteries, but only the northwestern unit was ultimately built. In March 1891 the design was submitted and approved. It closely followed the 1887 type plan, except the thickness of concrete and earth protection on the western (landward) side was reduced, and as the site was thought sufficiently inland, a ditch with flanking defense was also omitted. Work was initially done in 1891-1893. In 1895 it was sufficiently completed to allow mounting of the mortar base rings. Transfer was made on September 2, 1896 for a construction cost of $156,789.72. It was armed with sixteen 12-inch Model 1886 cast-iron mortars on Model 1891 carriages (#54/#38, #49/#55, #25/#69, #53/#53, #48/#68, #37/#44, #50/#70, #24/#49, #9/#30, #16/#28, #26/#27, #23/#29, #63/#17, #27/#19, #72/#20, and #61/#18). It was named on General Orders No. 194 of December 27, 1904 for Major General Benjamin Lincoln, American Revolution Continental Army officer. A severe accident in October 1904 killed several members of the crew. Premature firing dislodged a breechblock in one of the pits. Mortar #27 was destroyed; it being eventually replaced with like-model mortar #41. In 1906 the battery was administratively divided in two batteries, the western two pits retaining the Lincoln name, the eastern two became Battery Kellogg. These were named in General Orders No. 20 of January 25, 1906 for Brevet Colonel Sanford Kellogg, a Civil War cavalry officer. In 1910 it was decided to modernize, essentially entirely rebuild, the battery. In the early design the pits were too small and the ammunition storage and servicing were inadequate. Also, the poor Rosendale natural cement used in the early 1890s was showing serious deterioration. Finally, a new 30-foot-deep channel through Broad Sound to President Roads had left the coverage on the northern approaches to Boston inadequately provided. Several plans were advanced to rebuild the battery, the one selected used two of each pit's original walls, the other original walls were removed, thus expanding the pit footprint and considerably augmenting the internal magazine arrangement. The battery was rebuilt in halves, Battery Lincoln was rebuilt in 1911-1913 and cost $96,620. The armament was also changed. The rebuilt Battery Lincoln received more powerful (and more importantly longer-ranged) Model 1890M1 mortars on their Model 1896MII carriages. These were brought in from Battery Bagley as Fort Caswell, which received the older Model 1886s/ Model 1891s previously removed from Battery Lincoln. Mortars received were Niles #8/#213, Niles #9/#212, Watervliet #3/#216, Niles #7/#215, Bethlehem #37/#219, Builders #39/#217, Builders

BOSTON HARBOR, MASS.

GROVERS CLIFF

CABLE LINE.
AND DUCT LINE.

SERIAL NUMBER

EDITION OF MARCH. 4. 1914.
REVISIONS: DEC. 7, 1915; APR. 8, 1920.
FEB. 17, 1921; JAN. 29, 1925; AUG. 17, 1929.
JAN. 17, 1935; APR. 11, 1938

BATTERIES

LINCOLN	6-12"M.
KELLOGG	6-12"M.
WINTHROP	3-12"Dis

▲ ANTI-AIRCRAFT GUNS
FORT BANKS

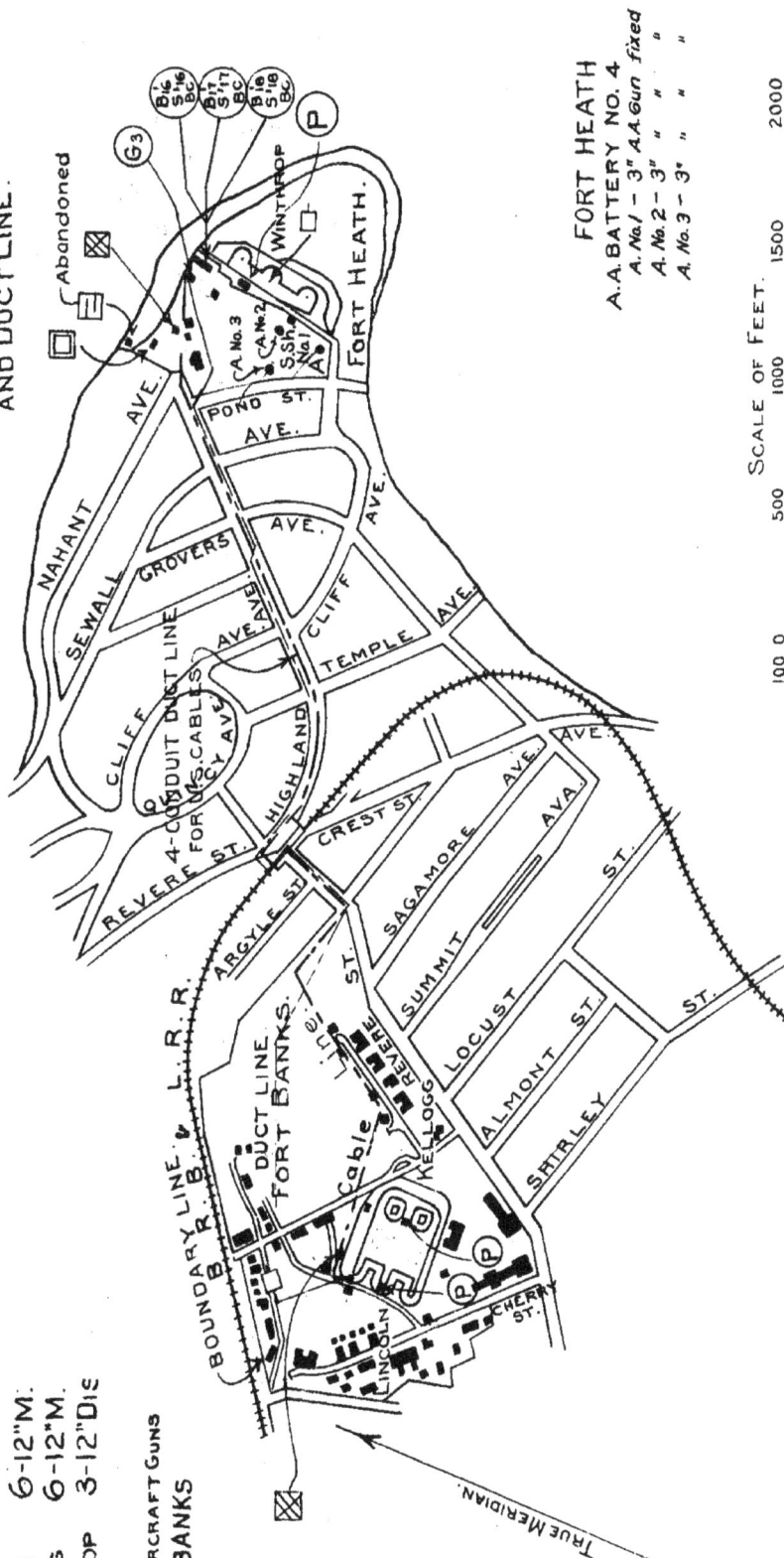

FORT HEATH

A.A. BATTERY NO. 4

A. No.1 – 3" A.A.6un fixed	
A. No.2 – 3" " " "
A. No.3 – 3" " " "

SCALE OF FEET.

100   0   500   1000   1500   2000

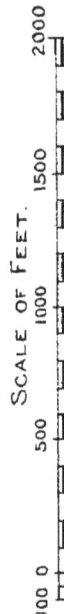

On Maintenance Status

NAHANT   AVE.
SEWALL   AVE.
GROVERS   AVE.
CLIFF   AVE.
POND ST.   AVE.
CLIFF   AVE.
TEMPLE   AVE.
HIGHLAND   AVE.
REVERE ST.
CREST ST.
SAGAMORE   AVE.
SUMMIT   ST.
LOCUST   AVA.
ALMONT   ST.
SHIRLEY   ST.
KELLOGG
ARGYLE ST.
CHERRY ST.
LINCOLN

WINTHROP
FORT HEATH.

Abandoned

4-CONDUIT DUCT LINE
FOR CABLES

BOUNDARY LINE B.R.B.   L.R.R.
DUCT LINE.
FORT BANKS.
Cable Line.
REVERE

TRUE MERIDIAN

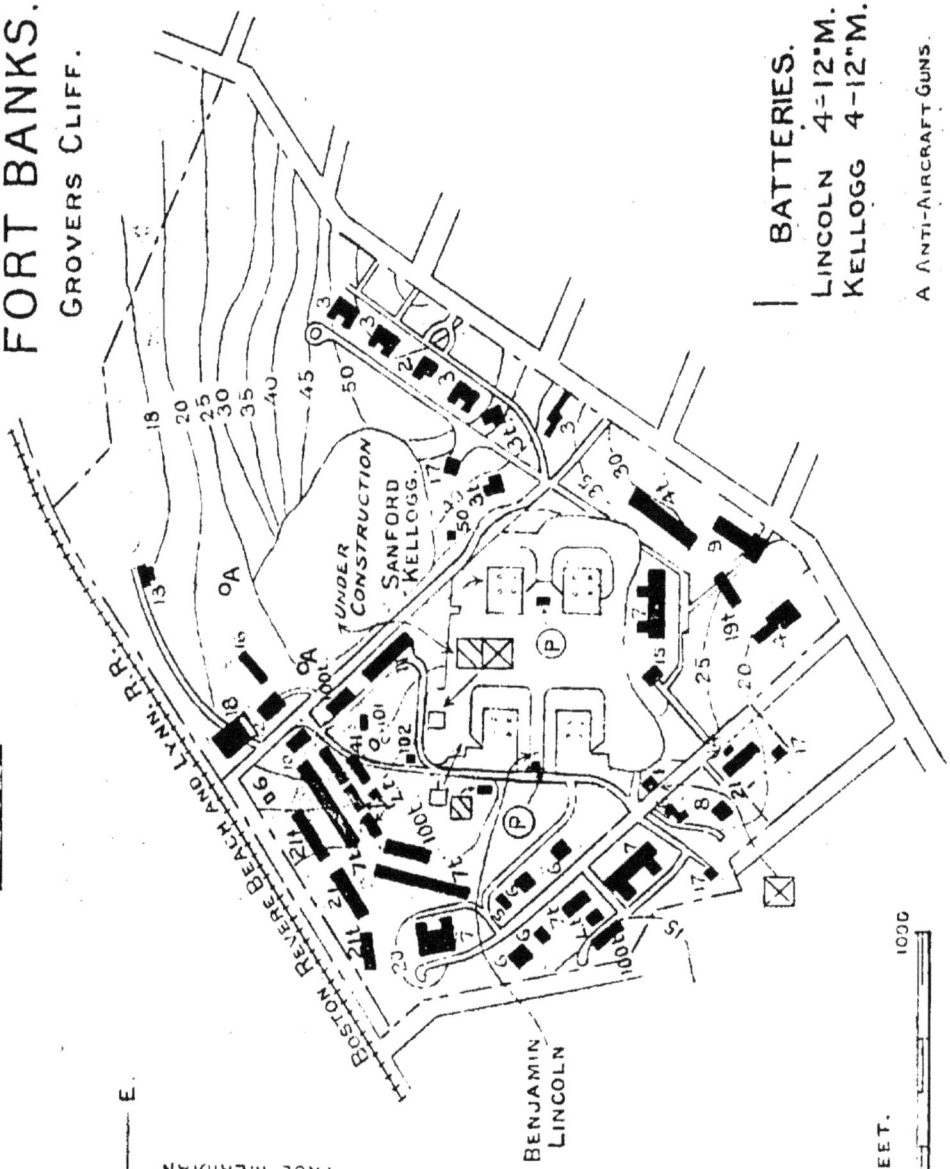

SERIAL NUMBER **124**

BOSTON HARBOR MASS.
FORT BANKS.
GROVERS CLIFF.

BATTERIES.
LINCOLN 4-12"M.
KELLOGG 4-12"M.

A ANTI-AIRCRAFT GUNS.

EDITION OF MARCH 4 1914.
REVISIONS: DEC 7, 1915;
NOV 3, 1916; APR. 8, 1920, FEB. 17, 1921.

LEGEND

1 ADMINISTRATION BLDG.
2 COMMANDING OFF. QRS.
3 OFFICERS QUARTERS.
3t. OFFICERS QUARTERS.
4 HOSPITAL.
4t. TEMP. HOSPITAL.
5 HOSPITAL STWD'S QRS.
6 N.C.O. QUARTERS.
7 BARRACKS.
7t. TEMP. BARRACKS.
8 GUARD HOUSE.
9 POST EXCHANGE AND
  GYMNASIUM.
10 BAKERY.
11 COAL SHED.
12 STABLE.
13 CREMATORY.
14 SCALES.
15 BARN.
16 WAGON SHED.
17 SHOP.
18 GARAGE.
19t. TEMP. NURSES HOME.
100t. TEMP. MESS.
21 Q.M. STOREHOUSE.
21t. TEMP. Q.M. ST. HO.
41 ENGINEER ST. HO.
50 PIGEON COTE.
101 GREENHOUSE.
102 OIL HOUSE.

TRUE MERIDIAN

BENJAMIN
LINCOLN

SANFORD
KELLOGG

UNDER
CONSTRUCTION

BOSTON REVERE BEACH AND LYNN R.R.

SCALE OF FEET.
100 0   500   1000

M

AA H

20 30 40

50

1A

TRUE

S

SCALE IN FEET

100 0 100 200 300 400 500

GROVERS
CLIFF

TRUE

WINTHROP

WINTHROP
BEACH

SCALE: 1/62,500

Snake I.

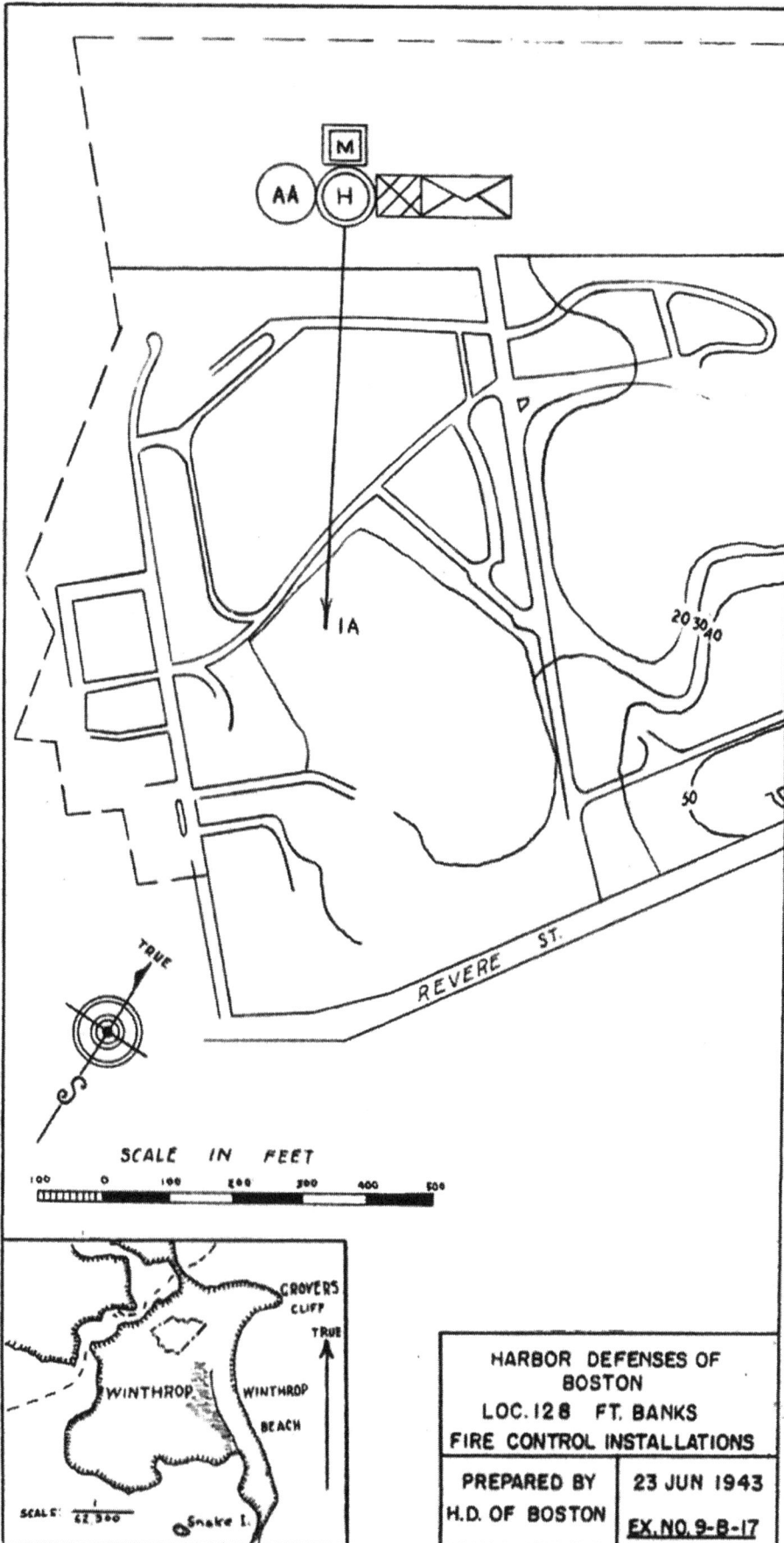

HARBOR DEFENSES OF
BOSTON
LOC. 128 FT. BANKS
FIRE CONTROL INSTALLATIONS

PREPARED BY	23 JUN 1943
H.D. OF BOSTON	EX. NO. 9-B-17

REVISED DATE 31 JAN 45

REVERE ST.

(VI50-885-0-101)3-7-41-2.P)(8.25-10.000)

Fort Banks and Fort Heath at Grovers Cliff Winthrop 1941 (NARA)

Fort Banks and Fort Heath at Grovers Cliff Winthrop 1920s (NARA)

Battery Lincoln-Kellogg Fort Banks (Terry McGovern)

Fort Heath Park (Terry McGovern)

#41/#218, and Builders #37/#214. Battery Kellogg was rebuilt between 1915 and 1916 at a cost of $111,210. It received its new mortars from Fort Washington's Battery Meigs- Model 1890M1 mortars and Model 1896MII carriages (all Watervliet tubes) #122/#169, #112/#170, #109/#171, #111/#173, #117/#172, #81/#175, #110/#208, and #125/#174. During World War I four guns (two rear guns from each of Kellogg's pits) were removed. They were shipped out on May 9, 1918 to Morgan Engineering for mounting on new railway carriages. The remaining armament of twelve mortars served during the interwar years, not being authorized for removal until November 6. 1942. After removal in mid-1943 Battery Lincoln was converted to the Boston Harbor Defense Command Post and Kellogg was used for ordnance storage. After the war the emplacement was partially destroyed for commercial development. However, some sections of the interior magazines and portions of pit walls remain on private property. The battery site is closed to the public.

Fort Heath (1895-1965) is located at Grovers Cliff in the city of Winthrop as a small reservation not far from Fort Banks. It was named in General Orders No. 43 on Apr. 4, 1900 for Maj. Gen. William Heath who served during the Revolutionary War. The fort's Endicott Program armament was three 12-inch disappearing guns organized into one battery. Later, a three-gun AA Battery Four was installed. During World War II, a battery of mobile 155mm GPF seacoast artillery was located here, followed by the installation of a 90mm AMTB battery in 1943. Became the U.S. Navy Field Test Station, Fort Heath (1946-1949), testing new fire-control systems, with a twin 5-inch naval gun turret and two 40mm guns emplaced. The site was used as a command center and radar site for the Nike missile program in the 1950s and 1960s. The fort was declared surplus in 1960s and sold to private interests. The site has been redeveloped with condominiums, as well as a city park, and nothing remains of the military structures today.

Fort Heath Gun Batteries

- **WINTHROP:** A battery for three 12-inch disappearing guns emplaced at the Grovers Cliff reservation of Fort Heath. Work was authorized for this battery with the National Defense Act funds of March, 1898. Submission of the battery plan was accomplished on April 18, 1898, although work had already begun on clearing the site. The cliff location was quite irregular, and considerable excavation was needed to create a level space for a three-gun battery. Work was done from 1898 to 1899 for transfer on October 4, 1901 for a construction cost of $202,480. It was generally of recommended type design. The battery was armed with three 12-inch Model 1888M1 Watervliet tube guns on disappearing carriages Model 1896 (#13/#12, #41/#20, and #15/#18). It was named on General Orders No. 43 of April 4, 1900 for Major Theodore Winthrop, U. S. Volunteers, who was killed in action at Big Bethel, VA in 1861. On June 25, 1907 the gun in emplacement No. 2 exploded, losing 6-8 feet of tis muzzle during practice firing. In July the damaged gun was replaced with Watervliet gun #2 (previously mounted at Battery Potter, Fort Hancock). The battery received several modifications between 1900 and 1916, including a BC station, loading platform changes, chain hoists, and modifications to the carriages to increase maximum elevation. The battery served as an important part of the north Boston defenses up until World War II. It was authorized for deletion in December 1942, scrapping of the armament was done the following year. The emplacement itself was destroyed in the early postwar years, no remains exist today.

- **AMTB-945:** A 1943 Program standard AMTB battery of two 90mm fixed and two 90mm mobile guns. The fixed blocks for the M3 90mm carriage were at the eastern tip of the reservation, close to the shoreline. It served until early postwar, when the guns were removed. The blocks were destroyed when modern construction was done at the reservation. No remains still exist.

BOSTON HARBOR, MASS.
FORT HEATH
GROVERS CLIFF.

SERIAL NUMBER

124

EDITION OF MARCH. 4. 1914.
REVISIONS: DEC. 7, 1915;
NOV. 8, 1916; APR. 8, 1920.
FEB. 17, 1921.

LEGEND
3. OFFICERS QUARTERS.
7. BARRACKS.
7t. TEMP BARRACKS.
10. SHED.
11. FIRE POINT (TOWER)
12. RIFLE BUTTS.
13t. TEMP. MESS.
14. STOREHOUSE.
41. ENGINEER ST. HO.

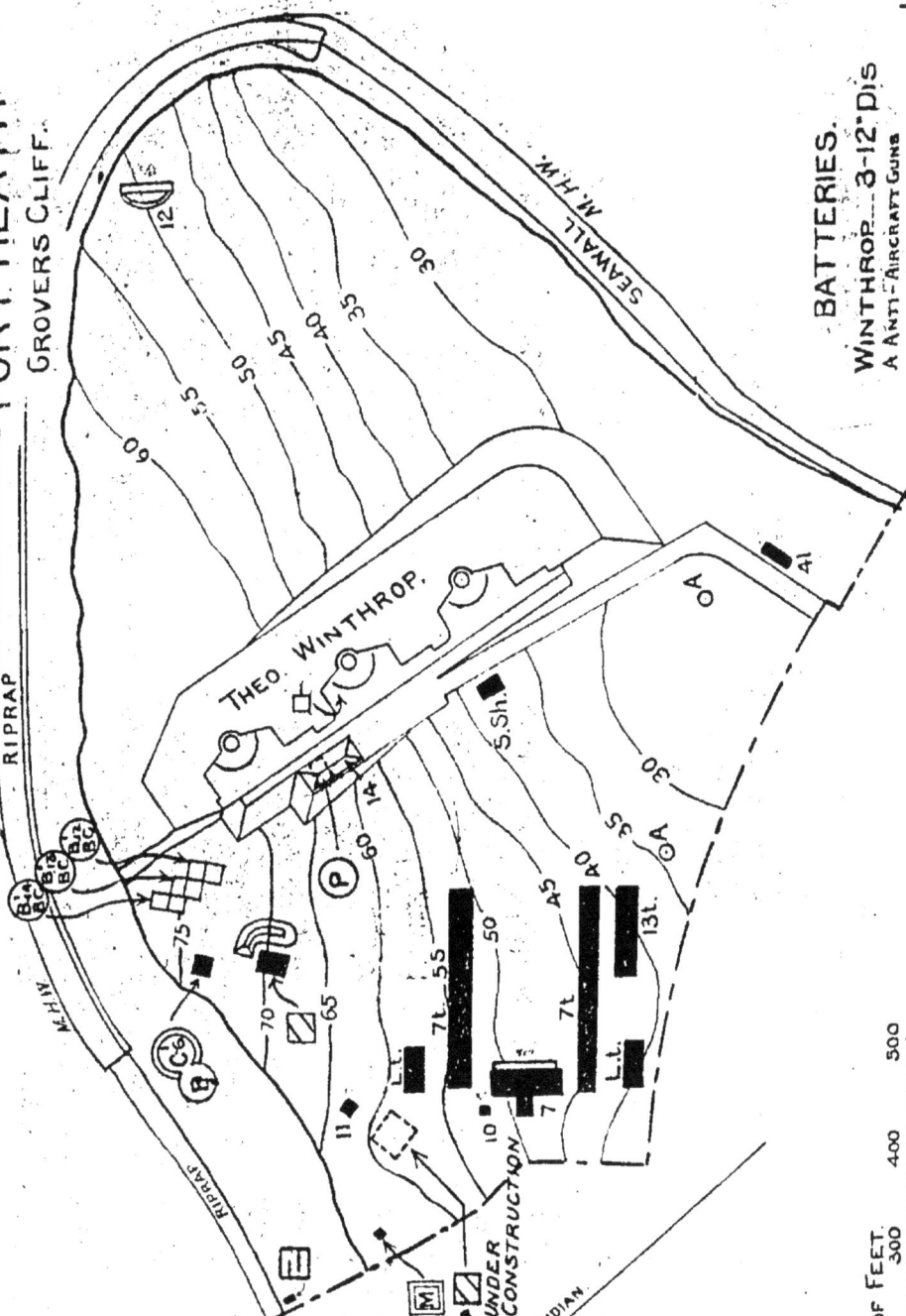

BATTERIES.
WINTHROP... 3-12" DIS
A ANTI-AIRCRAFT GUNS

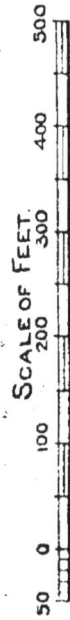

THEO. WINTHROP.

TRUE MERIDIAN

SCALE OF FEET.
50   0   100   200   300   400   500

UNDER CONSTRUCTION

SEAWALL M.H.W.

RIPRAP
M.H.W.

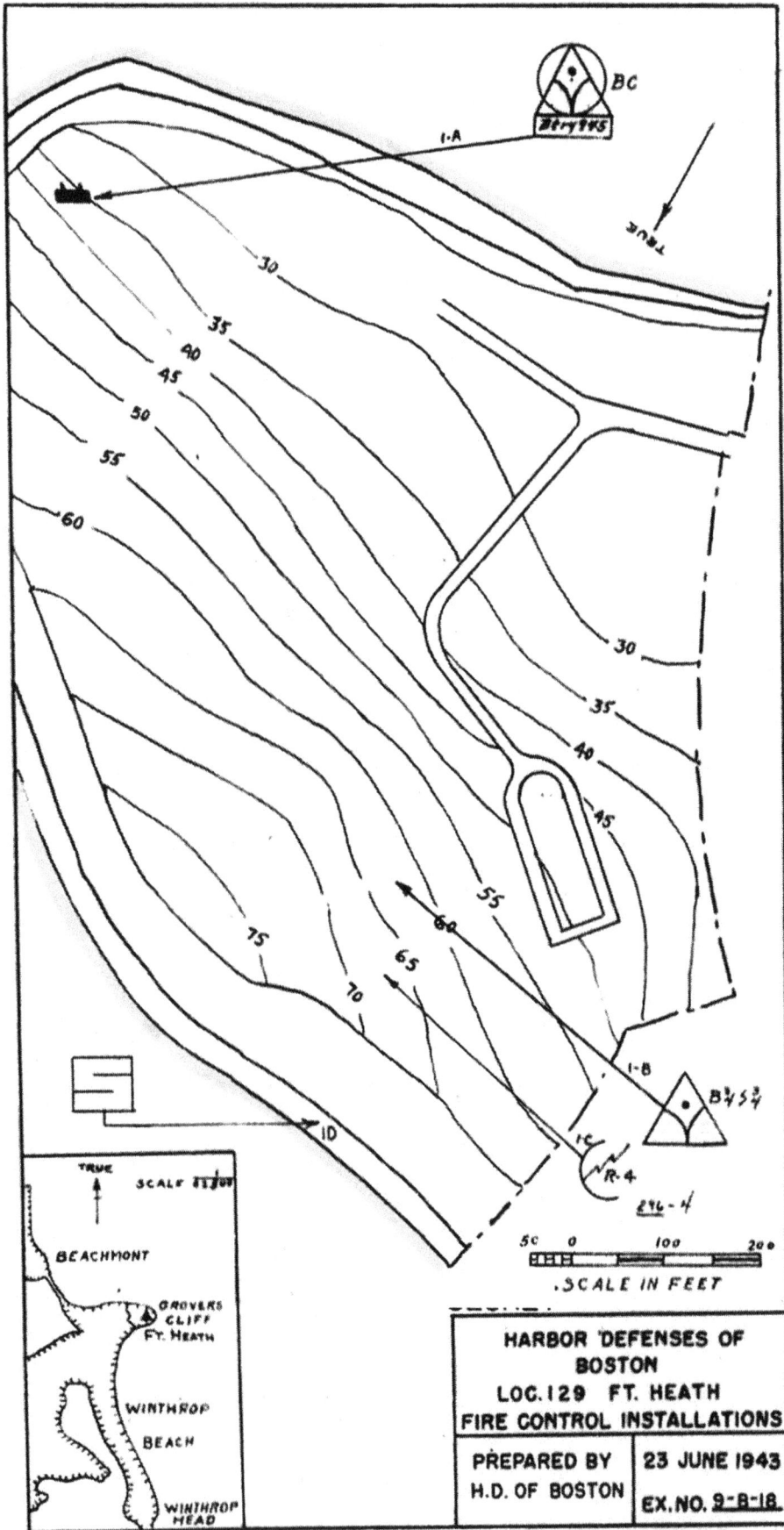

BC
Btry 445

I·A
TRUE

30
35
40
45
50
55
60

30
35
40
45

55
60
65
70
75

I·B
B⁴⁄₅₇

I·C
R·4

I·D

276·4

50   0   100   200
.SCALE IN FEET

TRUE
SCALE IN FEET

BEACHMONT

GROVERS
CLIFF
FT. HEATH

WINTHROP

BEACH

WINTHROP
HEAD

HARBOR DEFENSES OF
BOSTON
LOC. 129 FT. HEATH
FIRE CONTROL INSTALLATIONS

PREPARED BY
H.D. OF BOSTON

23 JUNE 1943
EX. NO. 9-B-18

REVISED DATE: 31 JAN 45

FORT HEATH
MASS.
ALT. 5000 FEET
PHOTOGRAPHED BY US ARMY AIR SERVICE
MAY 1924   NEG NO. 125
8TH PHOTO SECTION

Fort Heath 1930s (NARA)

Fort Dawes (1940-1963) was located on southern end of Deer Island in Boston Bay. Originally the site of fire control stations and searchlight during the Endicott Program, the fort was built as part of the 1940 Program to cover the northern approaches to Boston Harbor. It was named in General Orders No. 1 on Jan. 10, 1941 for William Dawes, a messenger who rode with Paul Revere on Apr. 18, 1775. In 1942 construction began on 16-inch casemated battery (#105 which was not armed) and 6-inch battery (#207 which was not armed). An AMTB battery was added in 1943, as well as controlled submarine mine casemate. Several fire control stations and HECP/HDCP were located with the fort's boundaries. Four 90mm AA guns were located here from 1952-57. Declared surplus by the Army in 1963, the Navy continued to use the reservation until transfer to the Metropolitan District Commission about 1980. Between 1985 and 2000, the entire reservation was leveled and converted to a regional wastewater treatment facility. None of the military fortifications remain at the site.

Fort Dawes Gun Batteries

- **Battery #105:** A 1940 Program dual 16-inch barbette battery emplaced on the Deer Island reservation of Fort Dawes. Deer Island had been used for many years as a site for fire control, but no armament was emplaced here until the 1940 Program. It was an ideal spot for a heavy battery with a clear field of fire to the east covering all of Boston Harbor. This work though, in the face of the existence of 16-inch Battery Long and the higher priority of Battery Murphy, was given a relatively low priority. It was assigned national priority #22 on September 11, 1940 and then advanced slightly to #20 on August 21, 1941. It had a completely conventional design following the type of configuration for standard 100-series batteries. Work was started with FY-1942 Budget, concrete construction being done from November 2, 1941 to August 31, 1944. It was transferred on September 28, 1944 at a cost of $1,085,369. The battery was never named, just known as Battery Construction No. 105. However shortly after being transferred all additional work was suspended. Both guns and carriages had been delivered, but not yet installed. It was to have mounted two 16-inch MkIIM1 guns on Model M1919M5 barbette carriages (#93/#60 and #98/#61). It remained unarmed to the end of the war, though recommended for completion under the 1946 Review. It was abandoned shortly thereafter. In 1989-1991 the emplacement was completely destroyed and removed for erection of the new Deer Island sewage treatment plant, no traces of it remain today.

- **Battery #207:** A 1940 Program dual 6-inch barbette battery emplaced at Fort Dawes at Deer Island. It was located just to the north of Battery #105, also firing to the east. It was funded under the FY-1942 Budget and initially assigned national priority #18. It was one of the four projected modern 6-inch batteries projected for the Boston defenses. Work was done from September 8, 1942 to October 31, 1943. Transfer was made on September 28, 1944 for a cost of $300,385. The battery was never named, simply being known as Battery Construction No. 207. It was a standard 200-series battery design. As soon as it was transferred work was suspended for lack of available gun tubes for armament. It was to receive two 6-inch M1(T2) guns on Model M3 barbette carriages. No gun tubes were ever delivered, though shields and carriages #4 and #5 were delivered (it is unclear if they were ever mounted). The battery remained incomplete in the postwar years. While it was recommended for completion in the 1946 Review, this was never done and the battery abandoned in 1948. The emplacement was destroyed in 1989-1991 with the construction of the Deer Island sewage facility. No traces remain of it today.

- **TAYLOR:** A relocated battery for two 3-inch pedestal guns made to a new emplacement at the southeastern tip of Fort Dawes. It served as an interim AMTB battery pending the completion of

BOSTON HARBOR, MASS.

DEER ISLAND.

SCALE OF FEET.

EDITION OF MARCH. 4. 1914.
REVISIONS: NOV. 8, 1916, APR. 8, 1920. FEB. 17, 1921.

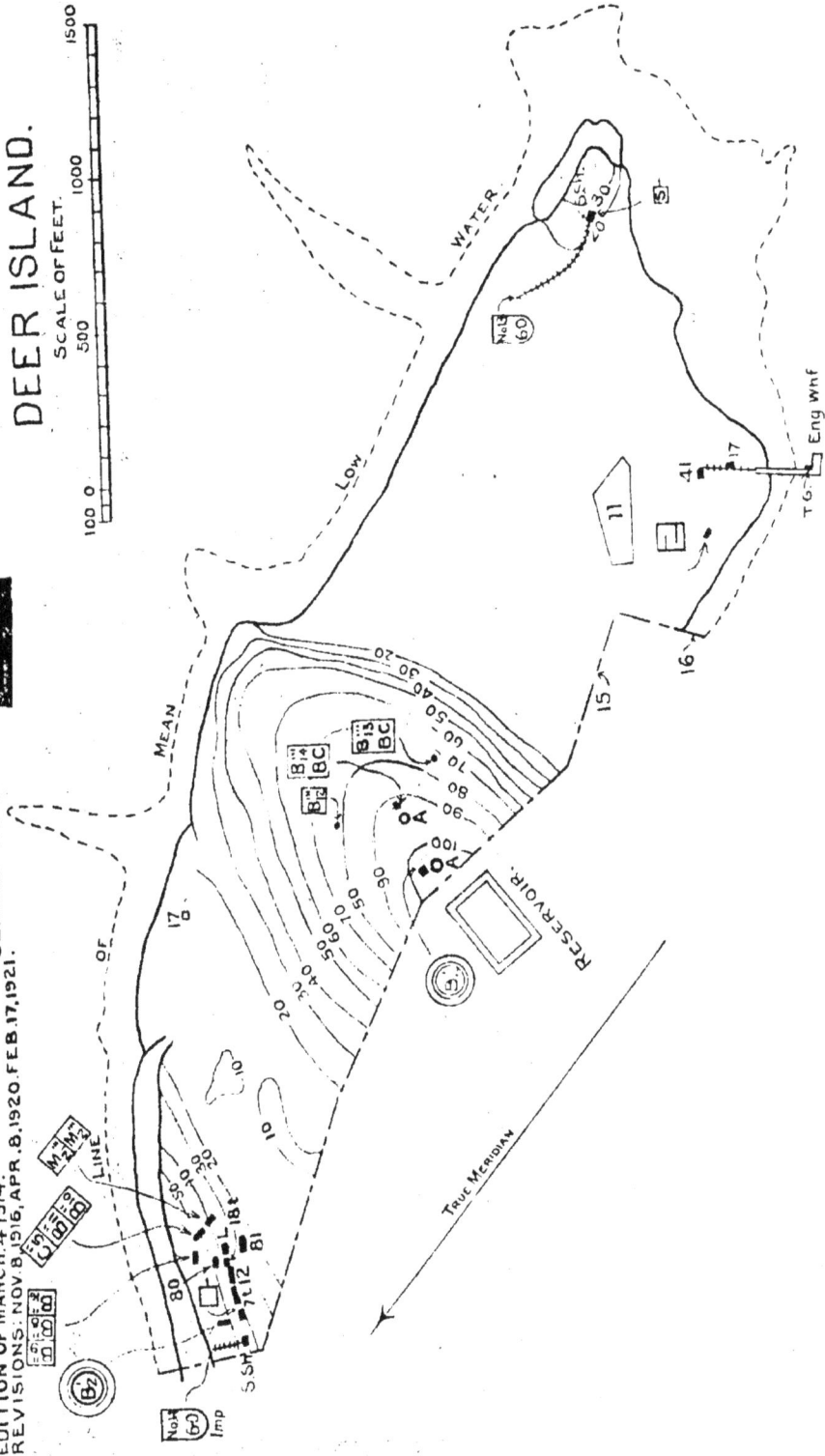

SERIAL NUMBER

WATER

Low

MEAN

OF

LINE

TRUE MERIDIAN

RESERVOIR

Eng Whf

T.G.

LEGEND.

7t BARRACKS.
11 CEMETERY.
12 F.C. DORMITORY.
15 CONCRETE WALL.
16 CATTLE FENCE.
17 WATCHMAN'S HOUSE.
18t LAVATORY.
41 ENGINEER ST. HO.
80 RADIO COMPASS STA.
81 " OPERATORS' QRS.

A - ANTI-AIRCRAFT
GUN FOUNDATIONS.

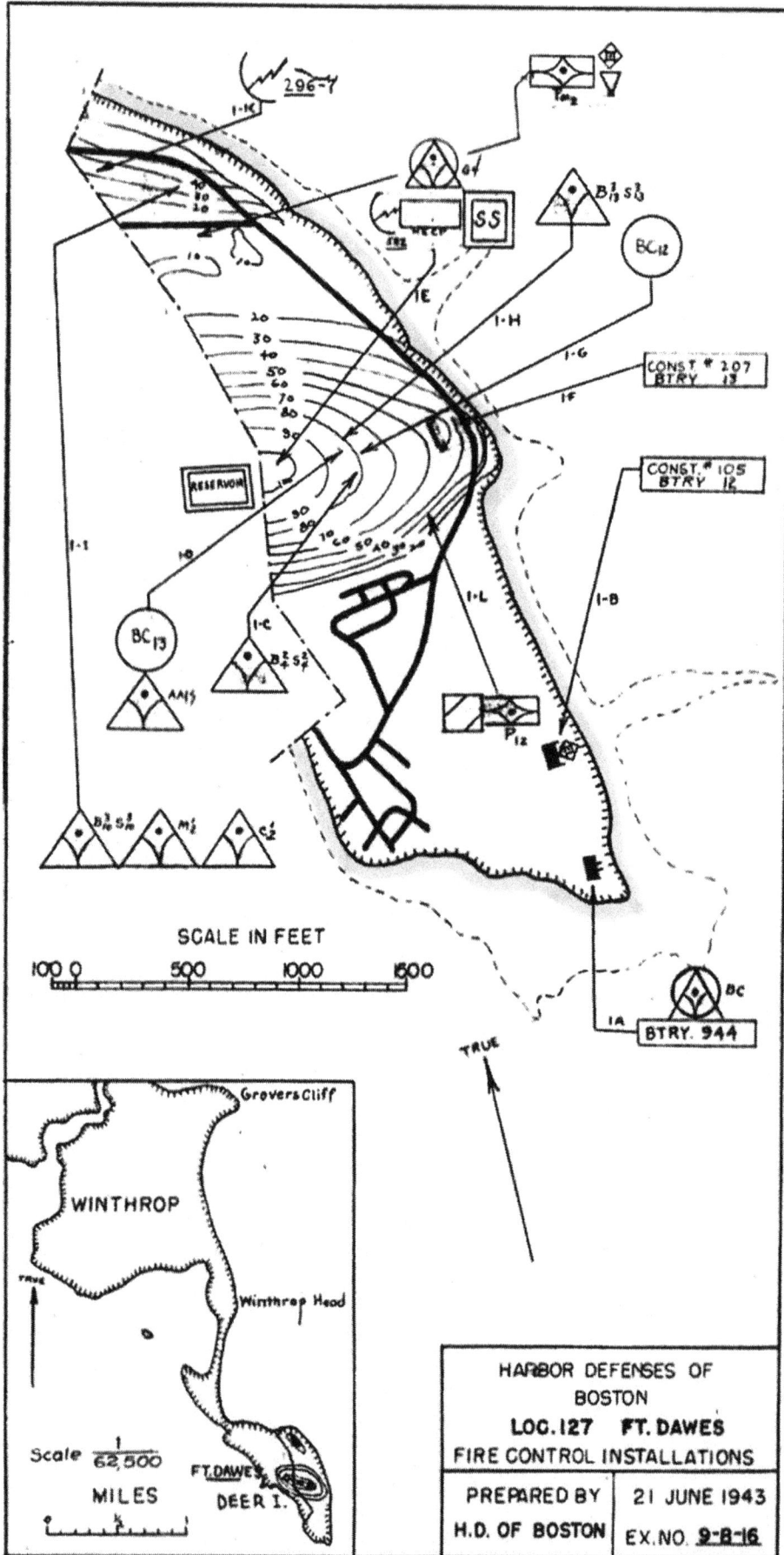

SCALE IN FEET

100 0   500   1000   1500

TRUE

Grovers Cliff

WINTHROP

TRUE

Winthrop Head

Scale 1/62,500

MILES

FT. DAWES

DEER I.

HARBOR DEFENSES OF
BOSTON
LOC. 127   FT. DAWES
FIRE CONTROL INSTALLATIONS

| PREPARED BY | 21 JUNE 1943 |
| H.D. OF BOSTON | EX. NO. **9-B-16** |

REVISED DATE: 31 JAN 45

Vicinity Map
Battery No. 207
Location No. 127, Site 1c

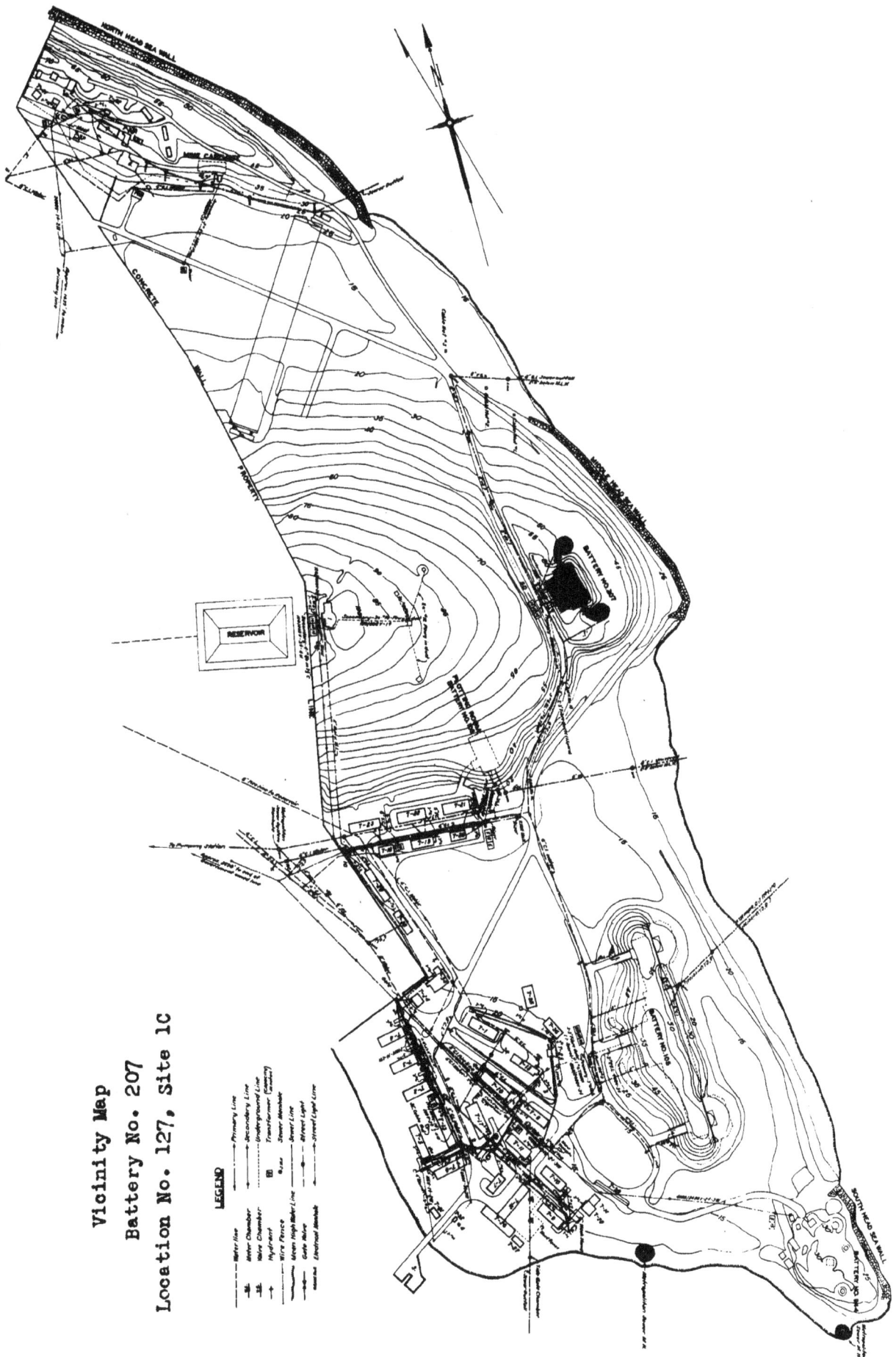

new 90mm batteries. Two 3-inch Model 1902M1 guns were relocated here from Battery Taylor at Fort Strong (guns and like-model pedestal mounts #45/#45 and #46/#46). The battery consisted of simple open gun blocks and a small concrete BC station. Work was done between June 12 and July 21, 1942 for transfer on August 17, 1942 at a cost of $7,808. Later AMTB-944 was constructed and replaced Taylor's function, though the guns remained in place as spares and the BC station reassigned to serve the new 90mm battery. It was probably disarmed shortly after war's end. All traces of the battery were destroyed in the early 1990s with the construction of the new Deer Island sewage plant.

* **AMTB #944**: A 1943 Program AMTB battery of two 90mm fixed and two 90mm mobile mounts. It was authorized on April 17, 1943 and built between April 26th and June 9, 1943. Transfer was made on September 10, 1943 at a cost of $7,625. It replaced 3-inch Battery Taylor in function and was built in almost exactly the same location at the southeastern tip of Deer Island. In fact, it utilized the small BC station and magazine from Taylor, though new blocks and hold-down bolts were built. It served during the balance of the war, being deleted and removed probably in 1946 or 1947. It was destroyed in the early 1990s with the construction of the new Deer Island sewage plant.

Calf Island Military Reservation (1941-1946) is located on Calf Island in Massachusetts Bay. During World War II, a searchlight station and an observation post, as well as an SCR-268 radar, were here. No remains today. Also, there were plans to build a dual 16-inch gun battery during the Interwar Period. The island is now part of the Boston Harbor Islands National Recreation Area and Boston Harbor Islands State Park.

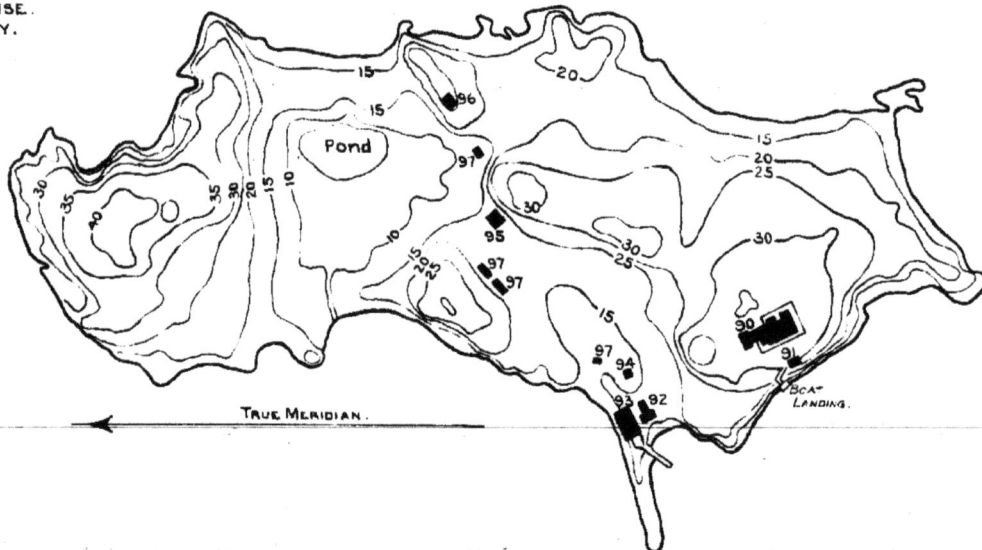

Calf Island M.R. Gun Battery

Calf Island was approved as a site for the construction of a modern 16-inch barbette battery under the 1915 Program. The island site was purchased in 1918. By August 1919 plans were prepared for this battery and a sister site near Scituate. It was to receive two of the newly developed army 16-inch guns. The primary mission of the battery was to protect Boston Harbor in general, and the South Boston shipyards in particular. However, by early 1920 opinion had changed against this installation, and plans were dropped for a battery here. The particularly isolated location and problematic wharf conditions probably figured heavily in this decision. The 16-inch barrels were subsequently emplaced at Fort Duvall in substitution for 12-inch long-range guns and carriages. The reservation itself remained in army hands until after World War II.

Outer Brewster Island Military Reservation (1943-1948) is located on Outer Brewster Island in Massachusetts Bay. Built as part of the 1940 Program to defend the entrance to Boston Harbor, its Battery Construction Number #209 had two shielded barbette 6-inch guns (later known as Battery Jewell) was built in 1942-1944. Due to its isolated location, the battery included a desalination plant to make fresh water. The island was declared surplus in 1947 and sold to a private owner in 1953. In 1973, the island was acquired by the Metropolitan District Commission. Today, former battery remains with only its concrete structures remaining. The island is now part of the Boston Island NRA and is accessible only by small watercraft.

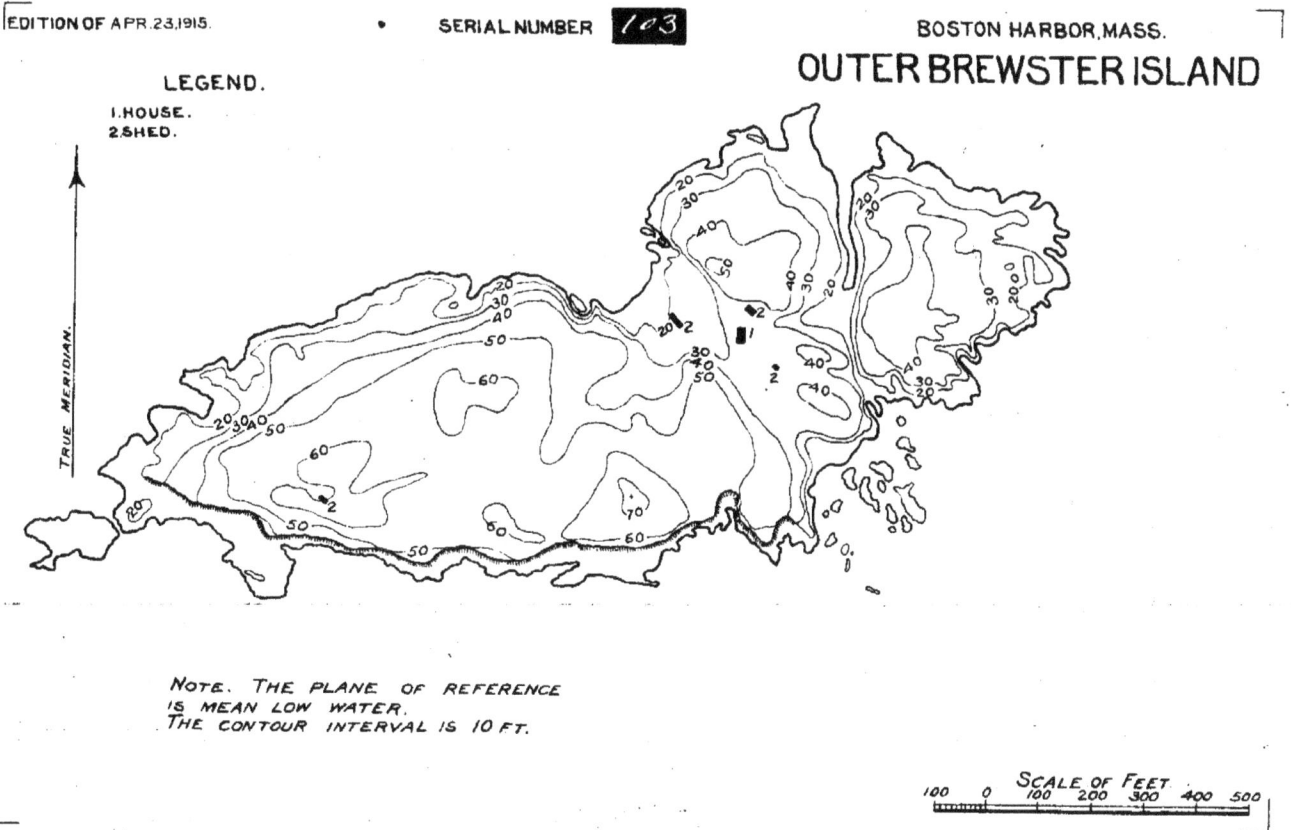

296·5
1·8

1A

SCALE OF FEET
100   0   100   200   300   400   500

Nantasket Roads

Outer Brewster

Gull Pt

Brewster Islands

TRUE

Scale 1/62500

Pt Allerton

HARBOR DEFENSES OF
BOSTON
LOC.126 OUTER BREWSTER
FIRE CONTROL INSTALLATIONS

PREPARED BY
H.D. OF BOSTON

23 JUNE 1943

EX. NO. 9-8-15

REVISED DATE: 31 JAN 46

Outer Brewster Island M.R. Gun Battery

- **JEWELL**: The highest priority 6-inch barbette battery of the 1940 Program authorized for Boston. While under construction it was known as Battery Construction No. 209. Planned and emplaced on Outer Brewster Island, it fired to the east. It was assigned national priority #12 in 1941 and funded under FY-1942 Budget. It was of conventional 200-series design. Work was somewhat delayed by the rocky nature of the site but was accomplished from June 8, 1942 to January 31, 1944. It was transferred in 1944. The battery was armed with two 6-inch guns Model M1(T2) on M1 barbette carriages (#1/#1 and #2/#2). The battery was named on Adjutant General's Memorandum of August 20, 1942 for Colonel Frank C. Jewell. It was retained in the immediate postwar years but finally deleted and disarmed probably in 1948. The emplacement still exists on Outer Brewster, part of the Boston Islands NRA. The battery is open to the public.

Greater Brewster Island Military Reservation (1917-1946) is located on Greater Brewster Island in Massachusetts Bay. During World War I, a searchlight was place on the island but removed at the end of the war. Built as part of the 1940 Program to defend the entrance to Boston Harbor, was AMTB Battery 942 (1943-1946), a mine casemate (1944) and a mine observation station (demolished). The island was declared surplus in 1946 and sold to a private owner in 1953. In 1973, the island was acquired by the Metropolitan District Commission. Today, former battery remains with only its concrete gun block remaining. The island is now part of the Boston Island NRA and is accessible only by small watercraft.

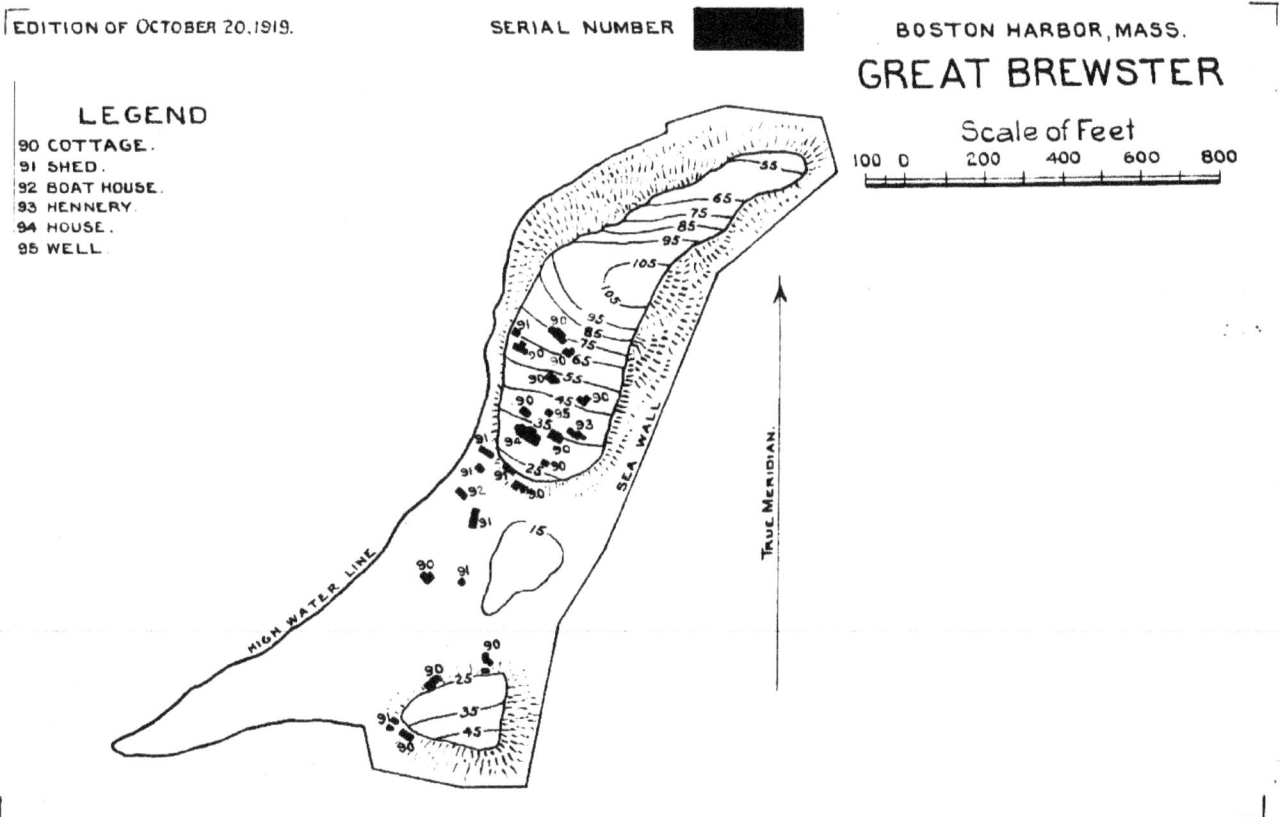

EDITION OF OCTOBER 20, 1919.   SERIAL NUMBER   BOSTON HARBOR, MASS.

GREAT BREWSTER

LEGEND

90 COTTAGE.
91 SHED.
92 BOAT HOUSE.
93 HENNERY.
94 HOUSE.
95 WELL.

Scale of Feet
100 0   200   400   600   800

Battery Jewell on Outer Brewster Island (Terry McGovern)

Battery Burbeck-Morris on Lovell's Island (Terry McGovern)

Greater Brewster Island M.R. Gun Battery

- **AMTB #942:** A 1943 Program AMTB battery of two 90mm fixes and two 90mm mobile guns emplaced on Great Brewster Island. It was authorized for construction on April 17, 1943. It was sited at the northern end of the island, on top of the island's highest drumlin. The island also had a mine casemate and mine station, as well as a searchlight position. The battery armament was removed postwar, some remain still exist on Great Brewster, part of the Boston Islands NRA. The battery site is open to the public.

Fort Standish (1900-1947) is located on Lovell's Island in Boston Bay. Construction of the major gun batteries was begun in 1899 as part of the Endicott Program. The fort's primary armament was four 10-inch disappearing guns organized into two batteries. Five rapid fire gun batteries (five 6-inch guns and nine 3-inch guns) provided coverage against small naval craft. It was named in General Orders No. 43 of 1900, for Myles Standish who arrived with the settlers on the Mayflower in 1620. In 1925, two of Battery Vincent's vacant positions were converted for a 3-inch antiaircraft battery. A 90mm AMTB battery was installed at the fort in 1943. The fort also was equipped with a mine casemate. The fort was declared surplus in 1947 and transferred to the State of Massachusetts in 1958. The island became owned by the Metropolitan District Commission and today is a part of the Boston Islands NRA. All garrison structures have been removed from island. The batteries at the southern end of the island have suffered from erosion, the 3-inch battery has been destroyed, and the 6-inch battery is breaking up. The island is a day use area, and overnight camping is allowed on a reservation basis. Refer to the Boston Harbor Islands website for ferry access to Lovell's Island.

Fort Standish Gun Batteries

- **BURBECK – MORRIS:** The 10-inch disappearing gun battery for Fort Standish on Lovell's Island. Plans were submitted on May 6, 1901 for four emplacements using the Model 1896 disappearing carriage (but ultimately changed to the M1901 carriage). Initially funds were available for building just the right-hand three emplacements, but soon the fourth and final unit was included. The battery was located on the central drumlin of the island, aligned parallel to the shoreline, firing to the northeast. In this position they covered both major channels into Boston Harbor. The plans were unique. Due to the low island terrain on either side (and deep water beyond) the battery was considered vulnerable to heavily enfilade fire from either flank. That prompted the use of very broad flank and internal traverses to offer the highest protection. In turn the plan could incorporate deep magazines in the traverses that could supply ammunition to the guns on the same level—avoiding the use of hoists. Work was done in 1901-1902. Transfer was made on December 19, 1907 for a cost of $256,903.84. It was named on General Orders No. 194 of December 27, 1904 for Brigadier General Henry Burbeck of Revolutionary War service. It was ultimately armed with four 10-inch guns Model 1900 on Model 1901 disappearing carriages (#8/#11, #10/#12, #5/#9, and #6/#10). The two northern guns were renamed as Battery Morris, a separate tactical battery, on December 13, 1909. This name was conferred by General Orders No. 245 for Brevet Major General William W. Morris, U.S. Artillery. Because of both the modernity of the gun types and the excellent field of fire, the emplacement remained during World War I and later. New BC stations were added to each battery in 1925. By the late 1930s they were slated for ultimate removal, however during the early years of the World War II Battery Morris was kept in reserve status. Battery Burbeck was disarmed in the 1935-38 period, Morris finally being declared obsolete under authority of November 6, 1942. For a while the emplacement was used for munitions storage for the local, smaller batteries Whipple

BOSTON HARBOR, MASS.

FORT STANDISH

LOVELLS ISLAND.

SERIAL NUMBER **124**

EDITION OF MARCH. 4 1914.
REVISIONS: DEC. 7, 1915.
NOV 8. 1916. APR. 8, 1920, FEB. 17, 1921.

LEGEND.

3t TEMP. OFFICERS QRS.
4t TEMP. HOSPITAL.
6 N.C.O. QUARTERS.
7t TEMP BARRACKS.
107 C. DORMITORY.
11 CISTERN.
12 C.
13 TARGET WAYS.
14 RIFLE BUTTS.
15 SHEDS.
16 OBSERVING TOWER.
17 COAL SHED.
18t TEMP. MESS.
19
100 STABLE.
101 WAGON SHED.
102 OFFICE.
103 LAVATORY.
21t TEMP. Q.M. ST. HO.
31 ORDNANCE ST. HO.

80 L.H. RESERVATION.
81 OIL HOUSE (U.S.L.H.E.)
82 L.H. TRESTLE WALK.

BATTERIES.

BURBECK..2-10" DIS.
MORRIS..2-10 " "
WHIPPLE..2-6" P.
* VINCENT...4-3" B.P.
WILLIAMS..3-3" B.P.
WEIR....2-3" P.
TERRILL..3-6" DIS.
▲ ANTI-AIRCRAFT GUNS.
* GUNS DISMOUNTED.

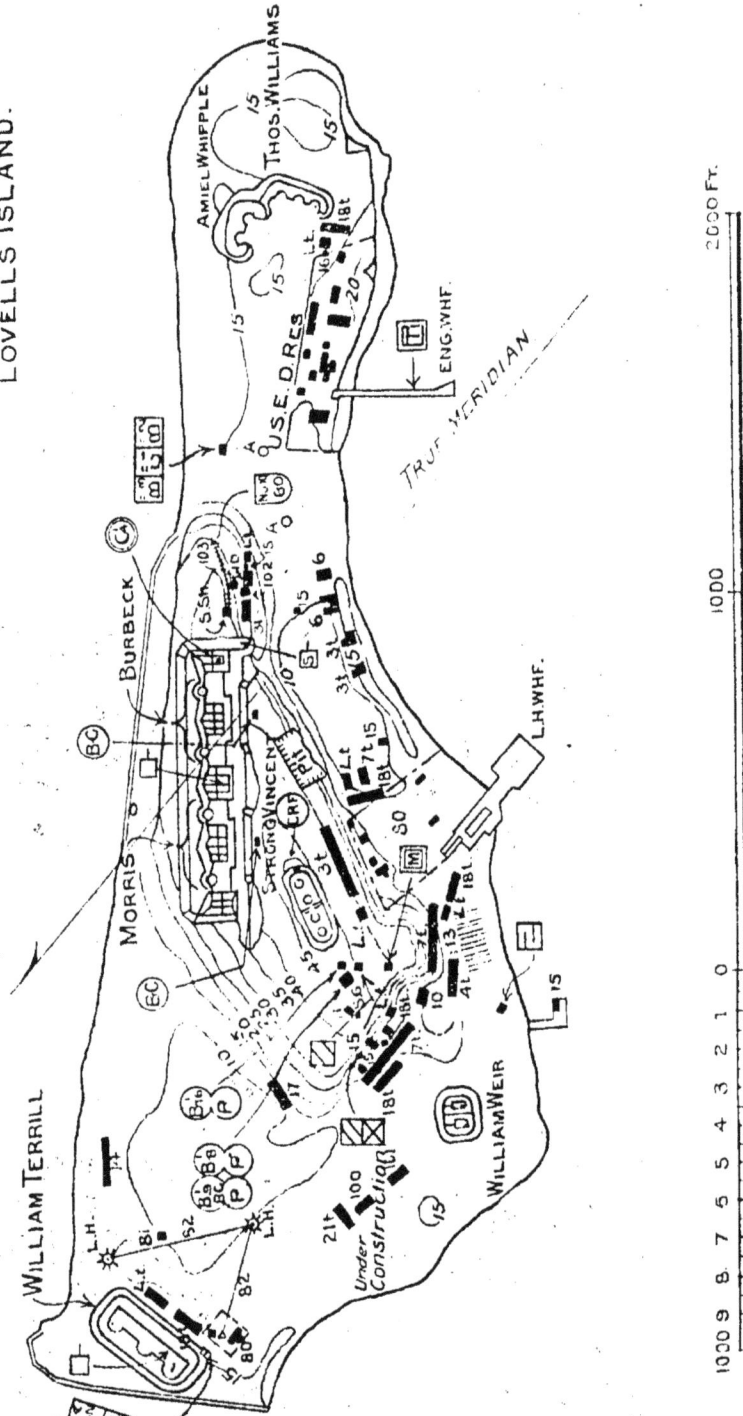

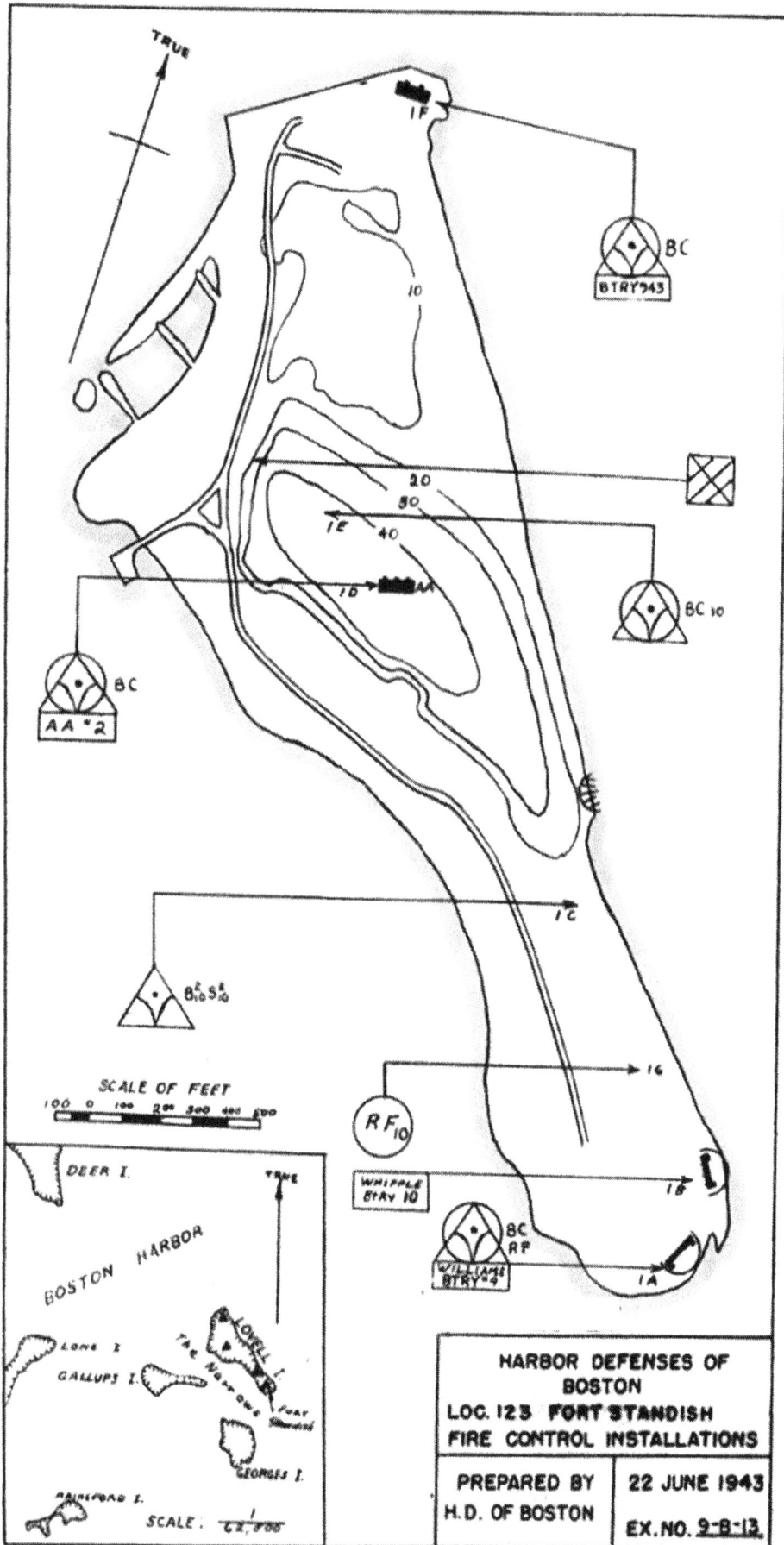

TRUE

1F

BC
BTRY 543

10

20
50
1E   40

1D   AA   BC 10

BC
AA #2

1C

B⁵₁₀ S⁵₁₀

SCALE OF FEET
100 0 100 200 300 400 600

16

RF₁₀

WHIPPLE
BTRY 10

1B

BC
RF
WILLIAMS
BTRY #4

1A

DEER I.
TRUE
BOSTON HARBOR
LONG I.
GALLUPS I.
THE NARROWS
LOVELL I.
FORT STANDISH
GEORGES I.
RAINSFORD I.
SCALE: 1/62,800

HARBOR DEFENSES OF
BOSTON
LOC. 123 FORT STANDISH
FIRE CONTROL INSTALLATIONS

PREPARED BY
H.D. OF BOSTON

22 JUNE 1943

EX. NO. 9-B-13

REVISED DATE: 31 JAN 45

FORT STANDISH
MASS.
ALT. 6600 FEET
PHOTOGRAPHED BY US ARMY AIR SERVICE
MAY 1924 NEG NO. 132
8TH PHOTO SECTION

Fort Standish 1930s (NARA)

and Terrill. The emplacements still exist at the Boston Harbor Islands National Park, Lovell's Island unit. The battery is open to the public.

- **TERRILL**: A battery for three 6-inch disappearing guns built at the low, northern end of Lovell's Island. The plan was submitted on May 16, 1899. It was located to cover Broad Sound and the proposed deep channel there. It fired to the north. The plan featured a shared magazine for the two emplacements on the east, and a single magazine for Emplacement No. 3 in the traverse between the two western units. Due to the danger of fire from the rear, a parados covered the back of the battery with a built-in concrete access entry in it. Work was done in 1899-1900 and transferred on November 21, 1902 for a construction cost of $86,673.33. It was named on General Orders No. 194 of December 27, 1904 for Brigadier General William R. Terrill who was mortally wounded at Perryville KY in 1862. The battery carried three 6-inch Model 1897M1 guns on Model 1898 disappearing carriages (#2/#15, #6/#16, and #10/#17). There appears to have been some early difficulties experienced, as the 1909 Report of Completed Works reports the battery condition as only fair. However, the battery was kept with its original armament in the war years. It was finally authorized for removal under the 1940 Program, permission to delete was granted on April 7, 1943. After disarmament the emplacement was used for ammunition storage for nearby AMTB-3, built on Terrill's parapet. The emplacement still exists at the Boston Harbor Islands National Park, Lovell's Island unit. The battery site is open to the public, but the emplacement has been damaged by shoreline erosion.

- **WHIPPLE**: An emplacement for two 6-inch rapid-fire pedestal guns emplaced at Fort Standish on Lovell's Island. It was located on the southern end of the island, with a field of fire almost due east. Plans were submitted on May 20, 1901. Due to concerns about enfilade fire disabling two guns sharing the same platform, the battery was built with two separate gun platforms and a central traverse covering a shared magazine—even before engineer mimeographs were changed to essentially recommend this configuration for all subsequent construction. It was fitted with light, hand-operated, shell lifts and a small power room in the traverse rooms. Three-inch Battery Williams would soon be built on the right flank of the battery. Work was done in 1900-1901. Transfer was made on December 29, 1904 for a cost of $12,000. It was named on General Orders No. 194 of December 27, 1904 for Major General Amiel W. Whipple who was killed in 1863 at the Battle of Chancellorsville. It was armed with two 6-inch Model 1900 guns and M1900 pedestal mounts (#11/#1 and #13/#30). There had been a considerable delay in mounting these, they were reported received at the fort in May 1902, but had to be returned to Watervliet Arsenal for breech modifications and then returned to the fort to be mounted in 1907. The tubes were dismounted in 1918 for potential use elsewhere but returned and remounted in 1920. The battery was retained in the 1932 Review. Modifications were made to the firing platforms and parapets in 1941-1942. It continued to serve throughout the war, not being removed until about 1947. The emplacements still exist at the Boston Harbor Islands National Park, Lovell's Island unit. The battery is open to the public.

- **VINCENT**: An emplacement four 3-inch pedestal guns. First plans were submitted on April 19, 1902 for three 3-inch, 15-pounder guns on Model 1898 masking parapet mounts for this site on the high ground that ran in a ridge across the center of the island to the west. Guns at this location would have the ability to fire either to the west or east, both having important water to cover with rapid-fire guns. In late 1901 a third gun was added to the plans, and the fourth shortly after that. Work was done in 1901-1902. It was transferred December 29, 1904 for a cost of $14,000. It was named on General Orders No. 194 of December 27, 1904 for Brevet Brigadier General Strong Vincent, U.S. Volunteers, mortally wounded at the Battle of Gettysburg in 1863. The battery was armed with four 3-inch Model 1898 guns on Model 1898 masking parapet mounts (Driggs-Seabury #1/#1, #8/#8, #64/#64, and #71/#71). In common with other balanced pillar batteries, it was disarmed in June of 1920. In subsequent years it was used to support Antiaircraft Battery No. 2, two platforms being converted out of emplacements No. 1 and No. 4 about 1925. The emplacements still exist at the Boston Harbor Islands National Park, Lovell's Island unit. The battery site is open to the public.

- **WEIR:** An emplacement for two 3-inch pedestal mount guns emplaced at the western end of Lovell's Island. Plans were submitted on October 7, 1902. It was positioned on low ground near the shore and had a field of five west and cover the Narrows. It was almost directly to the rear of Battery Terrill. It followed typical plans, except the magazine was pushed as far forward as possible for better protection. Work was done in 1902-1903, and transfer made on May 31, 1906 for a cost of $29,127. It was named on General Orders No.194 of December 27, 1904 for 1st Lieutenant William B. Weir, Ordnance Department, who was killed in action with Ute Indians near the White River Agency in 1879. It was armed with two 3-inch Model 1902 Bethlehem guns and pedestal mounts (#49/#49 and #50/#50). Unfortunately, it was very exposed on the beach and suffered severely from a series of storms and their erosion of the beach. It was abandoned by authorization of July 26, 1926, the removed armament being used elsewhere. Eventually it was totally destroyed.

- **WILLIAMS:** An emplacement for three 3-inch guns emplaced on the right flank of 6-inch Battery Whipple on the southern end of Lovell's Island. Plans were submitted on June 13, 1900 to use appropriations granted by the Act of May 25, 1900. As the most eastern of the Fort Standish batteries, it covered the main mine field offshore. Funds were granted to start the first two emplacements, and in April 1901 the third unit was added. It generally followed type plans in design for masking parapet emplacements. Work was done in 1901-1902. It was transferred on December 29, 1904 for a cost of $17,500. It was named on General Orders No. 194 of December 27, 1904 for Brigadier General Thomas Williams killed in action at Baton Rouge LA in 1862. It appears that three Model 1898 guns and pillars were actually shipped here. Ordnance records indicate guns and pillars #7, #12, and #95 were received at the post. However, they were shipped back out on May 8, 1906. For whatever reason, the battery never got these guns. Modifications were made to the platforms to mount Model 1903 guns and pedestals. It was armed around 1910-1912 with three 3-inch Model 1903 guns and pedestal mounts (#52/#82, #46/#81, and #45/#80). This new armament was retained during the following decades, not being removed after World War II, probably in 1946. The emplacements still exist at the Boston Harbor Islands National Park, Lovell's Island unit. The battery is open to the public, but one end is being eroded by the shoreline.

- **AMTB #943:** A 1943 Program AMTB battery for two 90mm fixed and two 90mm mobile guns. It was authorized on April 27, 1943 for the northern end of the island, utilizing the parapet of old Battery Terrill and its magazines for 90mm ammunition storage. The concrete blocks were built from March 29 to May 19, 1943. It was transferred on September 10, 1943 at a cost of $8,851.67. It was disarmed postwar, probably in 1946 or 1947. The two concrete blocks exist today at the Boston Harbor Islands National Park, Lovell's Island unit. The battery site is open to the public.

Fort Strong (1867-1947) is located on Long Island Head of Long Island in Boston Bay. Site of several 1870s Period fortifications, the major modern-era defenses on this island were begun in 1893 as part of the Endicott Program. The fort's primary armament was five 10-inch disappearing guns organized into two batteries. Four rapid fire gun batteries provided coverage against small naval craft. The fort also was equipped with a controlled submarine mine complex. The two-gun AA Battery Three was built in the 1920s, expanded to three guns in 1935, located near Battery Taylor. A mine casemate was built in 1906, which commanded the northern channel (President Roads) mine fields until replaced by Fort Dawes in 1944. It was named in General Orders No. 134 on July 13, 1899 for Maj. Gen. George C. Strong, USV for his Civil War Service. Disarmed by 1943, the fort was declared surplus in 1947. The island was also used as a Nike launch area from 1958 to 1963 on its western end. The City of Boston developed a chronic disease hospital on the western end of island along with an existing homeless shelter (1928) and a treatment center for alcoholics in 1941. Access to the island was by ferry until a bridge was built in 1951 from Moon Island, but this bridge to the mainland was declared structurally unsound in 2014 and was destroyed in 2015. The loss of bridge access resulted in closure of the Boston Health Commission's facilities on the island. The city built a new dock and a teenagers' day camp on Long Island called Camp Harbor View in March 2007 on part of the former garrison area of the fort. The City of Boston is trying to rebuild the bridge and reopen the medical facilities on the island. The fortifications remain in abandoned state. Access to the island is by boat only and arrangements must be made to visit the fortifications.

Fort Strong Gun Batteries

- **WARD – HITCHCOCK:** A battery for five 10-inch disappearing guns erected as the first modern emplacement at Long Island Head's Fort Strong. It was one of the very first Endicott emplacements for Boston. The initial plan for a single disappearing gun was made on October 11, 1892. It was to become the most easterly of what would be a continuous five-gun battery. It consisted of a single platform with an adjacent, lower magazine on its left flank, ammunition service being with hoists, with a field of fire to the north. It was on the side of the high drumlin at the north end of the island, just a little to the rear of the old 1870s earthen battery at the same fort. Concern over the stability of the earth and sand drumlin prompted the use of extra concrete in place of sand fill, and inclusion of large boulders in the concrete. Work was done in 1893, then suspended until 1895 when the Model 1894 10-inch disappearing carriage was adopted, and the racer and base ring could be laid. On July 2, 1896 three additional emplacements to the left were approved for the Model 1894 carriages. These were placed immediately adjacent to No. 1, though the crest was increased on these three; they were higher than emplacement No. 1. Work was done in 1896-1897. Finally, a fifth emplacement, this time for a Model 1896 carriage was added on the far left (No. 5 emplacement) in 1897. Also, as the emplacements progressed east to west, they were arranged in withdrawn echelon, progressively placed to the rear of the emplacement adjacent to the right. The four guns on the left were of the same crest height. Work was completed for transfer on October 21, 1899 at a cost of $269,633.53. All five emplacments were initially named Battery Ward on General Orders No. 78 of May 25, 1903 for Major General Artemas Ward of the Continental Army. In 1906 the battery was split tactically, the three western mounts becoming Barry Hitchcock. It was named in General Orders No. 20 of January 25, 1906 for Brevet 1st Lieutenant John Ford Hitchcock of who was killed in action at Stones River TN in 1862. The emplaced armament was for Battery Ward was two 10-inch Model 1888 Watervliet guns on Model 1894 disappearing carriages (#27/#30 and #39/#31). For Battery Hitchcock there were three Model 1888 Watervliet guns on two Model 1894 and one Model 1896 disappearing carriages (#53/#32, #43/#33 and with the M1896 #31/#12). The battery received numerous modernizing changes in 1900-1915, including adjustments to the

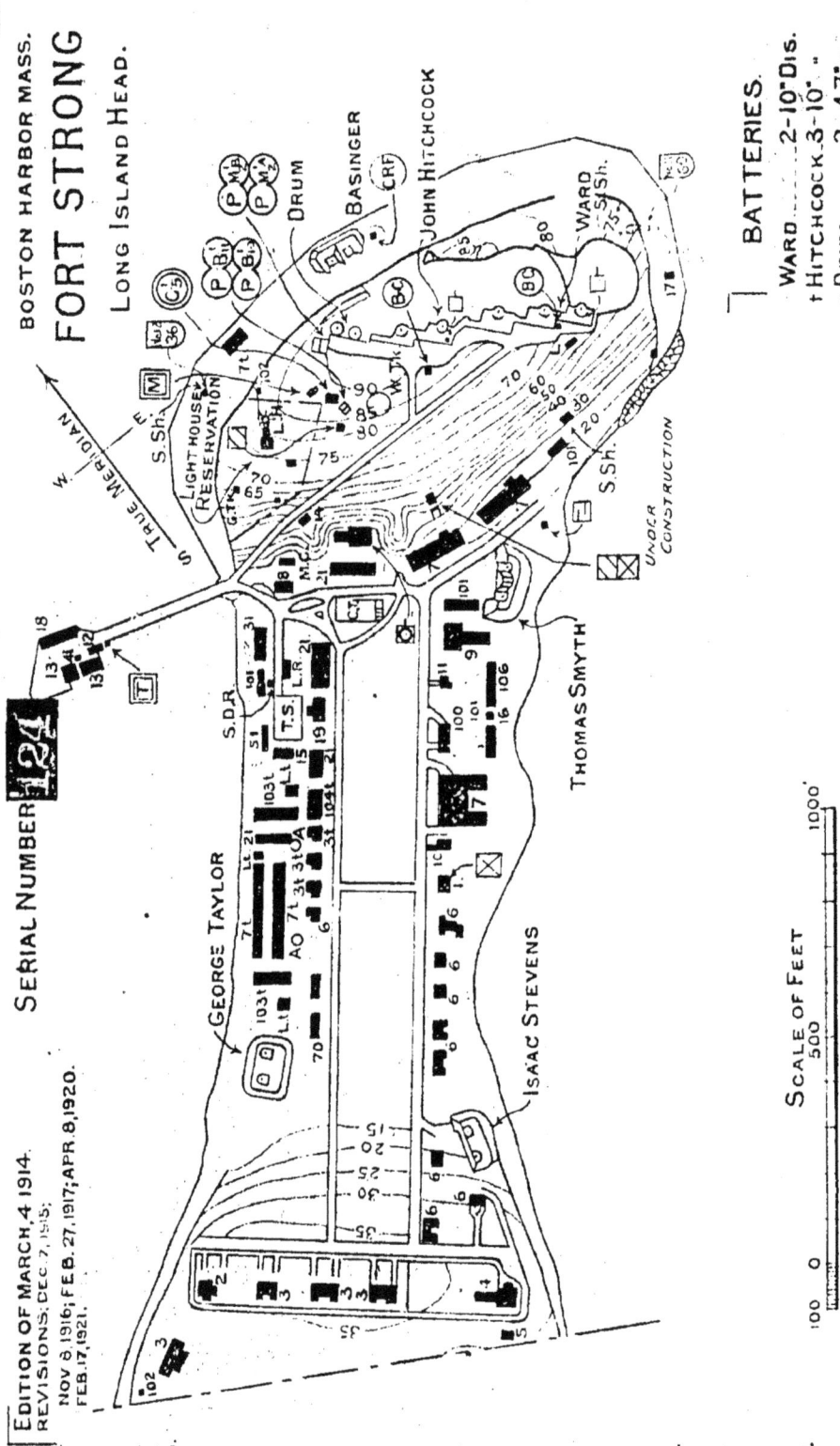

BOSTON HARBOR MASS.

FORT STRONG

LONG ISLAND HEAD.

SERIAL NUMBER 124

EDITION OF MARCH, 4 1914.
REVISIONS; DEC. 7, 1915;
NOV. 8, 1916; FEB. 27, 1917; APR. 8, 1920.
FEB. 17, 1921.

GEORGE TAYLOR

ISAAC STEVENS

THOMAS SMYTH

JOHN HITCHCOCK

BASINGER

DRUM

WARD

LIGHT HOUSE RESERVATION

UNDER CONSTRUCTION

SCALE OF FEET

100   0   500   1000'

BATTERIES.

WARD......2-10"DIS.
†HITCHCOCK..3-10"-
×DRUM......2- 4.7"-
×BASINGER..2- 3"B.P.
SMYTH.....2- 3" P.
STEVENS...2- 3" P.
TAYLOR....2- 3" P.

†*1 GUN ABSENT
* 2 GUNS ABSENT
A ANTI-AIRCRAFT GUNS

LEGEND

1 ADMINISTRATION BLDG.
2 COMMANDING OFF. QRS.
3 OFFICERS QRS.
31 TEMP OFFICERS QRS.
4 HOSPITAL.
5 HOSPITAL STWD'S. QRS.
6 N.C.O.QUARTERS.
7 BARRACKS.
7t BARRACKS.
8 GUARD HOUSE.
9 POST EXCHANGE.
10 FIRE APPARATUS.
11 SCHOOL HOUSE.
12 WAITING ROOM.
13 BOAT HOUSE.
14 ENGINE HOUSE.
15 MILITIA STOREHOUSE.
16 SCALES.
17 CREMATORY.
18 COAL SHED.
19 BAKERY.
100 STABLE.
101 SHEDS.
102 MINE PRACTICE BASE-
   END STATIONS.
103 MESS.
21 Q.M. & C.S. ST. HO.
31 ORDNANCE ST. HO.
41 ENGINEER DEPT. ST. HO.
51 SIGNAL CORPS. ST. HO.
70 SERVICE CLUB.
104 TAILOR SHOP.
105 CARPENTER AND
   BLACKSMITH SHOP.

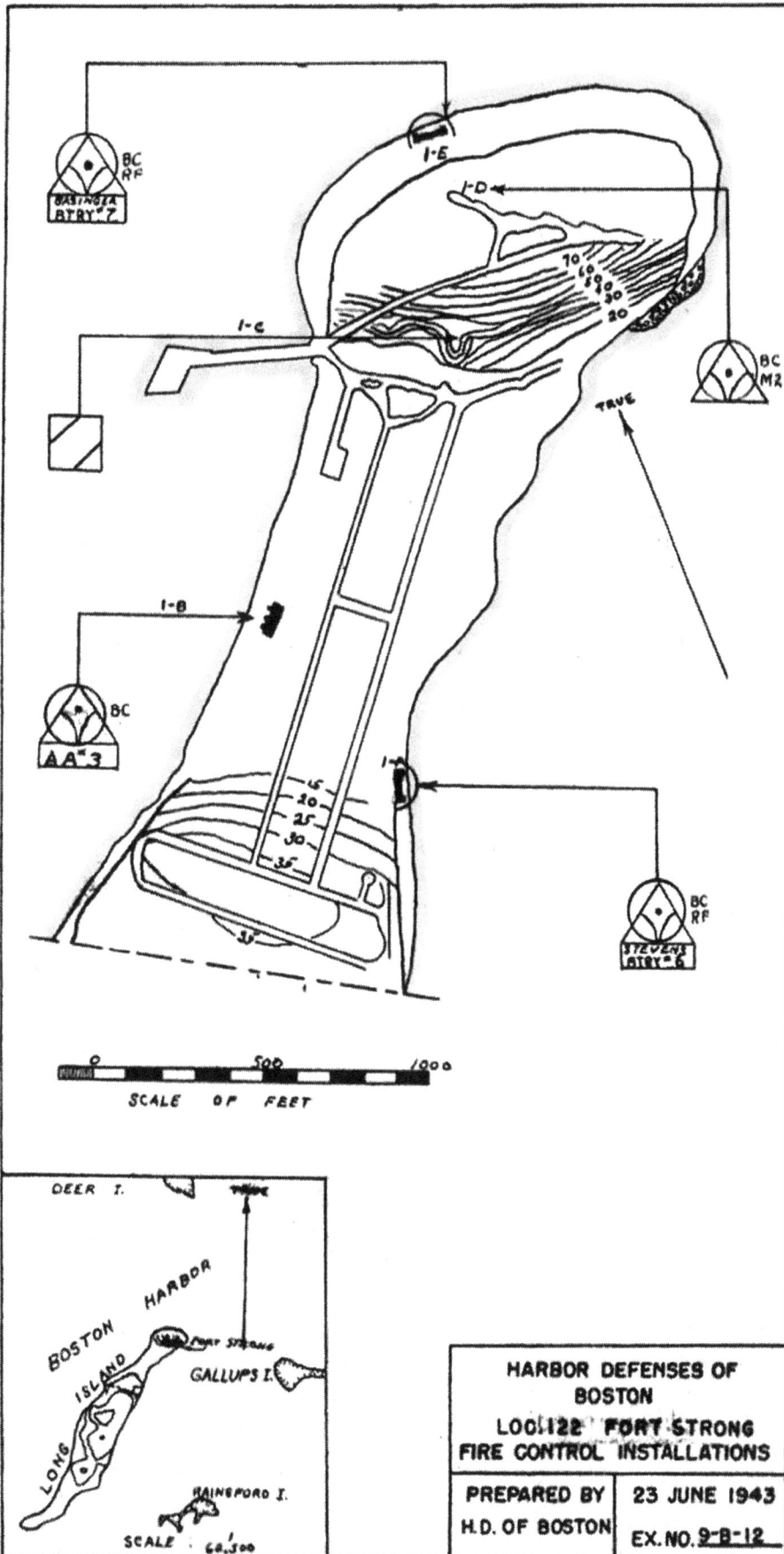

I-E

BC
RF
BASINGSA
BTRY 7

I-D

70
60
50
40
30
20

I-C

BC
M2

TRUE

I-B

BC
AA 3

5
10
25
30
35

37

I-A

BC
RF
STEVENS
BTRY 6

0   500   1000

SCALE   OF   FEET

DEER I.
TRUE

BOSTON HARBOR

FORT STRONG

LONG ISLAND

GALLUPS I.

RAINSFORD I.

SCALE   60,500

HARBOR DEFENSES OF BOSTON	
LOC. 122 FORT STRONG FIRE CONTROL INSTALLATIONS	
PREPARED BY H.D. OF BOSTON	23 JUNE 1943 EX. NO. 9-B-12

REVISED DATE: 31 JAN 45

Fort Strong 1920s (NARA)

Fort Strong 1930s (NARA)

loading platforms, hoists, electrification, and battery commander walks and stations. During World War I it was proposed to remove all five guns, but in fact only one was removed (Hitchcock No.1 carriage tube serial #53) on June 24, 1918. The batteries continued to serve until World War II as two batteries of two guns each. The battery was authorized for disarmament and abandonment about 1940 and the guns and carriages were eventually removed. The emplacement still exists, though of limited access on the land of the Boston Public Health Commission. The battery site is closed to the public.

- **DRUM**: An emplacement for two 4.72-inch rapid-fire guns emplaced at Strong a part of the urgent Spanish-American War defenses. Local engineers were instructed to plan and emplacement for two recently purchased Armstrong guns on April 18, 1898. A site was selected on the immediate left flank of 10-inch disappearing gun No. 5 on the drumlin at the north end of the island. It was built at the same crest height (105-feet) as these guns on the flank. Otherwise, the emplacement followed standard design criteria with two platforms and a central, lower-level magazine to serve the guns. It was built using the National Defense Act funding of March 1898, the work following that same summer. It was ready for transfer on October 21, 1899 for a cost of $14,737.39. The battery was named on General Orders No. 78 of March 25, 1903 for Captain Simon Drum, killed in action in 1847 during the attack on Mexico City. It was armed with two 4.72-inch Armstrong guns on pedestals (#11855/#10843 and #11856/#10842). These served until the guns and carriages were shipped out for use at the Sachuest Point Battery in 1917. The battery was not subsequently used for armament purposes. The emplacement still exists, though of limited access on the land of the Boston Public Health Commission. The battery site is closed to the public.

- **BASINGER**: An emplacement for two 3-inch guns on masking parapet mounts emplaced at Fort Strong. Needed to cover the main mine field to the north of the fort, the site selected was at the foot of the bluff near the water level at the northern tip of the island. Because of the need to cover the waters immediately adjacent, it had to be placed below the hill virtually at the seawall on the shoreline. Plans for the emplacement were submitted on April 25, 1898. It was an unusual design. As the bluff extended right down to the level of the platforms, access to the battery was only on the sides, through entrances cut through protected traverses on either flank. The guns were to be mounted on the roofs of the magazines, requiring the use of small hoists to bring full boxes of fixed rounds to the level of the gun platform. Work was done in 1898-1898 and transferred on March 6, 1901 for a construction cost of $18,780. It was armed with two 3-inch, 15-pounder Model 1898 guns on masking parapet mounts (Driggs-Seabury #34/#34 and #36/#36). It was named on General Orders No. 78 of May 25, 1903 for 2nd Lieutenant William E. Basinger, killed in action during the Seminole War in 1835. In 1918 the balanced pillar mounts were modified into non-retracting Model 1898M1 pedestals. The guns and carriages were declared obsolete and removed in mid-1920. Very soon this important battery location was rearmed with new guns. In 1921 two 3-inch Model 1902M guns and pedestals (#43/#43 and #44/#44) were transferred here from Battery Smyth also at Fort Strong. After slight modifications to accommodate the new bare rings, the new armament was up in November of 1921, and a new adjacent BC/CRF was also constructed. This now served through the 1930s and 1940s, being removed not until early postwar. The emplacement still exists, though of limited access on the land of the Boston Public Health Commission. The battery site is closed to the public.

- **SMYTH**: An emplacement for a pair of 3-inch rapid-fire, pedestals guns approved for the south shore of Long Island Head. It was sited at the base of the northern drumlin and the southern shore in the protected curve of the shoreline. It fired to the southeast, covering the Hubble Channel and

Fort Strong on Long Island (Terry McGovern)

Fort Warren on Georges Island (Terry McGovern)

the southern reaches of the bay. Plans were submitted on June 2, 1903. It was of modified type plans, needing strong retaining walls in front and rear due to the position relative to the sea on potential enemy fire. Work was done in 1903-1904, for transfer on May 31, 1906 at a cost of $16,000. It was named on General Orders No. 194 of December 27, 1904 for Brevet Major General Thomas A. Smyth who was mortally wounded near Farmsville, VA in 1865. It was armed with two 3-inch guns and pedestal mounts Model 1902 (#43/#43 and #44/#44). The battery was disarmed after World War I in order to re-locate the armament to the position of Battery Basinger at this same fort in November 1921. The emplacement still exists, though of limited access on the land of the Boston Public Health Commission. The battery site is closed to the public.

- **STEVENS**: One of two dual 3-inch emplacements submitted on June 24, 1903 for Fort Strong. It was located on the southeastern extreme of the reservation, firing to the east. It covered mine fields, but as importantly approaches for small enemy boat parties to the south of the island. There was enough room to allow the adoption of the standard plan for the design, but an extension of the seawall was needed in front of the battery. Work was done in 1903-1904, for transfer on May 31, 1906 at a cost of $21,500. It was named on General Orders No. 194 of December 27, 1904 for Brevet Major General Isaac I. Stevens killed in action at Second Bull Run in 1862. It carried two 3-inch Model 1902 guns on M1902 pedestal mounts of Bethlehem Steel manufacture (#47/#47 and #48/#48). This armament was carried throughout the service life of the battery, which was a long one. It was retained in the postwar reviews, and only finally deleted in 1946. The emplacement still exists, though of limited access on the land of the Boston Public Health Commission. The battery site is closed to the public.

- **TAYLOR**: The fourth and final 3-inch gun battery intended for Fort Strong at Long Island Head. It was to be emplaced on the lower west side of the island's neck, on the lowland near the reservation boundary. There was enough room to use the standard type plan for its design. The plan was submitted on June 24, 1903. Concrete work was done in 1903-1904. Transfer was made on May 31, 1906 for a cost of $21,000. It was named on General Orders No. 194 of December 27, 1904 for Brigadier General George W. Taylor mortally wounded at Second Manassas in 1862. It was armed with two 3-inch Model 1902M2 guns and M1902 pedestal mounts (#45/#45 and #46/#46). The armament was retained here for a number of years, eventually being relocated to a new Battery Taylor at the tip of Deer Island on the Fort Dawes reservation in 1942. The emplacement still exists, though of limited access on the land of the Boston Public Health Commission. The battery site is closed to the public.

Fort Warren (1833-1950) is located on Georges Island in Boston Bay. A two-level, pentagon-shaped granite fort of the Third System dominates the island. Begun in 1837 and completed in 1861, Fort Warren covered the main shipping channel into Boston Harbor. It was named in General Orders 32, on Apr. 18, 1853 for Maj. Gen. Joseph Warren, Continental Army killed in action at the Battle of Bunker Hill on June 17, 1775. The fort was modernized under the Endicott Program by the addition of new batteries in and around the old fort from 1892 to 1903. The fort's primary armament was two 12-inch disappearing guns and five 10-inch disappearing guns. The fort also hosted a controlled submarine mine complex and was the headquarters for the mine command for Boston's harbor defenses. Disarmed by 1944, the fort was declared surplus in 1958 transferred to the State of Massachusetts. Today the fort is the center piece of the Boston Harbor Islands NRA and is accessible by the Boston Harbor Islands Ferry. Fort Warren retains all its concrete emplacements, but access is restricted, as well as a number of its support buildings, including the powerhouse and the torpedo storehouse. The site hosts a number of Summer events and has an active interpretive program and a visitors' center.

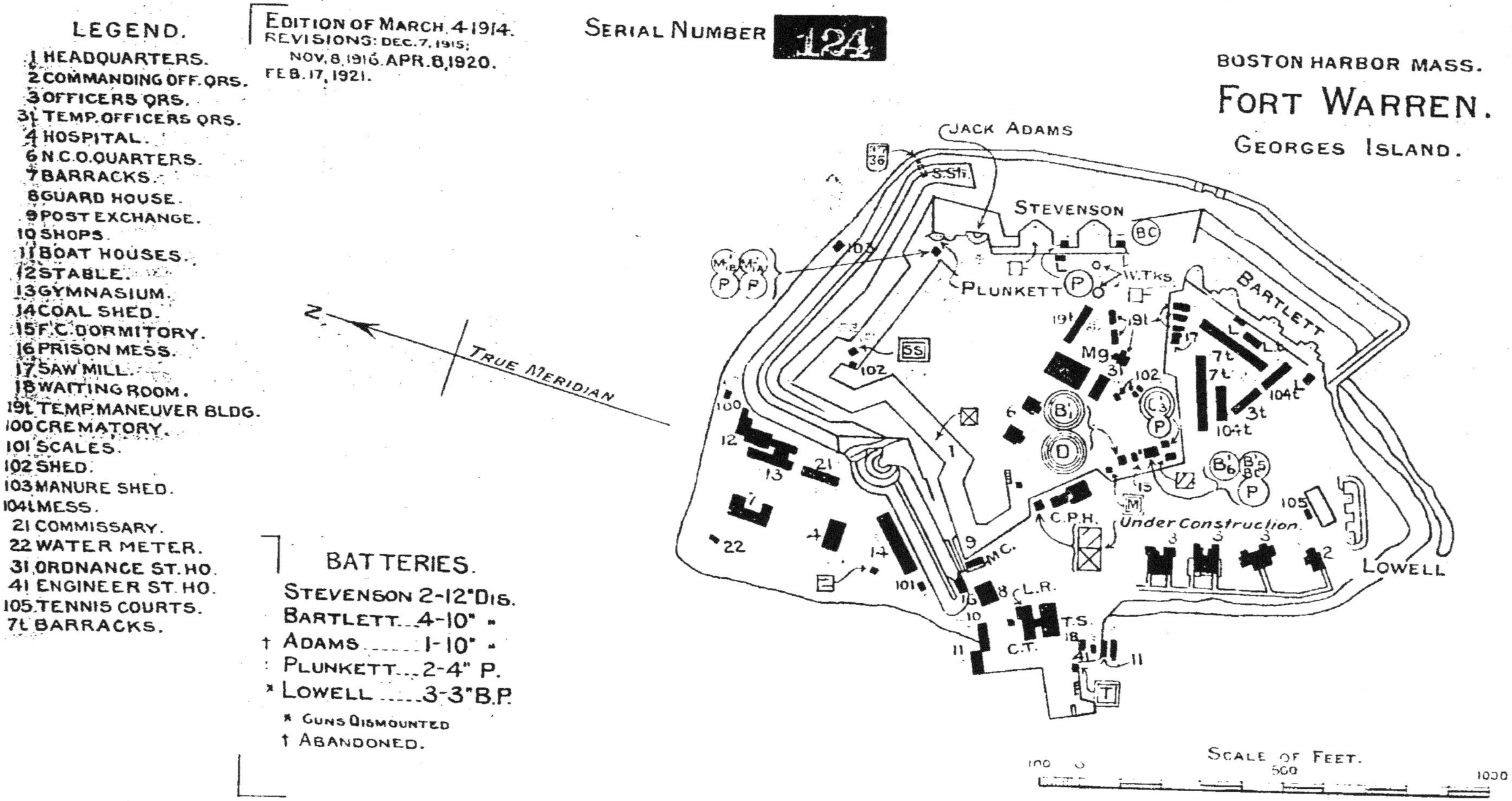
BOSTON HARBOR MASS.
FORT WARREN.
GEORGES ISLAND.
SERIAL NUMBER 124
EDITION OF MARCH 4 1914.
REVISIONS: DEC. 7, 1915; NOV. 8, 1916. APR. 8, 1920. FEB. 17, 1921.
LEGEND.
1 HEADQUARTERS.
2 COMMANDING OFF. QRS.
3 OFFICERS QRS.
3t TEMP. OFFICERS QRS.
4 HOSPITAL.
6 N.C.O. QUARTERS.
7 BARRACKS.
8 GUARD HOUSE.
9 POST EXCHANGE.
10 SHOPS.
11 BOAT HOUSES.
12 STABLE.
13 GYMNASIUM.
14 COAL SHED.
15 F.C. DORMITORY.
16 PRISON MESS.
17 SAW MILL.
18 WAITING ROOM.
19t TEMP. MANEUVER BLDG.
100 CREMATORY.
101 SCALES.
102 SHED.
103 MANURE SHED.
104t MESS.
21 COMMISSARY.
22 WATER METER.
31 ORDNANCE ST. HO.
41 ENGINEER ST. HO.
105 TENNIS COURTS.
7t BARRACKS.
BATTERIES.
STEVENSON 2-12" DIS.
BARTLETT 4-10" "
† ADAMS 1-10" "
PLUNKETT 2-4" P.
* LOWELL 3-3" B.P.
* GUNS DISMOUNTED
† ABANDONED.
N
TRUE MERIDIAN
JACK ADAMS
STEVENSON
PLUNKETT
BARTLETT
LOWELL
W. TKS.
C.P.H.
Under Construction.
L.R.
T.S.
C.T.
M.C.
S.Sh.
SCALE OF FEET.
500
1000

TRUE

AA1½

SS

G5
MINES

I·A

I·E

PM₃

I·D

I·B

SEA WALL

M1

M1

SCALE IN FEET

100   0   100   200   300   400   500

TRUE

BOSTON HARBOR

LOVELL IS.

THE NARROWS

GALLUPS IS.

LONG ISLAND

FORT
WARREN

GEORGES IS.

RAINSFORD IS.

SCALE: $\frac{1}{62,500}$

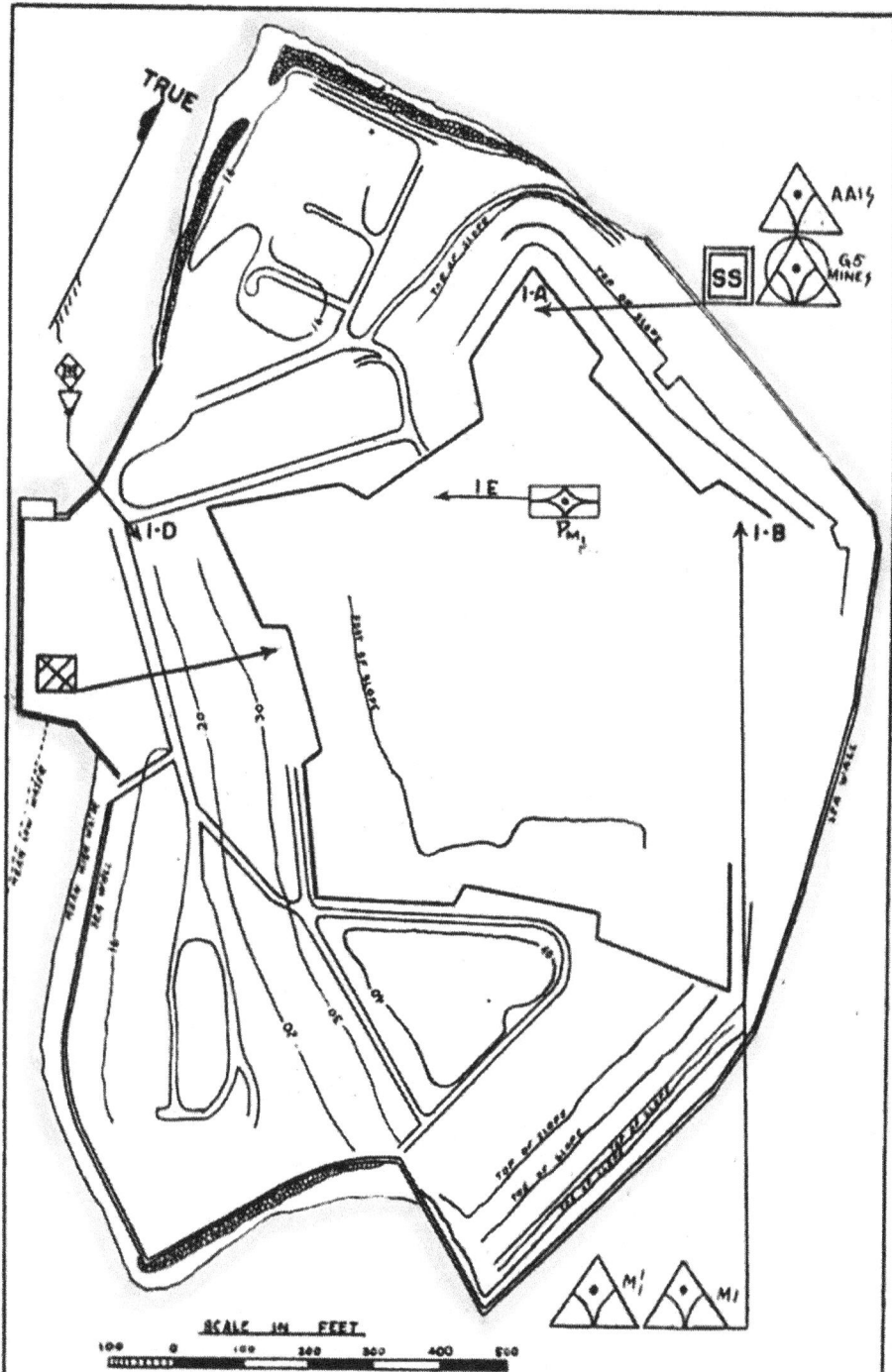

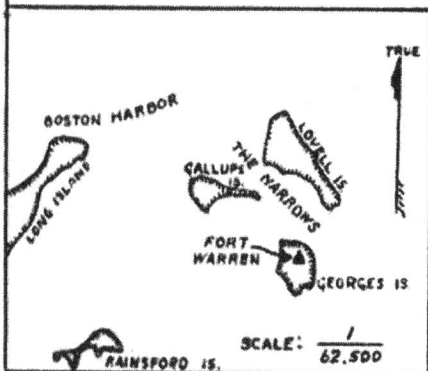

HARBOR DEFENSES OF
BOSTON
LOC. 121 FORT WARREN
FIRE CONTROL INSTALLATIONS

| PREPARED BY | 24 JUN. 1943 |
| H.D. OF BOSTON | EX. NO. 9-B-11 |

REVISED DATE: 31 JAN 45

FORT WARREN
MASS.
ALT. 4800 FEET
PHOTOGRAPHED BY US ARMY AIR SERVICE
MAY 1924   NEG. NO. 126
8TH PHOTO SECTION

Fort Warren 1930s (NARA)

Fort Warren Gun Batteries

- **JACK ADAMS:** An early emplacement for a single 10-inch disappearing gun emplaced at Fort Warren, George's Island, Boston Harbor. This was one of the first Endicott batteries to be funded, in fact it was originally started for an 8-inch disappearing gun as no approved carriage for any size gun of this type had yet to be adopted. Funds from the Act of August 18, 1890 were used to start this battery. Initial submission of a plan was made on November 12, 1890. At that time five disappearing guns were envisioned for Fort Warren—four along the line of the exterior ravelin, and one inside the old fort enceinte at bastion B. The latter was started first as it was deemed to have the more important field of fire. Designed when there were yet any mimeograph type plans, its plans were highly adapted to the site. The platform was constricted (only 52-foot across) and the magazines on the lower level poorly placed and served only by crane ammunition lifts. Major concrete work was done in 1891-1892, but it had to wait for its armament to become available and to receive its finishing work. It was eventually transferred on October 21, 1899 for a cost of $92,138.16. The armament was one 10-inch Model 1888 Watervliet gun on a Model 1894 disappearing carriage (28/#26). The battery was named on General Orders No. 16 of February 14, 1902 for Captain John G. B. Adams, a medal of honor recipient during the Civil War. In 1898 additional funding was requested to finish the ammunition handling facilities, pour concrete in the floors, and install doors and railings. The early design and poor quality natural, Rosendale cement led to poor serviceability, and the battery was soon essentially abandoned. Consequently, it was never modernized with extended platforms, chain hoists, or a plot and battery commander's station. It was authorized for deletion on January 13, 1914, and the armament removed before World War I. The original emplacement, in poor condition, still exists at the Boston Harbor Islands National Recreation Area. The battery site is closed to the public.

- **BARTLETT:** A battery for four 10-inch disappearing guns emplaced across the former ravelin and Bastion A of old Fort Warren. This work was an early planned element of the Boston defenses, being approved in the plans of 1890-1891. The first two emplacements were covered in a design plan submitted on December 2, 1891. The line of four guns was to extend along the eastern scarp line of the old ravelin, with the most northern two emplacements crossing the ditch and extending into Bastion A of the old masonry work. As less destruction was required for the building of the two southern emplacements, they were begun first. Work was done in 1892-1893, platforms were without anchor rings until an appropriate disappearing carriage was adopted. It followed conventional design plans. On April 21, 1898 plans were submitted for the second pair, the guns for emplacements No. 3 and 4. Funding was with the National Defense Act of March 1898. Much of the masonry had to be removed from the old fort, and the ditch filled in. The entire battery was turned over on October 21, 1899 for a construction cost of $236,105.68. It was named on General Orders No. 16 of February 14, 1902 for Brevet Major General William F. Bartlett of Civil War service. The first two emplacements were armed with 10-inch Model 1888 Watervliet guns on LF Model 1894 disappearing carriages (#26/#29 and #29/#25). The two northern emplacements were armed with 10-inch Model 1888 Bethlehem guns on LF Model 1896 disappearing carriages (#27/#36 and #26/#35). This armament was removed in 1917 (emplacements No. 2 and No. 4) and 1918 (emplacements No. 1 and No. 3). However just a year later, as soon as March 17, 1919 it was being rearmed, this time with all Model 1888 Bethlehem gun tubes (#15/#29, #16/#25, #23/#36, and #34/#35). This final armament was scheduled for deletion under the 1932 Review, but only older two emplacements with Model 1894 carriages were actually removed around 1936. Emplacements No. 3 and No. 4 were kept in reserve status until authorized for removal on November 4, 1942.

In later years emplacement No. 4 was destroyed by local park authorities in 1972 as a access way. The remaining three emplacements still exist at the park. The battery site is closed to the public.

- **STEVENSON**: A battery for two model 1897 12-inch disappearing gun carriages emplaced along the main seafront curtain wall of old Fort Warren. The plan was submitted on October 19, 1898 with funds of the Fortification Act of May 7, 1898. The emplacement was placed parallel to the old curtain wall between Bastions A and B, though it extended slightly beyond the scarp to the front. Final plans called for just two guns, versus three originally anticipated; this allowed a better fit closer to type plans for distance between platforms and arrangement of lower-level magazines served with hoists. Work was done in 1899-1900. Transfer was made on January 20, 1903 for a cost of $144,495. It was named on General Orders No. 16 of February 14, 1902 for Brigadier General Thomas G. Stevenson, killed in action at Spotsylvania in 1864. The armament was two 12-inch Watervliet tube guns Model 1895 on Model 1897 disappearing carriages (#33/#10 and #34/#11). This armament was retained throughout the battery's service life. Modifications to the platforms and hoists were made after construction, and the carriages were given increased elevation in 1916-1917. It was slated for removal with the completion of more modern 1940 Program elements, but its critical position kept it armed up until 1945. The emplacement still exists at the Fort Warren Park. The battery site is open, but interior is closed to the public.

- **PLUNKETT**: A battery for two rapid-fire guns emplaced during the Spanish American War emergency at Fort Warren. Originally Warren was allocated two of the recently purchased Armstrong 4.7-inch guns at the start of the 1898 conflict. However due to a shortage of this gun type, two substitute navy-type 4-inch Driggs-Schroder pedestal guns were emplaced on May 3, 1898. An emplacement plan was submitted on June 3, 1898. They were emplaced in Bastion B on the immediate left flank of 10-inch Battery Adams. Two platforms were built on the parapet at the same crest height as the adjoining disappearing battery. It had a single magazine between the two platforms. The battery fired to the east, covering Nantasket Roads and the Narrows. Work was quickly done, and transfer was made on October 21, 1899 for $9,083.90. It was named on General Orders No. 16 of February 14, 1902 for Sergeant Thomas Plunkett, a Civil War Medal of Honor awardee. It carried two 4-inch Model 1896 guns and pedestal mounts (#1/#1 and #2/#2). It was one of only two batteries built for this weapon. The emplacement was not entirely successful, a leaking, wet magazine plagued it for several years. The armament was declared obsolete in the early 1920s and was dismounted and displayed on the entryway to the fort during the 1930s and 1940s. At some point the guns were moved outside the main fort as ornaments and much later scrapped. The emplacement still exists at the Fort Warren Park. The battery site is open to the public.

- **LOWELL**: A battery for three 3-inch masking parapet guns emplaced at Fort Warren to protect the mine field in Nantasket Roads. The emplacement plan was submitted and approved on June 19, 1899. It was emplaced at the southern tip of the island, beyond the officers' quarters near the beach, firing to the south to cover the minefield. It followed type plans for masking parapet guns, with individual platforms and magazines at a lower level but still with hand-delivery ammunition service. The right flank, No. 1 position, was given a wider field of fire to the southwest. Work was done in 1899. Transfer was made on June 12, 1900 for a cost of $12,750. The battery was named on General Orders No. 16 of February 14, 1902 for Brigadier General Charles Russell Lowell, a Massachusetts cavalry officer mortally wounded at Cedar Creek, VA in 1864. It was armed with three 3-inch Model 1898 guns and balanced pillar mounts (#31/#31, #32/#32, and #33/#33). The pillars were modified in 1916 to become Model 1898M1 pedestal mounts. The armament was removed in 1920. The emplacement still exists at the Fort Warren Park. The battery is open to the public.

Point Allerton Military Reservation: Endicott-era fire control site

EDITION OF MARCH 4 1914.
REVISIONS; FEB. 17, 1921.

SERIAL NUMBER

BOSTON HARBOR, MASS.

POINT ALLERTON

CABLE LINE.
GENERAL MAP.

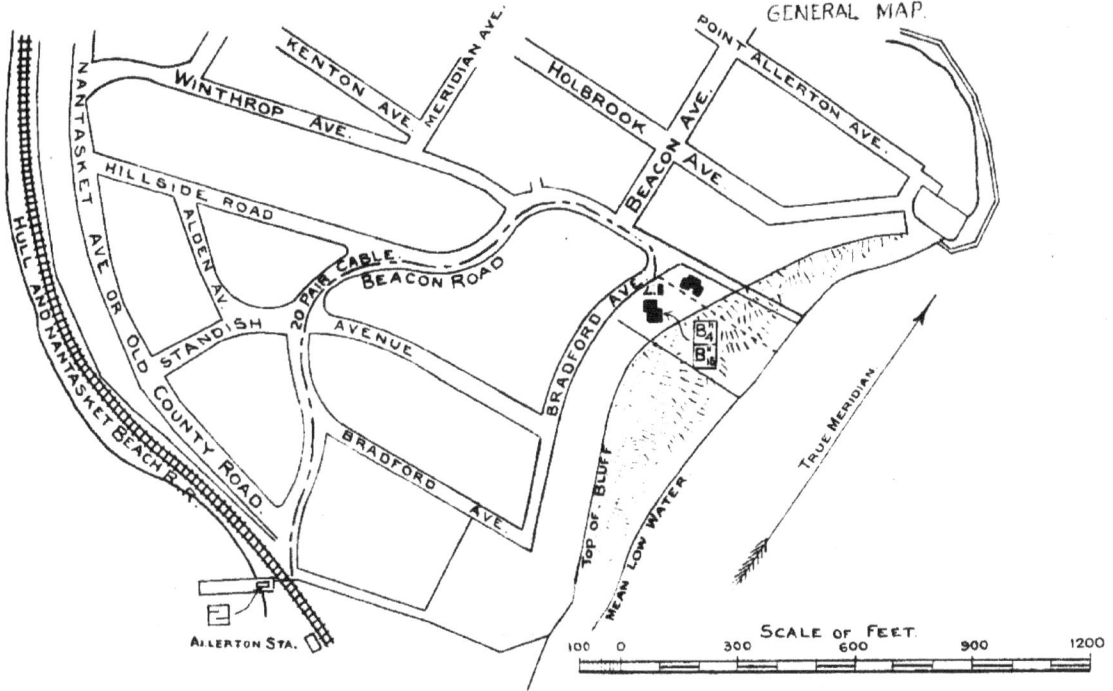

KENTON AVE.

MERIDIAN AVE.

HOLBROOK AVE.

POINT ALLERTON AVE.

WINTHROP AVE.

BEACON AVE.

NANTASKET AVE. OR OLD COUNTY ROAD

HILLSIDE ROAD

ALDEN AV.

20 PAIR CABLE.

BEACON ROAD

STANDISH AVENUE

BRADFORD AVE.

B"4
B"5

HULL AND NANTASKET BEACH R.R.

BRADFORD AVE.

TOP OF BLUFF

MEAN LOW WATER

TRUE MERIDIAN

ALLERTON STA.

SCALE OF FEET.

100 0 300 600 900 1200

EDITION OF MARCH 4 1914.
REVISIONS; FEB. 17, 1921.

SERIAL NUMBER

BOSTON HARBOR, MASS.

POINT ALLERTON D-1.

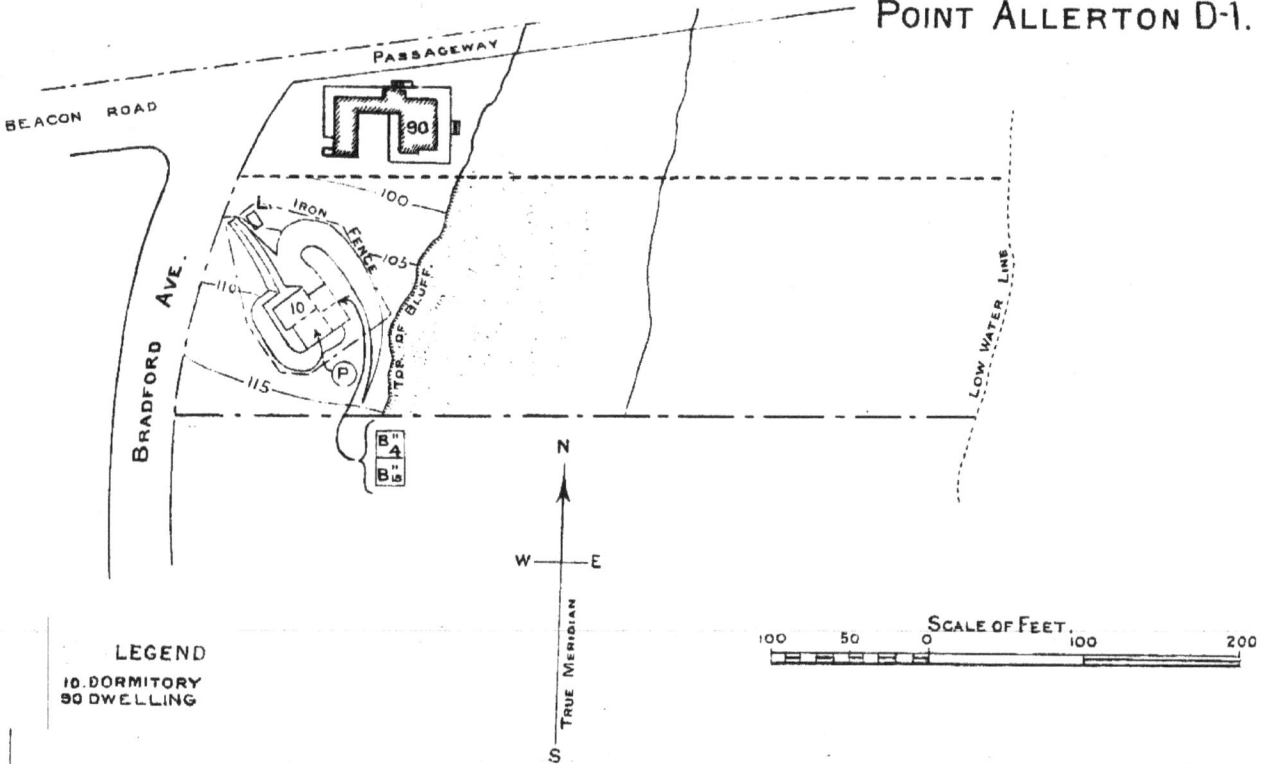

PASSAGEWAY

BEACON ROAD

90

100

IRON FENCE

105

BRADFORD AVE.

110

10

115

TOP OF BLUFF

LOW WATER LINE

B"4
B"5

N

W — E

TRUE MERIDIAN

S

LEGEND

10. DORMITORY
90 DWELLING

SCALE OF FEET.

100 50 0 100 200

Fort Revere (1900-1947) is located on Nantasket Head near the town of Hull. Site of several colonial defensive works; it was not until 1897 that the next defense work was constructed as part of the Endicott Program. It was named in General Orders No. 43 on Apr. 4, 1900 for Lt. Col. Paul Revere for his messenger ride and service in the Continental Army. The fort's primary armament was two 12-inch disappearing guns and six 6-inch disappearing guns (located on Telegraph Hill). The fort also had 5-inch balanced pillar battery (guns removed in 1917), and 3-inch antiaircraft battery installed in 1935. During World War II, a 90mm AMTB battery was added. The fort was declared surplus in 1947 and most of fort is now in private ownership. One officers quarters remains as a private residence. A small city park surrounds the 6-inch emplacements, the 12-inch and 3-inch emplacement has been buried.

Fort Revere Gun Batteries

- **RIPLEY**: An Endicott battery for two 12-inch barbette guns authorized with National Defense Act funds of March 1898. The plan was submitted on June 7, 1898. The battery, even though a little low with a crest elevation of just 96-feet, was assigned two of the new batch of relatively quickly-produced 12-inch barbette carriages. The location was on Cushing Hill, a high point in Hull. It was emplaced centrally on the reservation at Nantasket Head in the town. It was of fairly standard plan type with magazines below and on the left traverses of each gun platform, and it was sunk several feet to improve the earth protection in the front of the parapet. It fired almost directly north. Work was done in 1898-1899 and transfer made on January 26, 1901 for a construction cost of $110,645.58. It was named on General Orders No. 78 of May 25, 1903 for Major General Eleazer W. Ripley of War of 1812 service. It was armed with two 12-inch Model 1888 guns on Model 1892 barbette carriages (Bethlehem gun #8/carriage #25 and Watervliet #18/carriage #26). The battery had a long service life. It was retained for the 1932 Review and finally deleted under authorization of May 27, 1943. After the fort reservation was sold for private development, much of the emplacement was filled in and buried. Some of the lower magazines are used locally for storage. The battery remnants are on private property. The battery site is closed to the public.

- **FIELD**: An Endicott battery for two 5-inch guns on balanced pillar carriages. The battery plan was submitted on March 23, 1899. The site chosen was on the left side of 12-inch Battery Ripley. Its position allowed it to cover the mine fields of Nantasket Roads. Work was done in the summer of 1899, being reported completed by August 17 of that year. It was transferred with Battery Ripley of January 26, 1901 for a cost of $14,050. It was also named on General Orders No. 78 of May 25, 1903 for Captain George P. Field who was killed in action at Monterrey Mexico in 1846. It carried two 5-inch Bethlehem guns Model 1897 on Model 1896 balanced pillar carriages (#23/#3 and #24/#4). These guns were removed in September 1917 and the carriages scrapped in place in 1920. The emplacement was subsequently not used, but still exists partially destroyed and buried on private property. The battery site is closed to the public.

- **SANDERS – POPE**: A battery for six 6-inch disappearing guns emplaced around Telegraph Hill at the Fort Revere reservation. Plans for two guns (eventually becoming No. 5 and No. 6) were submitted on June 24, 1903. The plans for the final four (No. 1 through No. 4) were submitted less than a year later on May 31, 1904. These units were a continuous series, but there was a bend between No. 4 and No. 5 and the emplacement wrapped around the crest of Telegraph Hill. They closely followed the layout proscribed for Type 1903 disappearing batteries, with some rearrangement of entries into magazines and rooms due to the flank exposure to possible enemy enfilade fire. It generally fired to the north or northeast. Work was done in 1903-1904, with transfer be-

BOSTON HARBOR, MASS.

FORT REVERE

NANTASKET HEAD

SERIAL NUMBER 121

EDITION OF FEB. 17, 1921.

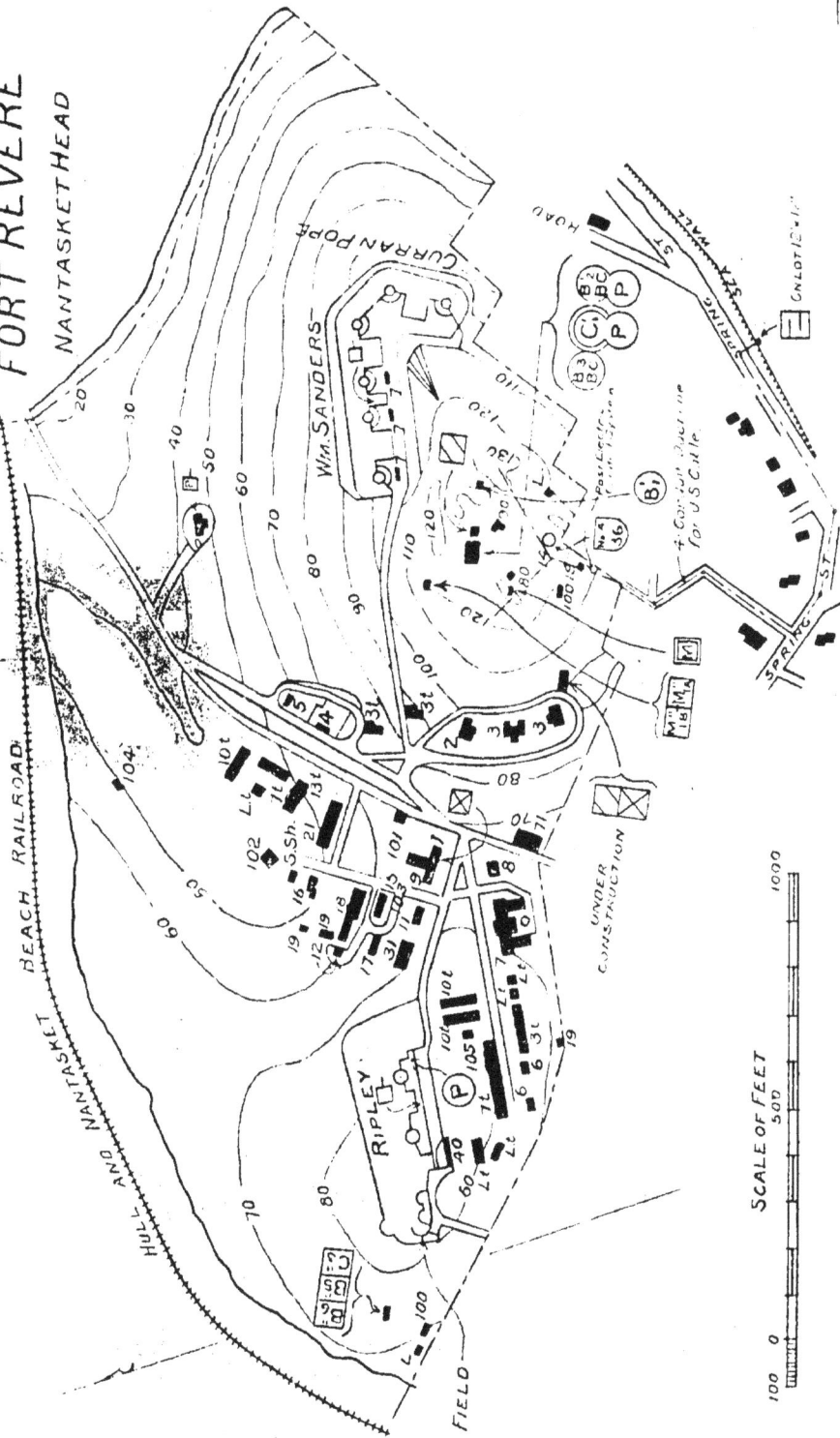

LEGEND

1. HEADQUARTERS.
2. COMMANDING OFFORS.
3. OFFICERS QRS.
3t. OFFICERS QRS. (TEMP.)
4. HOSPITAL
5. HOSPITAL STWD'S QRS.
6. N.C.O. QUARTERS.
7. BARRACKS
7t. BARRACKS (TEMP.)
8. GUARD HOUSE.
9. POST EXCHANGE.
10t. MESS HALL (TEMP.)
11. BAKERY.
12. PAINT SHOP.
13t. STORE ROOM (TEMP.)
14. STANDPIPE.
15. COAL SHED.
16. BARN.
17. SHOP.
18. STABLE.
19. SHED.
100. F.C. DORMITORY.
101. HOSE HOUSE.
102. GARAGE.
103. SCALES.
104. INCINERATOR.
105. ART. ENR. OFFICE.
21. ORDNANCE ST. HO.
31. COMMISSARY ST. HO.
40. ENG. DEPT. ST. HO.
71. GYM. SERVICE CLUB.
60. W.B. SIGNAL STAFF.
90. MARINE REPORTING STA.

BATTERIES

RIPLEY....2-12"N.DIS
SANDERS..3-6".DIS.
* POPE........3-6 "
* FIELD.....2-5"B.P.

* ARMAMENT REMOVED.

SCALE OF FEET

UNDER CONSTRUCTION

SCALE IN FEET

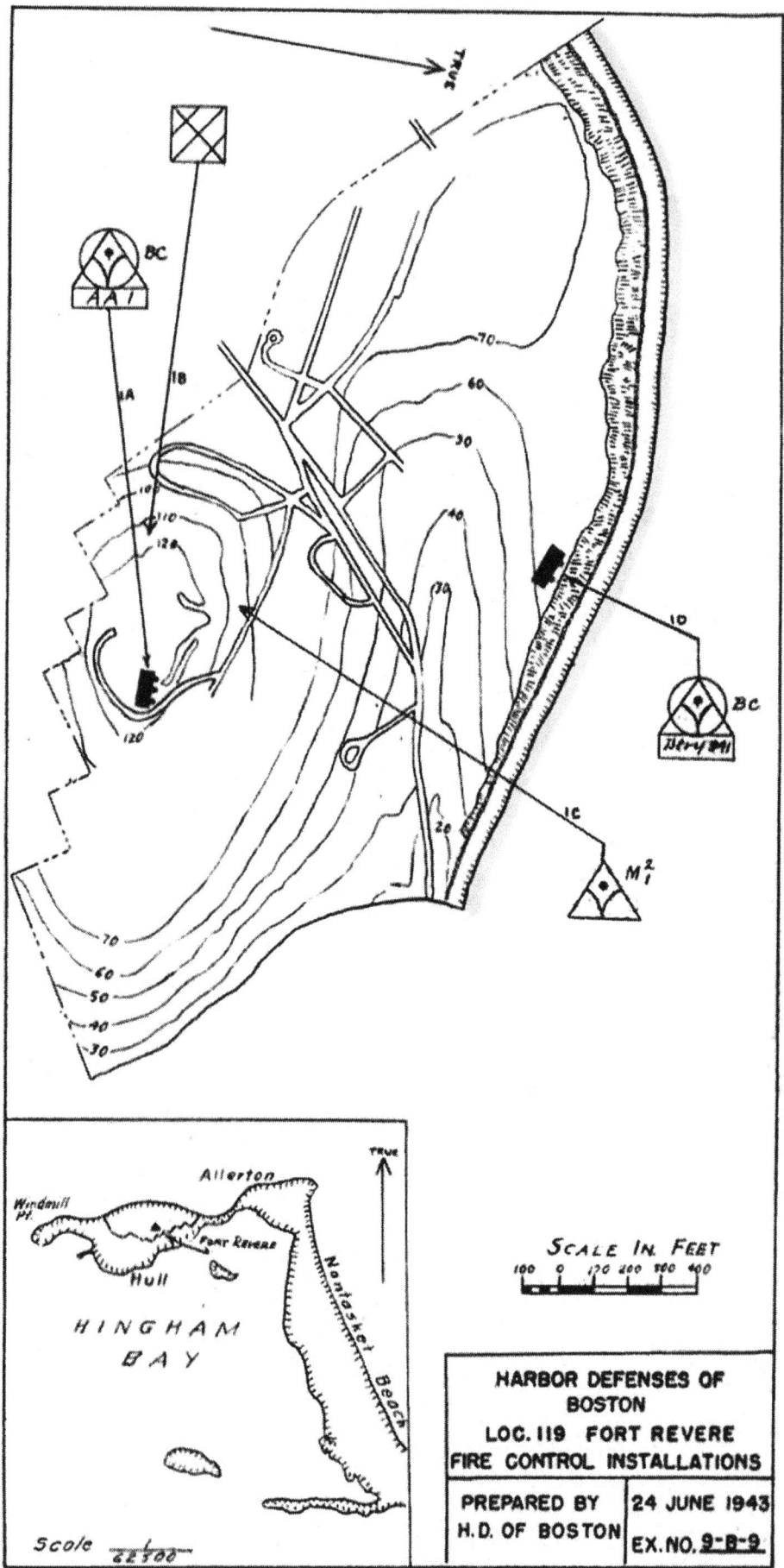

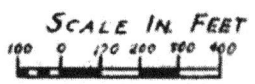

HARBOR DEFENSES OF
BOSTON
LOC. 119 FORT REVERE
FIRE CONTROL INSTALLATIONS

| PREPARED BY H.D. OF BOSTON | 24 JUNE 1943 |
| | EX.NO. 9-B-9 |

REVISED DATE: 31 JAN 45

FORT REVERE
MASS.
ALT. 6600 FEET
PHOTOGRAPHED BY US ARMY AIR SERVICE
MAY 1924 NEG. NO. 129
8TH PHOTO SECTION

Fort Revere 1930s (NARA)

Fort Revere Park, Nantasket Head, Hull (Terry McGovern)

Spinnaker Island (Fort Duvall) (Terry McGovern)

ing made on June 28, 1906 for a cost of $136,000. It was named on General Orders No. 194 of December 27, 1904 for Brevet Brigadier General William P. Sanders who was mortally wounded in Kentucky in 1863. The two eastern emplacements were tactically split off to form Battery Pope, named on General Orders No. 20 of January 25, 1906 for Colonel Curran Pope, 15th Kentucky Volunteer Infantry, who was mortally wounded at the Battle of Perryville, KY in 1862. The battery was armed with six 6-inch Model 1903 guns and disappearing carriages (Sanders: #14/#22, #47/#23, #57/#24, and #59/#25; Pope: #11/#62 and #13/#63). In 1917 the two guns from Pope and one from Sanders (#24) were removed for use on wheeled carriages. Sanders remained in service with its remaining three guns. These served until finally being authorized for removal on May 25, 1943. At that time the tubes were taken in order to use them in arming additional 1940 Program 6-inch batteries of the 200-series. The emplacement still exists at the town's Fort Revere Park. The battery is open to the public.

- **AMTB-941**: A 1943 Program AMTB battery emplaced at Fort Revere. It was authorized on April 17, 1943 and built in May and June of that year, transferring September 10, 1943 for a cost of $6856. The two fixed gun blocks were emplaced northeast of Battery Ripley near the shore, with platforms for the mobile guns outside the fixed pair. It served until disarmed in 1946. The fixed blocks still exist on private property. The battery site is closed to the public.

Fort Duvall (1920-1975) is located on Little Hog Island (now known as Spinnaker Island) just south of Hull. It was named in General Orders No. 13 on Mar. 27, 1922 for Maj. Gen. William P. Duvall U.S. Army. Constructed during 1920-1924, the battery, designed for 12-inch long-range barbette guns, was replaced with two 16-inch Army M1919 guns. These guns had "all around" fire to cover the waters of Massachusetts Bay. During World War II the battery was casemated. After World War II, the fort was used as a Nike integrated fire control site. In 1980, the 15-acre island was sold to private interest and renamed Spinnaker Island. Condominiums have been constructed over and around the battery. Fort Duvall's sole emplacement remains today in a highly modified form. The site is accessible to owners and guests only.

Fort Duvall Gun Battery

- **LONG**: A 1915 Board of Review Program dual 16-inch barbette battery for Hog Island, Boston Harbor. The original program for Boston in 1915 projected two dual 12-inch long-range gun or howitzer batteries for Boston—here at Hog Island and one at Fort Ruckman in Nahant. Also, a 16-inch barbette battery was to be built on Calf Island. Starting in 1920 the battery at Hog Island was begun. It was a standard battery of the system, for two 12-inch guns on Model 1917 barbette carriages in open positions and a large, protected traverse magazine. It had a closed back, with a small area for enlisted men quarters incorporated in the reverse gallery. When the Calf Island 16-inch battery project was cancelled in 1922, it was decided to convert the almost-completed 12-inch battery on Hog Island to emplace the larger 16-inch guns. New base rings were installed to hold the much larger and heavier carriage. However, only a few modifications were made to the traverse magazines consiting of larger handling block and tackle in the magazines and larger shell tables. The basic dimensions of the emplacement itself were not altered. The converted battery was ready for transfer on February 24, 1927. It was named on General Orders No. 13 of March 27, 1922 for 1st Lieutenant Frank S. Long who was killed in France in 1918. It was armed with two 16-inch guns Model 1919MII/MIII on Model 1919 barbette carriages (#3/#6 and #9/#1). The battery had a long service life as a key part of Boston's defenses. The emplacement was given heavy protected casemated gun houses over the gun blocks in 1941-1942. The guns were finally dismounted in

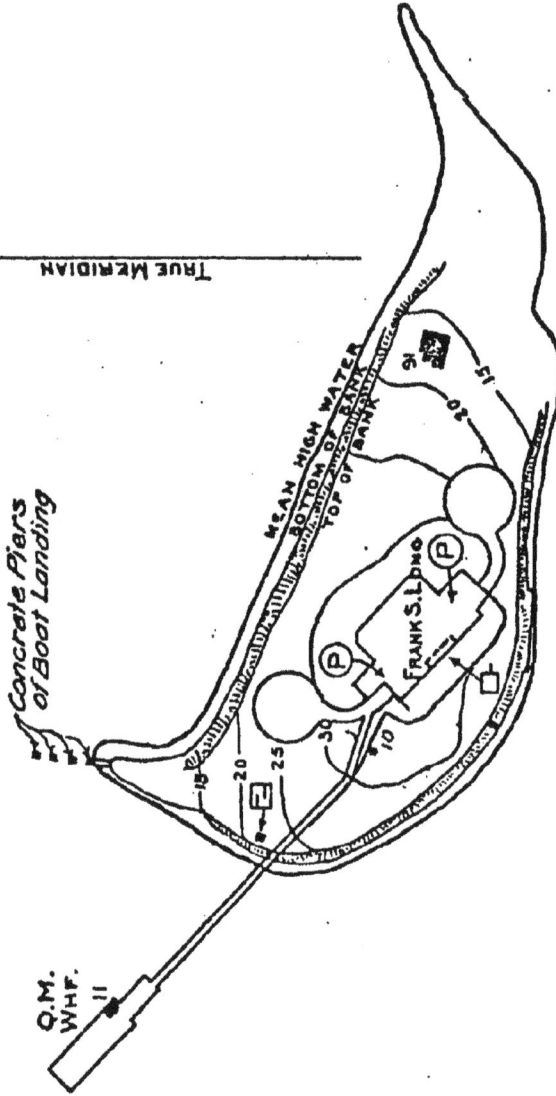

BOSTON HARBOR, MASS.
FORT DUVALL
LITTLE HOG ISLAND
Scale of Feet

LEGEND

EDITION OF OCTOBER 20, 1919.
REVISIONS: APR. 8, 1920; FEB. 17, 1921.
JAN. 29, 1925; MAY 6, 1929; JAN. 17, 1935;
APR. 11, 1938

SERIAL NUMBER

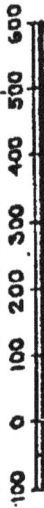

N.C.O. QUARTERS

HOIST HOUSE.
WHARFINGER'S HSE.

BATTERIES

LONG.............2-16" N Dⁱⁱ

TRUE MERIDIAN

Concrete Piers
of Boat Landing

MEAN HIGH WATER
BOTTOM OF BANK
TOP OF BANK

FRANK S. LONG

Q.M.
WHF.

This Reservation transferred and receipted for,
by Q.M.C. May 10, 1927. C. of E. 662.B. (Bos-Duvall) 8½2

On Maintenance Status.

SCALE IN FEET

TRUE

26 24 22 20 18 16 14 12

10

28

30

31

31

28

26 24

1-A

MEAN HIGH WATER

TOE OF BLUFF

FOOT OF BLUFF

1B

30

28

26

24

22

20

18

16

14

12

10

12

12

LONG BTRY'S

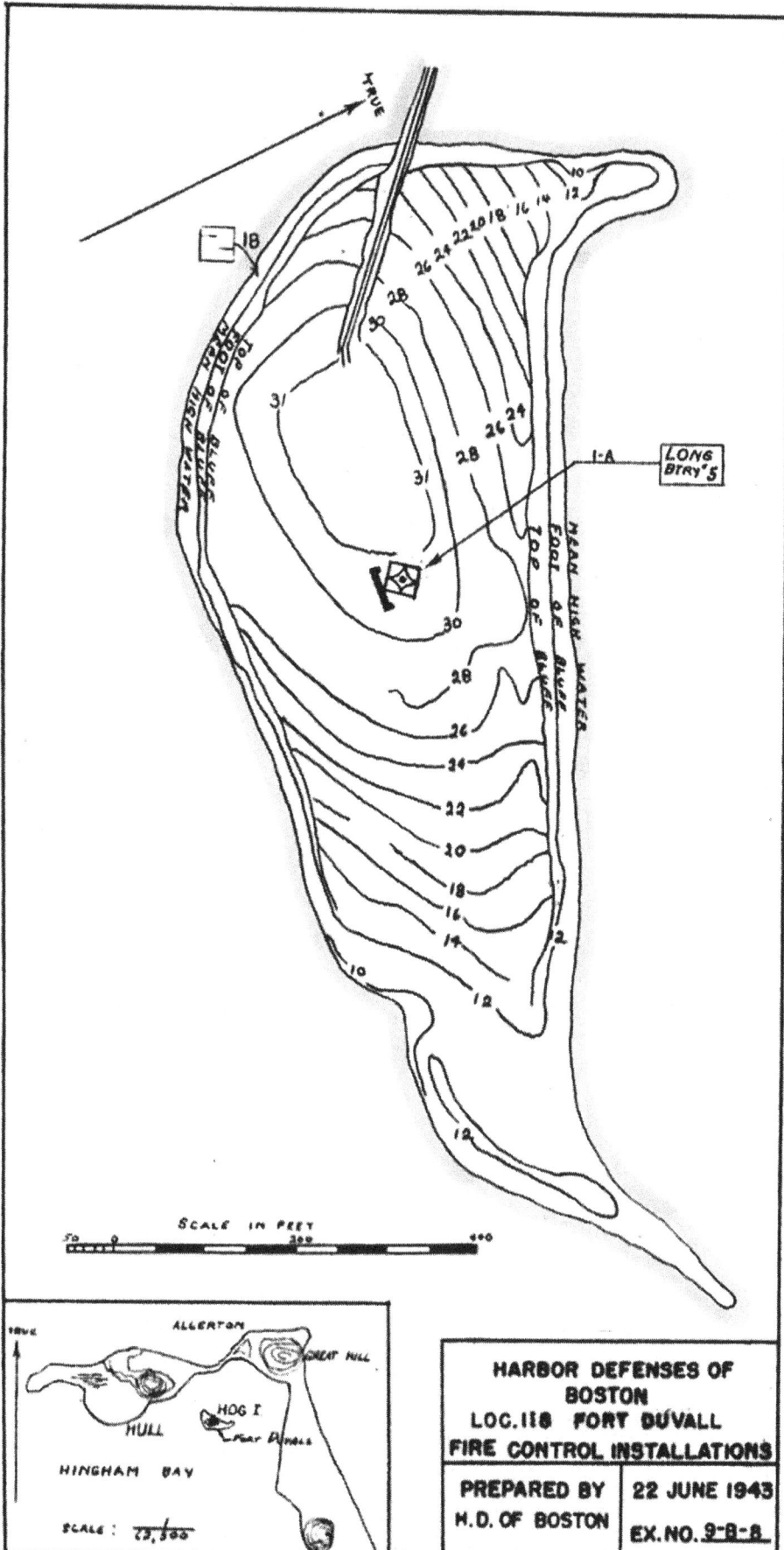

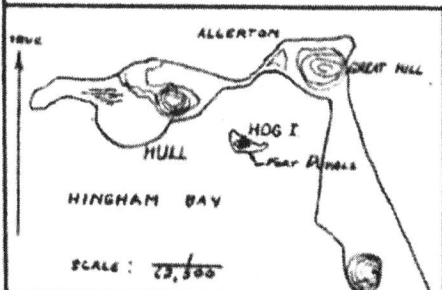

HARBOR DEFENSES OF
BOSTON
LOC. 118   FORT DUVALL
FIRE CONTROL INSTALLATIONS

PREPARED BY H.D. OF BOSTON	22 JUNE 1943
	EX. NO. 9-B-8

REVISED DATE: 31 JAN 45

ALLERTON

GREAT HILL

HULL

HOG I

FORT DUVALL

HINGHAM BAY

SCALE: 1:12,500

FORT DUVAL
MASS.
ALT 3000 FEET
PHOTOGRAPHED BY US ARMY AIR SERVICE
MAY 1924   NEG NO. 127.
8TH PHOTO SECTION

Fort Duvall, Mass.   7 - 10 - 24.   8th Photo Section.

Fort Duval 1924 (NARA)

1948, and the battery emplacement was used as part of the local air defense until abandoned in the 1960s. Subsequently the site was sold to private interests and condominiums were built around and on top of the emplacement. The island was renamed Spinnaker Island. Heavily covered over, but with magazine basically intact, important elements still exist on private property. The battery is closed to the public.

Fort Andrews (1898-1947) is located on the East Head of Peddocks Island in Boston Bay. This 88-acre fort was established in 1897. The fort's primary Endicott Program armament was two 12-inch mortar batteries (each with 8 mortars), and three two-gun rapid fire gun batteries which provided coverage of the channel between Peddocks Island and Georges Island. It was named in General Orders 43 of 1900 for Bvt. Maj. Gen. George Leonard Andrews, a professor at the U.S. Military Academy. This fort was the main garrison post for the Boston defenses. In 1913, Pit A of Battery Whitman had its M1890 mortars replace with M1908 mortars (the only battery on the East Coast to do so). The mortar batteries were declared surplus in 1942, and the fort became an Italian prisoner-of-war camp. The fort was declared surplus in 1947 and sold to private interest in 1957. In 1970 the fort became the property of Massachusetts Department of Conservation and Recreation. Today Peddocks Island is a component of the Boston Islands NRA with day use and camping facilities. A few of the old post buildings remain, many were knocked down in the 2000s for safety reasons (and these debris were dumped into Pit A of Battery Cushing). A ranger station has a limited amount of information on the site. The island is accessible by the Boston Harbor Islands Ferry.

Fort Andrews Gun Batteries

- **WHITMAN**: Fort Andrews on Peddocks Island was slated for a full battery of sixteen mortars in four pits, but funding from the Act of March 3, 1897 only authorized construction of half of the battery. Local engineers submitted a plan on July 16, 1897. It was located on the northwest bluff of the island, pointing to the north. The engineers recommended building the western half of a "quad" mortar plan of four pits. The full battery would be in a rectangle 140 by 126-feet. Work was done on the two pits and magazine gallery between the pits in 1897-1898. As it turned out, the design used was already obsolete and being phased out. The pits were too small and cramped, the ammunition rooms too small with a long a passage to the pits and shot storage in the laterals that were too constricting. From the start the battery was not considered successful. During construction, it was found that the sand of the drumlin on which it was built was unstable and presented problems in securing a firm foundation. Nonetheless, it was transferred on January 15, 1902 for a cost of $86,000. It was named Battery Cushing on General Orders No. 78 of May 25, 1903 for Brigadier General Thomas R. Cushing. With the completion of the adjacent mortar battery a couple of years later, names were exchanged, and this became Battery Whitman. That name was granted on General Orders No. 20 of January 24, 1906 for Major Frank H. Whitman of Spanish-American War service. The original armament consisted of eight 12-inch Model 1890M1 mortars on Model 1896 carriages (Builders #40/#210, Bethlehem #35/#104, Bethlehem #40/#209, Builders #43/#151. Bethlehem #31/#149, Bethlehem #33/#211, Bethlehem #32/#148 and Niles #12/#150). In August 1909 it was decided to entirely rebuild the battery. The plan was to enlarge the current two pits (remove and replace the side walls) and cut new entries into the magazines to enhance ammunition service. Work started in early 1910. At that point it was also decided to convert Pit A for the new Model 1908 mortars and carriages. These were to be diverted from intended emplacement at Battery Geary at Fort Mills in the Philippines, by rearrangement the displaced Model 1890M1/1896 mortars and carriages from Pit A would go to the Philippines instead. By July 1911 Pit B had received its replacement armament (Builders #40/#149, Builders #43/#148, Builders #35/#121, and

BOSTON HARBOR, MASS.

FORT ANDREWS

PEDDOCKS ISLAND.

SERIAL NUMBER 4124

EDITION OF MARCH 4, 1914.

REVISIONS JAN. 14, 1915; DEC. 7, 1915;
NOV. 8, 1916; APR. 8, 1920; FEB. 17, 1921.

BATTERIES.

CUSHING ... 4-12" M.
WHITMAN ... 6-12" M.
McCOOK ... 2-6" P.
RICE ... 2-5" P.
BUMPUS ... 2-3" P.

GUNS ABSENT.

SCALE OF FEET.

LEGEND.

1 ADMINISTRATION BLDG.
2 COMMANDING OFF. QRS.
3 OFFICERS QUARTERS.

4 HOSPITAL.
5 HOSPITAL STWD'S QRS.
6 N.C.O. QUARTERS.
7 BARRACKS.
7t TEMP. BARRACKS.
8 GUARD HOUSE.
8t TEMP. GUARD HOUSE.
9 POST EXCHANGE & GYM.
10 DORMITORY.
11 COMBINED ST. HO.
12 COAL SHED.
13 FIRE APPARATUS.
14 WAGON SHED.
15 STABLE.
16 CREMATORY.
17 CARPENTER AND
    PLUMBER'S SHOP
18 BAKERY.
19 GARAGE.
100 WIRELESS TEL. POLES.
101 WAITING ROOM.
102 OBS. PLATFORM.
103 COTTAGES.
104 OIL HOUSE.
105t TEMP. MESS.
106 SAW MILL.
107 SCALES.
108t TEMP. LAVATORY.
109 TRANSFORMER BLDG.
21 Q.M. STORE HOUSE.
31 ORDNANCE ST. HO.
40 ENG. DEPT. ST. HO.
110 PAINT SHOP.
111 SHEDS.
40 ENG. DEPT. ST. HO.

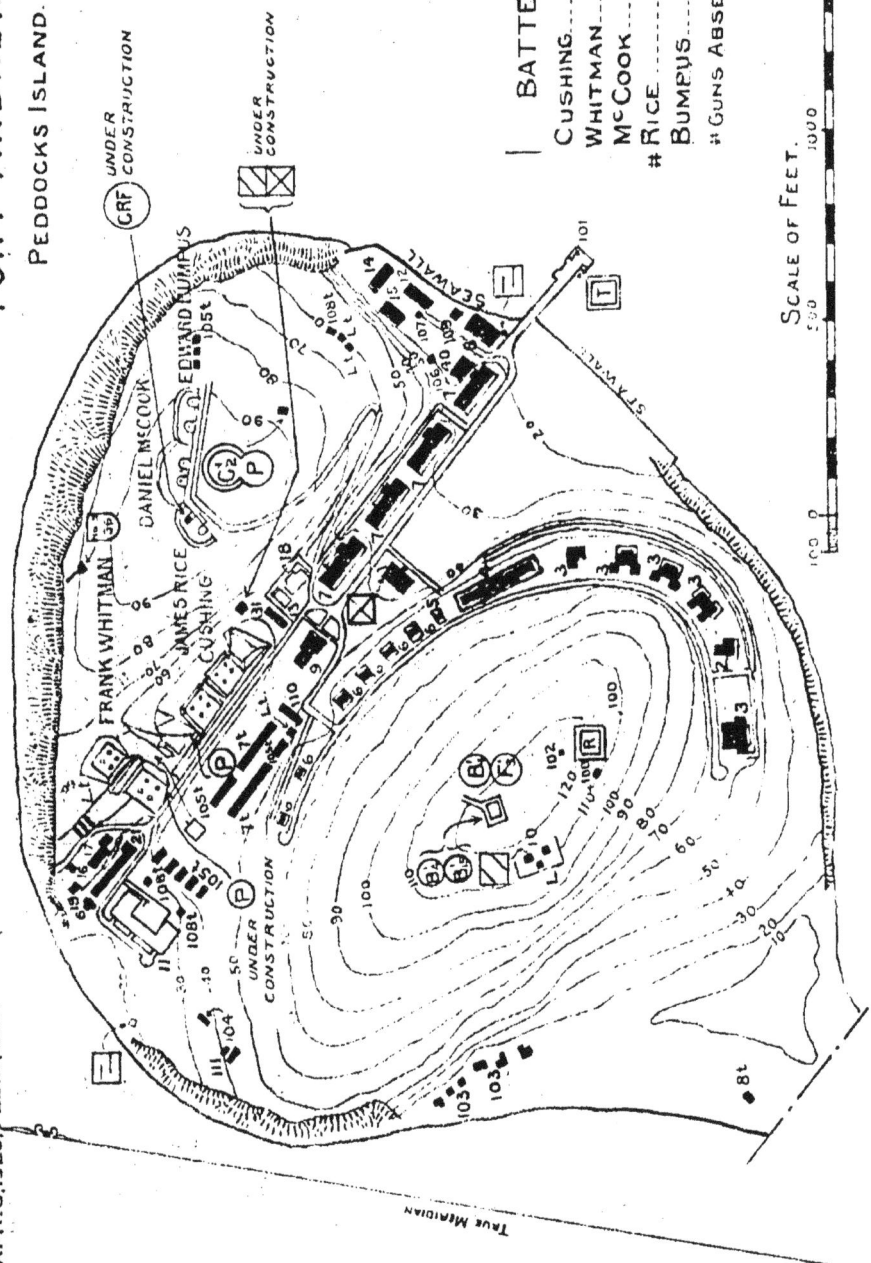

Fort Andrews 1920s (NARA)

Fort Andrews 1920s (NARA)

Niles #12/#125). Pit A received in 1913 and mounted four 12-inch mortars Model 1908 mortars on Model 1908 carriages (#1/#21, #3/#22, #4/#23, and #5/#24). This was the onlybattery of this mortar type in the Continental U.S. and important for training. In March 1918 two mortars from Pit B were sent away for use on railway carriages, their old carriages were scrapped in 1920, and the empty pits filled in. The final six mortars were eventually combined tactically with Battery Cushing and continued to serve up through World War II. They were finally authorized for removal on November 6, 1942. The emplacement exists at the Peddocks Island unit of the Boston Harbor Islands National Recreation Area. The battery is open to the public.

- **CUSHING**: The second mortar battery for Peddocks Island. Originally intended as the second half of Battery Whitman, when finally authorized in 1900 it was built as a separate unit from the earlier, problem-prone battery. Plans were submitted on August 29, 1900. This time it was planned with the wider pits of the later mortar plan design. Also to ease ammunition service it had open-backed pits and magazines under the front parapet and in flank and central traverses. The new battery was located to the immediate right (eastern) flank of Battery Whitman, but parallel to and with easy access to the road that ran by the battery. A new power plant was built into the space between the two batteries. Work was done in 1901-1902, and transfer made on December 29, 1904 for a cost of $150,578.31. Originally named Whitman (see naming citations for battery above) in 1906 the mortar battery names were exchanged, and it became Battery Cushing. The original armament was eight 12-inch Model 1890M1 mortars on Model 1896 carriages (#157/#287, #144/#286, #141/#285, #139/#294, #158/#297, Niles #2/#296, #162/#295, and #146/#288). This armament served until four mortars were removed on June 7, 1918 for use on railway carriages. Their carriages were left in place until scrapped in 1920. Around 1933 the battery was joined with Whitman to for a single 10-gun tactical unit called Battery Cushing-Whitman. It was retained until deleted under authority of November 6, 1942. The emplacement still exists at the Peddocks Island Park. The battery is open to the public.

- **McCOOK**: A rapid-fire battery for two 6-inch Model 1900 guns. Emplaced on the highest part of the Peddocks Island northern drumlin, it was on the eastern side of the reservation and fired to the north. It was the center of the series created by Batteries Rice, McCook, and Bumpus. The plan was submitted on June 6, 1901, and was similar to the plan made for the battery at Fort Standish. It was built late enough to be able to incorporate the changes and have each gun on its own platform rather than share a common platform. Work was done in 1901-1902. Transfer was made on December 29, 1904 for a cost of $27,000. It was named on General Orders No. 194 of December 27, 1904 for Brigadier General Daniel McCook, U.S. Volunteers who was mortally wounded at Kennesaw Mountain GA in 1864. It was armed with two 6-inch Model 1900 guns and pedestal mounts (#14/#11 and #20/#27). In early 1918 the guns were removed for use on wheeled, field carriages. However, by order of March 17, 1919 the guns were returned and remounted on their old pedestal carriages. The battery continued to serve as an important tactical unit up through World War II. The emplacement was substantially rebuilt from June to August 1943. The parapets were removed and the loading platforms extended. Opportunity was also taken to replace the worn gun tubes, the battery received guns #47 and #39 (previously at Battery Montgomery, Fort Monroe). It was finally disarmed postwar, probably in 1946 or 1947. The emplacement still exists at the Peddocks Island Park. The battery is open to the public, but overgrown.

- **RICE**: A battery for two 5-inch rapid-fire guns emplaced at the highest knoll of the drumlin at the northern end of Peddocks Island. Plans were submitted on November 23, 1899. The plan was for a type design in use for the proposed wire-wound, Brown type 5-inch gun (which was never actually produced). It had two separate platforms with a single magazine below the traverse to serve both guns. The compact design had hand-operated chain hoists to raise the ammunition to the loading level. Work was accomplished mostly in 1900. Transfer was made on December 29, 1904 for a cost of $23,800. It was named on General Orders No. 194 of December 27, 1904 for Brigadier General James C. Rice who was killed in action at Spotsylvania in 1864. There was a considerable delay in arming batteries built for the earlier type of 5-inch gun. The first gun was mounted on March 27, 1908 and the second one not until May. It held two 5-inch Model 1900 Watervliet guns on Model 1903 pedestals (#3/#1 and #5/#2). These guns were removed in late 1917 and moved to a temporary site at Cape Henry, Virginia. Later the emplacement was used as the site for the CRF station for 3-inch Battery Bumpus. The emplacement still exists at the Peddocks Island Park. The battery is open to the public, but overgrown.

- **BUMPUS**: The final rapid-fire battery put on the northwest drumlin of Fort Andrews. The plan was submitted on April 19, 1903. It was placed on the eastern end of the Rice-McCook line, extending that sequence. Work was done in 1903 for transfer made on December 29, 1904 for a construction cost of $15,500. It was named on General Orders No. 194 of December 27, 1904 for 1st Lieutenant Edward A. Bumpus, 9th U.S. Infantry, who was killed in action during the Philippine Insurrection in 1901. It was armed with two 3-inch Model 1902M1 guns on Model 1902 pedestal mounts (#52/#52 and #51/#51). This armament stayed with the battery throughout its service life. It served well into World War II, final removal coming after the war in 1946 or 1947. The emplacement still exists at the Peddocks Island Park. The battery is open to the public, but overgrown.

Fort Andrews barracks on Peddocks Island (Terry McGovern)

Strawberry Hill Military Reservation: Endicott-era fire control site

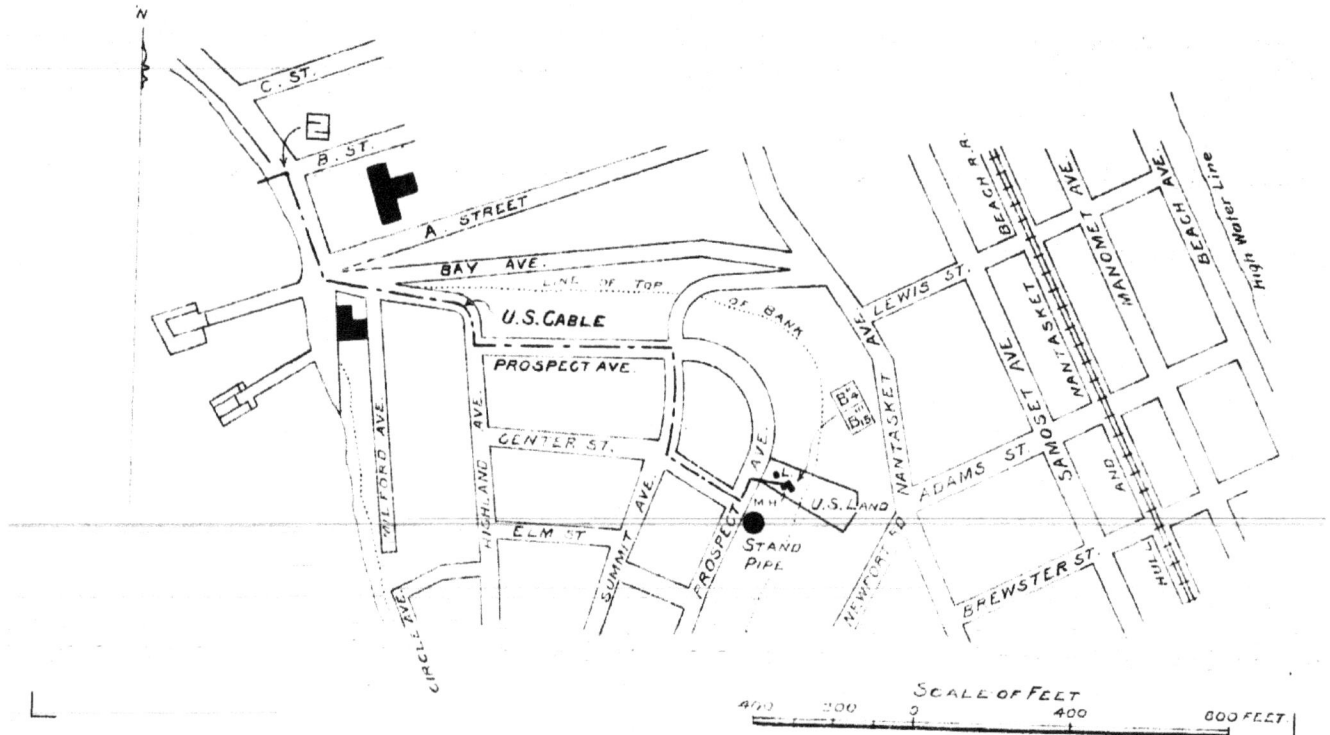

Fourth Cliff Military Reservation (1943-1948) is located on the bluff overlooking the mouth of the North and South Rivers north of Humarock, near Scituate, MA. A World War II era 6-inch battery (BCN 208) was built here as the southernmost Boston defense. A 16-casemated battery (BCN 106) was planned nearby but never built. Two disguised fire control stations remain at the site. Today the site is an Air Force recreation area, run by Hanscom Air Force Base, with all public access restricted. The 6-inch battery still retains its generators in its power room.

Fourth Cliff M.R. Gun Batteries

- **Battery #208**: One of the four dual 6-inch barbette batteries of the 1940 Program assigned to Boston Harbor. Battery Construction No. 208 was to be emplaced on the bluff above Fourth Cliff, on the southern flank of Boston Harbor. It was originally given a relatively low national priority and not funded until the FY-1943 Budget. It was of standard 1940 Program 200-series design. Work was done from September 8, 1942 to October 1, 1943 for transfer on November 9, 1944 at a cost of $334,000. It was armed with two 6-inch Model M1(T2) guns on Model M4 barbette carriages (#5/#53 and #3/#54). The battery was retained for a short while following World War II but was eliminated and disarmed in about 1947. The emplacement with its generators still exists at the Fourth Cliff Family Recreation Area operated by Hanscom Air Force Base. The battery site is not open to the public.

- *Battery #106* (planned): The third modern 16-inch barbette battery planned for Boston under the 1940 Program. It was to be emplaced at Flower Hill in Marshfield south of Boston Harbor. It would have had a conventional 100-series design structure. It was given the lowest national priority of the Boston batteries, ranking #30 on the September 11, 1940 list and with no priority status on August 11, 1941. Site preparation was begun, however the construction project was formally abandoned with authority to suspend on November 13, 1942.

Power plant in Battery #208 (Mark Berhow)

Battery Bartlett Fort Warren (Terry McGovern)

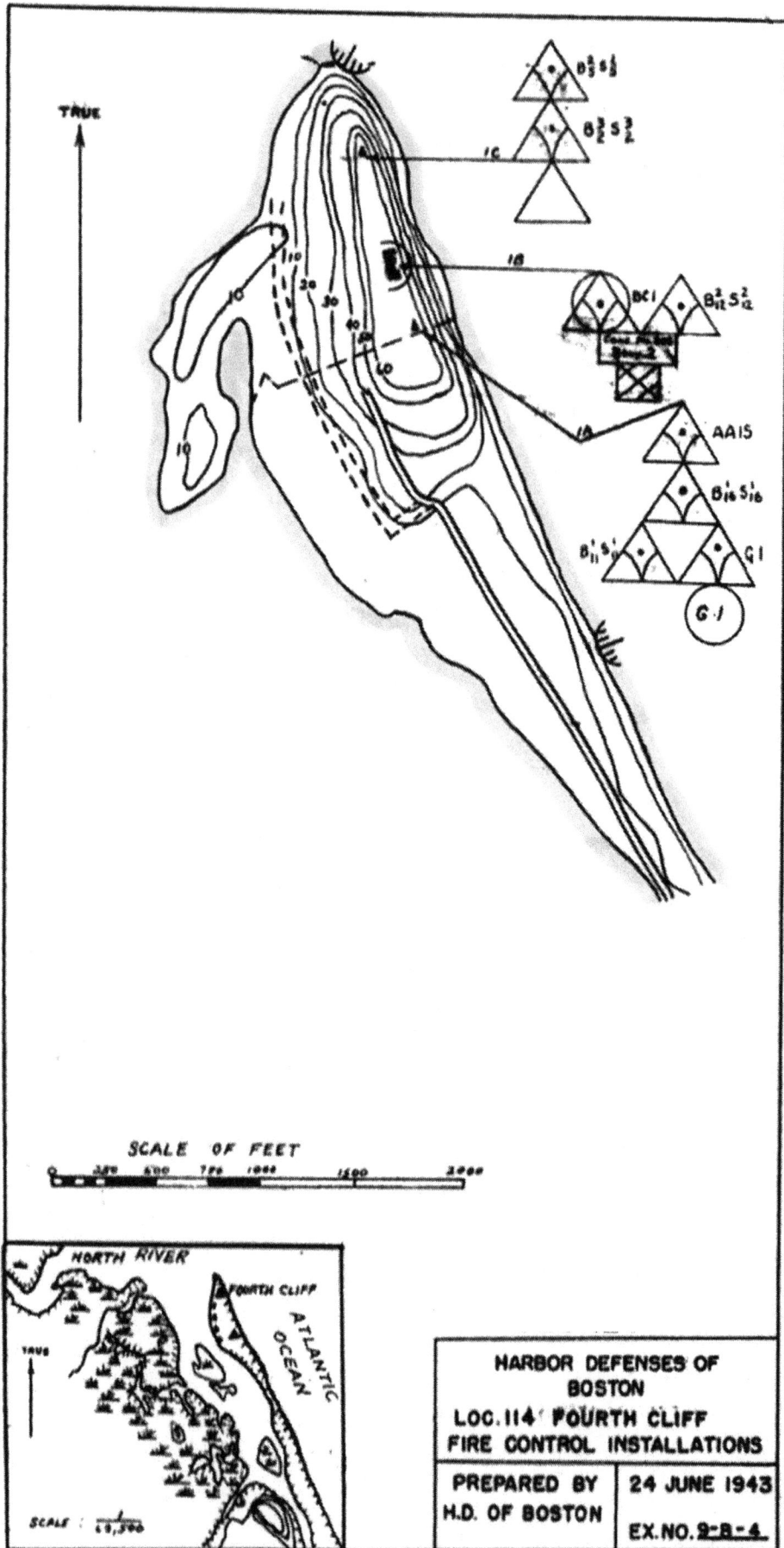

SCALE OF FEET

HARBOR DEFENSES OF
BOSTON
LOC. 114 FOURTH CLIFF
FIRE CONTROL INSTALLATIONS

PREPARED BY H.D. OF BOSTON	24 JUNE 1943
	EX. NO. 9-B-4

REVISED DATE: 31 JAN 45

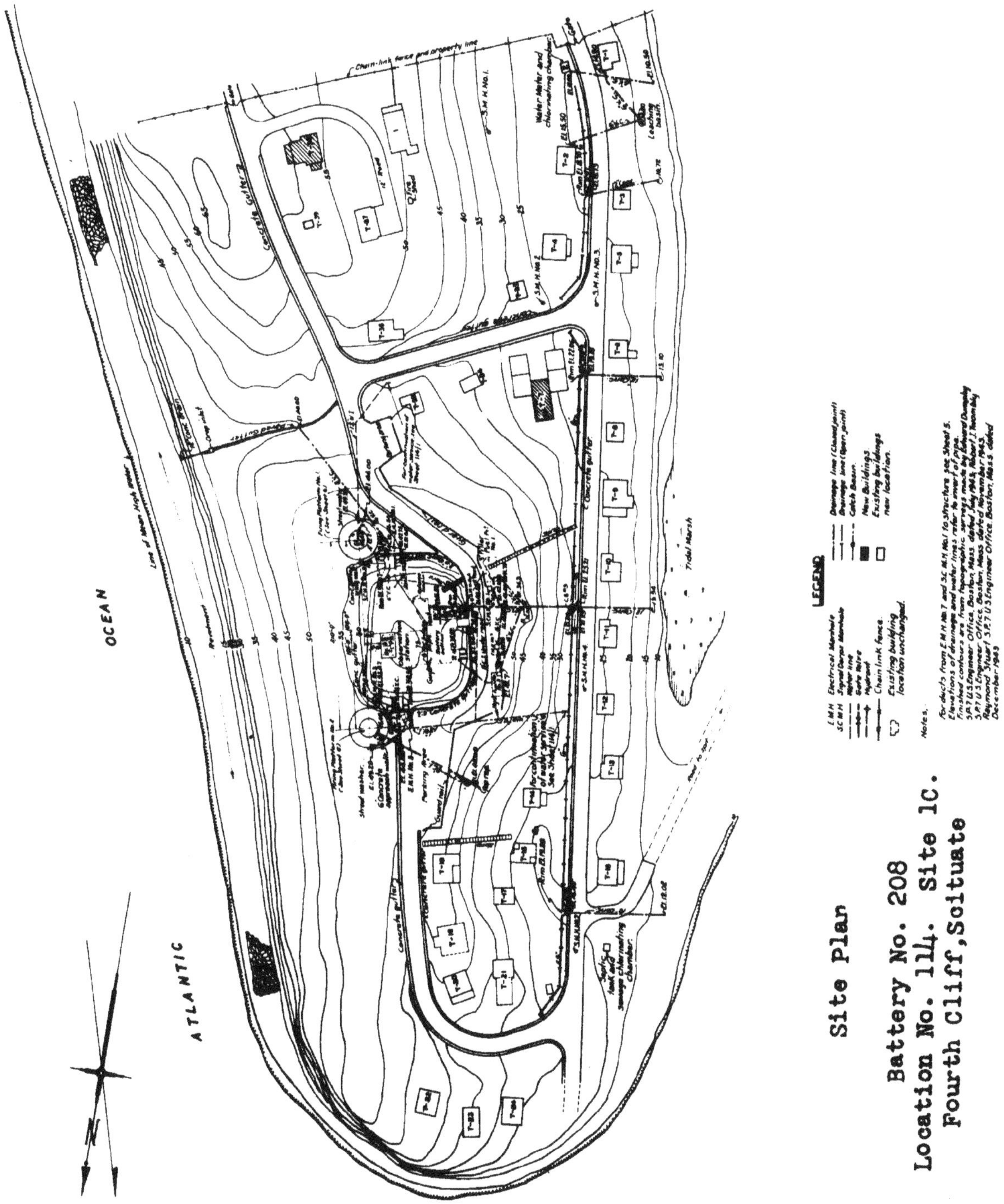

OCEAN

ATLANTIC

Site Plan

Battery No. 208
Location No. 114. Site 1c.
Fourth Cliff,Scituate

LOC. NO.137 CASTLE HILL	LOC. NO.136 HALIBUT POINT	LOC. NO.135 EMERSON POINT	LOC. NO.134A EASTERN POINT	LOC. NO.134 COOLIDGE POINT	LOC. NO.133 GALES POINT	LOC. NO.132 MARBLEHEAD NECK	LOC. NO.131 EAST POINT	LOC. NO.127 FORT DAWES	LOC. NO.120 POINT ALLERTON	LOC. NO.116 STRAWBERRY HILL	LOC. NO.115 STRAWBERRY POINT	LOC. NO.114 FOURTH CLIFF	LOC. NO.112A HOLLY HILL	LOC. NO.112 BRANT ROCK	LOC. NO.111 GURNET POINT
B_{12} S_{16}	C2 OP4	C2 OP3	B_{12} S_{12}	B_5 S_7	B_{12} S_{12}	C2 OP2	G6CP OP	HECP	CICP OP3	B_{10} S_{10}	C1 OP2	GICP OP	296 BTRY 2	B_2 S_2	C1 OP1
PORTSMOUTH	B_{16} S_{16}	B_{12} S_{12}	B_{16} S_{16}	B_{14} S_{14}	B_5 S_5	296 BTRY 15	G7CP OP	C2 OP1	BC 5	SL 9	G2CP OP	BTRY. 2		B_5 S_{11}	B_{12} S_{12}
PORTSMOUTH	SL 29	B_{16} S_{16}	SL 27	SL 26	B_{16} S_{16}	SL 28	BTRY 15	G4-AMTB	296 BTRY 4		B_2 S_2	296 BTRY 5		B_{12} S_{12}	SL 1
	PORTSMOUTH	SL 28	FCSB		SL 24	SL 24	BTRY 16	BTRY 13	SCR 682		B_{11} S_{11}	B_3 S_3		SL 3	SL 2
		FCSB			SL 25	SL 25	296 BTRY14	BTRY. 14	B_{4} S_{4}		B_{12} S_{12}	B_5 S_5		SL 4	
					FCSB	B_5 S_5	B_5 S_2	BTRY 944	B_{2} S_2		B_3 S_3	B_{12} S_{12}			
						B_{14} S_{14}	B_{11} S_{11}	296 BTRY 13	B_3 S_3		B_4 S_4	B_3 S_3			
						B_5 S_4	B_4 S_4	SCR 582	B_4 S_4		B_{14} S_{14}	B_3 S_3			
						B_{13} S_{13}	B_{12} S_{12}	MINE CASEMATE	B_2 S_2		SL 5&6	B_3 S_3			
						B_5 S_4	B_{11} S_{11}	B_4 S_4	B_3 S_3		FCSB	B_4 S_4			
						B_{16} S_{16}	B_4 S_4	B_4 S_4	B_5 S_5			B_5 S_{14}			
						SL 22	B_{15} S_{15}	B_{10} S_{10}	B_5 S_5			B_5 S_5			
						SL 23	B_4 S_4	SL 10&11	B_{16} S_{16}			B_{16} S_{16}			
							B_{13} S_{13}	M2				SL 7&8			
							SL 20	HECP-SL							
							SL 18								
							FCSB	SL AMTB 4							
								FCSB							

TRUE

10   0   10   20

THOUSANDS OF YARDS

LOC. NO.122 FORT STRONG	LOC. NO.121 FORT WARREN	LOC. NO.117 FORT ANDREWS	LOC. NO.119 FORT REVERE
BTRY. 6	G5CP OP	BTRY. 4	BTRY. 941
BTRY. 7	M1	BTRY. 3	BTRY. AA1
BTRY. AA3	M1 CP	CRF BTRY. 3	M_4^2
CRF BTRY. 6	SL 15	B_4 S_4	SL AMTB 1
CRF BTRY. 7	FCSB	SL 12	FCSB
M2 CP	SIGNAL STATION MINE CASEMATE	FCSB	LOC. NO.118 FORT DUVALL
SL 16			BTRY. 5
FCSB			

LOC. NO.123 FORT STANDISH
BTRY. 10
BTRY. 943
BTRY. 9
BTRY. AA2
B_{10} S_{10}
CRF BTRY. 10
CRF BTRY. 9
SL 17
SL AMTB 3
FCSB

LOC. NO.130 FORT RUCKMAN	LOC. NO.129 FORT HEATH	LOC. NO.128 FORT BANKS	LOC. NO.126 OUTER BREWSTER	LOC. NO.124 GREAT BREWSTER
C2 CP	BTRY. 945	HDCP	BTRY. 11	MINE CASEMATE
BTRY. 14	296 BTRY 10	G3-AACP	296 BTRY 11	BTRY. 942
BTRY AA4	B_4 S_4	FCSB	B_{11} S_{11}	M_2
B_4 S_4		MET STATION	SL 13 & 14	SL AMTB 2
B_{10} S_{10}			FCSB	
SL 19				
FCSB				

LOC. NO.131A PHILLIPS POINT — SL 21

H.D. OF BOSTON

LOCATION OF ELEMENTS

PREPARED BY. H D OF BOSTON	DATE 7-1-43
	EX. NO. 4A

REVISED DATE 31 JAN 45

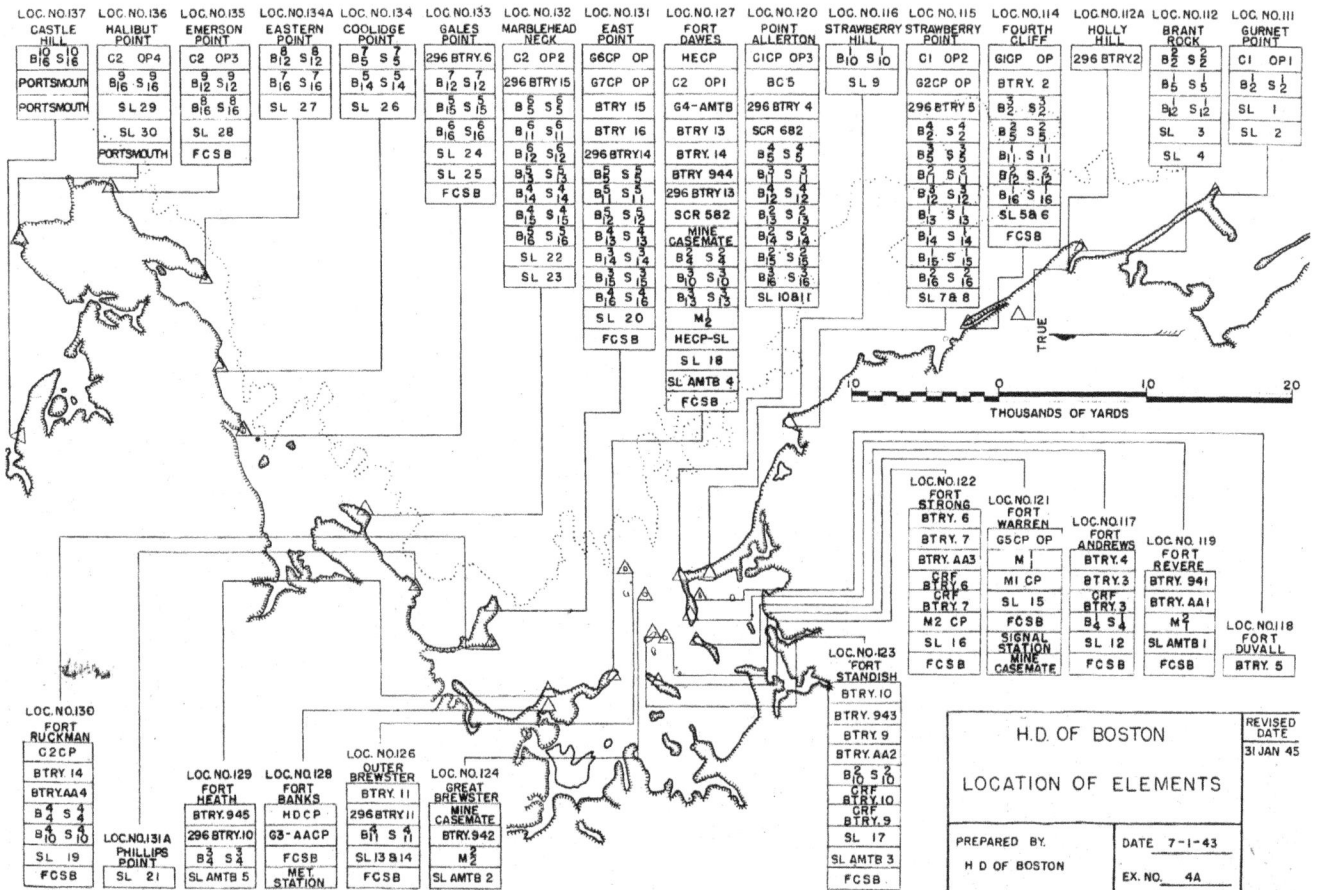

Fire Control Towers at East Point, Nahant (Terry McGovern)

Boston World War II-era Site Locations.   Stations housed in a single structure are connected by dashes (-)

location	Loc#	Purpose
Sagamore Beach	109	
Manomet Point	110	
Gurnet Point	111	C1-BS1/208, SL 1,2
Brant Rock	112	BS2/208-BS1/Long-BS1/105, SL 3,4
Holly Hill	112A	SCR296-208
Flower Hill	113	military reservation-planned
Fourth Cliff	114	Batt. Tact. #2 BCN 208, BC-BS3/208, BS7/105-BS2/Long, BS1/Jewell-BS1/Murphy-G1
Strawberry Point	115	G2-BS4/208-BS3/Long, C1-BS2/Jewell-BS2/Murphy, BS3/105-BS1/207, BS1/Gardner-BS2/206, SCR296-208, SBR
Strawberry Hill	116	BS1/Whipple, SL 9
Fort Andrews	117	Batt. Tac. #3 Bumpas, Batt. Tact. #4 McCook, BC/Bumpas, BC-BS1/McCook, SL 12
Fort Duval	118	Batt. Tact. #5 Long
Fort Revere	119	AMTB, SBR
Point Allerton	120	C1-BC-BS4/Long-BS2/Gardner, BS3/Jewell-BS2/207-BS2/206), BS4/105-BS3/Murphy, SCR682, SCR296-McCook, SL 10,11
Fort Warren	121	G5, M1/1, SL 15, SBR, MC, Mines
Fort Strong	122	Batt. Tact. #6 Stevens, Batt. Tact. #7 Basinger, BC/Stevens, BC/Basenger, M2CP, SL 17
Fort Standish	123	Batt, Tact. #9 Whipple, Batt. Tact. #10 Williams, BC-BS2/Whipple, BC/Williams, AMTB, SL17, SBR
Great Brewster Is	124	MC, AMTB, M2/2
Calf Island	125	SCR
Outer Brewster Is	126	Batt. Tact. #11 Jewell, BC-BS4/Jewell, SCR296-Jewell, SBR
Fort Dawes	127	Batt. Tact. #12 BCN 105, Batt. Tact. #13 BCN 207, HECP, C2, G4, BC/207, BC/105, AMTB, SCR582, MC, BS3/207, BS3/Whipple, BS3/McCook, M1/2, SL 18
Fort Banks	128	HDCP, G3, SBR, Met
Fort Heath	129	B3/McCook, SCR296-Whipple, AMTB
Fort Ruckman	130	Batt. Tact. #14 Gardner, C2CP, BC/Gardner, BS4/McCook, BS4/Whipple, SL10, SBR
East Point	131	Batt. Tact. #15 Murphy Batt. Tact. #16 BCN 206, G6-BS3/206-BS3/Gardner, G7-BS5/Long-BS5/Jewell-SCR296-Gardner, BC/206-BS4/Murphy-BC/Murphy-BS5/105-BS5/207, SL 20
Marblehead Neck	132	SCR296-206, C2-BS5/Jewell-BS5/Murphy, BS6/Long-BS4/Gardner, BS6/105, BS5/207-BS4/206, SL 22,23
Phillips Point		SL 21
Gales Point	133	BS6/Murphy-BS5/206-BS7/105, SL 24,25
Coolidge Point	134	BS7/Long-BS5/Gardner, SL 26
Eastern Point	134A	BS7/ Murphy-BS8/105, SL 27
Emerson Point	135	BS8/ Murphy-BS9/105-C2OP3
Halibut Point	136	BS9/ Murphy-C2OP4, SL 29,30
Castle Hill	137	BS10/ Murphy

THE HARBOR DEFENSES OF NEW BEDFORD – MASSACHUSETTS

Defended by seacoast forts since colonial times, New Bedford received a Third System fort in the late 1850s and was modernized with 5 new Endicott Program batteries after 1900, which protected the New Bedford harbor and the entrance to the newly opened Cape Cod Canal at Buzzards Bay. Upgraded with a new 12-inch battery in 1920 and a modern 6-inch battery in 1942, New Bedford's defenses were eventually integrated with the Narragansett Bay defenses.

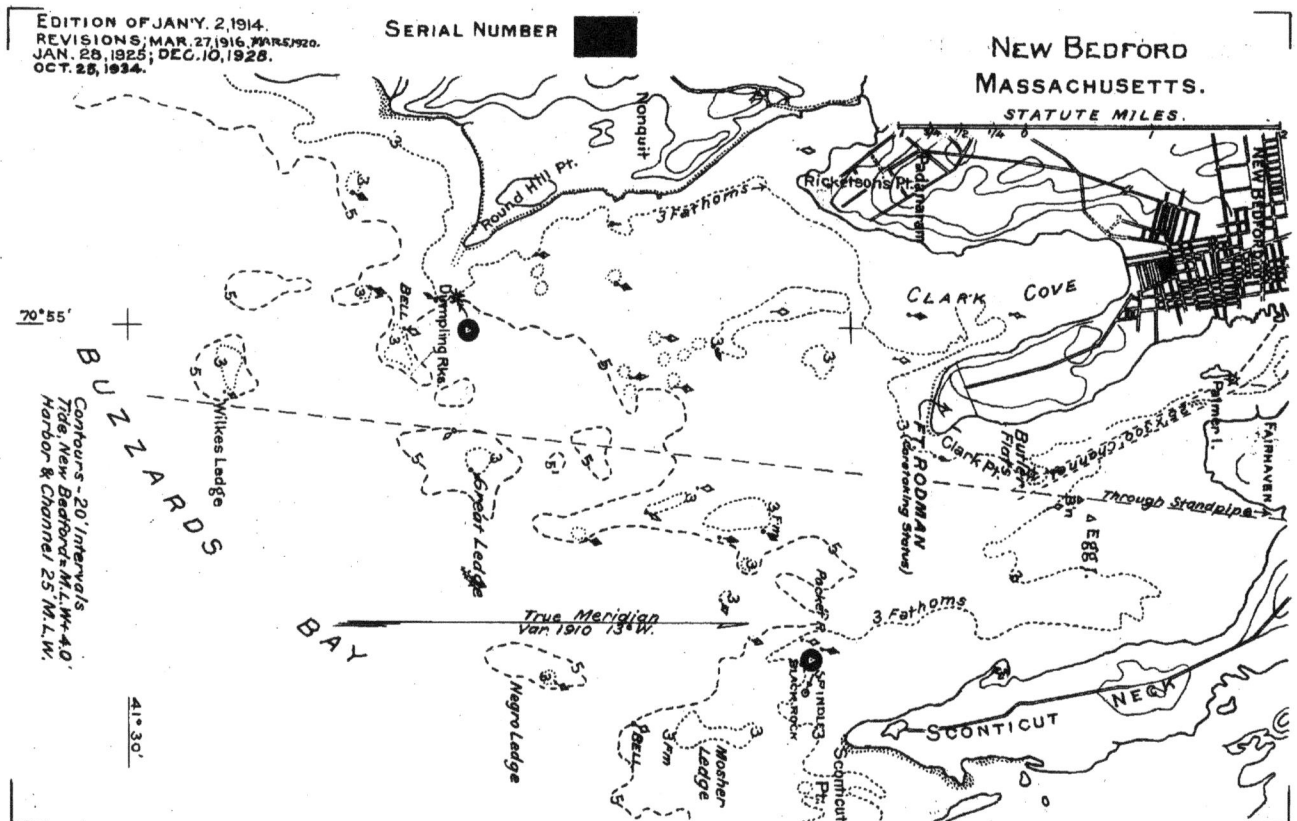

Fort Rodman (1840-1947) is located on Clarks Point, near the Fort at Clark's Point (a Third System fort). Fort Rodman was formally established in 1898 with five-gun Endicott Program batteries and a controlled submarine mine complex. It was named in General Orders No. 106 on July 23, 1898 for Lt. William Logan Rodman, Mass Vol. Inf. who was killed at Port Hudson, LA on May 27, 1863. The fort received a new 12-inch long-range barbette battery in the 1921 which was casemated during World War II. Two Panama mounts for mobile 155mm GPF artillery were emplaced here in 1938. The fort continued to be used as an U.S. Army and U.S. Navy Reserve Center after being deactivated as coast artillery fort. The reservation was eventually transferred to the City of New Bedford in the 1970s. The garrison area was transformed into a wastewater treatment facility and the campus of a local community college. The concrete batteries and the Third System fort are part of the city's Fort Taber Park. The emplacements are generally open to the public during daylight hours, the Third System fort is open for regular tours. Information can be obtained from the park's visitors center and shop.

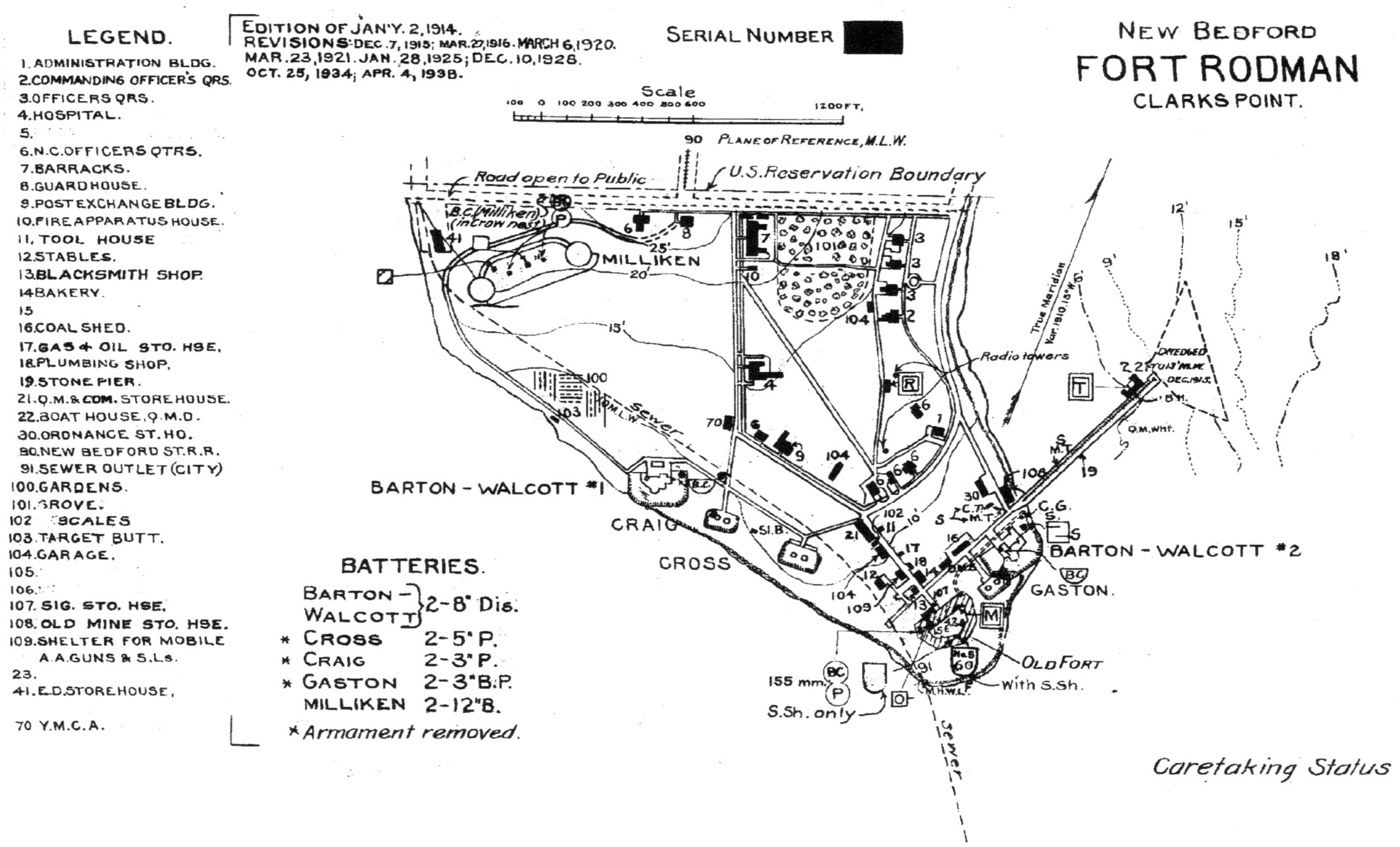

NEW BEDFORD

FORT RODMAN

CLARKS POINT.

SERIAL NUMBER

Scale

EDITION OF JAN'Y. 2, 1914.
REVISIONS: DEC. 7, 1915; MAR. 27, 1916. MARCH 6, 1920.
MAR. 23, 1921; JAN. 28, 1925; DEC. 10, 1928.
OCT. 25, 1934; APR. 4, 1938.

Caretaking Status

LEGEND.

1. ADMINISTRATION BLDG.
2. COMMANDING OFFICER'S QRS.
3. OFFICERS QRS.
4. HOSPITAL.
5.
6. N.C. OFFICERS QTRS.
7. BARRACKS.
8. GUARD HOUSE.
9. POST EXCHANGE BLDG.
10. FIRE APPARATUS HOUSE.
11. TOOL HOUSE
12. STABLES.
13. BLACKSMITH SHOP.
14. BAKERY.
15.
16. COAL SHED.
17. GAS & OIL STO. HSE.
18. PLUMBING SHOP.
19. STONE PIER.
21. Q.M.& COM. STOREHOUSE.
22. BOAT HOUSE, Q.M.D.
30. ORDNANCE ST. HO.
90. NEW BEDFORD ST.R.R.
91. SEWER OUTLET (CITY)
100. GARDENS.
101. GROVE.
102. SCALES
103. TARGET BUTT.
104. GARAGE.
105.
106.
107. SIG. STO. HSE.
108. OLD MINE STO. HSE.
109. SHELTER FOR MOBILE
   A.A. GUNS & S.L.S.
23.
41. E.D. STOREHOUSE.
70. Y.M.C.A.

BATTERIES.

BARTON - WALCOTT } 2-8" Dis.
* CROSS   2-5" P.
* CRAIG   2-3" P.
* GASTON   2-3" B.P.
  MILLIKEN   2-12" B.

* Armament removed.

Lat. 41°-35'-48"
Long. 70°-54'-41"

TRUE

N 70°10'55"E

RODNEY FRENCH BOULEVARD

BROCK AVE.

Lat 41°-35'-51"
Long. 70°-54'-13"

N 70°22'44"
N 70°32'36"E

O'REILLY AVE.

FRENCH AVE.

KEENE STREET

MILLS AVE.

HUNT AVE.

T

WHARF ROAD   WHARF

BS   M

R

S   4

2 1 0   2   4   6

HUNDREDS OF FEET

TRUE

FT. RODMAN

SCALE · THOUSANDS OF YDS

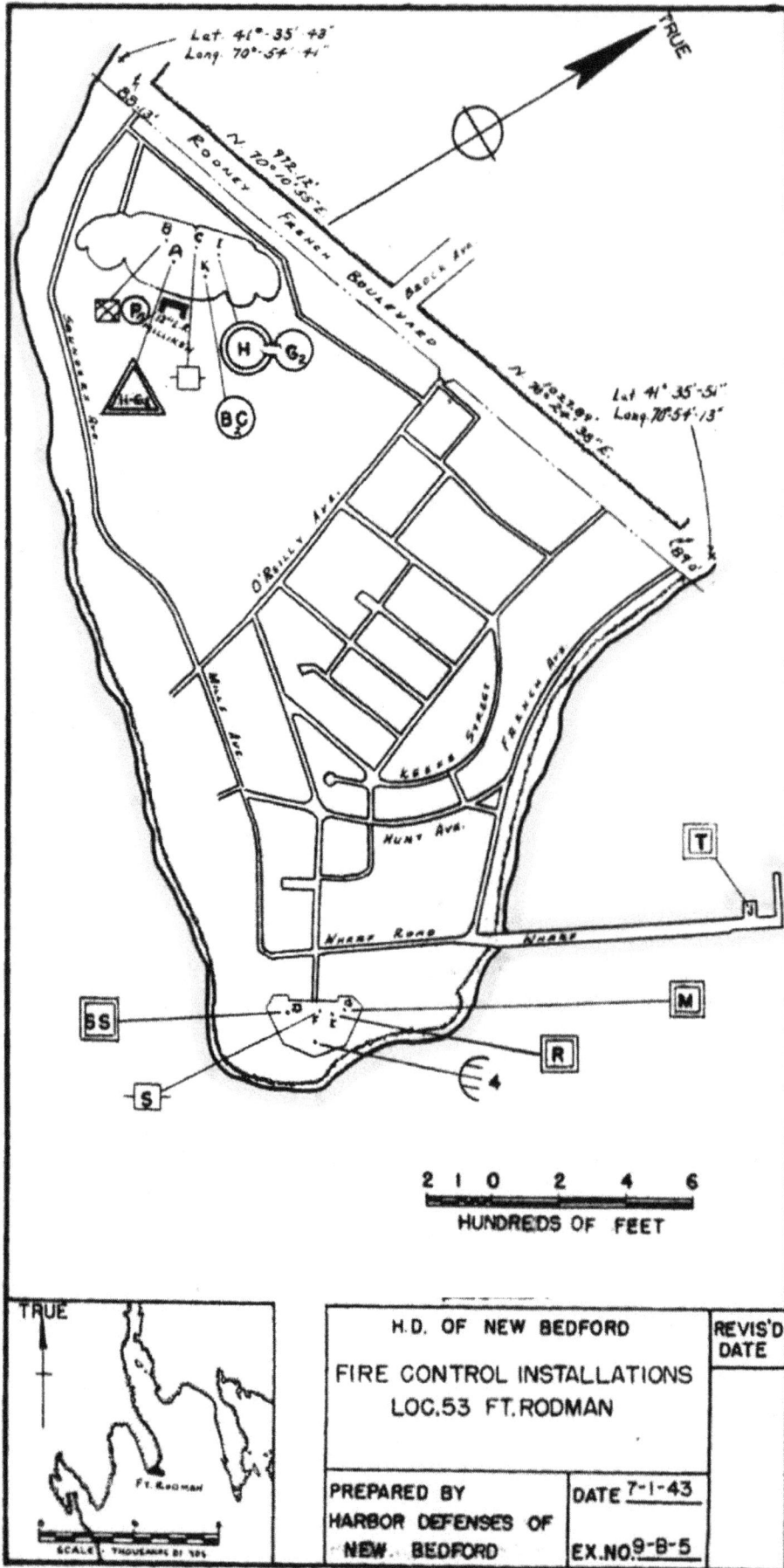

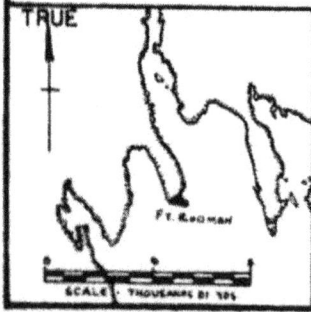

H.D. OF NEW BEDFORD	REVIS'D DATE
FIRE CONTROL INSTALLATIONS LOC.53 FT. RODMAN	
PREPARED BY HARBOR DEFENSES OF NEW BEDFORD	DATE 7-1-43
	EX.NO. 9-B-5

Fort Rodman Gun Batteries

- **BARTON:** A battery for a single 8-inch disappearing gun placed on the western side of the Fort Rodman reservation at Clark's Point to the northwest of the old masonry fort. It fired to the southwest. Plans were submitted on April 9, 1898. It was of conventional design, with just a single platform for the one gun and a lower-level magazine on the left flank. Ammunition service was by hoist. Work was done in 1898-1899. Transfer was made on July 31, 1899 for $51,812.50. It was named on General Orders No. 78 on May 25, 1903 for Colonel William Barton of Revolutionary War service. The battery was armed with one 8-inch Model 1888MII Bethlehem gun on a Model 1896 disappearing carriage (#12/#22). Modifications were subsequently made to the emplacement, including new hoists transferred in 1907. It was left armed during World War I. In the late 1930s batteries Barton and Walcott were combined tactically into a single battery, aptly named Battery "Barton-Walcott". As a single unit they served until salvaged under authority of November 6, 1942. The emplacement still exists at the Fort Taber Park. The battery is open to the public.

- **WALCOTT:** A battery for a single 8-inch disappearing gun. It was placed on the most eastern part of the reservation, close to the shore and adjacent to the pier. It fired to the southeast. Plans were submitted on April 9, 1898. It was conventional in design, with its single platform and a lower-level magazine on the right flank and ammunition service by hoist. It was practically a mirror image of Barton. Work was done in 1898-1899. Transfer was made on July 31, 1899 for $51,812.50. It was named on General Orders No. 78 on May 25, 1903 for Captain William H. Walcott of Civil War service. The battery was armed with one 8-inch Model 1888MII Bethlehem gun on a Model 1896 disappearing carriage (#15/#18). Modifications were subsequently made to the emplacement, including new hoists transferred in 1907. It was left armed during World War I. In the late 1930s batteries Barton and Walcott were combined tactically into a single battery, aptly named Battery "Barton-Walcott". As a single unit they served until salvaged under authority of November 6, 1942. The emplacement still exists at the Fort Taber Park. The battery is open to the public.

- **CROSS:** A battery for two 5-inch pedestal guns emplaced near the reservation's western shore, between Walcott and the masonry fort. It also fired to the southwest. Plans were submitted on August 20, 1898. Its location was about 500-feet from Battery Walcott. It was of conventional design for this generation of rapid-fire batteries, conforming to current mimeograph specifications. It had two separate platforms and magazine on the lower left flank of each platform. Work was done in 1899-1900, for transfer on December 13, 1902 at $11,610.58. It was named on General Orders No. 78 of May 25, 1903 for Captain Charles E. Cross, killed in action at Fredericksburg in 1862. After the usual delay awaiting the replacement of the original wire-wound 5-inch guns with later models, the battery was armed with two 5-inch Model 1900 guns on Model 1903 pedestal mounts (#16/#3 and #17/#4). Trunnion height was 23.28-feet. The armament was removed for a short while during World War I (removed May 17, 1918) for use on an army transport but returned and remounted in 1919. Then in 1920 it was permanently removed. The emplacement still exists at the Fort Taber Park. The battery is open to the public.

- **GASTON:** A battery for two 3-inch, 15-pounder guns emplaced on the left flank of Battery Walcott, firing to the southeast. Plans were submitted on June 29, 1900. Local engineers wanted to place both Gaston and Craig on the top tier of the old masonry fort, but Washington rejected this due to the vulnerability of this location. It was moved to its site on the southeast even though that location didn't have as clear a field of fire. It was of type plan, with two platforms and lower magazines on the right flank of each, ammunition service was by hand. Work was done in 1900-1901.

Transfer was made on December 29, 1902 for a construction cost of $18,900 (which included Battery Craig). It was named on General Orders No. 78 of May 25, 1903 for 2nd Lieutenant William Gaston, 1st U.S. Dragoons, who was killed in action with the Spokane and Coeur D'Alene Indians in Washington Territory in 1858. It was armed with two 3-inch Model 1898 Driggs-Seabury guns and masking parapet mounts (#83/#83 and #84/#84). These were modified about 1915-1916 to pedestal M1898M1 mounts. This was removed in June 1920 in common with other 3-inch balanced pillar batteries. The emplacement still exists at the Fort Taber Park. The battery is open to the public.

- **CRAIG**: A battery for two 3-inch, 15-pounder guns emplaced on the left flank of Battery Barton, close to the shore and firing to the south. Plans were submitted on June 29, 1900. The first suggested location for placing the guns atop the old masonry fort at Clark's Point was rejected by Washington, and eventually the site next to Battery Barton was chosen even if it meant a restriction in coverage due to the interposing bulk of the older work. It was of type plan, with lower level magazines on the left flank and ammunition service by hand. Work was done in 1900-1901. Transfer was made on December 29, 1902 for a construction cost of $18,900 (which included Battery Gaston). It was named on General Orders No. 78 of May 25, 1903 for 2nd Lieutenant Presley O. Craig, killed in action at Bull Run in 1861. It was armed with two 3-inch Model 1898 Driggs-Seabury guns and masking parapet mounts (#81/#81 and #82/#82). These were modified about 1915-1916 to pedestal M1898M1 mounts. This was removed in June 1920 in common with other 3-inch masking parapet batteries. In the late 1920s the northwest emplacement (No. 1) was converted to take a 3-inch Model 1917 anti-aircraft gun, which served until into World War II. The emplacement still exists at the Fort Taber Park The battery is open to the public.

- **MILLIKEN**: A 1915 Board of Review Program battery for two 12-inch long-range barbette guns emplaced at Fort Rodman for coverage over Buzzard Bay. It was located on the northwest corner of the reservation, firing to the southeast and over the Endicott batteries and old masonry fort. It closely followed type plans, with standard distance between gun pads and arrangement of magazines. It was an open-back type of plan. It did have a reduced-sized embankment on the flanks that were not endangered by incoming fire. Plans were submitted on February 20, 1917. Concrete work was done between 1917 and 1919. It was transferred on May 3, 1921 for a cost of $326,616.54. The battery was named on General Orders No. 15 of March 10, 1920 for 2nd Lieutenant Alfred S. Milliken, 6th Engineers, who was killed in France in 1918 during World War I. The battery was armed with two 12-inch Model 1895M1A4 guns on Model 1917 barbette carriages (#49/#15 and #59/#14). It was retained during the Interwar period. In 1942 authorization was issued to provide heavy overhead casemates to the guns. New gun houses were built and connected to the battery structure. The armament served until the early postwar period, apparently being dismounted in 1946. The emplacement still exists at the Fort Taber Park. The battery site is open, but the interior is closed to the public.

Fort Rodman, New Bedford (Terry McGovern)

Battery Millikan, Fort Rodman Park, New Bedford (Terry McGovern)

Mishaum Point Military Reservation (1943-1948) is located in Dartmouth, Bristol County. A 155mm GPF two-gun battery on Panama mounts was completed and transferred to the Coast Artillery for use in 1943. Deactivated in 1945 as Battery BCN 210 came online with two 6-inch M1 guns on long-range shielded mounts. The site was also had a radar tower and the harbor entrance control post for New Bedford and served as the examination battery for HD New Bedford. Declared surplus after World War II, the military reservation was sold to a private owner. A large private home now sits atop the battery. Access to the point is restricted.

Mishaum Point M.R. Battery

- **Battery #210**: A 1940 Program modern 6-inch barbette battery, the only new modern battery for the fixed defenses of New Bedford. It was originally to be located at Fort Rodman but was soon relocated on a small reservation at Mishaum Point, to the southwest of Fort Rodman by letter authority of July 27, 1942. Funding came under the FY-1942 appropriations. It was originally assigned national priority #21 under the list of September 1940. Work was done in 1943-1944 and was completed in 1945. It was never named, just known during construction and service as Battery Construction No. 210. It was ultimately armed with two 6-inch guns M1(T2) on model M3 barbette carriages (#7/#5 and #6/#6). These were removed postwar, probably in 1946-1947. The emplacement still exists but is privately owned with a residence built on top of it. The battery is closed to the public.

Elizabeth Islands Military Reservations (1940-1945) Defending the passage to New Bedford between Dartmouth and Cuttyhunk Island were two batteries of 90mm dual-purpose guns, one at Barneys Joy Point Military Reservation and one on Cuttyhunk Island. These were Batteries 931 and 932, respectively. The AMTB batteries had an authorized strength of four 90 mm guns, two on fixed mounts and two on towed mounts. An additional 90mm battery, AMTB Battery 933, was on Nashawena Island, just east of Cuttyhunk Island. Protecting the southern entrance to the Cape Cod Canal was a two-gun 155mm battery on Panama mounts, replaced in 1943 by the 90mm AMTB Battery 934, at Butler Point Military Reservation in Marion. Most of these locations are on private islands with no public access.

Barney's Joy M.R. Battery

- **AMTB #931**: An AMTB battery for two 90mm fixed and two 90mm mobile guns authorized on April 17, 1943. It was subsequently built and armed, then disarmed after the end of the war. One mount is covered. The battery site is closed to the public.

Cuttyhunk Island M.R. Battery

- **AMTB #932**: An AMTB battery for two 90mm fixed and two 90mm mobile guns authorized on April 17, 1943. It was subsequently built and armed, then disarmed after the end of the war. The blocks still exist on private land. The battery site is closed to the public.

Nashawena Island M.R. Battery

- **AMTB #933**: An AMTB battery for two 90mm fixed and two 90mm mobile guns authorized on April 17, 1943. It was built on Nashawena Island. Reportedly the blocks for the fixed guns still exist on private land. The battery site is closed to the public.

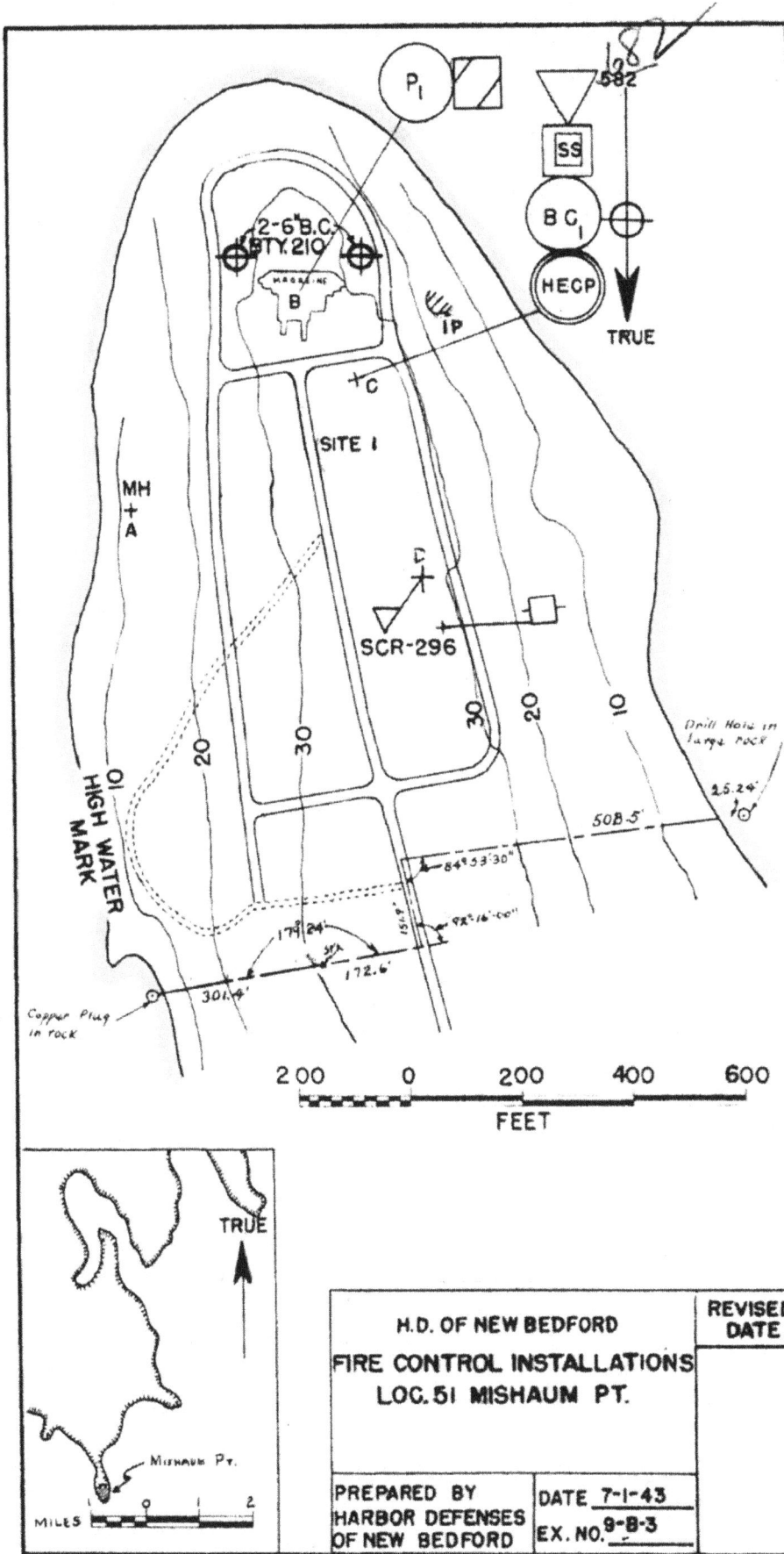

P₁

SS

B C₁

HECP

TRUE

582

2-6 B.C.
BTY. 210

MAGAZINE

B

IP

+C

SITE 1

MH
+A

D
+

SCR-296

30

20

10

Drill Hole in large rock

25.24'

OI
HIGH WATER
MARK

20

30

50B·5'

84°53'30"

156·9'

9X2·16'·con'

179·24'

172·6'

301·4'

Copper Plug
in rock

200   0   200   400   600

FEET

TRUE

MISHAUM PT.

MILES   0   1   2

H.D. OF NEW BEDFORD

FIRE CONTROL INSTALLATIONS
LOC. 51 MISHAUM PT.

REVISED
DATE

PREPARED BY
HARBOR DEFENSES
OF NEW BEDFORD

DATE 7-1-43

EX. NO. 9-B-3

Butler's Point M.R. Battery

- **AMTB #934:** An AMTB Battery for two 90mm fixed and two 90mm mobile guns authorized on April 17, 1943. This 90mm battery replaced two 155mm GPF on Panama mounts. It was built later in that year at Butler's Point. It was disarmed postwar. Mounts for both batteries still remain. The battery site is closed to the public.

Battery Walcott Fort Rodman (Mark Berhow)

Battery #210 Mishaum Point (Terry McGovern)

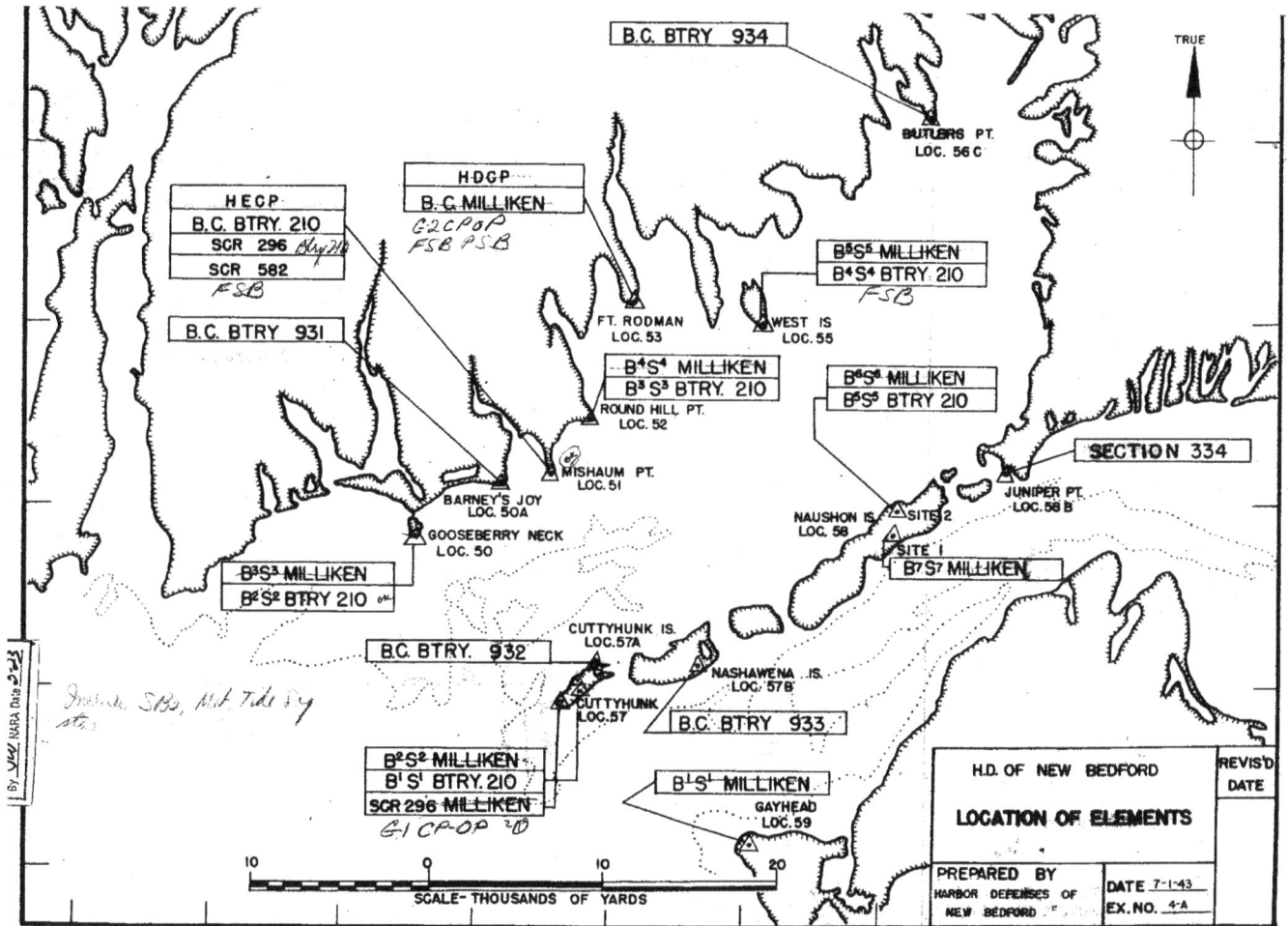

New Bedford World War II-era Site Locations. Stations housed in a single structure are connected by dashes (-)

location	Loc#	Purpose
Goosebury Neck	50	BS3/Millikan-BS3/210
Barney's Joy	50A	AMTB
Mishaum Point	51	Batt. Tact. #1 BCN 210, HECP-HDOP, BC/210, SCR296-210, SCR582
Round Hill	52	BS3/210
Fort Rodman	53	Batt. Tact. #2 Millikan, HDOP-BC/Millikan-BS4/210
West Island	55	BS5/Millikan-BS4/210
Pease Point	56D	SL
Butler's Point	56C	AMTB
Cuttyhunk	57	BS2/Millikan-BS1/210, SCR296-Millikan, AMTB
Nashawena Isl	57B	AMTB
Naushon Island	58	BS4/Millikan-BS2/210
Juniper Point	58B	
Gayhead	59	BS1/Millikan

THE HARBOR DEFENSES OF NARRAGANSETT BAY – RHODE ISLAND

Narragansett Bay is a large set of bays and estuaries that provide a good natural harbor. Initially colonized in 1638 from the Massachusetts colony and chartered as the Rhode Island colony in 1643. Several small seacoast forts were built along its shores in colonial times. The United States established several seacoast fortifications under the First and Second Systems. A large Third System fort was built in Brenton Point, just south of Newport between 1824 and 1857. New brick and earthwork batteries were constructed after 1870 at four locations but not completed. In 1869, the Naval Torpedo Station was founded on Goat Island. A naval station was established at Newport in 1883. Over 20 new concrete gun batteries were built at 5 locations to control the main entrances to the bay during the Endicott Program, as well as controlled mine fields for the East and West Passages. The defenses were extensively upgraded beginning in 1939 with several repositioned guns, three new 16-inch casemated gun batteries, one casemated 8-inch gun battery, three new 6-inch gun batteries, and five new 90mm AMTB batteries. Three of these forts served as POW Camps later in the war. Several of the reservations were used by the U.S. Navy after 1950.

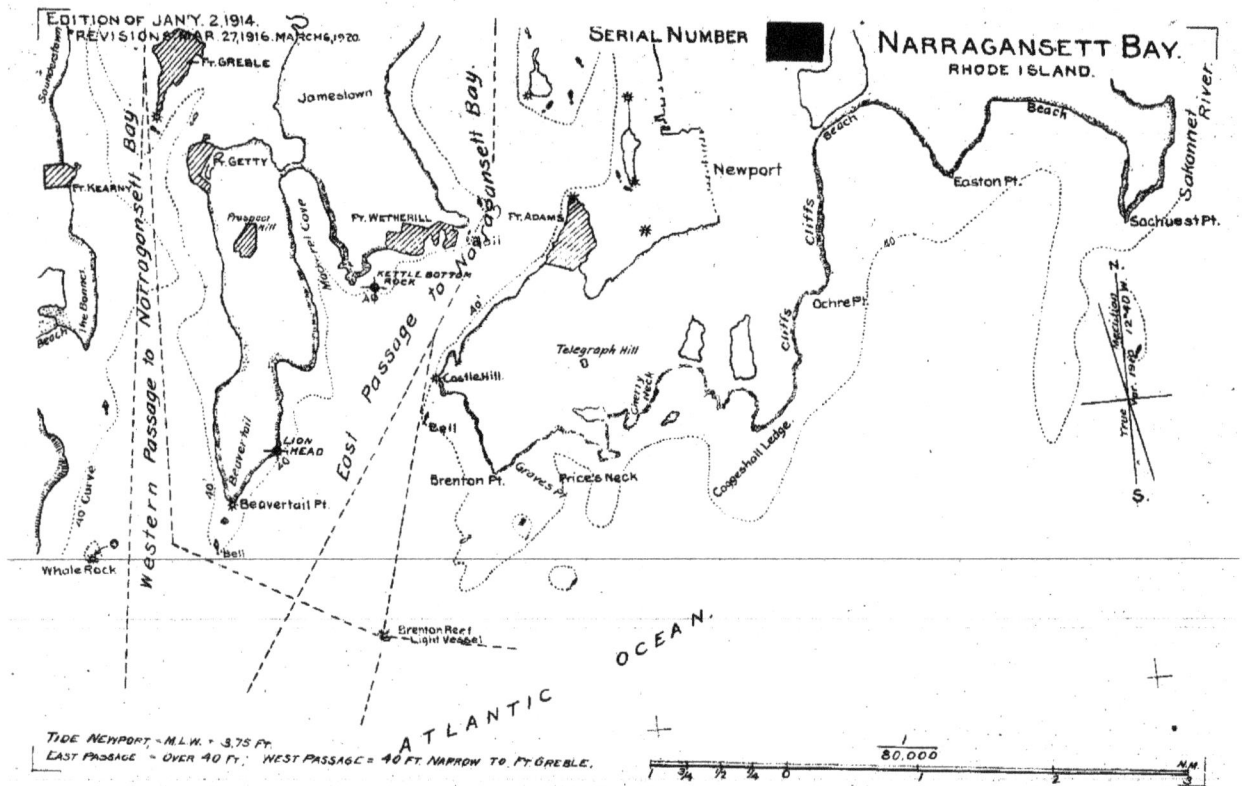

Fort Church (1940-1948) is located near Sakonnet Point off State Highway 77, south of the town of Little Compton. It was named on General Orders 5 of 1940 for Col. Benjamin Church of the Continental Army. The 1940 Program fort was divided into three parcels of land. The North Reservation was the location of a 16-inch casemated battery built in 1941-42, known at Battery Gray (#107). The East Reservation contained Battery Reilly; an 8-inch casemated battery built in 1941-42. This unique battery used two U.S. Navy 8-inch guns on U.S. Army barbette carriages in a scaled-down version of 16-inch 100-Series construction. The South Reservation was located on high ground overlooking Sakonnet Point and the ocean and was the location of shielded 6-inch Battery #212 and two Panama mounts for 155mm guns (1942-1943), which may still remain buried. All locations had some associated fire control structures. Today all the reservations of Fort Church are privately owned. Battery Grey is partially buried on the grounds of the Sakonnet Golf Club. Battery Reilly is buried on private land. Battery #212 has house built atop it and one of its gun blocks is a site of a swimming pool. There is no public access to these sites.

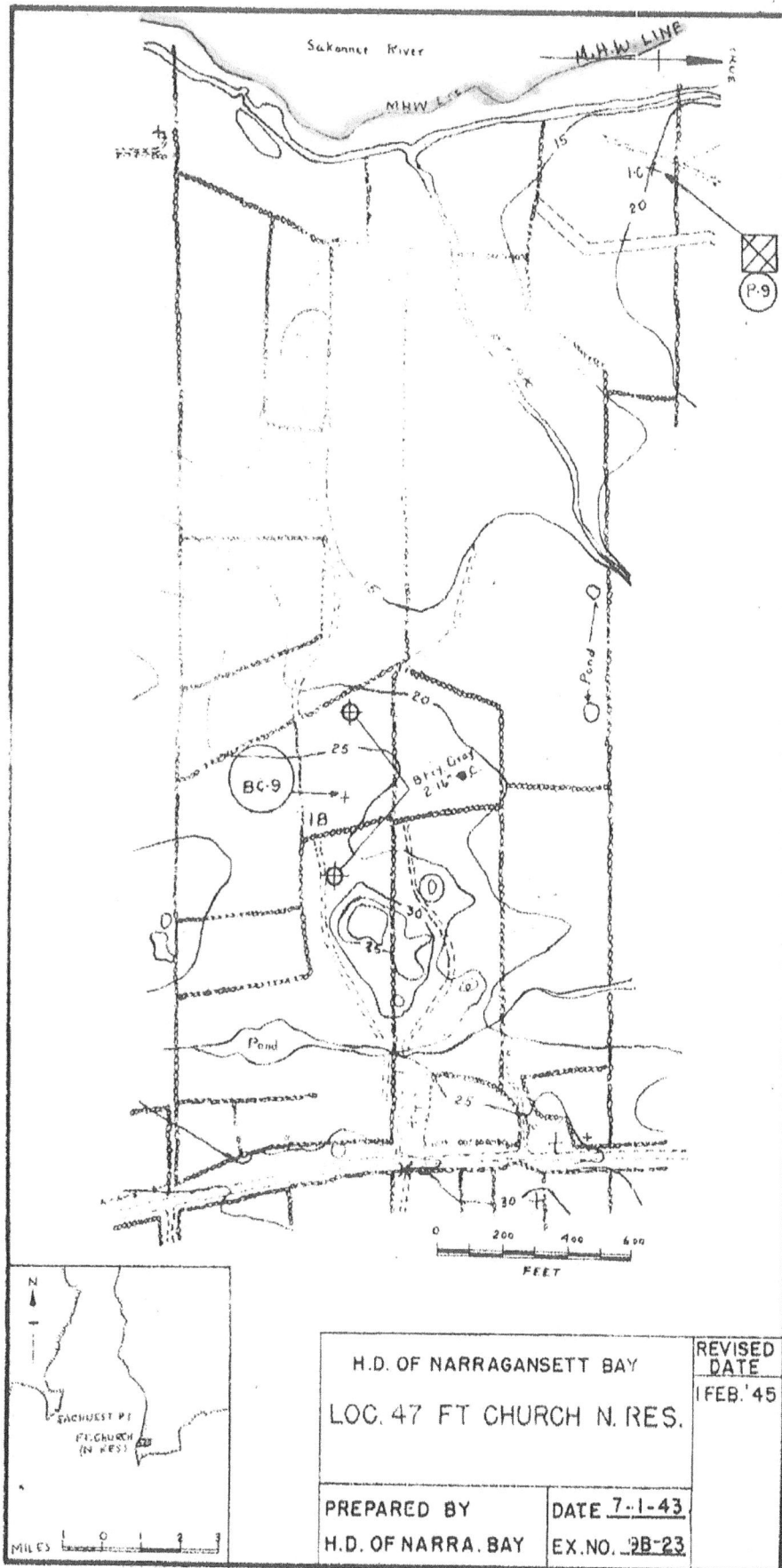

Sakonnet River

M.H.W. LINE

MHW L.

15

10

20

P.9

20

25

BC.9

18

Ott Pond

30

25

Brit Gun
2 16" C

Pond

25

30

0 200 400 600

FEET

N

SACHUEST PT

FT CHURCH
(N RES)

MILES 1 0 1 2 3

H.D. OF NARRAGANSETT BAY

LOC. 47 FT CHURCH N. RES.

REVISED
DATE
1 FEB. '45

PREPARED BY
H.D. OF NARRA. BAY

DATE 7-1-43

EX. NO. 9B-23

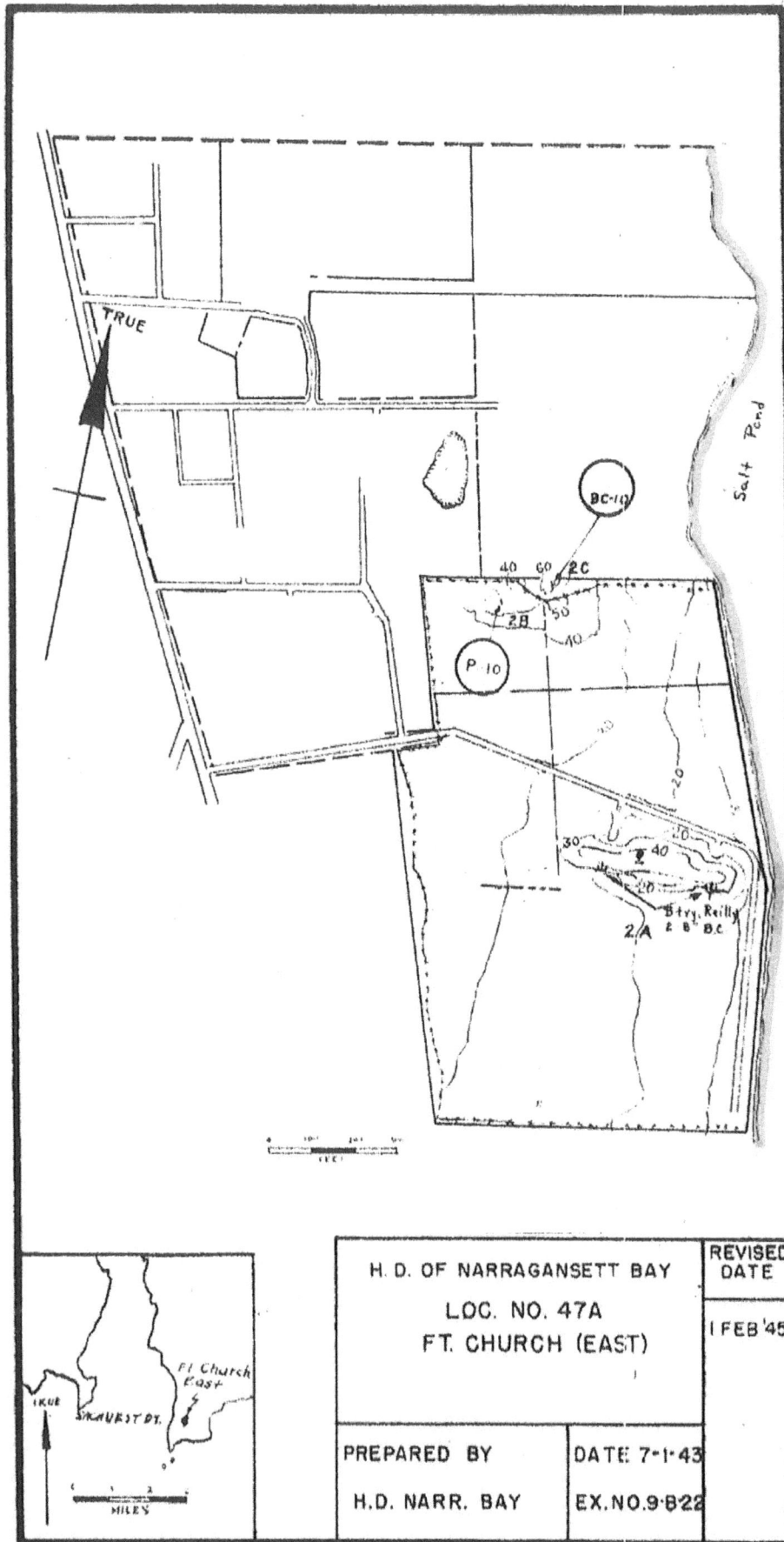

H. D. OF NARRAGANSETT BAY

LOC. NO. 47A
FT. CHURCH (EAST)

REVISED
DATE

I FEB '45

PREPARED BY

H.D. NARR. BAY

DATE 7-1-43

EX. NO. 9-B22

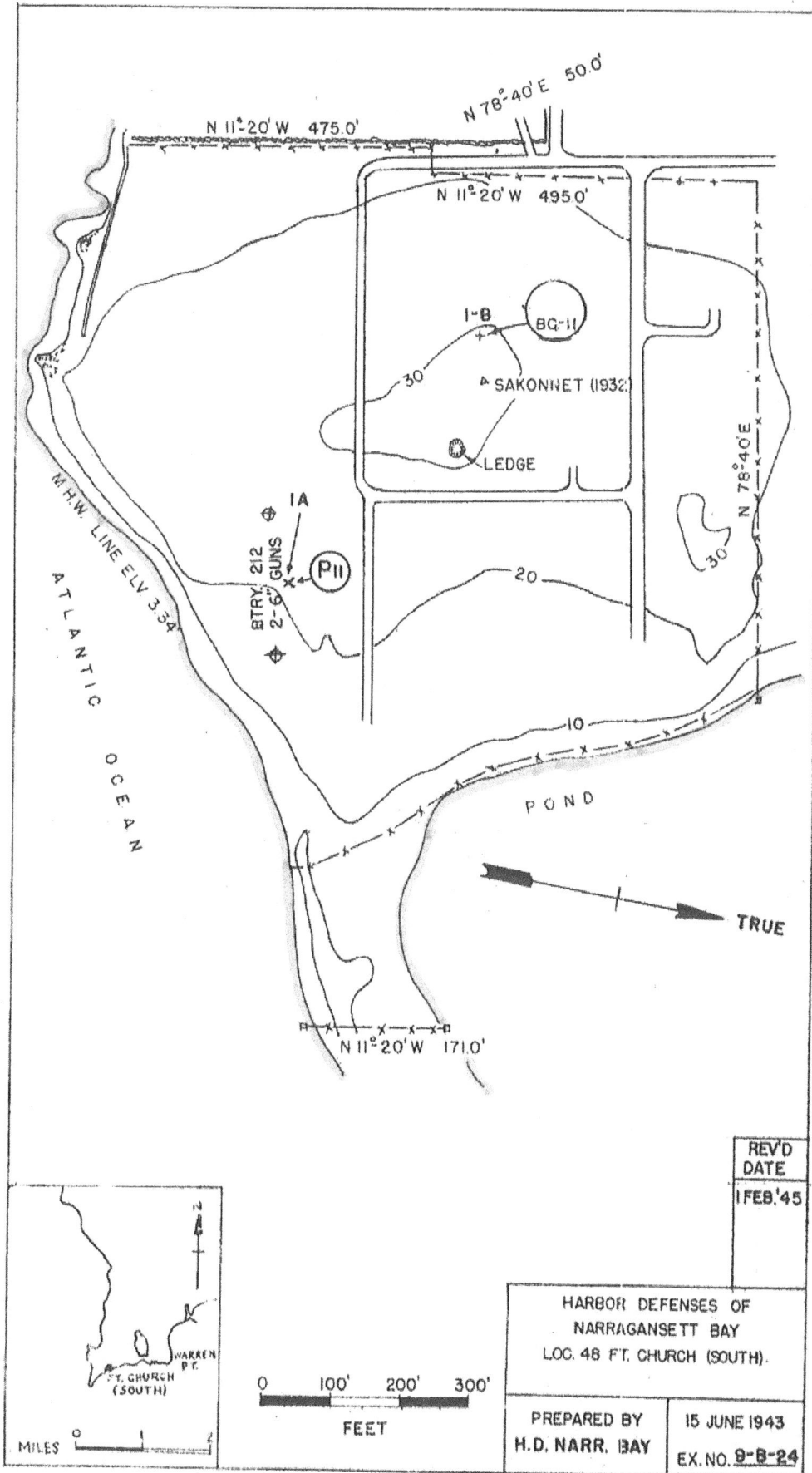

N II°-20' W 475.0'

N 78°-40' E 50.0'

N II°-20' W 495.0'

I-B

BC-11

30

SAKONNET (1932)

LEDGE

IA

BTRY. 212
2-6" GUNS

P II

M.H.W. LINE ELV. 3.34

ATLANTIC OCEAN

N 78°-40' E

30

20

10

POND

TRUE

N II°-20' W 171.0'

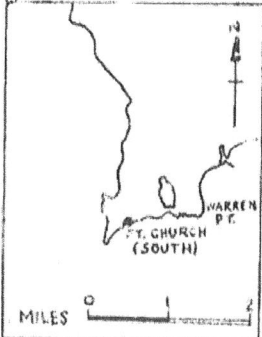

N

WARREN
PT.
FT. CHURCH
(SOUTH)

MILES 0   1   2

0   100'   200'   300'

FEET

REV'D
DATE

1 FEB. '45

HARBOR DEFENSES OF
NARRAGANSETT BAY
LOC. 48 FT. CHURCH (SOUTH).

PREPARED BY	15 JUNE 1943
H.D. NARR. BAY	EX. NO. 9-B-24

Fort Church Gun Batteries

- **GRAY**: Installation of heavy caliber guns had been contemplated for Sakonnet Point on the east side of Narragansett Bay for a decade before World War II. 14-inch railway guns were proposed at one time, but in 1934 authority was approved to emplace a pair of modern 16-inch gun batteries here. Funding during the depression was delayed but eventually work on the Sakonnet site was started on September 27, 1939 and completed on May 15, 1942. This was one of the five pre-1940 Program dual batteries, similar to sister-battery Battery Hamilton built at Point Judith just across the bay. In fact, it was one of the first of the modern series, predated only by San Francisco's two prototypes. It uses the design structure of the early types, with main gallery recessed and having two separate entries into the gun rooms. The battery was located considerably inland, on the western segment of the reservation. It fired on a bearing 5-degrees east of south. A separate plotting room to the northeast was built, and ten base-end stations were built and assigned. When the 1940 Program was adopted, this battery (then actually under construction) was given the designation Battery Construction No. 107. In the September 1940 national priority list, it was assigned #3 and funded with the FY-1940 and FY-1941 Budgets. It was completed for transfer on May 15, 1942 at a cost of $943,000. The battery was armed with two 16-inch guns Model MkIIM1 on Model 1919M3 barbette carriages (#79/#17 and #94/#18). It was named in 1941 for Major Quinn Gray, a Coast Artillery Corps officer. It served as one of the primary defensive batteries during the war. It was removed from service and disarmed under authority of May 14, 1948. The emplacement still exists on the property of the Sakonnet Golf Course, and it has been partly covered with earth. The battery site is closed to the public.

- **REILLY**: This was a battery for 8-inch barbette guns built to complement the long-range fire of 16-inch Battery Gray. It also pre-dates the 1940 Program and had a unique prototype casemate design. Essentially it resembles a scaled-down version of the larger, pre-1940 16-inch dual battery. Gun centers are at 240-feet, the complete gun houses on the same axis line, and heavily covered magazines and power room are in the protected traverse between the gun rooms, though there is no central rear entry. It was proposed for construction as early as 1940, but funding was not available immediately. Work was finally undertaken on February 2, 1941 and completed at year's end. It was transferred on January 15, 1942 for a cost of $248,000. Another $40,000 was spent on the separate protected plotting room to the north. The expense of effort and dollars were considered inappropriate for guns of only 8-inch bore size, and no further batteries of this design were ever built, war time 8-inch gun emplacements were much simpler and less expensive alternates. It was named on General Orders No. 13 of November 24, 1941 for Captain Henry Joseph Reilly, killed on August 15, 1900 during the China Relief Expedition, the name transferred from the battery at Fort Adams. Guns were sent here on July 21, 1941. It emplaced two 8-inch ex-navy MkVIM3A2 guns on Model M1 barbette carriages (#205L2/#3 and #111L2/#4). The battery served throughout the war, being finally removed and disarmed by authority of March 9, 1947. The emplacement survived but was basically filled in and covered with earth by private owners in the early 2000s, so the emplacement today is just a hill.

- **Battery #212**: A 1940 Program dual 6-inch barbette battery located on the southern section of the Fort Church reservation, near, but east of Sakonnet Point. It was funded under the FY-1942 Budget, and in 1940 assigned the very high national priority of #3. It was of conventional dual 6-inch design, with gun centers at 210-feet, and magazines in the protected traverse between the guns. Work was done between November 10, 1941 and October 15, 1942. Transfer was made on August 6, 1943 at a cost of $202,905. The battery was never named, it was just known as Battery

Sakonnett Golf Club (partially buried Battery Gray (Fort Church North) (Terry McGovern)

The site of the buried Battery Reilly (Fort Church East) (Terry McGovern)

Construction No. 212. It was armed with two 6-inch Model 1903A2 guns on Model M1 barbettes carriages (#22/#94 and #65/#95). The emplacement was retained at the end of the war, but then deleted and disarmed in 1948, The emplacement still exists, but it is privately owned with a residence built on top of it. The battery is closed to the public.

Fort Adams (1824-1958) is located on Breton Point to the south of the city of Newport. Fort Adams is one of the largest Third System forts in the United States. The construction of this large fort, which was to mount almost 500 cannons, was begun in 1824 and its final form was completed in 1857. It was named after the second President of the United States, John Adams. During the Endicott Program, six batteries were constructed within the Fort Adam's reservation. Of special interest is a Third System redoubt, located south of the main fort, that was used as the HDCP during World War II and was also the site of a 3-inch anti-aircraft battery. After World War II, the fort reservation was used for U.S. Navy housing, the officers' quarters and some new quarters are still used by the U.S. Navy today. The rest of Fort Adams is now a Rhode Island state park and home to many festivals and events during the year. Fort Adams Trust, a non-profit organization, runs the Third System fort for the state. The Trust runs regular tours, and they maintain a visitors' center and gift shop. An active volunteer organization has been clearing the vegetation on the fort outer works and the Endicott batteries. They have cleaned up and restored the Advanced Redoubt and it is open for semi-regular tours.

Fort Adams Gun Batteries

- **EDGERTON – GREENE:** The mortar battery for the Fort Adams reservation. Along with the battery at Fort Monroe, it was among the first of the "in-line" type of emplacements with the four pits arranged in a line as opposed to the previous 2 x 2 quadrangular design. For protection the battery was situated away from the main front on the eastern side of the reservation, south of old Fort Adams and east of the old redoubt. Plan submission was made on August 29, 1896. The plan was unique. The four small-dimension pits were separated from each other only by a splinter-proof concrete passageway. All magazines were under the parapet in front of the pits, and each pit had its own set, designed to hold 173 rounds per mortar. The slope available allowed the overhead magazine protection to be of almost entirely earth, a large depth of concrete was not required. Work was carried out from September 1896 to June 1898. It was transferred in May of 1898 for a cost of $126,700. The battery was named on General Orders No. 43 of April 4, 1900 for Continental Army Major General Nathaniel Greene. The battery was armed with sixteen 12-inch Model 1890M1 mortars (one was a Model 1890) on Model 1896 carriages (Watervliet tubes #10/#29, #12/#17, #19/#26, #21/#22, #6/#16, #9/#23, #11/#13, #14/#15, Builders tubes #5/#24, #8/#27, #1/#21, #2/#19, and #7/#20 and Bethlehem tubes #7/#25, #29/#28, #7/#25, #23/#18 and #29/#28). In its early years the emplacement suffered some severe problems with wetness in the magazines and passageways. Several attempts were made to line the magazines with various materials, some more successful than others. While partially solved, the battery was still always prone to moisture problems. In 1906 with the reorganization of U.S. mortar batteries, the two eastern pits were split tactically to form Battery Edgerton, named on General Orders No. 20 of January 25, 1906 for Lieutenant Colonel Wright P. Edgerton, a professor at the Military Academy who died in 1904. Two mortars were ordered removed from each pit (8 in total) on May 3, 1918 for use on railway carriages. The eight remaining mortars served up into World War II. In 1932 the battery was placed in a reserve category. In 1940 Battery Greene was renamed Battery Gilmore when the Greene name was transferred to the new fort near Point Judith in General Orders No. 9 of September 11, 1940 for Colonel John C. Gilmore Jr. who had died on July 8, 1934. Final authority to remove the armament came on December 15,

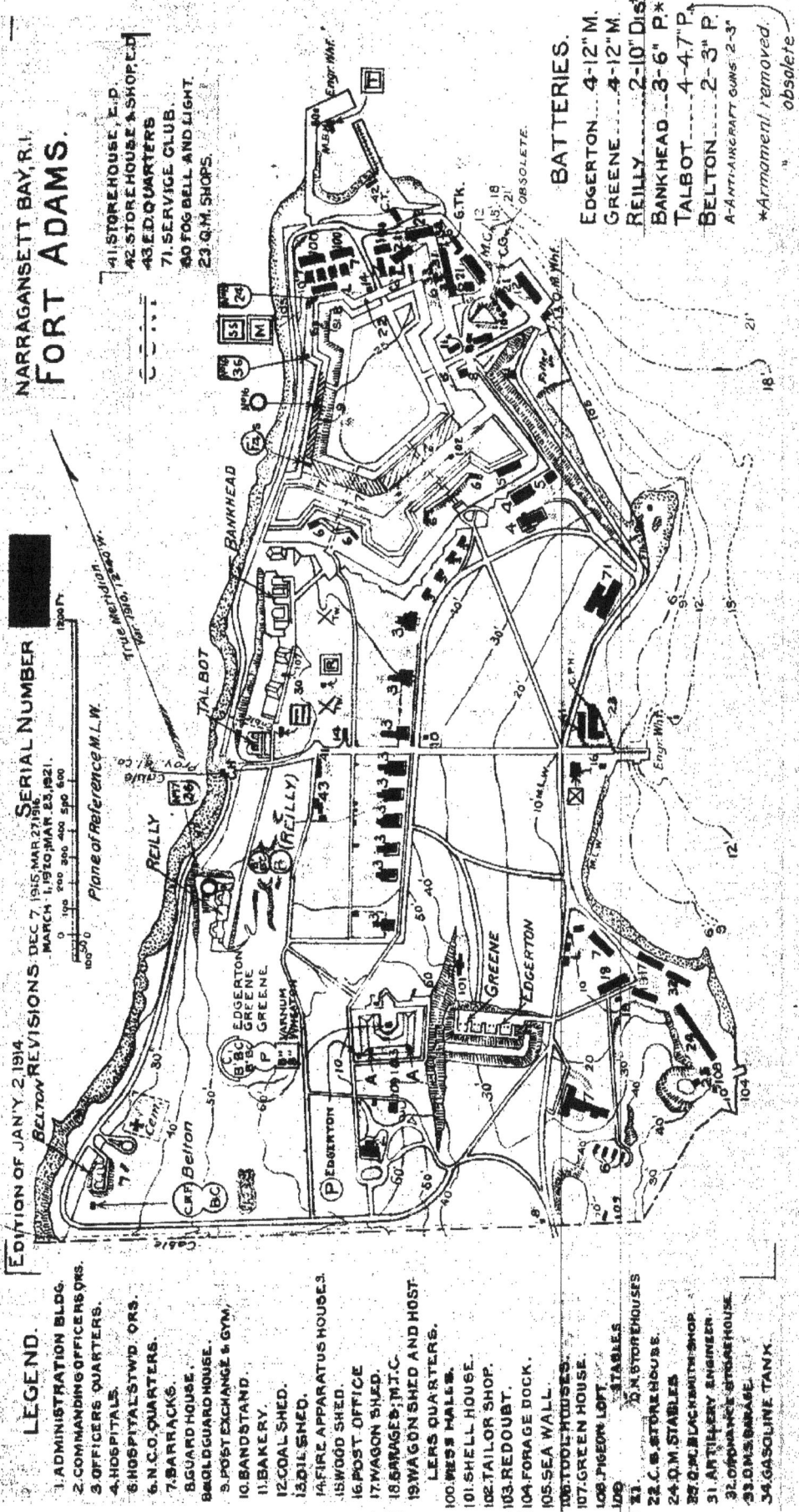

NARRAGANSETT BAY, R.I.

FORT ADAMS

Edition of Jan'y. 2,1914.

SERIAL NUMBER

REVISIONS: DEC. 7, 1915, MAR. 27, 1916, MARCH 1, 1910; MAR. 23, 1921.

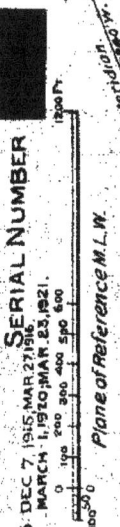

Plane of Reference M.L.W.

LEGEND.

1. ADMINISTRATION BLDG.
2. COMMANDING OFFICERS QRS.
3. OFFICERS QUARTERS.
4. HOSPITALS.
6. HOSPITAL STW'D QRS.
6. N.C.O. QUARTERS.
7. BARRACKS.
8. GUARD HOUSE.
8½. OLD GUARD HOUSE.
9. POST EXCHANGE & GYM.
10. BANDSTAND.
11. BAKERY.
12. COAL SHED.
13. OIL SHED.
14. FIRE APPARATUS HOUSE. 3.
15. WOOD SHED.
16. POST OFFICE
17. WAGON SHED.
18.5. GARAGE; M.T.C.
19. WAGON SHED AND HOST-LERS QUARTERS.
100. MESS HALLS.
101. SHELL HOUSE.
102. TAILOR SHOP.
103. REDOUBT.
104. FORAGE DOCK.
105. SEA WALL.
106. TOOL HOUSES.
107. GREEN HOUSE.
108. PIGEON LOFT.
STABLES.
21. STOREHOUSE
22. C.S. STOREHOUSE.
24. Q.M. STABLES.
25. Q.M. BLACKSMITH SHOP.
31. ARTILLERY ENGINEER
32. ORDNANCE STOREHOUSE.
33. Q.M. GARAGE.
34. GASOLINE TANK.

41. STOREHOUSE, E.D.
42. STOREHOUSE & SHOP, E.D.
43. E.D. QUARTERS
71. SERVICE CLUB.
50. FOG BELL AND LIGHT.
23. Q.M. SHOPS.

BATTERIES.

Edgerton....4-12" M.
Greene.....4-12" M.
Reilly.....2-10" Dis.
Bankhead...3-6" P.*
Talbot....4-4.7" P.
Belton....2-3" P.

A- ANTI-AIRCRAFT GUNS 2-3"

*Armament removed.

obsolete.

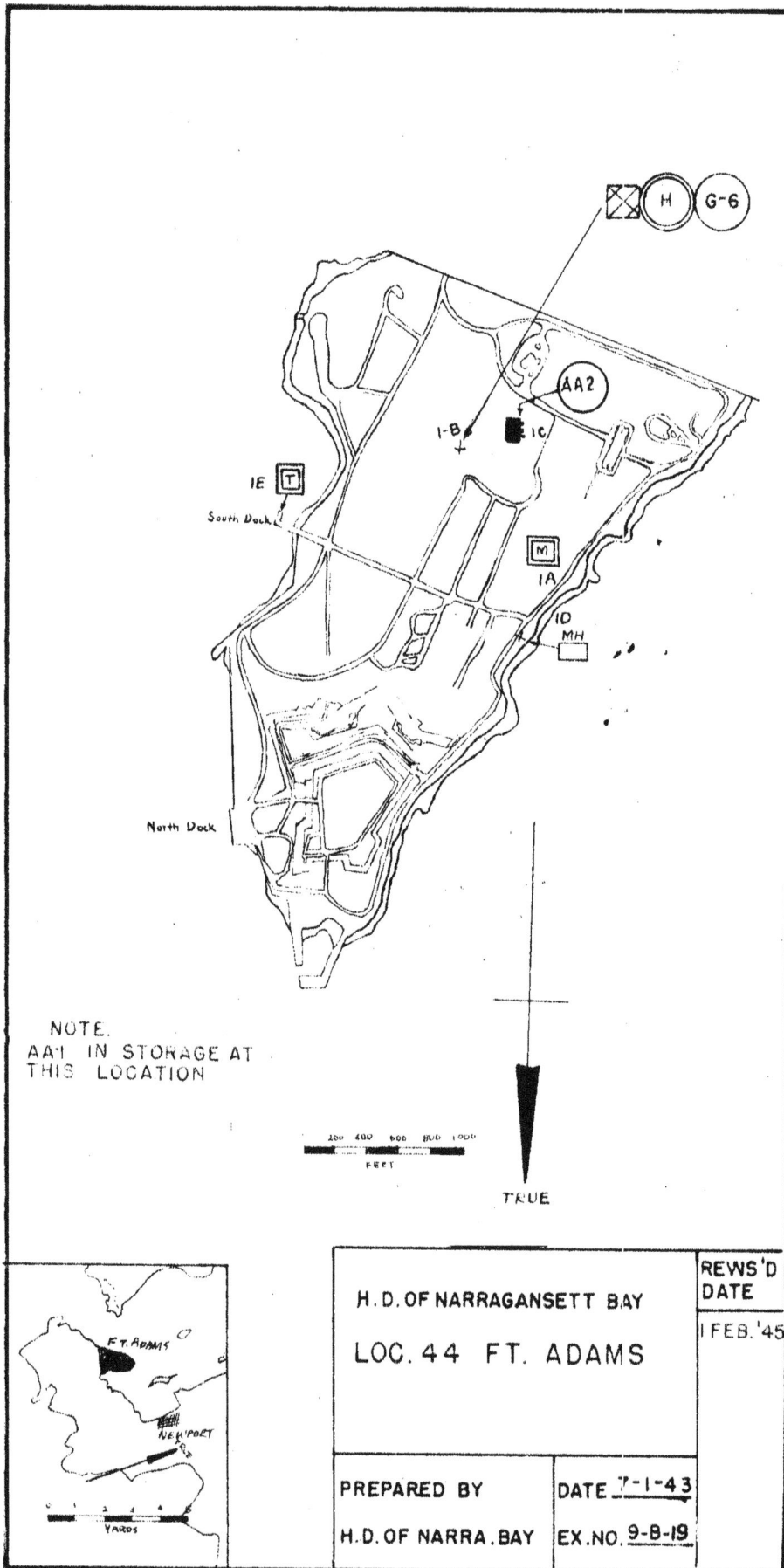

NOTE.
AA1 IN STORAGE AT
THIS LOCATION

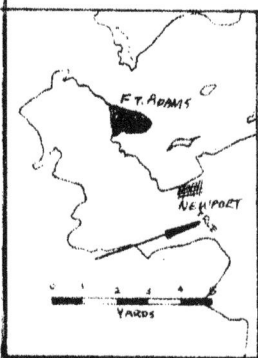

South Dock

North Dock

1E [T]

1-B

1C

AA2

H G-6

M
1A

1D
MH

TRUE

100 400 600 800 1000
FEET

FT. ADAMS

NEWPORT

YARDS

| H. D. OF NARRAGANSETT BAY | REWS'D DATE |
| LOC. 44 FT. ADAMS | 1 FEB. '45 |

| PREPARED BY | DATE 7-1-43 |
| H. D. OF NARRA. BAY | EX. NO. 9-B-19 |

Fort Adams 1941 (NARA)

Fort Adams 1930s (NARA)

1942. The emplacement still exists at the Fort Adams State Park. The battery is open to the public.

- **REILLY**: A battery for two 10-inch disappearing guns emplaced south of the masonry work in the old outworks, firing to the west. Plans were submitted on March 26, 1898. It followed the type plans of mimeograph No. 22, with two separate platforms, No. 1 being a flank type, No. 2 and internal one. Magazines were on the lower left flank of each gun, service with lifts and later chain hoists. Work was done from April 1898 to September 1899. It was transferred on June 15, 1899 for a cost of $92,517.57. It was named on General Orders No. 10 of March 19, 1902 for Captain Henry J. Reilly, 5th U.S. Artillery, who was killed in Peking, China in 1900. It was armed with two 10-inch Model 1888MII Bethlehem guns on Model LF 1896 disappearing carriages (#16/#46 and #23/#41). The battery was given extended loading platforms, a new BC station and hoist modifications during its first decade. The guns were removed for potential railway serve in 1917, one carriage being removed from the emplacement in 1920 and the other in 1925. The battery name was rescinded on General Orders No. 13 of November 24, 1941 to allow it to be used at another new battery at Fort Church. By World War II the old battery BC station was being used as the Harbor Defense Met station. The emplacement, in recent years in rather poor condition, still exists on the property of the Fort Adams State Park. The battery site is open to the public.

- A Spanish-American War emergency emplacement for a single 8-inch gun on modified 15-inch Rodman carriage emplaced temporarily at Fort Adams. On May 2, 1898 the fort was directed to prepare an emplacement using a new Model 1888 8-inch tube with a modified, reinforced 15-inch Rodman carriage which was already at the fort. Work was done without a lot of formality in May, and on the 17th an allocation of $2500 was provided to increase the earth cover over the magazine. It was emplaced in the 1870s water battery which extended just outside the masonry fort on its southwest corner. An already existing 15-inch Rodman platform and adjacent brick magazine were repurposed for this battery. It utilized platform No. 404, the second most northerly of the 1870s platform in this outwork. Bethlehem 8-inch Model 1888 tube #5 was shipped here and received on July 19, 1898. The gun served only a short time, being removed in about 1900 when more modern work took place in other places of the battery. It certainly was gone by 1905. The actual platform and magazine used still exist at the Fort Adams State Park. The battery is open to the public.

- **BANKHEAD**: An unusual battery for three 6-inch guns built into the old 1870s earthwork battery just to the south of old Fort Adams. The fort had been allocated a battery of three 6-inch disappearing guns, but this changed to a battery for 6-inch British Vickers guns before construction. One gun and carriage became available in July 1903 with the decision to build a new 6-inch disappearing battery at Fort Greble (Battery Mitchell) which required the dismounting of the existing Vickers gun and destruction of the simple old emplacement. It was decided to relocate to Adams two similar guns from other forts and complete a new emplacement for three guns, at the site of the old earthen battery. Formal plans were submitted on September 22, 1903. The emplacement closely resembled the features and layout of the most modern 1903 type of battery for 6-inch disappearing guns but modified to hold elevated pedestals for the British RF mounts. A single magazine was in the traverse between No. 1 and 2, and a shared, dual magazine between No. 2 and 3. Work was done during 1904-1905. Transfer was made on July 31, 1907 for a cost of $63,036.46. It was named on General Orders No. 194 of December 27, 1904 for Brevet Brigadier General James Bankhead, U.S. Army officer who served in the War of 1812 and Seminole War. The guns came from Battery Logan at Fort Moultrie, Battery Backus at Fort Screven, and the unnamed 6-inch battery at Fort Greble. It was armed with three 6-inch Vickers guns and pedestal mounts (#12138/#11160, #12137/#11158, and #12134/#11159). These guns and carriages are sometimes mistakenly described as Armstrong guns. Some of the serial numbers are reported differently in other period documents, but the listed serials are those verified in 1907 by careful inspection by local officers. The battery served only a half-dozen years. In 1913 the armament was removed and shipped to Benicia Arsenal prior to going to arm new batteries on Ford Island in Pearl Harbor, Oahu (along with the gun from Battery Hobart at Fort Williams). The emplacement's north traverse magazine was then used for the post fire control telephone switchboard. The emplacement still exists at the Fort Adams State Park. The battery is open to the public.

- **TALBOT**: A battery for two rapid-fire 4.7-inch guns installed during the Spanish American War emergency at Fort Adams. The post was alerted on April 13, 1898 that two of these guns recently purchased in England were assigned to the fort, and to prepare emplacement plans along the design as those for 5-inch RF guns. The emplacement was sited on the southern end of the old 1870s outwork battery and caused the demolition of some of that older work. Concrete work was done in just a couple of months. Transfer was made on January 27, 1899 for a cost of $16,883.76. It was named on General Orders No. 30 of March 19, 1902 for Lieutenant Colonel Silas Talbot of Revolutionary War service. The battery was armed with two 4.7-inch Armstrong guns and pedestals (#12123/#10981 and #12124/#10982). It served with this armament until removed in 1920. Eventually the guns were donated as municipal memorials in Newport and Westerly, Rhode Island. The battery was not subsequently reused for armament. It still exists at the Fort Adams State Park. The battery is open to the public.

- **BELTON**: A battery of two 3-inch guns emplaced on the southern side of the reservation, just to the south and in line with Battery Reilly. Plans were submitted on May 19, 1903, using funds from the Fortification Act of March 3, 1903. It was of standard 3-inch mimeograph type design, with two platforms separated by 62-feet. The central traverse between the guns covered two protected magazines and storeroom. Work was done from 1903 to 1904. Transfer was made on July 31, 1907 for a cost of $15,800. It was named on General Orders No. 194 of December 27, 1904 for Colonel Francis S. Belton of Mexican War service. It was armed during the summer of 1908 with two 3-inch Model 1903 guns on Model 1903 pedestal carriages (#31/#24 and #30/#25). This armament served through World War I but was removed on April 12, 1925 and used to rearm Battery Crittenden at Fort Wetherill. The emplacement still exists at Fort Adams State Park. The battery is open to the public.

Brenton Point Military Reservation (1942-1945) is located at the southwestern tip of Aquidneck Island in the city of Newport, Rhode Island. During World War II, the Budlong home was condemned and purchased by the government in order to erect a coastal defense battery. One of the original four circular concrete Panama mounts, built for towed 155 mm guns, remains in place. Radar and searchlights were also in the area. An anti-motor torpedo boat battery, AMTB 923, was at Brenton Point from July 1943 until moved to Fort Wetherill in July 1944. This battery had four 90 mm guns and two 37 mm guns. At war's end, the battery was dismantled, and the estate returned to the Budlong's, who allowed the property to fall into disrepair. The heavily vandalized manor house was partially destroyed by fire in 1960 or 1961, then torn down in 1963. The state took possession in 1969, and it was then dedicated as a state park in 1974.

Brenton Point M.R. Gun Batteries

- **AMTB #923**: A 1943 Program AMTB battery of two fixed and two mobile 90mm guns built at Brenton Point Military Reservation. Battery #923 was moved in 1944 to Fort Wetherill. Construction was done in 1943 on the site of the middle of the existing 155mm GPF battery on Panama mounts. It served until disarmed in June 1944. The battery site is open to the public.

Fort Adams State Park (Terry McGovern)

Fort Adams State Park (Terry McGovern)

Prospect Hill Military Reservation (1916-1945) was an Endicott and World War II-era fire control reservation located on Conanicut Island. The site is administered by the Jamestown Historical Society and contains the ruins of a revolutionary war fort.

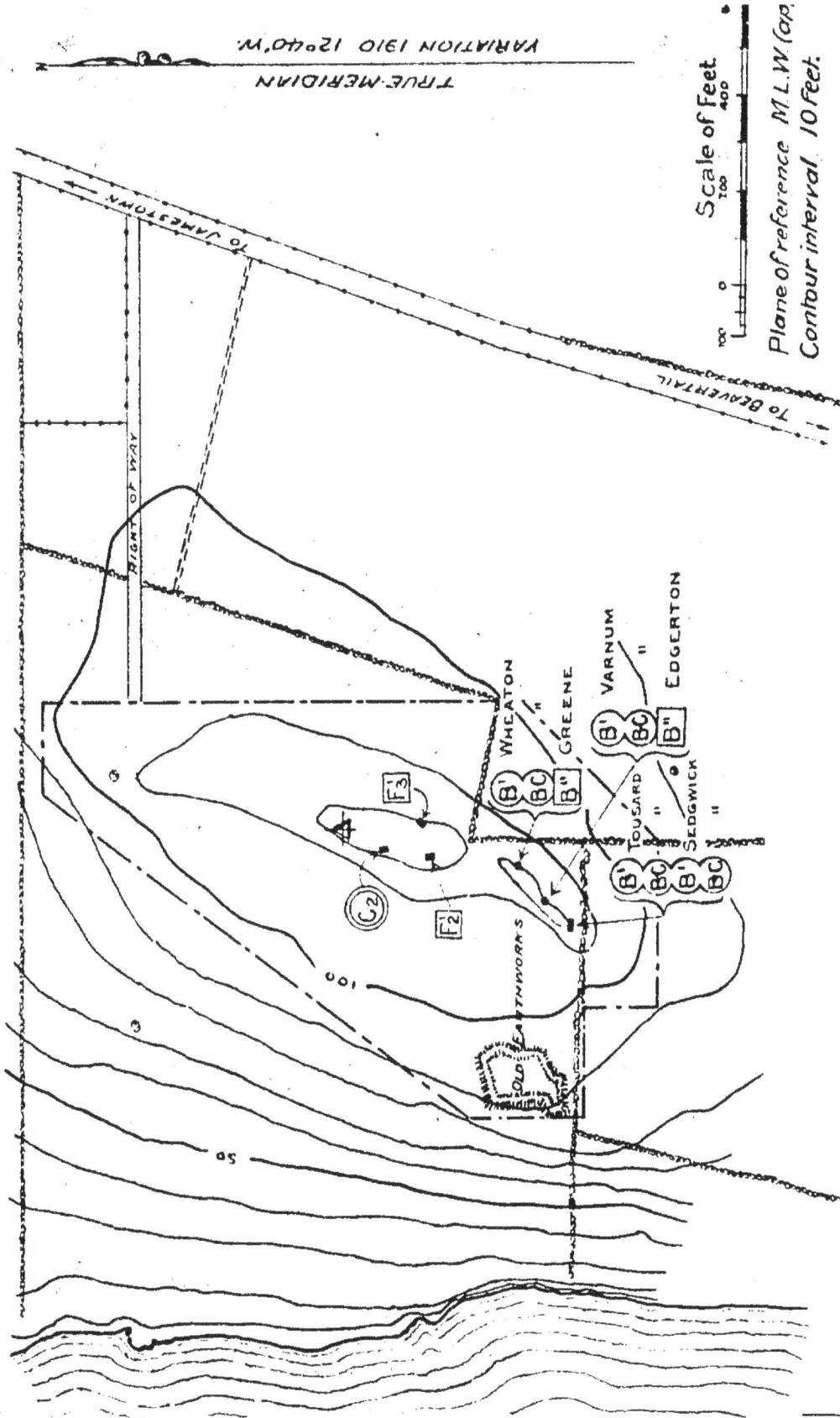

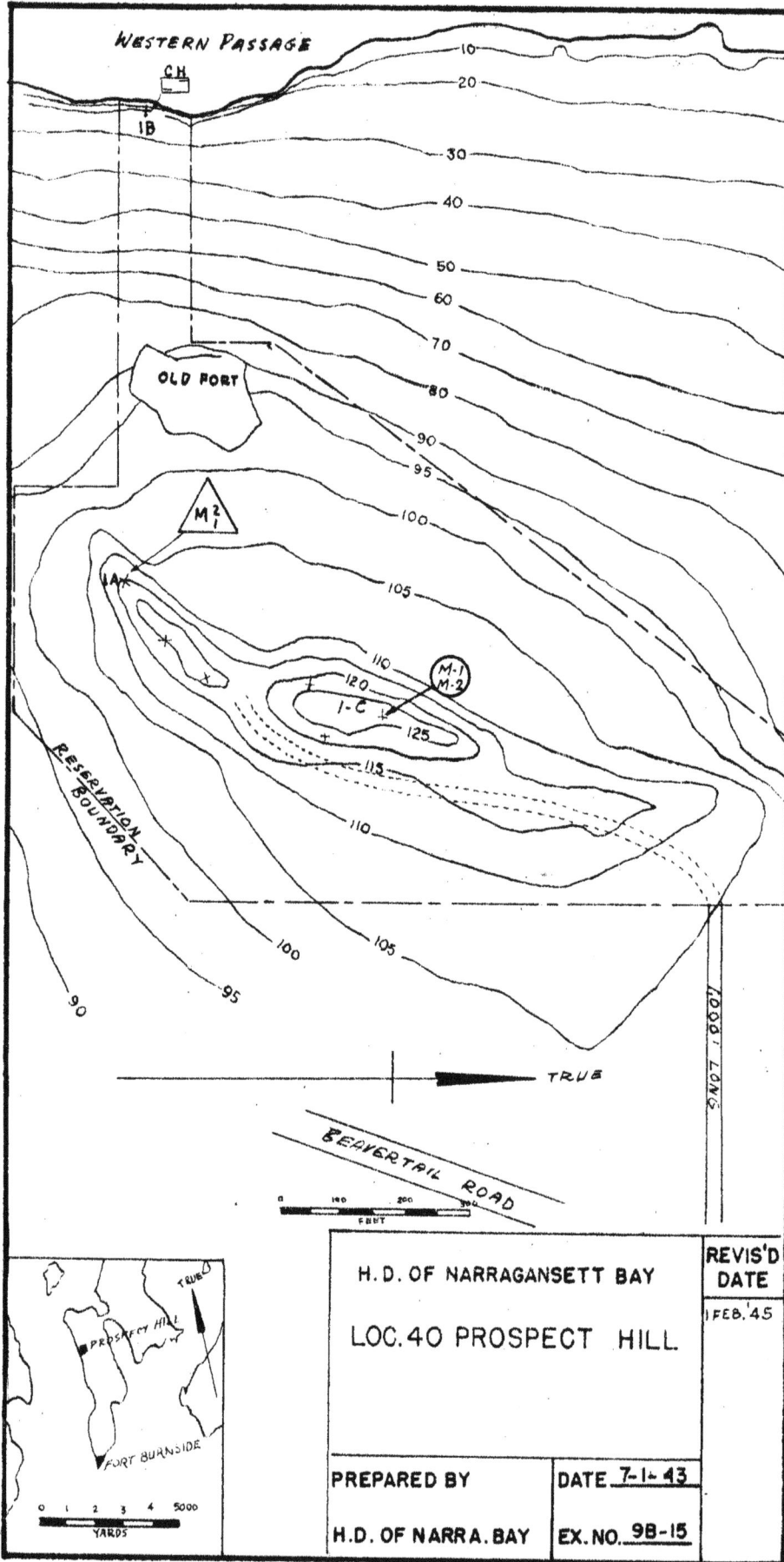

WESTERN PASSAGE

C.H.

IB

10
20
30
40
50
60
70
80
90
95

OLD FORT

M²₁

1A

100

105

110

120

M-1
M-2

1-C

125

115

110

105

100

95

90

RESERVATION BOUNDARY

TRUE

LONG. 1000

BEAVERTAIL ROAD

0   100   200   300
FEET

PROSPECT HILL

TRUE

FORT BURNSIDE

0 1 2 3 4 5000
YARDS

H.D. OF NARRAGANSETT BAY

LOC. 40 PROSPECT HILL

REVIS'D DATE

I FEB. '45

PREPARED BY

H.D. OF NARRA. BAY

DATE 7-1-43

EX. NO. 9B-15

Fort Wetherill (1896-1951) is located on the east side of Conanicut Island at the "Dumplings" across the East Passage from Fort Adams. The location of Fort Wetherill contained batteries from the Colonial Period and First System. The post was developed with three large caliber batteries, four small caliber batteries, and a mine facility during the Endicott Program but had few garrison buildings. It was named in General Orders No. 43 on Apr. 4, 1900 for Capt. Alexander M. Wetherill, U.S. Infantry, killed in action on July 1, 1898 at San Juan, Santiago, Cuba. During World War II a 90mm AMTB battery was located on the post. Anti-submarine nets were set across the East Passage to Fort Adams during Wrld War II. After 1943, its primary activity was as a German prisoner-of-war camp. Used by the U.S. Navy in the 1960s the fort was eventually turned over to the State of Rhode Island in 1972 and became a state park, while the mine complex has been renovated and is used by the Rhode Island State Division of Marine Fisheries. This site is a compact complex consisting of two mine storehouses, a cable tank, a loading room, a mine casemate, the loading wharf, and the remains of a double mine station. This section has Battery Varnum, one of the few older batteries that mounted 12-inch guns in non-disappearing carriages. The western portion of Fort Wetherill has an impressive collection of linked gun batteries (known as a "gun line") with 12 emplacements mounting guns ranging from 12-inch to 3-inch in size which face directly down the East Passage into Narragansett Bay. They offer what is perhaps the longest linear concrete gun line in the coast defenses of New England. This section is an undeveloped and somewhat neglected park area open to the public. Some of the batteries have been partly buried to the loading platforms with many batteries suffering from spalling of exterior concrete surfaces as well as being covered with graffiti.

Fort Wetherill Gun Batteries

- **VARNUM:** A battery for two 12-inch barbette guns emplaced on the rocky southeast coast of the Dumplings, just in front of the old seacoast defense tower on the site. Location selection for the battery was submitted on June 13, 1898 for what was known at the time as the Dumplings Reservation on Conanicut Island. Formal plans and estimates were submitted on July 1, 1898. A low site like this would not normally have been selected for barbette carriages, but the relatively immediate availability of carriages of this type was made from funds for the Spanish American War and this was one of the sites selected to take advantage of this opportunity. The layout of the battery was consistent with type plans, but the exposed situation on the peninsula point required some special considerations. The hill rising behind it forced construction of an epaulement immediately behind the battery which was used as space for supporting and power rooms. Also, the old tower on the Dumplings was destroyed as it offered too tempting an aiming point for enemy fire. Magazines were on the lower right level of each platform, hoists used for ammunition service. The left (No. 2) platform parapet was of unusual shape in order to maximize fire to the left flank to cover the inner reaches of the passage. Work was done from June 1898 to August 1899 for transfer on February 15, 1901 at a cost of $146,703.50. It was named on General Orders No. 78 of May 25, 1903 for Brigadier General James M. Varnum of the Continental Army. It was armed with two 12-inch Model 1888 Watervliet guns on Model 1892 barbette carriages (#5/#32 and #14/#24). In the early 1900s changes were made to the platforms and hoists. It served until final authority to remove armament came on May 27, 1943. The emplacement still exists at the Fort Wetherill State Park. The battery is open to the public.

- **WALBACH:** A battery for three 10-inch disappearing guns emplaced on the western end of the Fort Wetherill reservation. It fired to the south. Plans for two emplacements were submitted on November 14, 1900, soon to be joined by the third gun of the emplacement submitted on July 16, 1901. All three emplacements were in a line, with magazines on the lower level to the left. Am-

NARRAGANSETT BAY, R.I.
FORT WETHERILL.

SERIAL NUMBER

EDITION OF MAR. 27,1916.
REVISIONS; MAR. 1,1920; MAR. 23, 1921.

True meridian
von 1910, 12°40'W.

BATTERIES.

VARNUM __ 2-12"N.DIS.
WHEATON _ 2-12" DIS.
*WALBACH _ 2-10' -
‡ZOOK _____
DICKENSON _ 2-6"P.
CRITTENDEN 2-3"P.
‡COOKE _____ 2-3"P.
*No.3 dismounted.

† Obsolete.
A-Anti-aircraft guns 2-3"

SCALE OF FEET

PLANE OF REFERENCE M.L.W.

LEGEND.
1.
2.
3. OFFICER'S QUARTERS.
4. HOSPITAL.
5.
6. N.C.O. QUARTERS.
7. BARRACKS.
8. GUARD HOUSE.
9. POST EXCHANGE.
10. STOREHOUSES.
11. MESS HALLS.
12. CAMP BUILDINGS.
13. GARAGE.
14. FIRE APPARATUS.
21. Q.M. STONE CRUSHER.
41. E.D. TOOLHOUSE.
42. E.D. STOREHOUSE.
43. E.D. STOREHOUSE AND SHOP.
71. Y.M.C.A.
44. E.D. STONE CRUSHER.

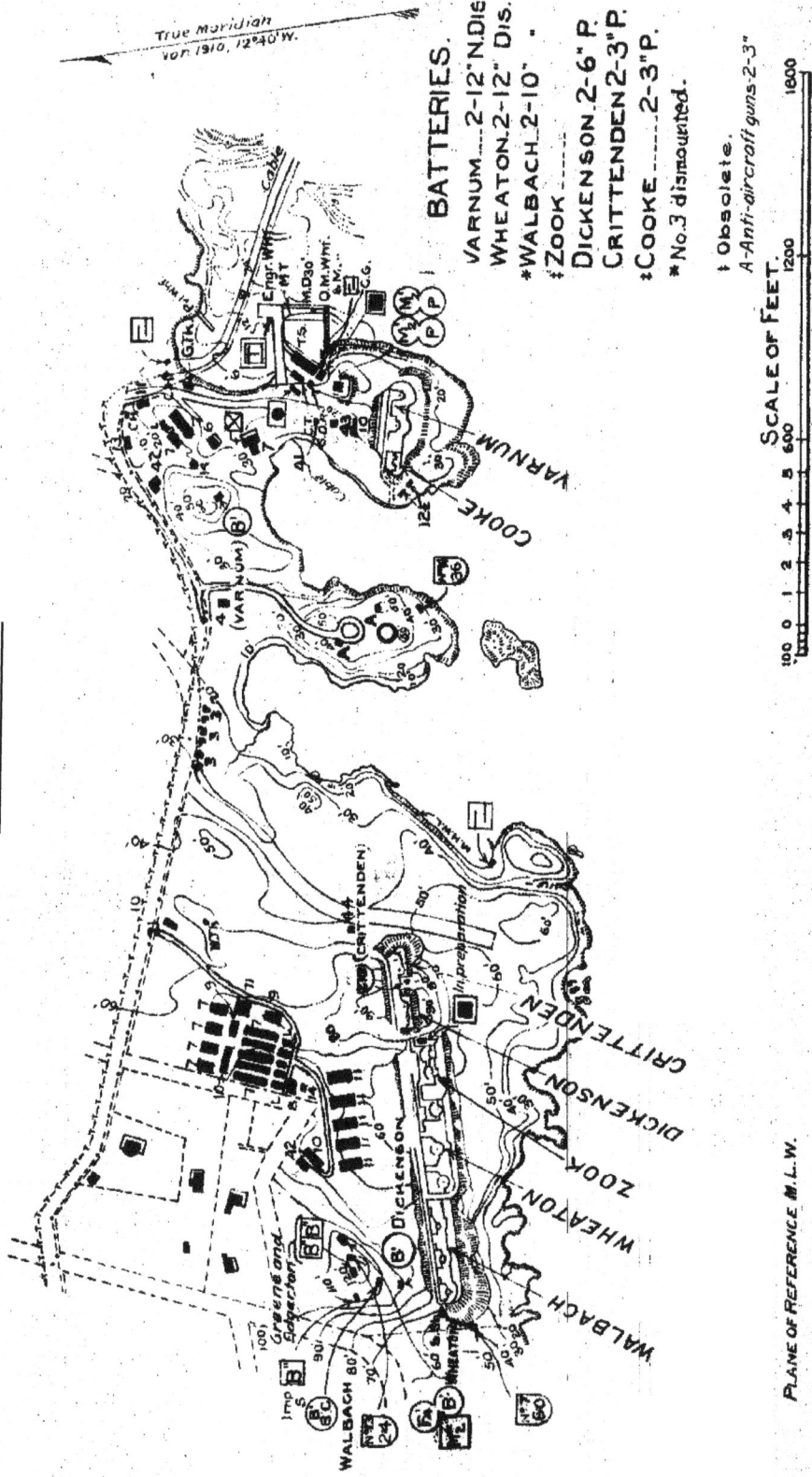

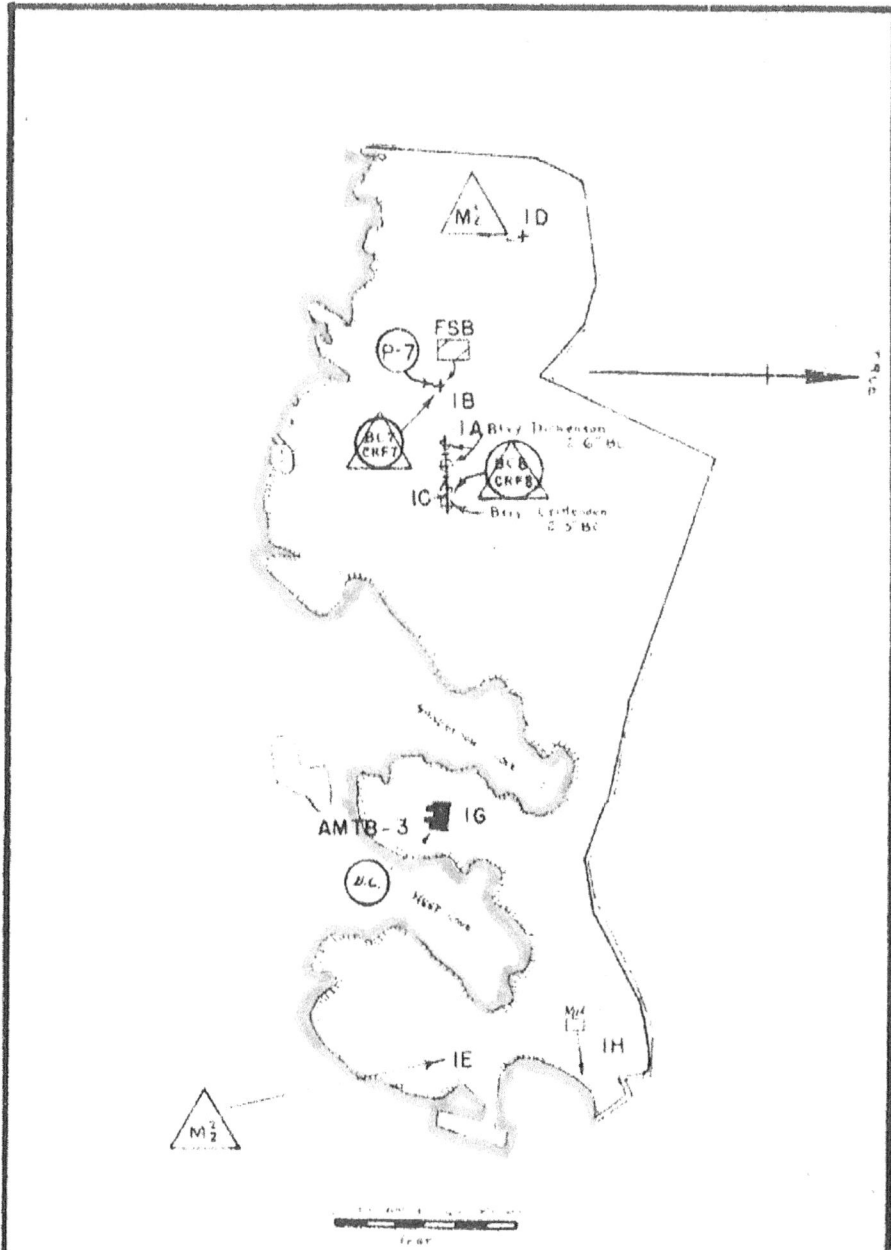

NOTE:
AA-1 IN STORAGE AT LOC. 44

H. D. OF NARRAGANSETT BAY	REVISED DATE
LOC NO. 43 FT. WETHERILL	1 FEB. '45

PREPARED BY H.D. NARR. BAY	DATE 7-1-43
	EX. NO. 9-B-18

(O-165-886-152)(3-5-41-10:25A)(10-600)WETHERILL,JAMESTON

Fort Wetherill 1941 (NARA)

FT. WETHERILL, R.I.

Fort Wetherill 1930s (NARA)

Fort Weatheriill State Park (Terry McGovern)

Fort Wetherill mine complex (Terry McGovern)

munition service was with hoists, though the original ones were changed to electric chain hoists in 1916. While most of the battery closely follows type plans, the No. 3 emplacement differs in its platform—it was designed for a Model 1901 disappearing carriage (rather than the Model 1896 in the other two), and was also given clearance to allow a wider field of fire so it could bear on the eastern channel if required. Work was done from May 1901 to June 1904, for transfer on May 7, 1908 at a cost of $119,170.70. It was armed with three 10-inch guns Model 1888MII Bethlehem guns on two Model 1896 and one Model 1901 disappearing carriage (#41/#3, #45/#2, and #50/#1). It is somewhat surprising that a later Model 1901 or even Model 1895 10-inch gun wasn't selected to go with the most modern Model 1901 carriage. It was named on General Orders No. 78 of May 25, 1903 for Colonel John Walbach, 4th U.S. Artillery who served in the Revolutionary War and War of 1812. An accident occurred at the battery when testing a 10-inch gun on December 7, 1904, when the gun in emplacement No. 1 had a premature detonation. All of the crew were knocked down (but no casualties), the parapet partly blown away, the gun dismounted and the carriage damaged. The carriage took a year to repair at Watertown Arsenal but was eventually returned to its emplacement and the gun remounted. All the guns were removed in 1918 for use on railway carriages, but then the guns in emplacements No. 1 and 2 (guns #41 and #45 on the Model 1896 carriages) were returned and remounted. These final two guns continued to serve until removed in 1936. The carriages remained until authorized for scrapping under authority of December 7, 1942. The emplacement still exists at the Fort Wetherill State Park. The battery is open to the public, but parts are fenced and filled-in.

- **WHEATON**: A battery for two 12-inch disappearing guns emplaced just to the east and adjacent but a little forward of Battery Walbach. It was one of the latest large disappearing batteries built in the continental U.S. Instructions were issued on September 24, 1902 to being planning for the battery to be built with funds from the Act of June 6, 1902. Formal submission of the design was made on December 2, 1902. It was on the knoll just to the east of Battery Walbach. Extensive excavation had to be done, and another factor that affected the construction cost of this, and in fact all of the other batteries here, was the lack of earth—sand or dirt for parapets and fill had to be trucked in from elsewhere at a high cost. Because of the limited space, the plan adopted was the type plan in most respect, with reduced dimensions in order to fit into the space available. Magazines were on the lower left of each battery, served each by two chain hoists. Because of the experience with numerous wet or damp magazines in the construction here at Newport, each magazine has a completely separate interior room with walls and ceiling separated by air passages from the main structure. Work was done from March 1903 to July 1904, for transfer on May 7, 1908 for a cost of $200,000. It was named on General Orders No. 194 of December 27, 1904 for Major General Frank Wheaton of Civil War and Spanish American War service. The battery was armed with two model 12-inch Model 1900 Watervliet guns on Model 1901 disappearing carriages (#11/#2 and #10/#11). The carriages were modified in 1916-18 to allow an additional 5-degrees of elevation and thus increased range. The armament continued to serve for many years. It was slated for replacement by the 1940 Program construction, but actual disarmament and scrapping of the guns and carriages did not occur until 1945. The emplacement still exists at the Fort Wetherill State Park. The battery is open to the public, but parts are fenced off and filled-in.

- **DICKENSON**: A battery for two 6-inch pedestal guns erected on the hilltop of the western peninsula, but to the east of the heavy batteries. Plans were submitted on May 7, 1901. Local engineers rejected the type plan for placing the two guns together on one shared platform and opted for two platforms and two separate lower-level magazines, with shell service by hoists. As a result, it varies somewhat from other current type plans. Work was done from July 1901 to July 1902 for transfer on May 7, 1908 for a cost of $36,500. It was named on General Orders No. 78 of May 25, 1903 for 1st Lieutenant George Dickenson who was killed in 1862 during the Civil War. The battery was armed with two 6-inch Model 1900 guns and pedestal mounts (#36/#36 and #40/#35). It was used extensively for training purposes, during fire exercises in 1922 it sustained some structural damage from a firing accident. In the 1930s-40s it was extensively modified, getting widened platforms, an adjustment for a wider field of fire, with commander's and fire control stations. The battery served until removed from service and the armament scrapped by authority of March 9, 1947. The emplacement still exists at the Fort Wetherill State Park. The battery is open to the public.

- **ZOOK**: A battery for three 6-inch disappearing guns erected in the constrained space between Battery Dickenson and Wheaton. At the time of its construction space was running out on the reservation, but as the cost of additional land to the west was too high, every effort was made to shorten the footprint of the battery. In this case it meant that the traverse between guns No. 2 and 3 was entirely omitted, the battery plan had magazines only between guns No. 1 and 2 and on the left flank of No. 3. Plans were submitted on April 11, 1902, but the work was not funded until passage of the Fortification Act of June 6, 1902. Work was done between May 1903 and November 1904. It was transferred on May 7, 1908 at a cost of $81,475. It was named by General Orders No. 194 of December 27, 1904 for Brevet Major General Samuel K. Zook who died at Gettysburg in 1863. The battery was armed with three 6-inch Model 1903 guns on Model 1903 disappearing carriages (#2/#15, #4/#17, and #29/#18). The gun tubes were removed in 1918 for use on wheeled field carriages. It was not subsequently used for armament, but it was utilized as the site for a mining casemate for a number of years. The emplacement still exists at the Fort Wetherill State Park. The battery is open to the public.

- **COOKE**: A battery for two 3-inch masking parapet guns placed on the eastern peninsula of the fort, immediately adjacent to Battery Varnum. Plans were submitted on March 3, 1899. The site was considered a prolongation of the main works, but special care was exercised to allow the guns to fire over part of the parapet of adjacent Battery Varnum in order to secure coverage of the mine fields. Work was done from April to September 1899. It was transferred on February 15, 1901 for a cost of $11,454.90. The battery was named on General Orders No. 78 of May 25, 1903 for 1st Lieutenant William W. Cooke, 7th Cavalry, killed in action at the Little Big Horn in 1876. The battery was armed with two 3-inch, 15-pounder Model 1898 Driggs-Seabury guns and balanced pillar mounts (#54/#54 and #56/#56). The carriages were modified to Model 1898M1 pedestals about 1916. In common with other such model guns, the armament was declared obsolete and removed in 1920. The emplacement still exists at the Fort Wetherill State Park. The battery is open to the public.

- **CRITTENDEN**: A battery for two 3-inch, rapid-fire guns placed at the eastern end of the main sequence that started with Battery Walbach on the flank. Like the other batteries of this line, it fired to the south. The battery was built from June 1901 to July 1902. The design plan was for a standard type battery intended for 3-inch Model 1898 masking parapet guns, with square, internal platforms with barrel niches, gun centers 29-feet apart, and lower-level magazines on the right flank of each platform. In September 1901 a decision was made to change the mounts to pedestals, and the construction was modified. While the pit shape wasn't changed, the hold-down bolts for the Model 1902 pedestal mounts were adapted. The battery was completed for transfer on May 7, 1908 for a cost of $15,050. It was named on General Orders No. 78 of May 25, 1903 for 2nd Lieutenant John J. Crittenden, 20th U.S. Infantry, killed in action at the Little Big Horn in 1876. It was armed with two 3-inch Model 1902 Bethlehem guns and M1902 pedestal mounts (#28/#28 and #29/#29). In the mid-1920s the battery armament was switched to Model 1903 guns and pedestal mounts that were previously mounted at Battery Belton at Fort Adams (#30/#24 and #31/#25). The removed M1902 guns and mounts, after some time in storage, were installed in the new Exam battery at Fort Story during World War II. Battery Crittenden disarmed after Wold Wa II, authority coming on March 14, 1946. The emplacement still exists at the Fort Wetherill State Park. The battery is open to the public.

- **AMTB #923**: A 1943 Program AMTB battery of two fixed and two mobile 90mm guns built at Fort Wetherill. Originally Battery #923 was scheduled for emplacement here, but a reorganization before construction moved it to Brenton's Point. In June 1944 it was relocated here from Brenton Point. Construction was done in 1944 on the site of the middle lobe or peninsula of the fort, the gun blocks being converted from two 3-inch anti-aircraft blocks previously at the same site. It served until disarmed in late 1945. The blocks still exist at the Fort Wetherill State Park. The battery site is open to the public.

Fort Getty (1900-1948) is located on a small headland on the west side of Conanicut Island due west from Fort Wetherill. It was named on General Orders 78 of 1903 for Bvt. Maj. Gen. George Getty, USV for his Civil War service. This fort in conjunction with Fort Greble and Fort Kearny defend the West Passage into Narragansett Bay during the Endicott Program. This small 31-acre reservation had three Endicott Program batteries and a mine casemate. In 1942, its guns were either scrapped or relocated, however in 1943 a AMTB Battery #922 with four 90 mm guns arrived. Two of the guns were on fixed mounts and two were on towed mounts. The fort was also used during World War II as a German prisoner-of-war camp. Today Fort Getty is a City of Jamestown Park and open to the public. All buildings have been removed, and the two large main batteries have been filled with earth. The park has 83 seasonal RV sites and 24 tent sites, as well as a public dock and pavilion.

Fort Getty Gun Batteries

- **TOUSARD**: A battery for three 12-inch disappearing guns emplaced as the primary armament for Fort Getty on Fox Hill. Plans were submitted on June 15, 1901. The battery was sited on the south central part of the fort, firing to the southwest. The entire reservation was a rocky prominence, requiring bringing in earth for parapets and fill. The plan conformed to standardized mimeo types, with lower-level magazines and hoists. The three-gun sequence was bent, rather than straight, with each emplacement having a slightly different bearing. Work was done from July 1901 to early 1903. Transfer was made on June 7, 1910 for a cost of $222,000. It was named on General Orders No. 78 of May 25, 1903 for Lieutenant Colonel Louis Tousard, of Continental Army service. It was one of the few batteries built to carry the latest model guns and disappearing carriages: three 12-inch Model 1900 Watervliet guns on Model 1901 disappearing carriages (#4/#8, #8/#7, and #7/#9). These were not reported mounted until 1910. In 1918 new back-delivery hoists were added, and about the same time the disappearing carriages were modified to allow another 5-degrees of elevation. The armament served until authority for removal issued on September 7, 1942, though this appears to not have been accomplished until August 14, 1943. The emplacement still exists at the Fort Getty Park. The battery site is open to the public, but the emplacement buried to the top of parapet.

- **HOUSE**: A battery for two 6-inch, rapid-fire pedestal guns emplaced on the immediate right flank of Battery Tousard, also firing in generally the same southwestern direction. Plans were submit-

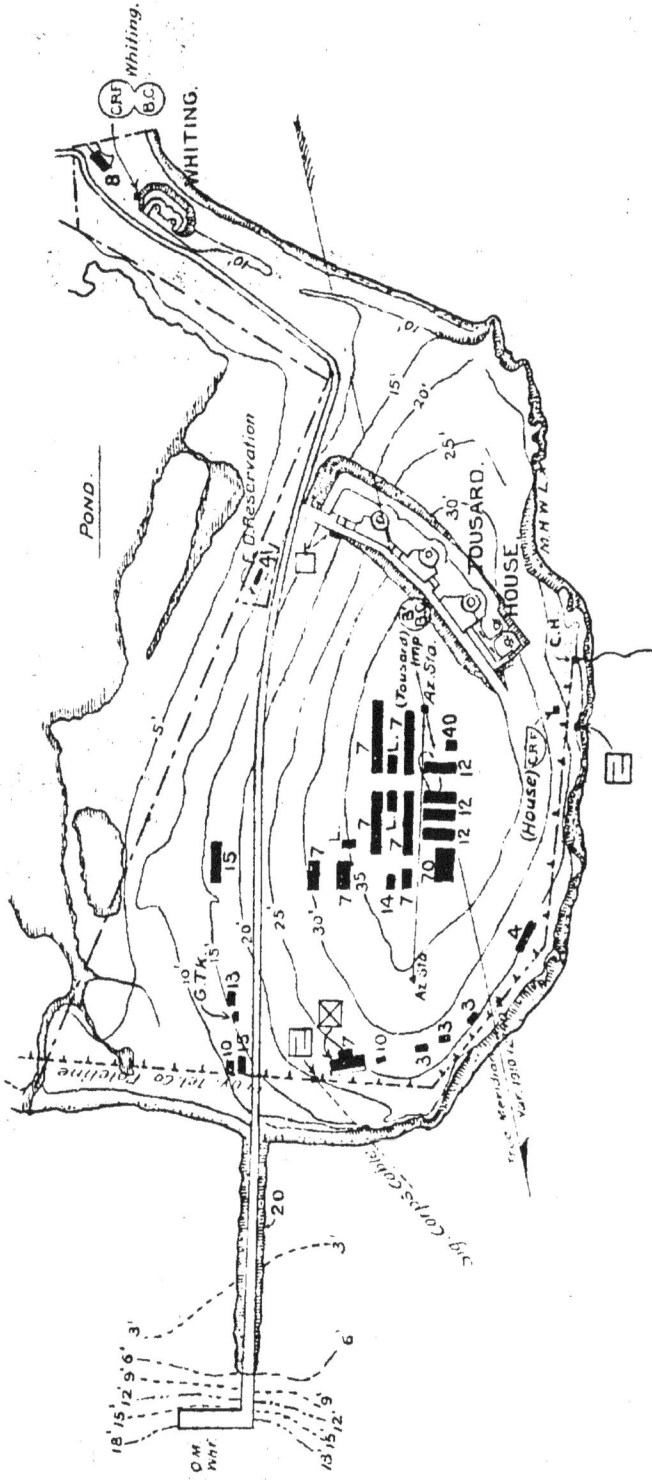

NARRAGANSETT BAY, R.I.
FORT GETTY.

EDITION OF JAN'Y 2, 1914. SERIAL NUMBER **124**
REVISIONS: DEC. 7, 1915; MAR. 27, 1916.
MAR. 1, 1920; MAR. 23, 1921.

Plane of Reference, M.L.W. (approx.)

LEGEND.

1.
2.
3. OFFICERS QRS.
4. HOSP'TAL.
5.
6.
7. BARRACKS.
8. GUARD HOUSE.
9.
10. WOOD SHED.
12. MESS HALLS.
13. GARAGE.
14. FIRE APPARATUS HO.
15. STORE HOUSE.
20. ROCK FILL EMBANK-
       MENT.
40. OFFICE, E.D.
41. STORE HOUSE, E.D.
70. Y.M.C.A.

BATTERIES.

Tousard ..... 3-12" Dis
House ...... 2-6" P.
Whiting ...... 2-3" P.

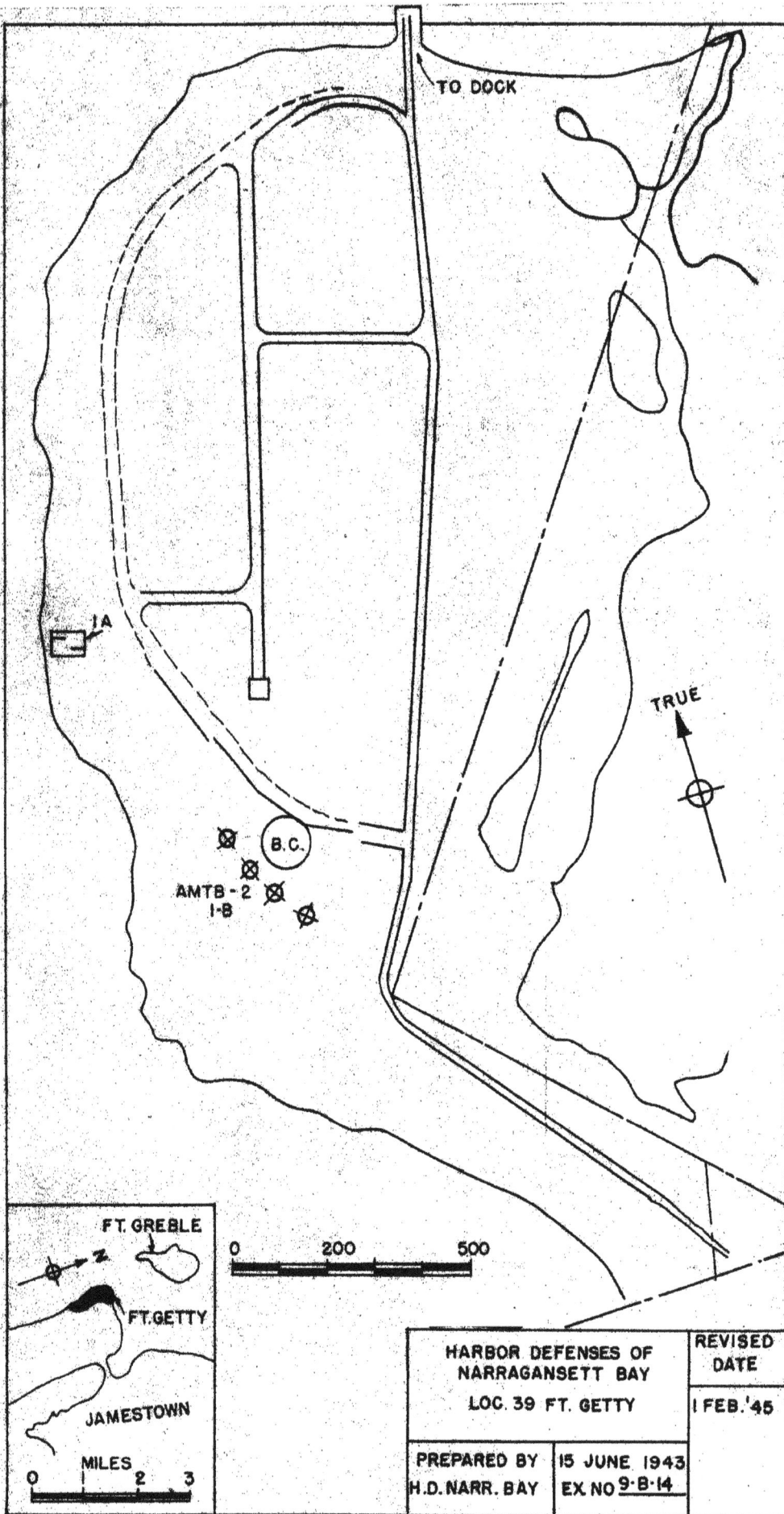

TO DOCK

IA

TRUE

B.C.

AMTB-2
I-B

FT. GREBLE

FT. GETTY

JAMESTOWN

MILES
0   1   2   3

0   200   500

HARBOR DEFENSES OF NARRAGANSETT BAY	REVISED DATE
LOC. 39 FT. GETTY	1 FEB. '45

PREPARED BY	15 JUNE 1943
H.D. NARR. BAY	EX NO 9-B-14

Fort Getty 1924 (NARA)

Fort Getty 1941 (NARA)

Fort Getty Park (Terry McGovern)

Dutch Island Wildlife Management Area (Fort Greble) (Terry McGovern)

ted on August 9, 1901. Initially the battery was begun with the older type plans having two guns share a single platform. However, it was modified during construction to a plan with two separated platforms. Hoists were incorporated to aid shell handling to the guns. Work began immediately in September 1901 and was mostly done by early 1902. Transfer was only made on June 7, 1910 at a cost of $35,292.02. Guns were mounted in 1906 and consisted of two 6-inch Model 1900 guns on pedestal mounts (#30/#28 and #32/#29). The battery was named on General Orders No. 194 of December 27, 1904 for Brevet Brigadier General James House, U.S. Army 1799-1834. The armament was removed for a short period of time from 1917 to 1918 (it is possible that they didn't even leave the reservation). The battery then continued with its armament until World War II. The guns were moved on July 31, 1942 to arm a new emplacement at Fort Varnum. The emplacement still exists at the Fort Getty Park. The battery site is open to the public, but the emplacement is buried to the top of the parapet.

- **WHITING**: An emplacement for two 3-inch pedestal guns for Fox Hill. There was difficulty in finding a location that could cover the minefield and fit in with other works. The original site on the left flank of Battery Tousard was found too rocky and irregular, and the position was fixed further to the south near the shore practically on the approach causeway to the reservation. Plans were submitted on December 15, 1902. The design is unusual, with a single shared magazine in the center traverse. Closeness to the shore required an encircling front retaining wall. Work was done from January to September 1903, and transfer made on June 7, 1910 for a cost of $13,715.13. It was armed with two 3-inch Model 1903 guns and pedestal mounts (#34/#31 and #43/#32). These were mounted in October 1908. It was named on General Orders No. 194 of December 27, 1904 for Lieutenant Colonel Levi Whiting of War of 1812 service. A new CRF station for the battery was added in 1922 just to the southeast of the emplacement. The battery had a long service life, but early in World War II it was decided to relocate these guns to a new position at Fort Burnside. Authority was granted in 1942 for this transfer, and the guns were moved on July 20, 1942. Somewhat damaged, the emplacement still exists at the Fort Gerry Park. The battery is open to the public.

- **AMTB #922**: A 1943 Program AMTB battery of two 90mm fixed and two 90mm mobile guns. It was emplaced in front of Tousard, on the earthen parapet of that battery. The blocks for the fixed guns was built from June 25 to August 6, 1943, transfer being made on December 23, 1943 at a cost of $4,651. Two simple concrete gun blocks and a wooden BC station nearby comprised most of the physical work. The guns were mounted in late 1943 and then dismounted after the war. The blocks still exist at the Fort Getty Park. The battery site is open to the public.

Fort Greble (1860-1947) is located on Dutch Island off the west side of Conanicut Island. The 102-acre island received several earthen batteries during the American Civil War. The 1870s Period saw positions for 15-inch muzzle-loading cannons constructed on the southern end of Dutch Island. The island was rearmed during the Endicott Program with a mortar battery, three-gun batteries, and a controlled submarine mine facility. It was named on General Orders 59 of 1898 for 1st Lt. John T. Greble who was killed in action at Big Bethel, VA, on June 10, 1861. The post had a small set of garrison buildings. Fort Greble was often placed on caretaking status except for use during World War I and World War II. The fort's armament was removed during World War II and the fort was used as a German prisoner-of-war camp. Following the war, it was used as a training facility for the Rhode Island National Guard. The island is now owned by the State of Rhode Island and is designated as a wildlife management area by the state's Department of Environmental Management (DEM). All the garrison buildings have been demolished. In 2016, the U.S. Army Corps of Engineers completed a project to mitigate safety hazards on the island. The island is only accessible by boat and may require formal permission to visit.

Fort Greble Gun Batteries

- **SEDGWICK:** The mortar battery was emplaced on the northern side of Dutch Island, to the north and behind the central hill. Plans were submitted on July 27, 1898. The design was according to mimeo types, with two adjacent wide pits, magazines on each flank and the central traverse rather than under the front parapet. Concerns with possible fire from the rear by ships that might pass through either channel prompted the erection of a large earth parados in the rear of the battery protecting the open pits on the north side. Work was done from September 1898 to November 1900. On December 20, 1900 to work was reported ready for transfer. Transfer was made on January 23, 1901 for a cost of $140,827.37. It was named on General Orders No. 43 of April 4, 1900, for Major General John Sedgwick who was killed in action at Spotsylvania in 1864. The original armament was mounted in July 1901 and consisted of eight 12-inch mortars model 1890M1 on Model 1896 carriages (Watervliet tubes: #66/#182, #68/#177, #69/#181, #88/#195, #91/#176, #97/#180, #99/#179 and Builders tube: #55/#178). Orders to remove four guns and carriages were issued on May 3, 1918, and this was soon accomplished. Removed were #69/#181, #88/#196, #91/#178 and #99/#179. The vacant pits were filled in with concrete in May 1925. The other four mortars served throughout the 1920s and 30s. The battery was finally disarmed with authority granted on November 12, 1942. The armament was removed and scrapped. The emplacement still exists at the Rhode Island Wildlife Management Area of Dutch Island. The battery is open, but the island is closed to the public.

- **HALE:** A battery for three 10-inch disappearing guns emplaced on the southern side of Dutch Island's central hill, firing to the south/southwest basically covering the entrance to the western side of Narragansett Bay. Plans for the first two emplacements were submitted on October 28, 1896. The emplacements followed the trace of a Civil War earthen battery in this location, being to the rear of it by about 50 feet. The No. 1 emplacement was a right flank type, the No. 2 was an internal emplacement, both were built for the Model 1896 disappearing carriage. On December 2, 1897 the third emplacement plans were submitted for the unit on the left flank. While in most regards similar in plan to the other two, it was turned more to the south and fired to the south. All the emplacements had lower-level magazines on their left flanks, ammunition service being by hoists. Difficulties were encountered with the construction. Work was done from March 1897 to May 1898 by a contractor for transfer on June 6, 1898 for a cost of $99,034.05. It was named on General Orders No. 43 of April 4, 1900 for Captain Nathan Hale, a Revolutionary War hero. The original armament emplaced in 1898 was three 10-inch guns Model 1888MII on Model LF 1896 disappearing carriages (tubes Bethlehem Steel #9/#15, #12/#16, and #14/#17). In 1903 the battery was reported to be in very poor shape, being very wet and with inoperable hoists. The usual repairs and modifications to the loading platforms, hoists, and communications were done in the early 1900s. Emplacement No. 3 was disarmed early, with authority for removal coming on May 25, 1918. It was rearmed when the gun was replaced with Model 1888 Watervliet tube #50 in mid-1919 coming from Battery Walbach at Fort Wetherill. The 1932 Review recommended the abandonment of Battery Hale, but final authority to delete it did not come until November 12, 1942. The emplacement still exists at the Rhode Island Wildlife Management Area of Dutch Island. The battery is open, but the island is closed to the public.

- A temporary Spanish-American War battery for a single 6-inch Armstrong pedestal guns emplaced on the western exposure of the main island hill, firing to the southwest. A simple emplacement

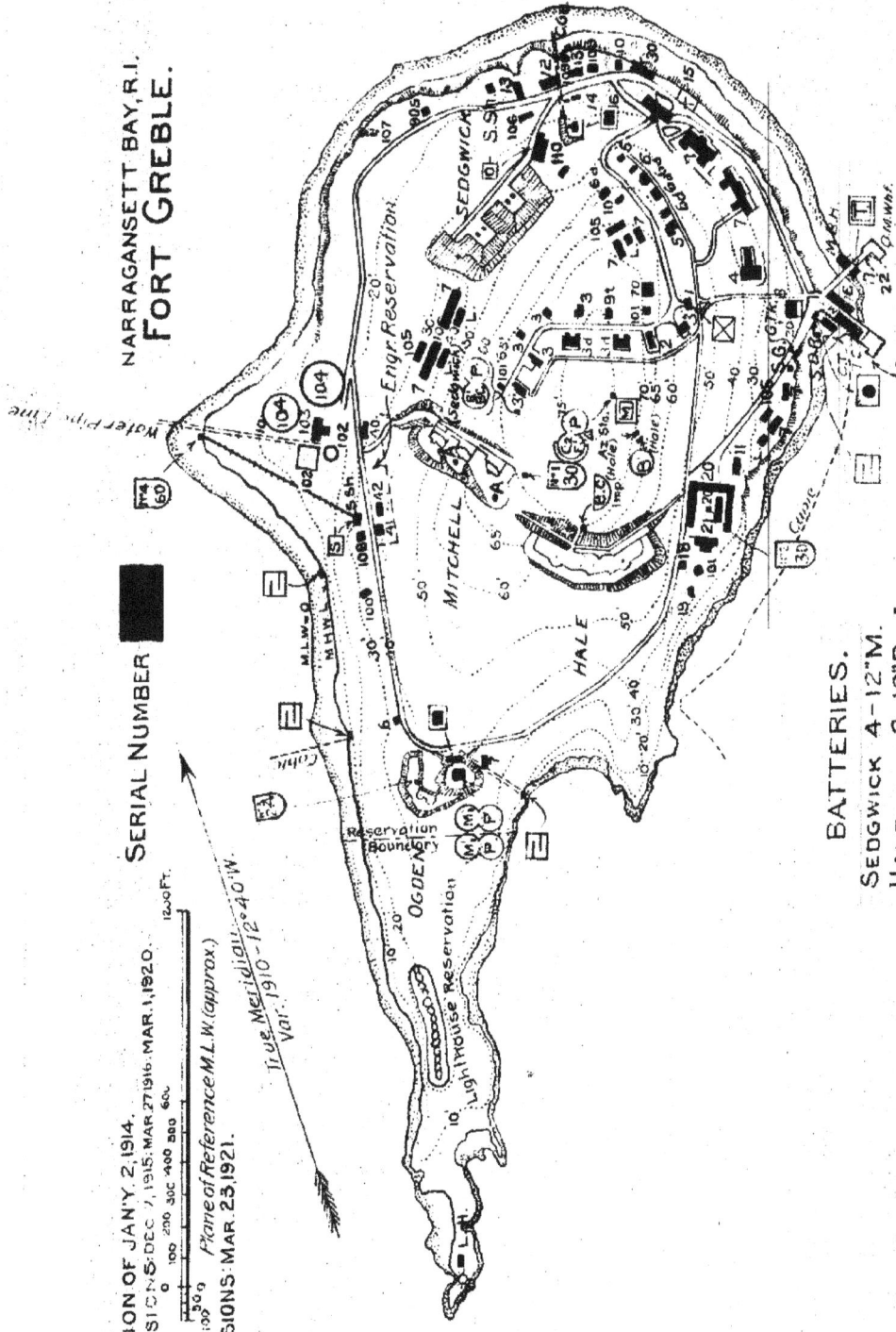

NARRAGANSETT BAY, R.I.
FORT GREBLE.

SERIAL NUMBER

EDITION OF JAN'Y. 2, 1914.
REVISIONS: DEC. 7, 1915. MAR. 27, 1916. MAR. 1, 1920.
REVISIONS: MAR. 23, 1921.

Plane of Reference M.L.W. (approx.)

True Meridian
Var. 1910-12°40'W.

BATTERIES.

SEDGWICK 4-12"M.
HALE   3-10"Dis
*MITCHELL 3-6" "
*OGDEN   2-3" P.
A-ANTI-AIRCRAFT GUNS 2-3"

*Dismounted.

LEGEND.

1. ADMINISTRATION BLDG.
2. COMMANDING OFF. QRS.
3. OFFICERS QUARTERS.
3d. DOUBLE OFFICERS QRS.
4. HOSPITAL.
5. "   STW'D QRS.
6. N.C.O. QUARTERS.
6d. DOUBLE N.C.O. QRS.
7. BARRACKS.
8. GUARD HOUSE.
9. POST EXCHANGE.
10. FIRE STATION.
11. BAKERY.
12. STABLE.
13. WAGON SHED.
14. SPECIAL DUTY MENS QRS.
15. CEMETERY.
16. COAL SHED.
17. WAITING ROOM.
18. BLACKSMITH SHOP.
19. OIL SHED.
100. PLUMBERS QRS.
101. GARAGE.
102. RESERVOIR SHELTERS.
103. PUMP HOUSE.
104. COVERED RESERVOIR.
105. MESS HALL.
106. MILITIA EQUIPMENT
     STOREHOUSE.
107. TARGET BUTT.
108. QUARTERS ENGR.
     OF PUMPING STATION.
109. WOOD SHED.
20. Q.M. AND C.S. ST. H'O
21. QM.ST.H'O. & ART.ENGRS.
22. FERRY SLIP.
30. ORDNANCE ST. HOUSE.
40. OFFICE, E.D.
41. BLACKSMITH SHOP E.D.
42. E.D. CARPENTER
     SHOP & ST. HOUSE
70. Y.M.C.A.
90S. GARAGE.
110 CARPENTER SHOP

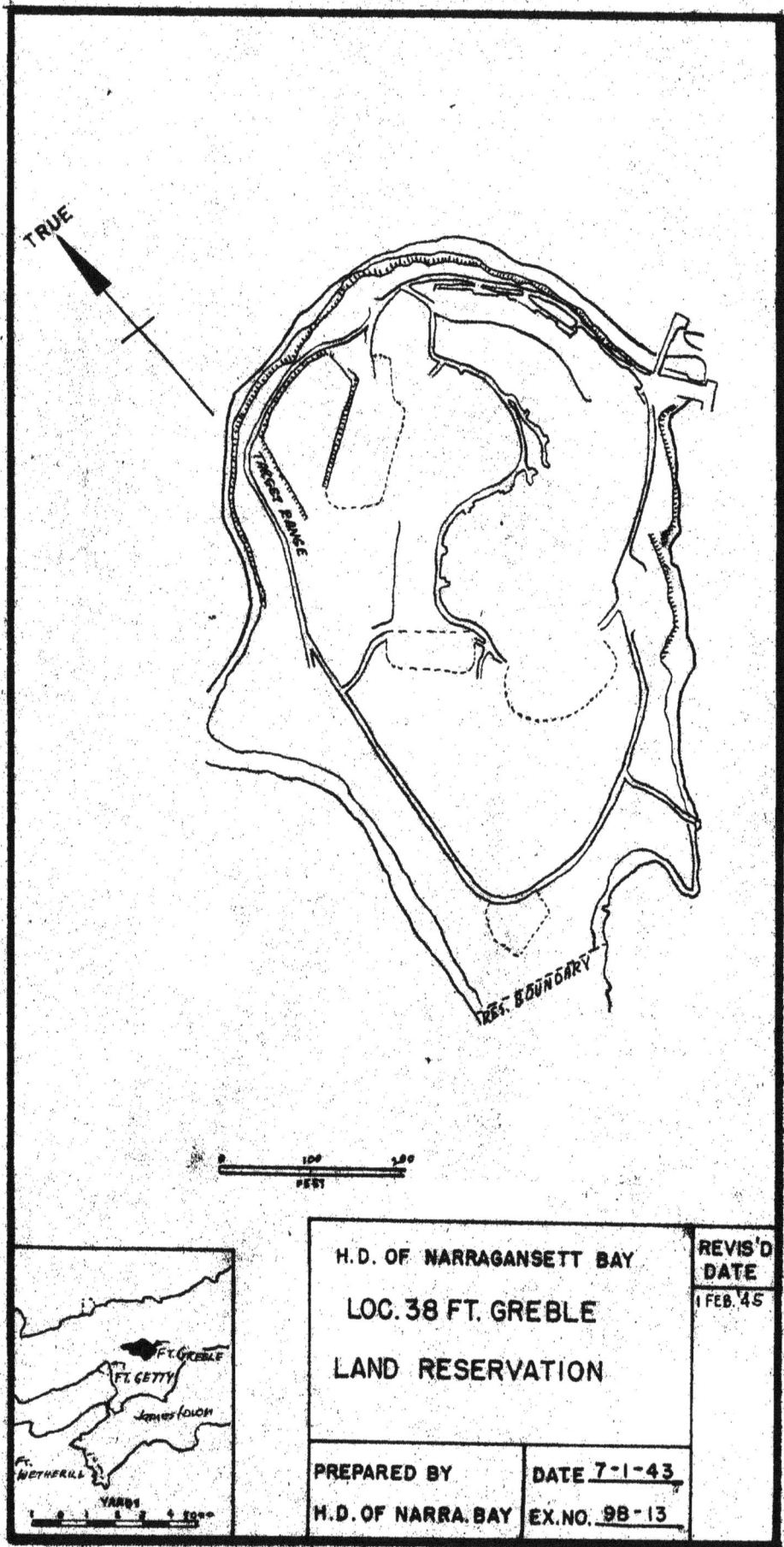

TRUE

TARGET RANGE

RES. BOUNDARY

0   100   200
FEET

H.D. OF NARRAGANSETT BAY

LOC. 38 FT. GREBLE

LAND RESERVATION

REVIS'D
DATE

1 FEB. 45

FT. GREBLE

FT. GETTY

JAMESTOWN

FT.
WETHERILL

YARDS

PREPARED BY

H.D. OF NARRA. BAY

DATE 7-1-43

EX. NO. 9B-13

with just a platform and adjacent magazine was authorized on May 27, 1898 with notification that a British Armstrong purchased 6-inch gun on pedestal mount would be made available. Work was done in 1898, the gun mounted in August, and transfer to service troops made in February 1899. It carried one 6-inch gun and pedestal ($12134/#11163). Apparently, the hurried construction work was not impressive. In early 1903 the emplacement was reported to be in poor condition and recommended for complete rebuilding. In addition, the plans for an expanded (3-gun) 6-inch disappearing gun in the vicinity overtook this effort. On May 27, 1903 approval was granted to dismount and move the gun to Fort Adams and to destroy the emplacement to allow the complete construction of Battery Mitchell, the single emplacement for an Armstrong gun was never named and was completely obliterated in 1903.

- **MITCHELL:** A battery for three 6-inch disappearing guns emplaced on the southwest side of the island's central rise, roughly as a prolongation of the western flank of Battery Hale. In 1897 defense plans dropped the project for two 8-inch disappearing guns on this side of the hill and substituted a battery for three 6-inch guns. After some discussion, and the decision to remove the Spanish American War temporary emplacement for a 6-inch Armstrong gun, plans for a battery of modern Model 1903 disappearing 6-inch carriages were submitted on September 11, 1903. The standard plans were modified to adapt to the sloping site. The three emplacements are in line, and all are standard internal emplacements. However, each is 10-feet different in elevation from its neighbor, sloping downhill to the No. 1 emplacement. Also, each has its own magazine in the traverse on its left flank, there is not the usual shared magazine between pairs of emplacements. Work was done between August 1903 and December 1904. Transfer was made on December 29, 1905 at a cost of $85,360. The battery was named on General Orders No. 78 of May 25, 1903 for Captain David D. Mitchell, 15th U.S. Infantry who was killed in 1900 during the Philippine Insurrection. The armament mounted on January 2, 1906 were three 6-inch guns Model 1903 on Model 1903 disappearing carriages (#32/#19, #37/#20, and #38/#21). In common with many other batteries of this type, it was disarmed during World War I to release tubes for use on wheeled field artillery. The emplacement was subsequently used for fort support—as a mine casemate, switchboard, and two antiaircraft blocks were placed on the front parapet after World War I. The emplacement still exists at the Rhode Island Wildlife Management Area of Dutch Island. The battery is open, but

Fort Greble 1924 (NARA)

Dutch Island Wildlife Management Area (Fort Greble) (Terry McGovern)

HECP-HDCP in Beavertail State Park (Fort Burnside) (Terry McGovern)

the island is closed to the public.

- **OGDEN**: An emplacement for two 3-inch rapid-fire guns emplaced towards the southern end of the island. It was located within the structure of a Civil War and 1870s earthen "Upper Battery", and almost 500-feet from the main gun line on the hill to the rear. Plans were submitted on June 6, 1899. The plan used was typical for the 15-pounder masking parapet type. The left emplacement was rounded as a flank emplacement for wider fire, the right-hand one was an internal emplacement. Platforms were separated by 29-feet, and trunnion height was reported at 47.5-feet. Magazines were below and on the left flank of each platform, ammunition service was by hand. The battery fired to the southwest to cover the minefield in the western channel. Work was done from May to October 1900. Transfer was made on October 24, 1900 for a cost of $9,576.86. It was named on General Orders No. 30 of March 19, 1902 for 1st Lieutenant Frederick C. Ogden killed in action at Trevilian Station VA in 1864. It was armed with two 15-pounder, 3-inch Model 1898 guns and balanced pillar mounts (Driggs-Seabury #53/#53 and #52/#52). The pillars were modified to the M1898M1 pedestal standards about 1916. The armament was declared obsolete and removed in June 1920. The emplacement still exists at the Rhode Island Wildlife Management Area of Dutch Island. The battery is open, but the island is closed to the public.

Fort Burnside (1940-1948) is located on Beavertail Point, the southern end of Conanicut Island covering 118 acres. It was named on General Orders 65 of 1942 for Maj. Gen. Ambrose E. Burnside, U.S. Army. A 1940 Program 6-inch battery (#213) and a relocated 3-inch battery (Battery Whiting) were constructed during World War II to cover both the East and West Passages into Narragansett Bay. A fourth 16-inch casemated battery (#110) was planned but not built. A large HECP was constructed to appear as a summer house. Fort Burnside's military use ended in 1974, and it is currently Beavertail State Park and open to the public, though the HECP/HEDP structure is not open. The site is largely overgrown.

Fort Burnside Gun Batteries

- *Battery #110* (planned): A proposed 1940 Program dual 16-inch barbette battery to be emplaced at the tip of Beavertail in Narragansett Bay. With the completion of the earlier 16-inch batteries at forts Greene and Church, this unit was assigned a low priority for construction. On the initial program priority list of September 11, 1940, it was given priority #25, and a year later on August 11, 1941 #26. Funding was never authorized for construction, the entire project was deleted under authority of November 13, 1942.

- **Battery #213**: A 1940 Program dual 6-inch barbette battery built at the Beavertail reservation of Fort Burnside. It was funded under the FY-1943 Budget, but work proceeded relatively quickly. It was of standard 200-series type design. It was placed centrally on the reservation, firing to the southeast. Work was done from February 19, 1942 to June 30, 1943 for transfer on December 23, 1943 at a construction cost of $239,623. It was never named, being known during construction and service at Battery Construction No. 213. It was armed with two 6-inch guns Model 1905A2 on Model M1 barbette carriages (#18/#62 and #25/#63). The emplacement served in the latter stages of the war and was approved for deletion postwar in 1947. The emplacement still exists at the Beavertail State Park. The battery is open to the public.

- **NEW WHITING**: The defenses of Narragansett Bay included some unique transfers of older Endicott guns to substantial new emplacements. Two 3-inch guns from old Battery Whiting at Fort Getty were relocated in 1942 to a new position at Fort Burnside. It was on the southeast side of

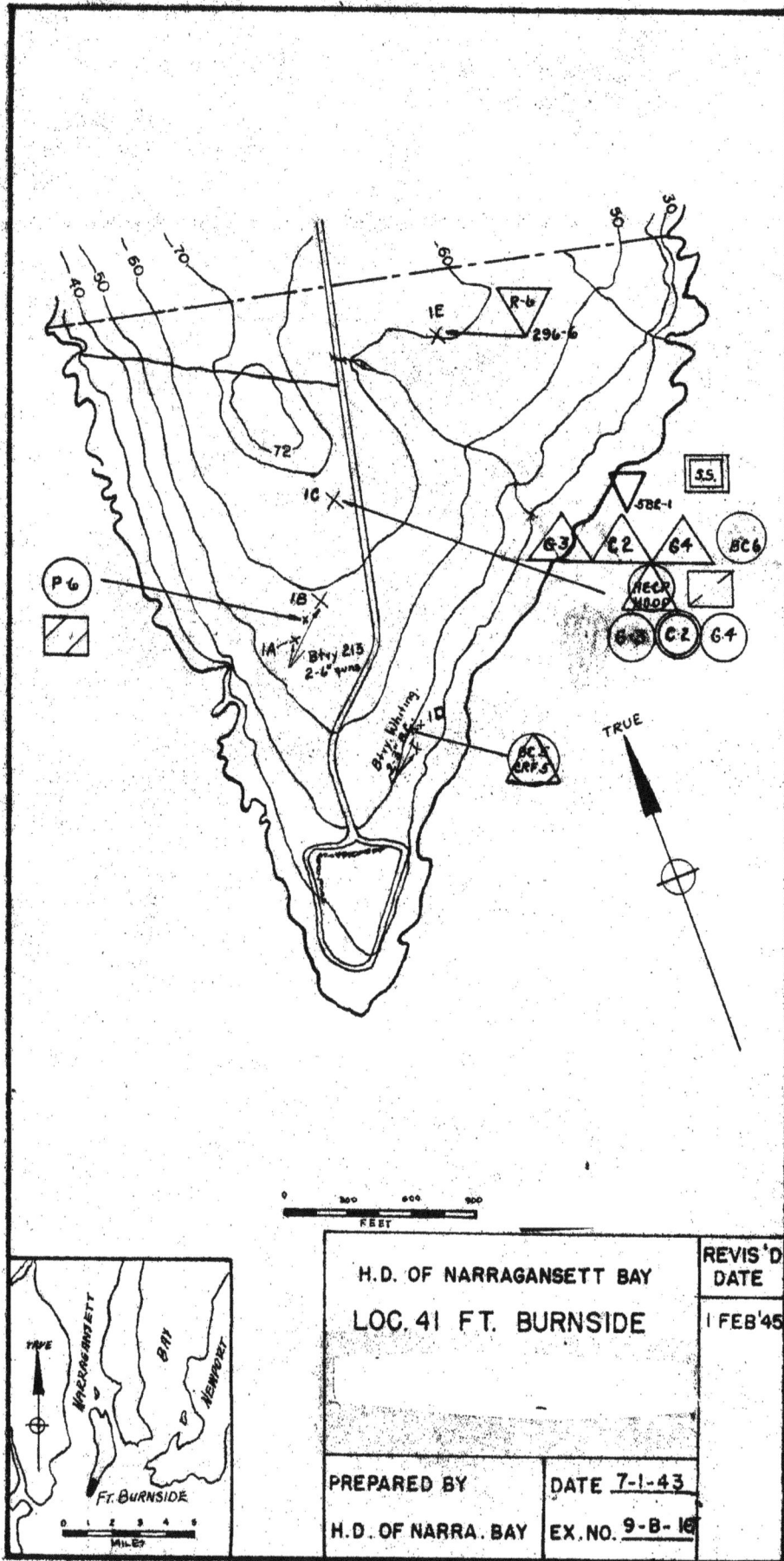

H.D. OF NARRAGANSETT BAY

LOC. 41 FT. BURNSIDE

	REVIS'D DATE
	I FEB '45

PREPARED BY	DATE 7-1-43
H.D. OF NARRA. BAY	EX. NO. 9-B-18

the peninsula, to the southeast of Battery #213. They were to function in the anti-motor torpedo boat role, pending development and construction of new positions for 90mm guns. This battery was a complete and substantial design, with two-gun blocks, adjacent, protected traverse magazine and BC station. Work was done from May 5 to September 15, 1942. It was transferred on August 29, 1942 for a cost of $30,980 with two 3-inch Model 1903 guns and pedestals mounts (#34/#31 and #43/#32). These served until removed under authority of March 14, 1946. The emplacement still exists at the Beavertail State Park, though heavily overgrown and partially filled in. The battery site is open to the public.

Fort Kearny (1906-1946) is located at South Ferry overlooking the East Passage into Narragansett Bay. The fort was established in 1906, and three Endicott Program batteries were constructed. It was named on General Orders 194 of 1904 for Maj. Gen. Philip Kearny, who was KIA at the Battle of Chantilly, VA on Sept. 1, 1862. Fort Kearny was placed on caretaking status soon after completion and remain so except for World War I and World War II. During World War II it was used as a German prisoner-of-war camp until 1946. The former fort is now used by the University of Rhode Island. Of special interest is Battery French, a 6-inch battery that mounted four guns. A research nuclear reactor has been built upon the battery's two center emplacements, while Batteries Cram and Armistead have been refurbished as office space. Parts of the former batteries can be visited by the public.

Fort Kearny Gun Batteries

- **FRENCH:** A battery for four 6-inch disappearing guns and the first Endicott battery for this reservation at Boston Neck. It was placed about in the center of the reservation and fired to the southeast. Plans were submitted on May 19, 1903, using funds from the act of March 3, 1903. All four emplacements were placed in line. It closely followed the mimeograph type plans. Each end emplacement was designed for a flank fire type, with two interior types between. There was a shared magazine between emplacements No. 1 and 2, and another between No. 3 and 4. A powerplant was in the smaller traverse between platforms No. 2 and 3. Work was done from April 1904 to November 1906. Transfer was made on May 7, 1908 at a cost of $102,427.64. It was named on General Orders No. 194 of December 27, 1904 for Major General William H. French of Mexican and Civil War service. The battery was armed with four 6-inch Model 1905 Watervliet guns on Model 1903 disappearing carriages (#23/#70, #24/#10, #25/#9, and #26/#69). To release the tubes for mounting on wheeled field carriages, the battery was disarmed in 1917, and the carriages scrapped in place in 1922. The emplacement was not subsequently used. After the war the property went to the university, and the emplacement was heavily modified for new structures. A research nuclear reactor was built in the center of the battery. Still, remains of the emplacement exist at the University of Rhode Island Bay Campus. The battery site is open, but interior is closed to the public.

- **CRAM:** A battery for two 6-inch disappearing guns, located just on the northeast flank of Battery French, firing to the southeast. Plans were submitted on July 16, 1904. It followed typical Type 1903 plans for 6-inch disappearing batteries, with two flank emplacements and a single shared magazine in the central traverse between the platforms. Work was done from October 1904 to November 1906 and was transferred with the other two batteries at the fort on May 7, 1908 for a construction cost of $65,580. It was named on General Orders No. 194 on December 27, 1904 for Brevet Major General Thomas J. Cram of Topographical Engineers and Civil War service. It was armed with two 6-inch Model 1905 guns on Model 1903 disappearing carriages (#14/#8 and #15/#82). Unlike Battery French, the armament was retained during World War I. The battery

NARRAGANSETT BAY, R.I.
FORT PHILIP KEARNY

EDITION OF JANY. 2 1914.   SERIAL NUMBER **124**
REVISIONS. DEC.7, 1915; MAR.27,1916.
ADMINISTRATION BLDG. MAR.1,1920; MAR.23,1921.

LEGEND
1. ADMINISTRATION BLDG.
2.
3. OFFICERS QRS.
4. HOSPITAL.
7. BARRACKS.
8. GUARD HOUSE.
10. STORE HOUSE.
11. OLD WELL.
12. MESS HALL.
13. GARAGE.
14. FIRE APPARATUS HO.
15. PUMP SHELTER.
16. STABLE.
20. PUMP HOUSE. Q.M.
41. OLD STONE ST.HO. E.D.
43. OLD SHOP E.D.
90.

BATTERIES.
*FRENCH......4-6"DIS.
CRAM..........2-6" "
ARMISTEAD 2-3" P.
*DISMOUNTED.

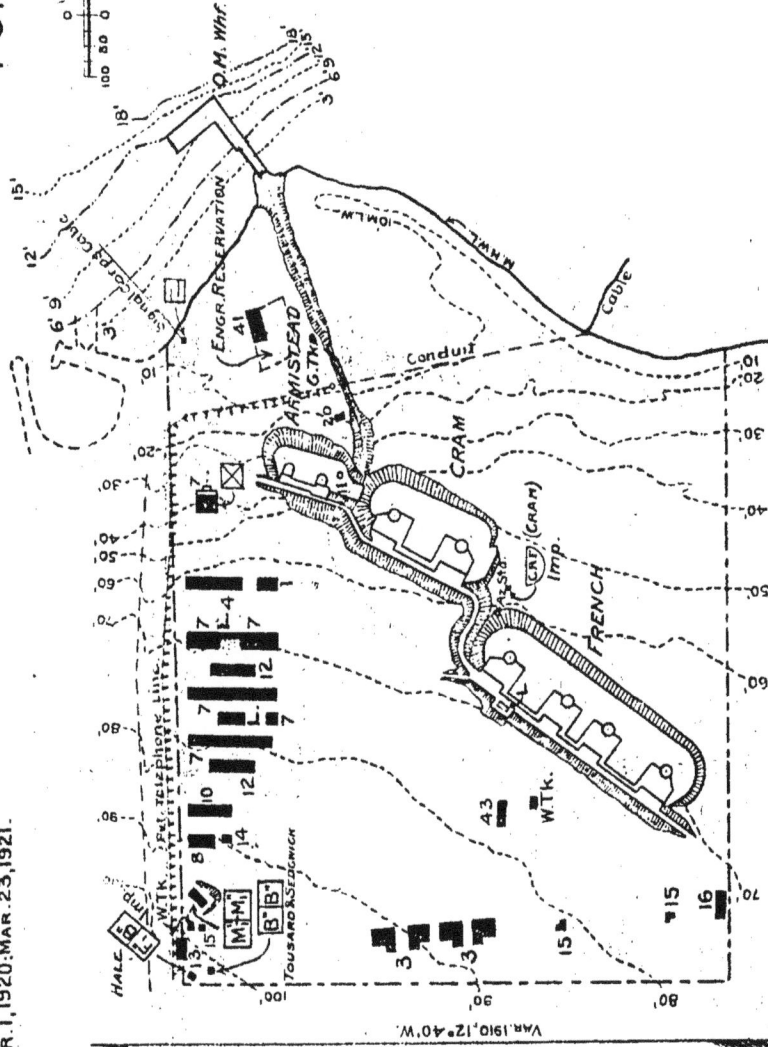

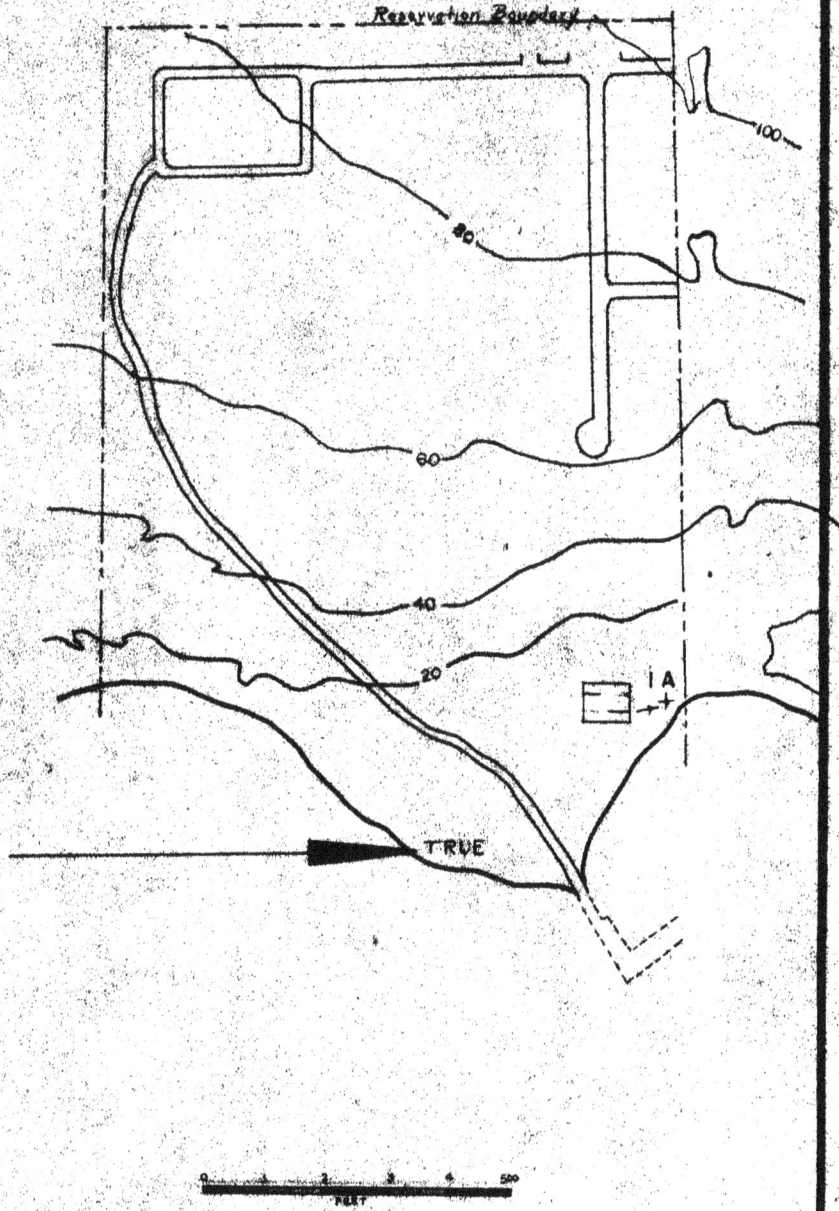

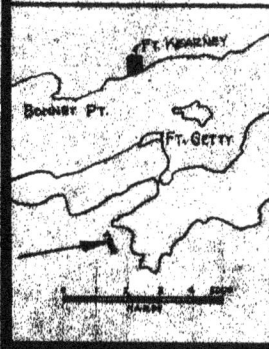

Reservation Boundary

100

80

60

40

20

A

TRUE

FEET

FT. KEARNEY

BONNET PT.

FT. GETTY

H.D. OF NARRAGANSETT BAY

LOC. 37 FT. KEARNEY

LAND RESERVATION

| REVIS'D DATE |
| 1 FEB '45 |

| PREPARED BY | DATE 7-1-43 |
| H.D. OF NARRA. BAY | EX. NO. 98-12 |

Fort Kearny 1941 (NARA)

Fort Kearny 1941 (NARA)

University of Rhode Island Campus (Fort Kearny) (Terry McGovern)

Fort Varnum (Terry McGovern)

served throughout the 1920s and the guns were still in place in 1940. Under authorization of May 25, 1943, the guns were finally removed for use in the modernization program and were actually dismounted on June 7, 1943. The emplacement still exists at the University of Rhode Island Bay Campus. The battery site is open, but the interior is closed to the public.

- **ARMISTEAD:** A battery for two 3-inch pedestal guns located adjacent to and just to the north of Battery Cram. Plans were submitted on May 19, 1903 using funds from the Fortification Act of March 3, 1903. It utilized the standardized emplacement plan for late 3-inch pedestal batteries, with two platforms separated by 62-feet, and a central traverse covering two magazines and a single storeroom. Work was done from August 1904 to October 1906. It was transferred on May 7, 1908 for a cost of $16,600. It was named on General Orders No. 194 of December 27, 1904 for Captain Lewis G. A. Armistead, who was killed in action during the Canadian campaign of the War of 1812 in 1814. Original armament was two 3-inch Model 1903 guns and pedestal mounts (#44/#34 and #42/#33). This was carried throughout the battery's service life. The guns and mounts were re-sited to Fort Varnum early in World War II under authority of July 20, 1942. The emplacement still exists at the University of Rhode Island Bay Campus. The battery site is open, but interior is closed to the public.

Fort Varnum (1942-1948) is located off State Highway 1A on Boston Neck which is north of the town of Narragansett. This 34-acre fort was established during World War II to cover the West Passage into Narragansett Bay. It was named on General Orders 61 of 1942 for Brig. Gen. James M. Varnum of the Continental Army. A 6-inch battery and 3-inch battery were relocated here from Fort Getty and Fort Kearny in 1942. Later, a 90mm AMTB battery (#921) replaced the older guns. The fort is now the property of the Rhode Island National Guard and is now known as Camp Varnum. Most of the World War II structures have been retained allowing visitors to see the fort much as it was constructed in the 1940s. Several buildings are actually two-level fire control structure disguised as summer cottages. The camp is closed to the public.

For Varnum Gun Batteries

- **NEW HOUSE:** The relocated position for two 6-inch pedestal guns from Battery House of Fort Getty. Authority was issued in early 1942 for the movement of these guns to the new reservation of Fort Varnum at Boston Neck. It was located on the southeast side of the main reservation, firing to the southeast. The battery consisted of simple open gun blocks with an adjacent concrete magazine and plotting room. Work was done from April 18 to September 15, 1942 for transfer on August 29, 1942 at a total cost of $53,955. It was armed with two 6-inch Model 1900 guns on pedestal mounts (#30/#28 and #32/#29). It served until deleted under authority of March 9, 1947. The gun blocks still exist at the Rhode Island National Guard Camp Varnum facility. The battery site is closed to the public.

- **NEW ARMISTEAD:** The relocated position for two 3-inch pedestals guns moved from Battery Armistead at Fort Kearny. Authority was granted for this project in early 1942, after at least one cancellation and then reinstatement of the move. It consisted of simple separate gun blocks, magazine, and BC station. Work was done from April 18 to July 15, 1942. Transfer came on August 29, 1942 at a cost of $22,185. It was armed with two model 1903 guns and pedestal mounts (#42/#33 and #44/#34). It served until deleted after the war on authority of March 14, 1946. The gun blocks still exist at the Rhode Island National Guard Camp Varnum facility. The battery site is closed to the public.

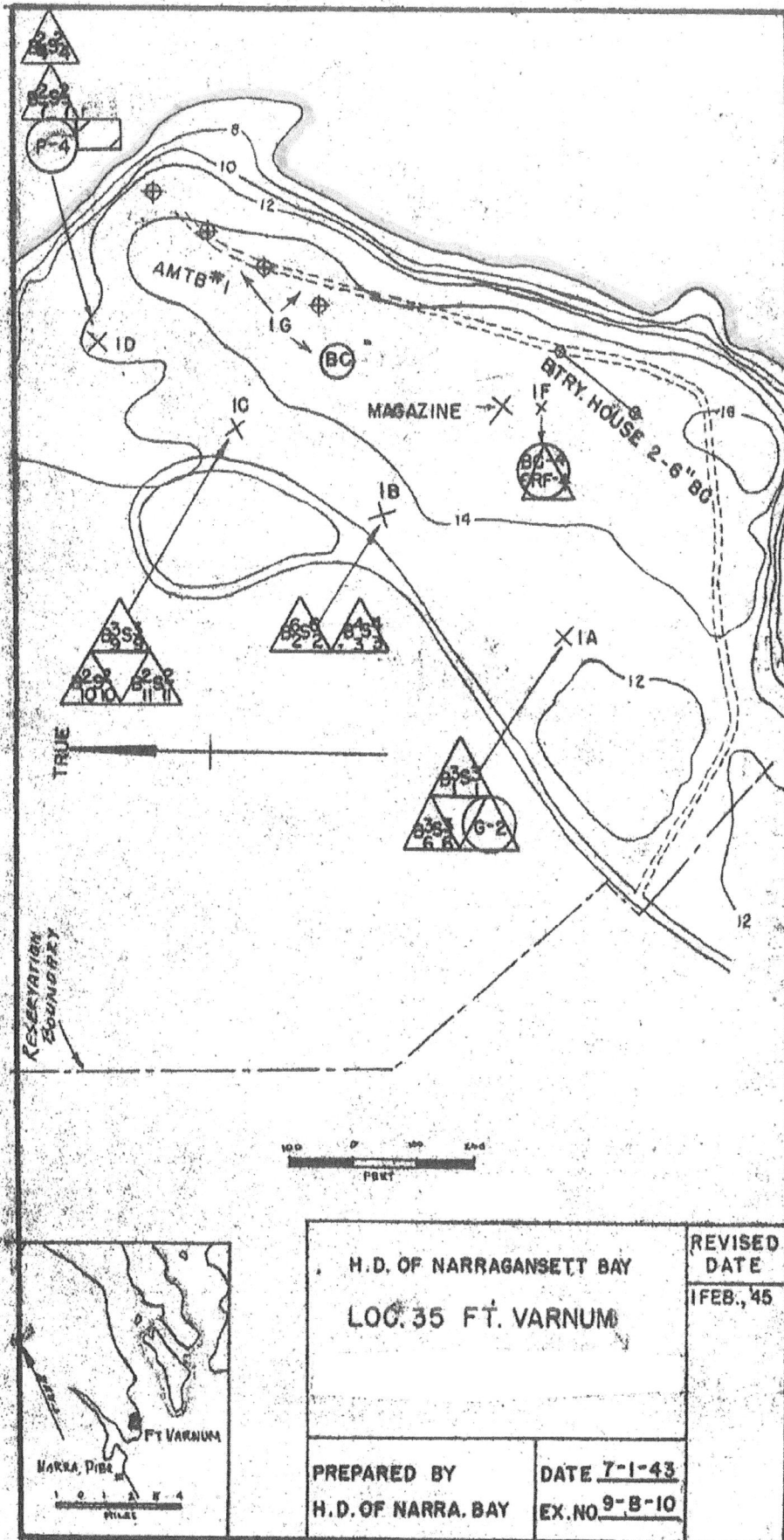

H.D. OF NARRAGANSETT BAY

LOC. 35 FT. VARNUM

REVISED DATE

1 FEB., 45

PREPARED BY
H.D. OF NARRA. BAY

DATE 7-1-43

EX. NO. 9-B-10

- **AMTB #921**: A 1943 Program AMTB battery for two 90mm fixed and two 90mm mobile guns. It was built from April 1 to May 2, 1943 for transfer on July 22, 1943 at a cost of $9,141. It consisted of two fixed gun blocks and a simple wood and earth magazine. It was located near the earlier emplacement of Battery New Armistead. It served until deleted under authority of November 29, 1947. The gun blocks still exist at the Rhode Island National Guard Camp Varnum facility. The battery site is closed to the public.

Fort Greene (1934-1947) is located on Point Judith off State Highway 108, north of the town of Point Judith. The fort was first established in 1934 as the Point Judith Military Reservation. In 1940 it was designated Fort Greene as part of the 1940 Program to the protect the approaches to Narragansett Bay. It was named on General Orders 8 of 1940 for Gen. Nathaniel Greene of the Continental Army with the name transferred from the mortar battery at Fort Adams. The fort had three parcels of land. The East Reservation was the location of Battery Hamilton (#108), a 16-inch casemated battery built in 1940-42. This battery is on the grounds of the Fort Greene U.S. Army Reserve Training Center. The West Reservation contained Battery #109; a 16-inch casemated battery built in 1942-43. This battery was never armed and is now part of the Fisherman's Memorial State Park. Of special interest is the battery commander's station that camouflaged to be a farmer's grain silo (now used the park's offices). The South Reservation was located near the Point Judith Lighthouse. It contained a standard #200 Series 6-inch battery and an unnamed four-gun 155mm battery on Panama mounts that are mostly buried under the beach. Today, Battery #211 is slowly being undermined by the sea. Located nearby were Panama mounts for four 155mm guns during World War II. The South Reservation is now in mixed use as housing and park land.

Fort Greene Gun Batteries

- **HAMILTON**: A modern 16-inch dual barbette battery emplaced at the new Point Judith reservation of Fort Nathanael Greene on the western side of Narragansett Bay. Modern defenses here had been considered for a number of years after World War I. Construction was authorized in 1934, but funding was delayed until several years later. It was one of the five pre-1940 program major batteries constructed and followed the early design plan with recessed main corridors and dual-entry gun houses. A separate PSR to the north and twelve base-end stations were also constructed. It was on the eastern segment of the reservation, pointed to the south, southeast (12-degrees east of south). Construction was done with funds from the FY-1941 Budget ($625,000) and FY-1942 (another $625,000). Work was done from September 20, 1940 to November 1, 1942. Transfer was made on April 19, 1943 for a cost of $1,183,715. While under construction it was given Battery Construction No. 108. In the 1940 Program priority assignments, it was given national priority #4 on the September 11, 1940 list. It was named on General Orders No. 13 of November 24, 1941 for Brigadier General Alston Hamilton of World War I service. Proof firing was conducted on August 6, 1943. It was armed with two 16-inch guns MkIIM1 on barbette carriage M1919M4 (#107/#19 and #137/#20). The battery served out the war and early postwar period, not being authorized for deletion and disarmament until May 14, 1948. The emplacement still exists on Rhode Island National Guard property. The battery site is closed to the public.

- **Battery #109**: A 1940 Program dual 16-inch barbette battery intended to augment Battery Hamilton's coverage from the western section of the Point Judith Fort Nathanael Greene military reservation. As part of the later program, it had the simplified plan with a straight main corridor and standardized room arrangement, Funds came from the FY-1942 and FY-1943 budgets, and it was assigned priority #18 on the national list of September 11, 1940 and #17 a year later on August

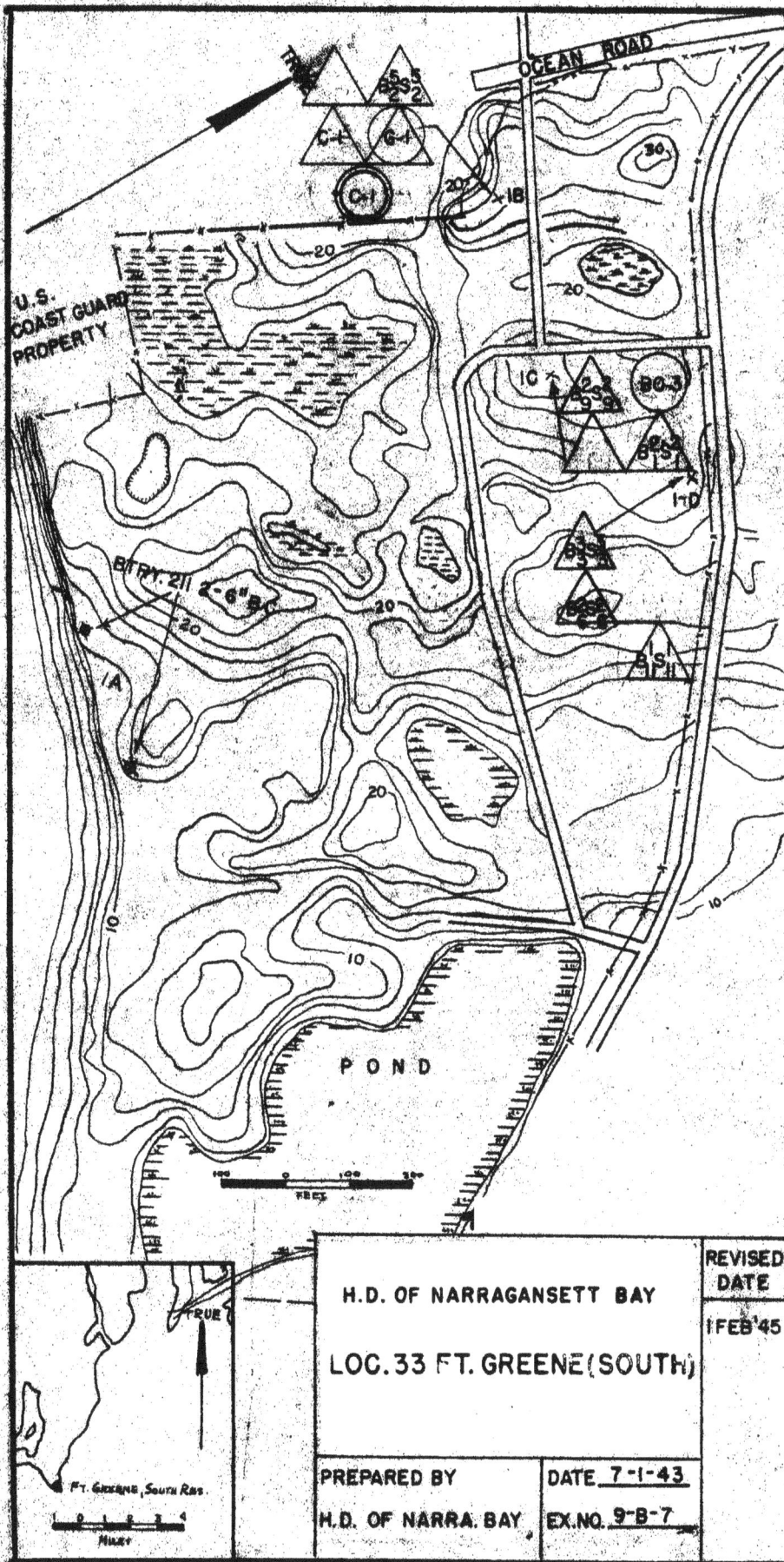

OCEAN ROAD

U.S. COAST GUARD PROPERTY

POND

H.D. OF NARRAGANSETT BAY

LOC. 33 FT. GREENE (SOUTH)

REVISED DATE
1 FEB '45

PREPARED BY	DATE 7-1-43
H.D. OF NARRA. BAY	EX. NO. 9-B-7

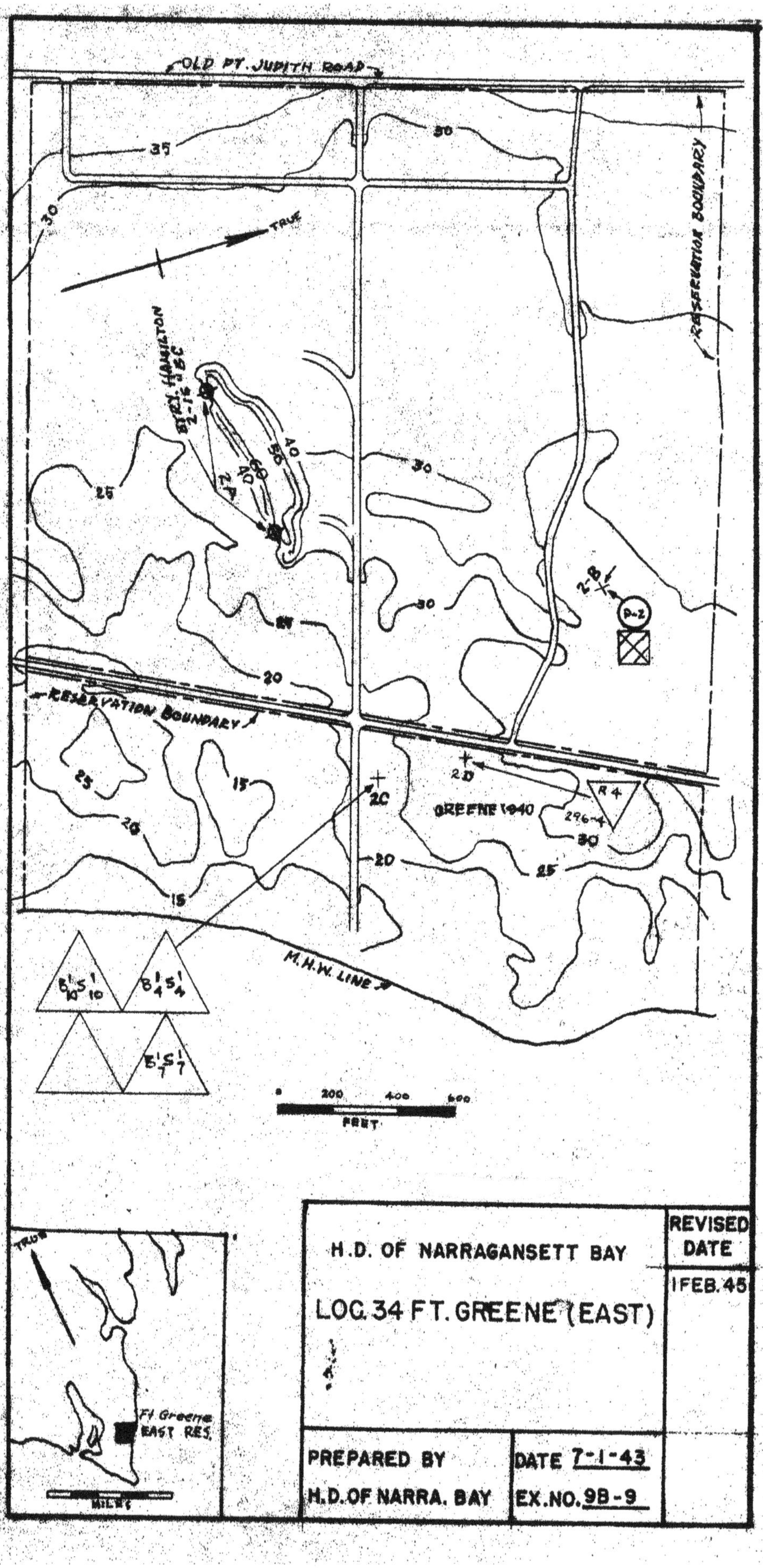

OLD PT. JUDITH ROAD

RESERVATION BOUNDARY

TRUE

BTRY. HAMILTON
2-16 B.C.

RESERVATION BOUNDARY

M.H.W. LINE

GREENE 1940

B' S'
10 10

B' S'
4 4

B' S'
7 7

200   400   600
FEET

H.D. OF NARRAGANSETT BAY

LOC. 34 FT. GREENE (EAST)

	REVISED DATE
	1 FEB. 45

PREPARED BY	DATE 7-1-43
H.D. OF NARRA. BAY	EX. NO. 9B-9

TRUE

Ft. Greene
EAST RES.

MILES

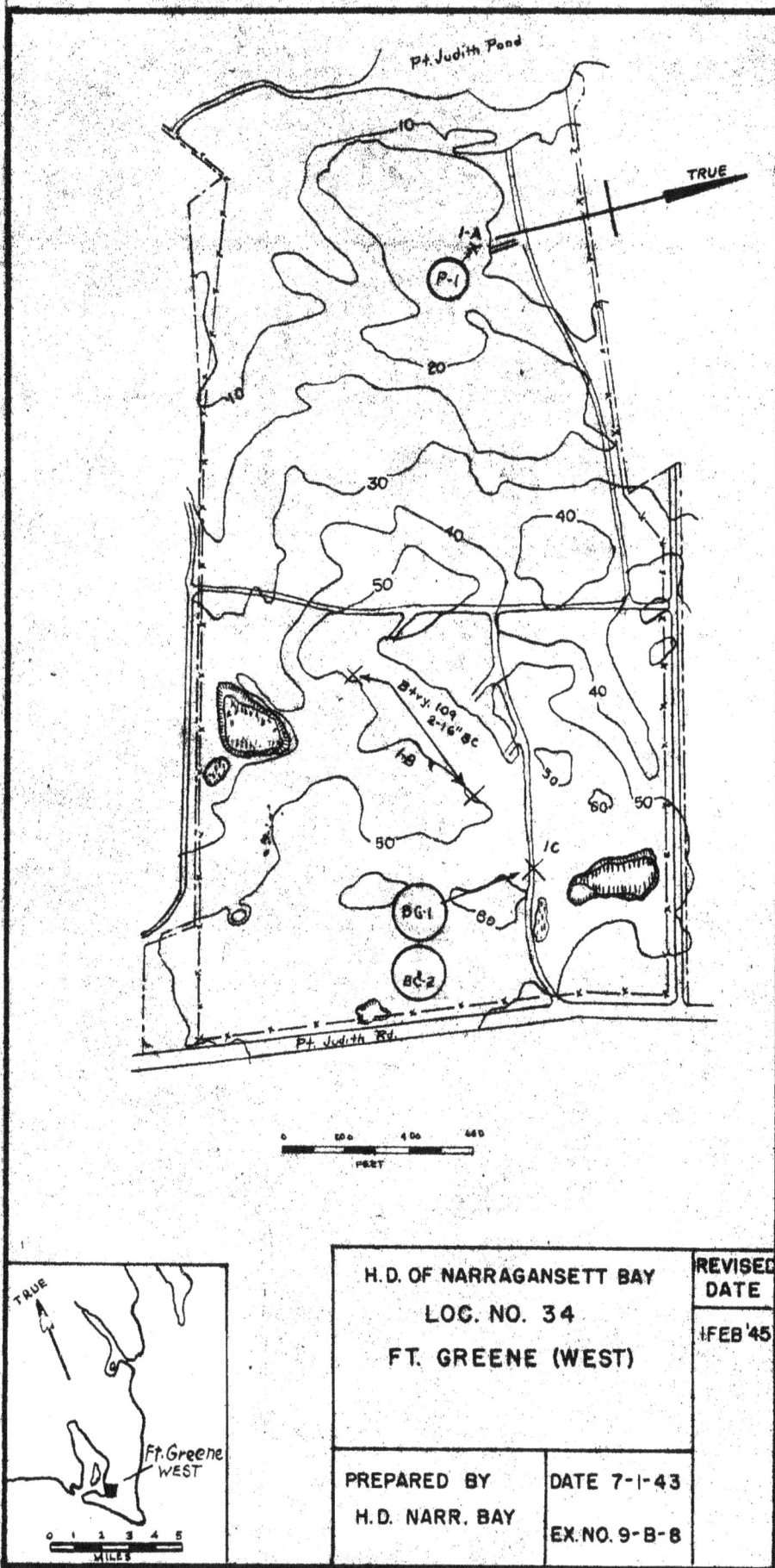

Pt. Judith Pond

10

TRUE

I-A

P-I

20

40

30

40

50

40

Btry. 109
2-16" BC

I-B

50

50

60

50

IC

BG-I

60

BC-2

Pt. Judith Rd.

0   50.0   100   150
FEET

TRUE

Ft. Greene
WEST

0 1 2 3 4 5
MILES

H.D. OF NARRAGANSETT BAY

LOC. NO. 34

FT. GREENE (WEST)

REVISED
DATE

1 FEB '45

PREPARED BY

H.D. NARR. BAY

DATE 7-1-43

EX. NO. 9-B-8

11, 1941. Actual work was done between September 24, 1942 and October 31, 1943 for transfer on September 18, 1944 for a cost of $1,705,667. A PSR to the west and ten base-end station assignments completed the structural work. The armament assigned was two 16-inch MkIIM1 guns with Model M4 barbette carriages (#56/#5 and #50/#47). These items were received at the post (the tubes about November 1, 1943), but never actually mounted. The guns were stored in the emplacement after arrival. The battery was never named. Authority was granted to delete and abandon the site on March 9, 1947. The emplacement still exists at the Fishermen's Memorial State Park. The battery site is open, but the interior is closed to the public.

- **Battery #211:** A 1940 Program dual 6-inch barbette battery emplaced near the southern shore of Point Judith, on the south section of Fort Nathanael Greene. It was one of the earliest 6-inch batteries, being assigned priority #2 in the 1940 list and funded in the FY-1942 Budget. It was a standard 200-series design type. Work proceeded rapidly, being done between December 29, 1941 and July 22, 1942. Transfer was made on June 15, 1943 for a cost at completion of $257,723. The battery was never named, simply being referred to as Battery Construction No. 211. It was armed with two 6-inch Model 1903A2 guns on M1 barbettes (#21/#15 and #46/#16). As completed it fired to the south. It was retained throughout the war but eventually deleted and disarmed in 1948. Though partially buried and impacted by shoreline erosion, the emplacement still exists. The battery site is open, but the interior is closed to the public.

Battery #211 at Point Judith (Terry McGovern)

Schroder, Walter K. *Defenses of Narragansett Bay in World War II.* Rhode Island Bicentennial Foundation. Newport, RI, 1980.

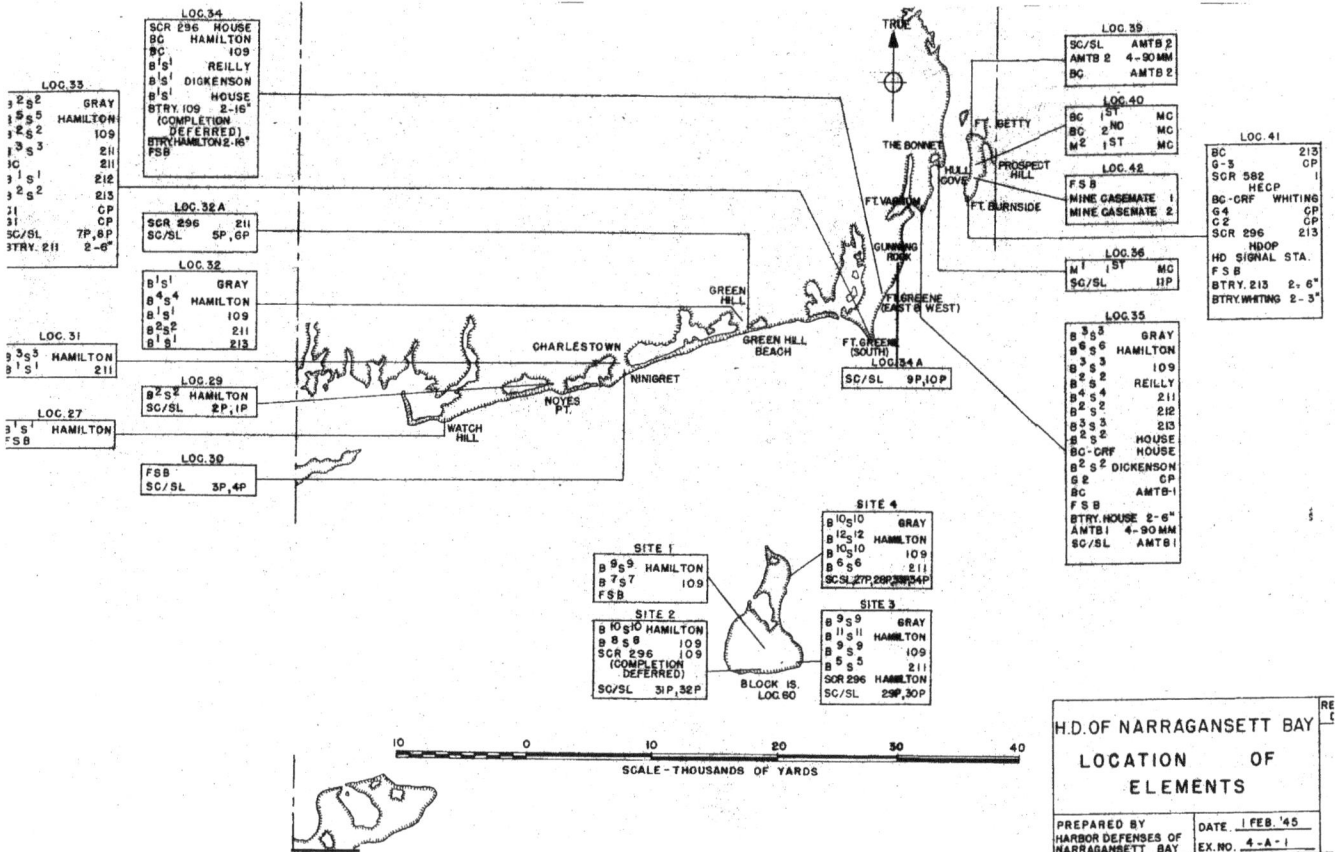

H.D. OF NARRAGANSETT BAY
LOCATION OF ELEMENTS
PREPARED BY HARBOR DEFENSES OF NARRAGANSETT BAY
DATE 1 FEB. '45
EX. NO. 4-A-1

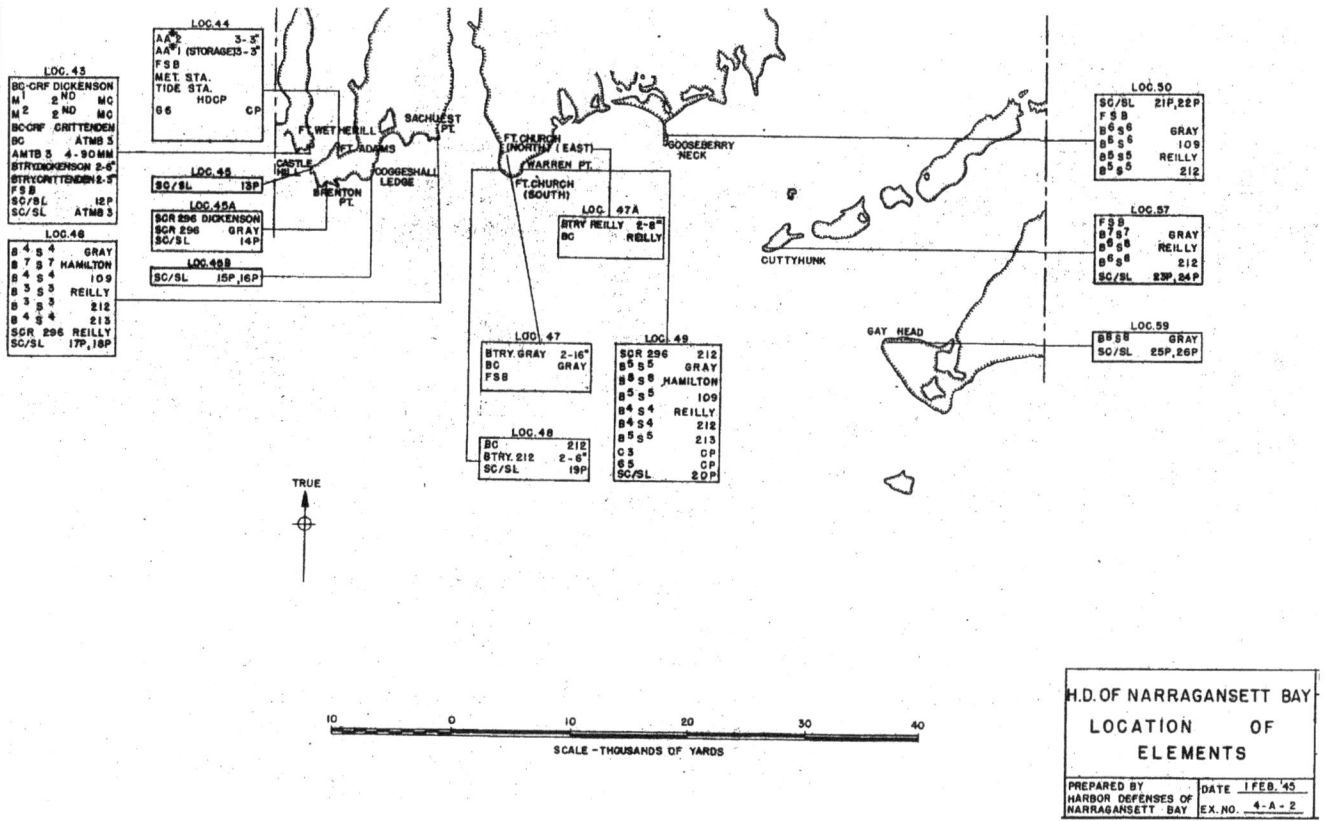

H.D. OF NARRAGANSETT BAY
LOCATION OF ELEMENTS
PREPARED BY HARBOR DEFENSES OF NARRAGANSETT BAY
DATE 1 FEB. '45
EX. NO. 4-A-2

Narraganset Bay World War II-era Site Locations. Stations housed in a single structure are connected by dashes (-)

location	Loc#	Purpose
Watch Hill Pt	27	BS1/Hamilton, SBR
Noyes Point	29	BS2/Hamilton, SL 1,2
Ninigret	30	SBR, SL 3,4
Charlestown	31	BS3/Hamilton, BS1/211
Green Hill	32	BS1/Gray-BS1/213, BS12/111-BS1/109, BS4/Hamilton-BS2/211"
Green Hill Beach	32A	SCR296-211
Fort Greene West	34	Batt. Tact. #1 BCN 109, BC/109
Fort Greene East	34	Batt, Tact, #2 Hamilton, BC/Hamilton, BS1/Reilly, BS1/Dickenson, BS1/House, SBR
Fort Greene South/ Point Judith	33	Batt. Tact. #3 BCN 211, BS2/Gray, BS5/Hamilton, BS2/109, BC-BS3/211, BS1/212, BS2/213, G1, C1, SL 7,8
Fort Varnum/ Boston Neck	35	Batt. Tact. #4 House, BS3/Gray, BS6/Hamilton, BS3/109, BS2/Reilly, BS4/211, BS2/212, BC-BS2/House, BS2/Dickenson, G2, AMTB
The Bonnett	36	M1, SL 11
Fort Kearney	37	SBR, SL 3,4
Fort Greble	38	military reservation
Fort Getty	39	AMTB
Prospect Hill	40	M2
Fort Burnside Beavertail	41	Batt. Tact. #5 Whiting, Batt. Tact. #6 BCN 213, BC/213, G3, SCR582, HECP-HDCP-HDOP, BC/Whiting
Hull Cove	42	M1, M2
Fort Wetherill	43	Batt, Tact. #7 Dickenson, Batt. Tact. #8 Crittenden, BC/Dickenson, BC/Crittenden, M1/2, M2/2, AMTB3, SL 12, SBR
Fort Adams	44	Met, T, HDOP, G6
Breton Point	45A	SCR296-Dickenson, SCR296-Gray, SL 13,14
Sachuest Point	46	BS4/Gray, BS7/Hamilton, BS4/109, BS3/Reilly, BS3/212, BS4/214, SCR296-Reilly, SL 17,18
Fort Church North	47	Batt. Tact. #9 Gray, BC/Gray
Fort Church East	47A	Batt. Tact. #10 Reilly, BC/Reilly
Fort Church South/ Sakonnet Point	48	Batt. Tact. #11 BCN 212, BC/212, SL 19
Warren Point	49	SCR296-219, BS4/Reilly, BS4/212, BS5/213, BS5/Grey, BS8/Hamilton, BS5/109, C3, G5, SL 20
Goosebury Neck	50	BS6/Gray, BS6/109, BS5/Reilly, BS5/212, SBR, SL 21,22
Cuttyhunk Island	54/57	BS7/Gray, BS6/212, BS6/Reilly, SL 23,24
Gayhead	59	BS8/Gray, SL 25,26
Block Island	60A	BS9/Hamilton, BS7/109
Block Island	60B	BS10/Hamilton, BS8/109, SCR-109 SL 31,32
Block Island	60C	BS9/Gray, BS11/Hamilton, BS9/109, BS5/211, SCR296-Hamilton, SL 29,3
Block Island	60D	BS10/Gray, BS12/Hamilton, BS10/109, BS6/211, SL 27-34

THE HARBOR DEFENSES OF LONG ISLAND SOUND – NEW YORK

The wide northern entrance to Long Island Sound led to several major ports in Connecticut, which were protected by a variety of local forts from the First Second and Third Systems. The Endicott Program built nearly 30 new concrete gun batteries on the string of 3 major islands that stretched from Napatree Point in Rhode Island to Orient Point on Long Island in New York. This effectively closed off the entrance to Long Island Sound. The defenses were upgraded by the 1940 Program with three new 16-inch casemated gun batteries (with a fourth battery planned), four new 6-inch gun batteries, and 6 new 90mm AMTB batteries. During World War II these defenses were expanded to include Montauk Point in New York and Block Island in Rhode Island.

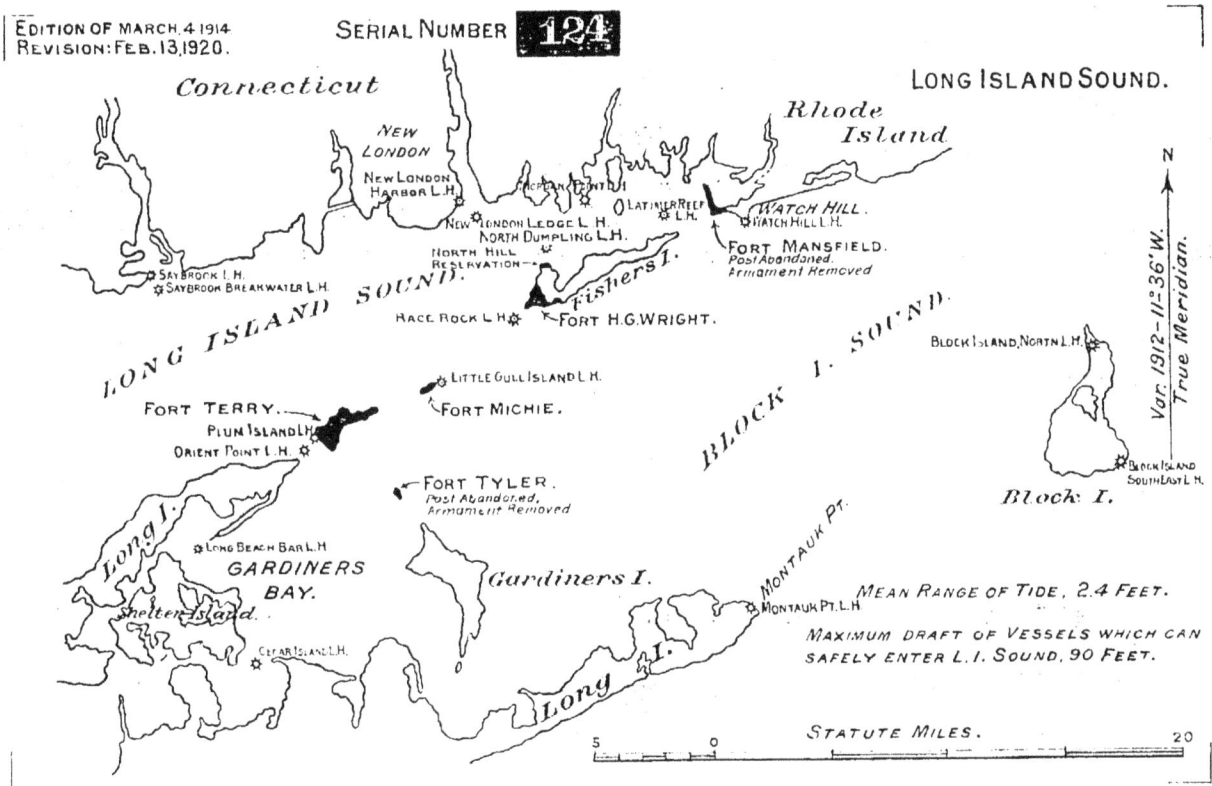

Fort Mansfield (1898-1926) is located on Napatree Point which is near Watch Hill, Rhode Island. Napatree Point is sand split which marks the eastern end of Fishers Island Sound and Long Island Sound. The 70-acre fort was established in 1900, and three Endicott Program batteries were constructed. It was named on General Orders 43 of 1900 for Maj. Gen. J.K.F. Mansfield, USV, mortally wounded during the Battle of Antietam on Sept. 18, 1862. The post suffered from hurricane damage and eventually was inactivated, declared surplus by 1926 and sold. Subsequent hurricanes have removed most remains of the fort. Only two Endicott Program batteries remain today. The spit is now part of the Napatree Point Conservation area and only accessible by foot. The parking area is controlled by the Napatree East Beach Association and may not be open to the general public.

Fort Mansfield Gun Batteries

- **WOOSTER:** A battery for two 8-inch disappearing guns erected as the primary armament of the Fort Mansfield reservation on Napatree Point, Rhode Island. It was sited to the east of the dock, on the central part of the reservation, firing to the south. It was submitted on July 3, 1898 with funding

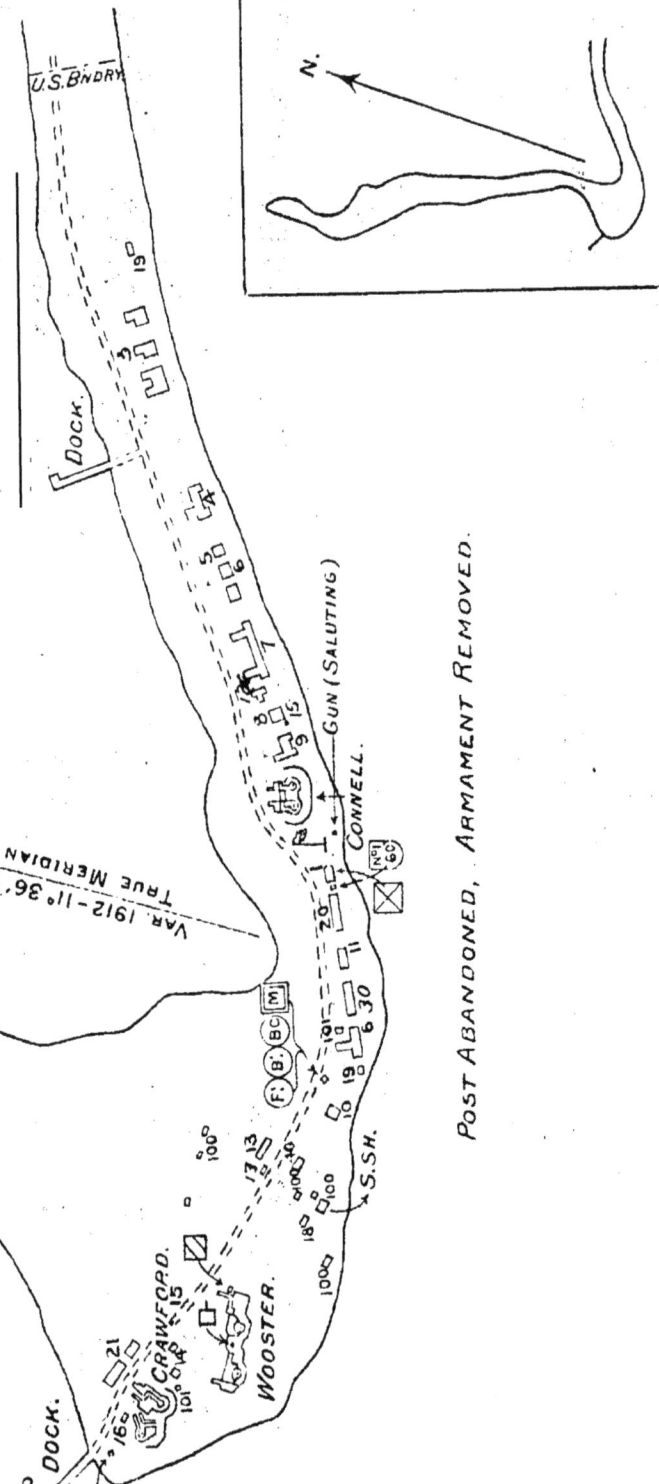

LONG ISLAND SOUND,

FORT MANSFIELD

NAPATREE POINT, R.I.

SERIAL NUMBER **124**

EDITION OF AUG. 28, 1914.
REVISIONS: DEC. 7, 1915.
JUNE 9, 1916.
FEB. 13, 1920.

VAR. 1912-11°36'W.

TRUE MERIDIAN

N.

POST ABANDONED, ARMAMENT REMOVED.

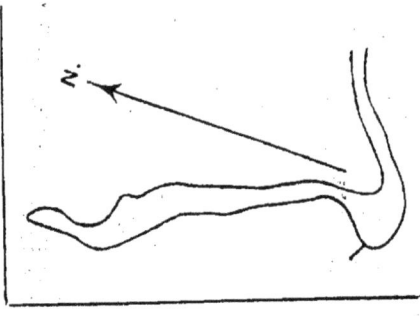

GUN (SALUTING)

U.S. BNDRY

DOCK.

S.SH.

CRAWFORD.

WOOSTER.

CONNELL.

DOCK.

LEGEND.

1 ADMINISTRATION BLDG.
2
3 OFFICERS QUARTERS.
4 HOSPITAL.
5 HOSPITAL STW'D. QRS.
6 N.C.O. QUARTERS.
7 BARRACKS.
8 GUARD HOUSE.
9 POST EXCHANGE.
10 BAKERY.
11 SHOP.
12 LAVATORY.
13 COAL SHED.
14 OIL HOUSE.
15 SENTRY BOX.
16 TEAMSTERS BUILDING.
17 SCALES.
18 WOOD SHED.
19 BATH HOUSE.
100 CAMP BUILDINGS.
101 HOSE HOUSE.
20 Q.M. & COM. STOREHOUSE.
21 Q.M. STABLES.
30 ORDNANCE ST. HOUSE.
40 ENGINEER BUILDING.
80 FOG SIGNAL.

BATTERIES.

WOOSTER __ 2-8"D.
CRAWFORD __ 2-5"B.
CONNELL __ 2-5"P.

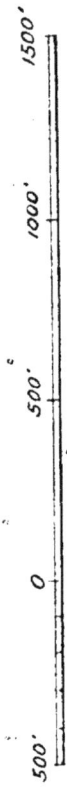

500'   0   500'   1000'   1500'

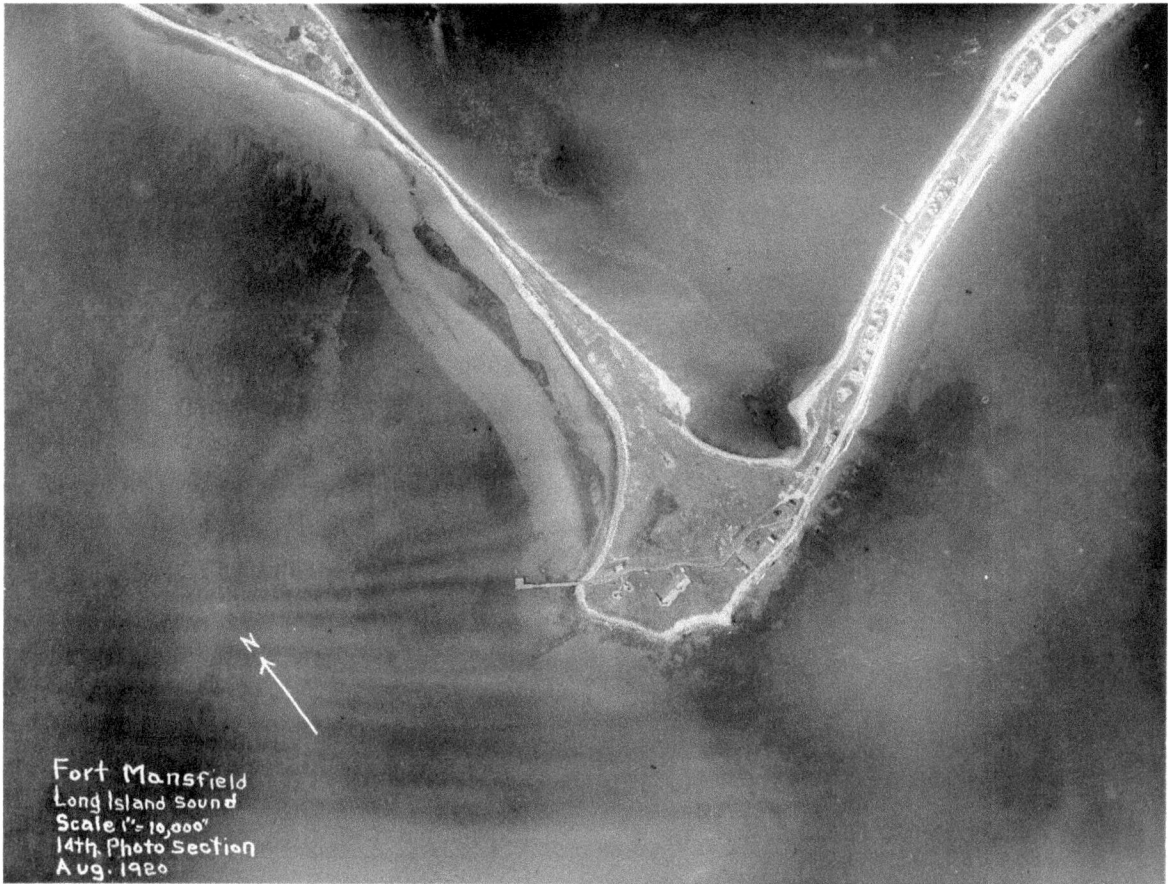

Fort Mansfield 1920 (NARA)

Fort Mansfield 1924 (NARA)

from the Act of May 7, 1898. The plan featured differences from the standard mimeographs—the left (No. 2) emplacement was withdrawn in echelon to give better coverage to the eastern flank. Also, the battery was emplaced "half-sunk" to provide it more coverage and less visibility on an otherwise low elevation. Work was done in 1898-1899 and reported as completed on July 1, 1900. It was transferred on February 18, 1901 for a cost of $75,000. It was named on General Orders No. 30 of March 19, 1902 for Brigadier General David Wooster, a Continental Army war hero who was mortally wounded in 1777. The battery carried two 8-inch Model 1888 Watervliet guns on disappearing carriages Model LF 1896 disappearing carriages (#12/#35 and #13/#34). Changes were made to loading platforms and hoists in the subsequent years. The armament was authorized for removal in 1917. The fort's location had been demonstrated as vulnerable to rearward fire in the maneuvers of 1907 and was no longer considered critical. The reservation was disposed of in 1927. The emplacement still exists on unmanaged property of the Napatree Point Conservation Area. The battery is open to the public, accessible by a long beach walk.

- **CRAWFORD:** A battery for two 5-inch balanced pillar gun mounts emplaced 300-feet to the northwest of Battery Wooster, between that emplacement and the dock. The emplacement plan was submitted on August 20, 1898. Plans called for two virtually separate, but immediately adjacent single emplacements, composed of a gun platform and lower, flank magazine. The two emplacements were centered to fire in different directions, giving the battery an unusually wide field of fire. Also, their position on a slight rise, along with the sunken construction of the 8-inch battery allowed them the ability to fire over the other emplacement. Work was done in 1898-1899. It was turned over with the other two batteries here on February 18, 1901 for a cost of $11,600. It was named on General Orders No. 30 of March 19, 1902 for Captain Emmet Crawford, 3rd U.S. Cavalry mortally wounded pursuing hostile Indians in Mexico in 1886. It was armed with two 5-inch Model 1897 Watervliet guns on Model 1896 balanced pillar mounts (#3/#8 and #7/#7). This armament was removed in 1917-1918, and the carriages were scrapped in place shortly after. The emplacement was not used again for armament and was abandoned in the 1920s. It still exists on unmanaged property of the Napatree Point Conservation Area. The battery is open to the public, but requires a long beach walk.

- **CONNELL:** A battery for two 5-inch pedestal guns emplaced on the spit of land to the east of the fort's main works. It fired to the southeast. With funds provided in the Fortification act of July 7, 1898 the plan was submitted on July 18, 1898. It called for a standard type emplacement of two flank emplacements with a central traverse covering magazines. It was originally intended and built to carry two 5-inch Brown wire-wound guns, which never were produced. Eventually, though after several years, replacement Model 1900 guns and Model 1903 pedestal mounts were substituted. Work was done in 1898-1899, transfer being made at the same time as all the fort's batteries on February 28, 1901 for a cost of $14,000. The battery was named on General Orders No. 30 of March 19, 1902 for Captain James W. Connell, 9th U.S. Infantry, killed during the Philippine Insurrection in 1901. It carried two 5-inch Model 1900 Watervliet guns on Model 1903 pedestals (#4/#5 and #6/#6). This armament was removed in 1917. The empty emplacement was gradually destroyed by surf erosion, only fragmentary remains of the gun blocks persist at the Napatree Point Conservation Area.

Napatree Point Preserve (Fort Mansfield) (Terry McGovern)

Napatree Point Preserve (Fort Mansfield) (Terry McGovern)

Goshen Point Military Reservation (1943-1946) is located Goshen Point, near Waterford, CT. Anti Motor Torpedo Boat Battery 914 was built on the Harkness Estate in 1943. The battery site is now covered over as part of the Harkness Memorial State Park.

Goshen Point M.R. Gun Battery.

- **AMTB-914**: A 1943 Program battery of two 90mm fixed and two 90mm mobile guns emplaced at Goshen Point, on land of the Harness Estate. It was built from April 32 to June 19 of 1943. Transfer was made on December 20, 1943 for a cost of $14,251. The armament was mounted and served until removed after the end of the war. The gun blocks still remain on the Harkness Memorial State Park but are covered. The battery site is open to the public.

Pine Island Military Reservation (1943-1946) is located off the coast of Groton in waters historically known as "The Eastern Chops of New London Harbor." Pine Island and the Plant estate at Avery Point eventually were auctioned off and sold to the State of Connecticut in 1939. In 1941 the state turned both over to the U.S. government in support of the federal defense of the coast. A Coast Guard station was created on the mainland property, and a AMTB battery was installed on the western end of Pine Island. Anti Motor Torpedo Boat Battery 915 was abandoned in 1945. In 1968, after the Coast Guard operations ceased at Avery Point estate and Pine Island, the properties reverted to the State of Connecticut and were granted to the University of Connecticut as a campus branch. The island is currently owned by the University of Connecticut.

Pine Island M. R. Gun Battery

- **AMTB-915**: A 1943 Program battery of two 90mm fixed and two 90mm mobile guns emplaced on Pine Island in 1943. It served until removed right after the end of the war, probably in late 1945. The blocks still exist on this small island, but access is difficult as a boat is required. The battery site is open to the public.

Fort H.G. Wright (1898-1958) is located on western end of Fishers Island, New York in Long Island Sound. It was named for Maj. Gen. Horatio Gouverneur Wright, USV, for his Civil War service. This 334-acre fort on Race Point grew to become the headquarters of Harbor Defense of Long Island Sound in 1913. The Endicott Program defenses consisted of two major caliber disappearing gun batteries, four 6-inch gun batteries, two 3-inch gun batteries and a two-pit mortar battery. There was a large garrison area and submarine mine complex on Silver Eel Pond. Of special interest is the emplacement for an experimental Dynamite Gun battery at the tip of Race Point. The ferry landing is in Silver Eel Pond which was the center of the controlled mine complex and cantonment area. Southwest of Silver Eel Pond is the "gun line" for the fort. The defenses of fort were supplemented during World War II by new batteries located around Fishers Island. On Race Point, near the Dynamite battery, a #200 Series battery was constructed in 1943. To the northeast of the main fort at Wilderness Point, a #100 Series battery for two 16-inch guns and a #200 Series battery for two shielded 6-inch guns were installed. On Mount Prospect is a complex of fire control stations and a HECP were located. North Hill on the northeastern point of Fisher Island a battery of 3-inch guns was installed during World War II. Three 90mm AMTB batteries added in 1943 at Race Point, Goshen Point, and Pine Island. The U.S. Army placed Fort H.G. Wright in caretaker status in 1949 after scrapping the remaining coast artillery. In 1958 the main fort was sold by the GSA to the corporation that oversees Fishers Island. The military airfield located in the center portion of the fort was converted into a general aviation airport. The mortar battery and some of the gun batteries have been repurposed for trash

LONG ISLAND SOUND

FORT H.G. WRIGHT

FISHERS ISLAND, N.Y.

GENERAL MAP

SERIAL NUMBER 124

EDITION OF MAR. 28, 1921.
REVISIONS:

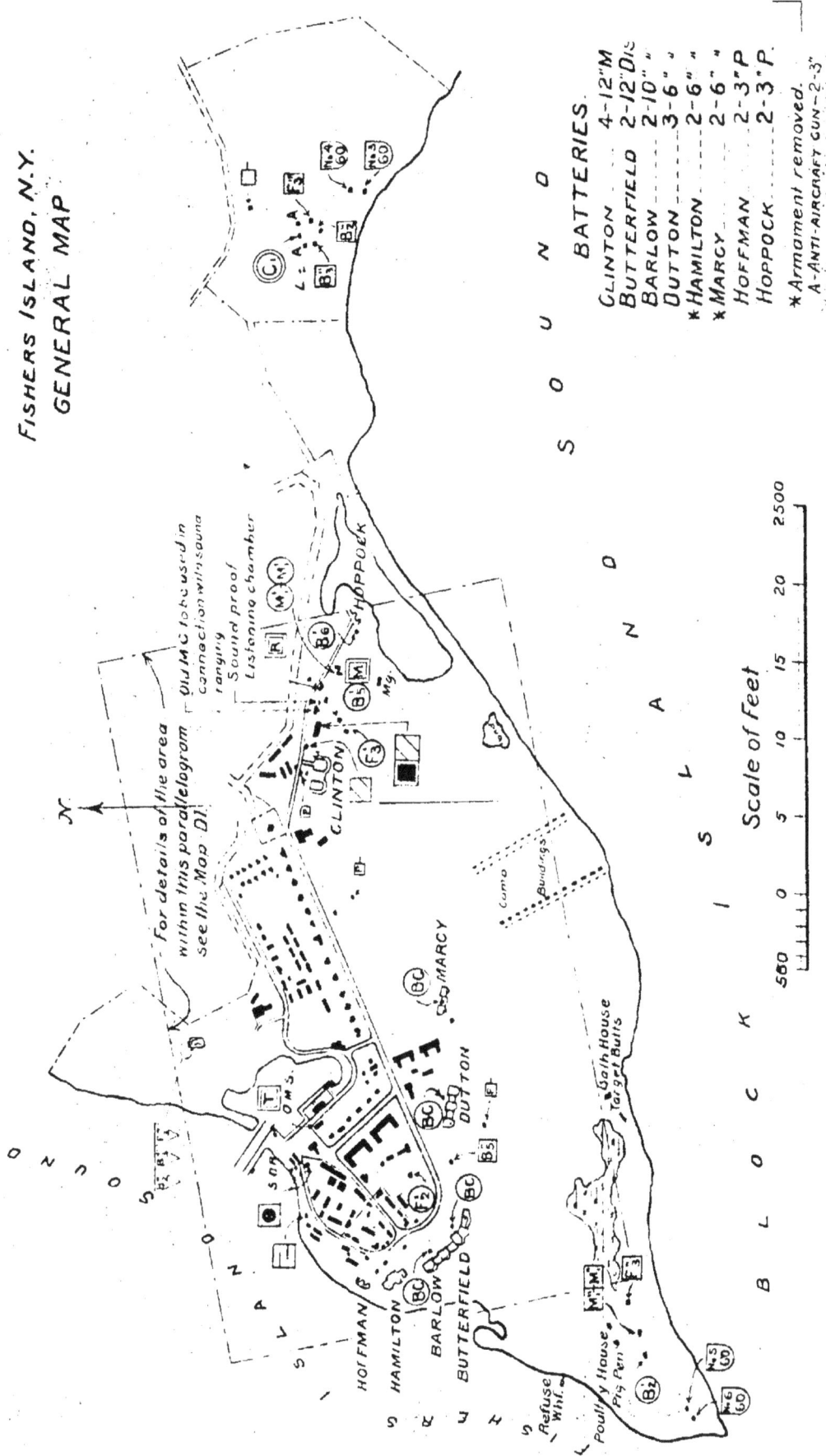

BATTERIES.

CLINTON .... 4-12"M
BUTTERFIELD .. 2-12"D¹⁵
BARLOW .... 2-10" "
DUTTON .... 3-6" "
*HAMILTON .... 2-6" "
*MARCY ...... 2-6" "
HOFFMAN ... 2-3"P
HOPPOCK .... 2-3"P.

* Armament removed.
A-Anti-aircraft gun-2-3".

Scale of Feet

500   0   5   10   15   20   2500

LONG ISLAND SOUND

FORT H.G. WRIGHT-D1.

FISHERS ISLAND, N.Y.

Scale of Feet

SERIAL NUMBER **124**

EDITION OF MAR. 28, 1921.
REVISIONS:

Old M.G. tube used in connection with sound ranging. Sound proof Listening chamber.

Engr. Reservation

SOUTH BEACH POND

BATTERIES.
CLINTON ------- 4-12"M.
BUTTERFIELD -- 2-12"Dis.
BARLOW ------- 2-10" *
DUTTON ------- 3-6" "
*HAMILTON ---- 2-6" "
*MARCY ------- 2-6" "
HOFFMAN ------ 2-3"P.
HOPPOCK ------ 2-3"P.
* Armament removed

CAMP
BUILDINGS

SILVER EEL POND. O.M.S.

HOFFMAN
HAMILTON
BARLOW
BUTTERFIELD
DUTTON
MARCY
HOPPOCK
CLINTON

LEGEND
1. ADMINISTRATION BLDG.
1a. NEW ADMIN. BUILDING.
2. COMM. OFFICERS QRS.
3. OFFICERS QUARTERS.
4. HOSPITAL.
5. HOSPITAL CORPS QRS.
6. N.C.O. QUARTERS.
7. BARRACKS.
8. GUARD HOUSE.
9. POST EXCHANGE AND GYMNASIUM.
10. WIRELESS POLES.
11. CANTONMENT BLDGS. (OFFICERS QUARTERS.)
12. BLACKSMITH SHOP.
13. DRAINAGE PUMP.
14. BAKERY.
15. OFFICE OF ARTY. ENGR.
16. BAND QUARTERS.
17. SHOP.
18. WAGON SHED.
19. BOAT HOUSE.
100. GARAGE.
101. BAND STAND.
102. FIRE ENGINE HOUSE.
103. HOSE HOUSE.
104. RESERVOIR.
105. MARRIED SOLDIERS QRS.
106. COAL OFFICE.
107. SCALES.
108. COAL SHED.
109. WOOD SHED.
110. OIL HOUSE.
111. TEAMSTERS QRS.
112. CANTONMENT BLDGS.
113. BALLOON HANGAR.
114. GREENHOUSE.
21. Q.M. & COM. ST. HOUSE.
22. Q.M. STOREHOUSE.
23. Q.M. BARRACKS.
24. Q.M. STABLE.
25. STONE CRUSHER.
30. ORD. STOREHOUSE.
40. ENGR. BUILDINGS.
70. SERVICE CLUB.

80. FOG SIGNAL.
92. WAITING ROOM.

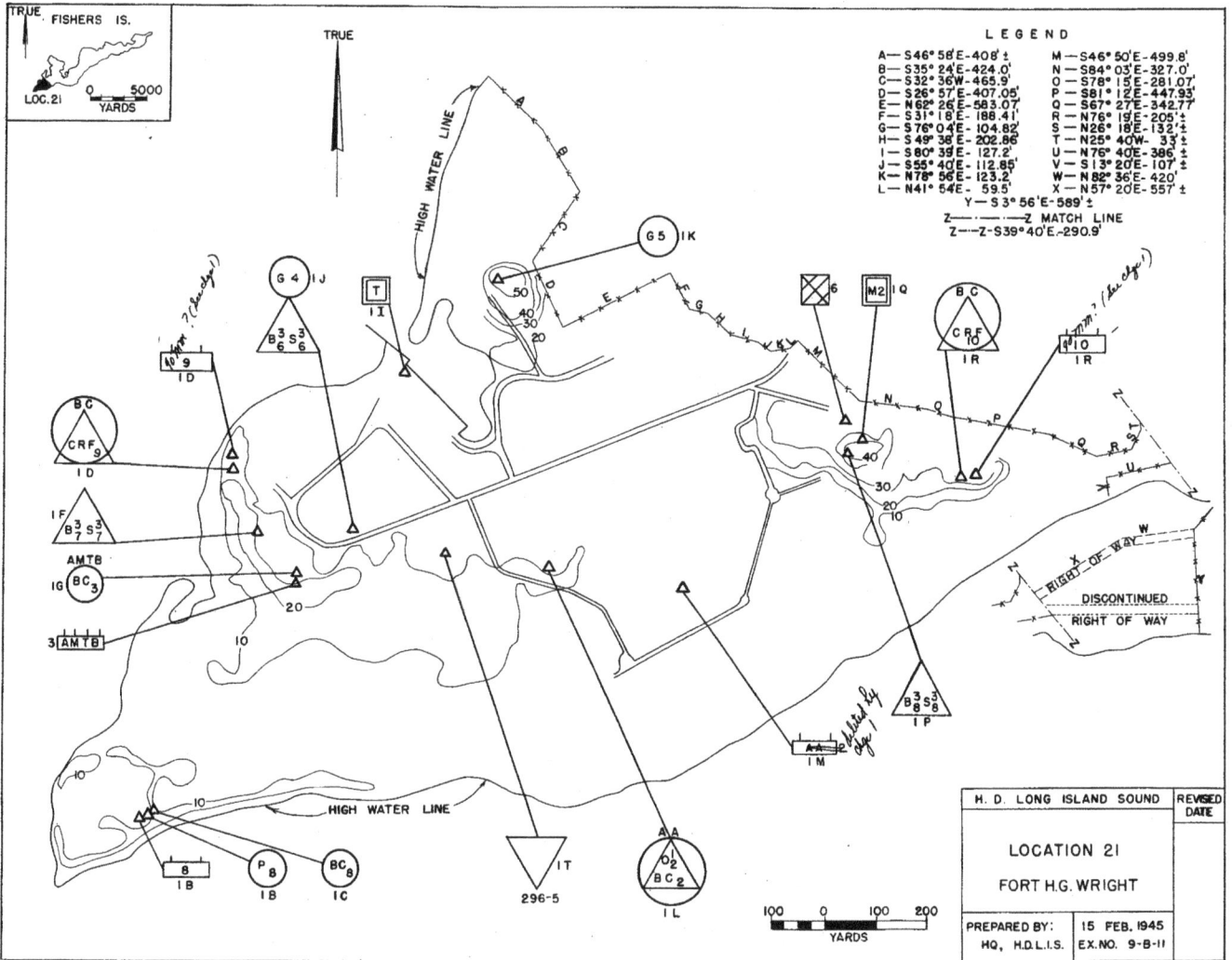

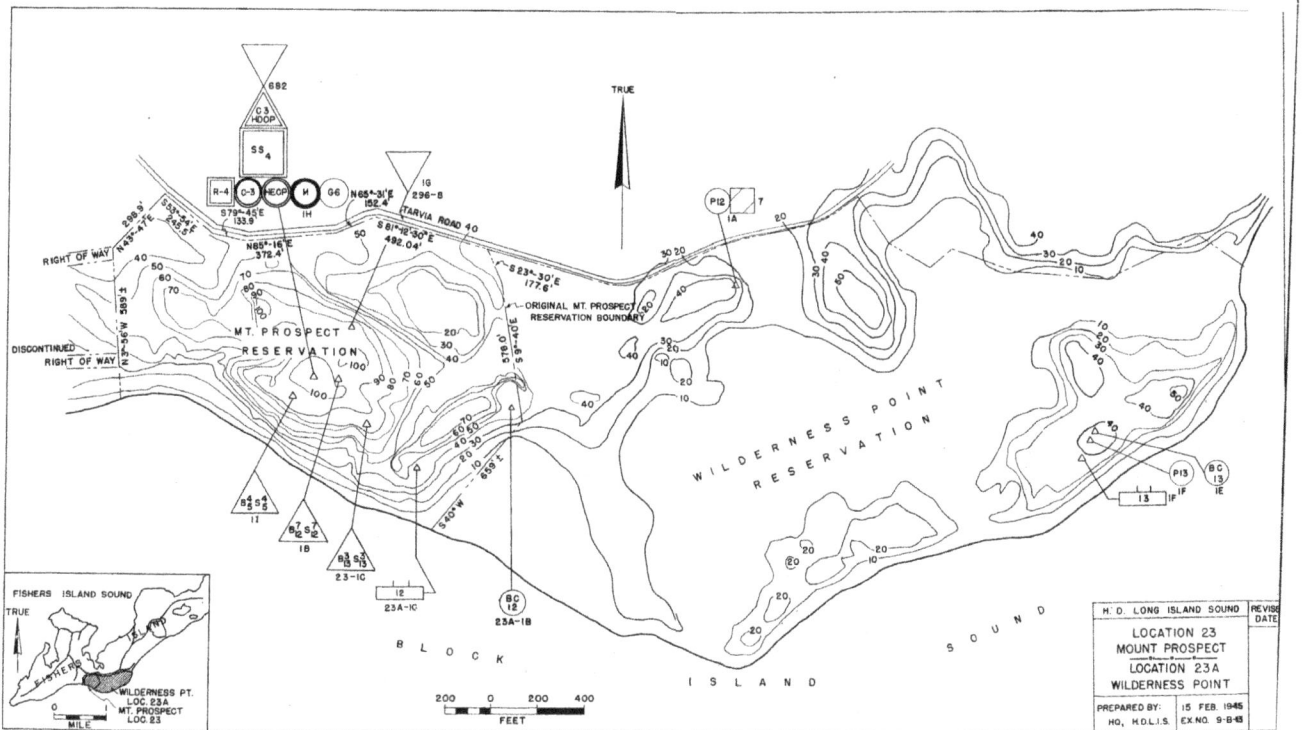

Fort H.G. Wright 1930 (NARA)

Fort H.G. Wright 1924 (NARA)

Fort H.G. Wright, Fishers Island (Terry McGovern)

Fort H.G. Wright, Fishers Island (Terry McGovern)

Battery Clinton (upper left) Fort H.G. Wright, Fishers Island (Terry McGovern)

Battery #111 Wilderness Point Fishers Island (Terry McGovern)

storage/maintenance yard use and are partially buried. Many of fort's structures have been converted into new uses. The island fort is reached by a ferry service from New London, Connecticut. While Fisher's Island is open to the public, public visitation is not generally encouraged as most of the island is privately owned.

Fort Wright Gun Batteries

- **BUTTERFIELD:** A battery for two 12-inch guns on disappearing carriages emplaced along with Battery Barlow as the heavy sequence of guns for Fishers Island. It was located on a low knoll on the western edge of the reservation, with a field of fire to the southwest. The battery was built immediately adjacent to 10-inch Battery Barlow. However, the eastern-most emplacement (No. 2) swung right a further 10-degrees to get a wide field of fire to the southeast. Like To the rear of the emplacement was a retaining wall with casemated storage rooms. Otherwise, the battery conformed to the recommended layout of current mimeographs. The battery plan had been submitted on July 7, 1898. Work was done in 1898-1900. Transfer was made on March 7, 1901 for a construction cost of $110,000. The battery was armed with two 12-inch Model 1895 Watervliet guns on Model 1897 disappearing carriages (#30/#21 and #31/#22). It was named on General Orders No. 30 of March 19, 1902 for Major General Daniel Butterfield of Civil War Service. Between 1905 and 1915 the battery had the usual modifications to its chain hoists, loading platforms, and commander's station. During 1911-12 the gun tubes were removed and replaced with two new Watervliet tubes (#38 and #2). In 1916-17 the Model 1897 carriages were altered to allow 5-degrees of additional elevation. The guns remained emplaced during World War I. However, around 1932 the second set of tubes were again replaced (probably with relined Watervliet tube #7 and Bethlehem tube #3). The battery was declared surplus under the 1932 Review, but apparently nothing was done about removal until World War II. Final authority for abandonment came on October 15, 1944. The emplacement still exists but is partly filled in and damaged due to use as a waste transfer area. The battery site is open to the public.

- **BARLOW:** A battery for two 10-inch guns on disappearing carriages on the western edge of the Fort H. G. Wright reservation, with a field of fire to the west. Plans were submitted on May 23, 1898 for the two-gun emplacement. Within just a couple of months 12-inch Battery Butterfield was positioned contiguously on the left flank, making in effect one large emplacement for four guns. Characteristics of loading platforms and magazines for Barlow followed mimeograph recommendations. However, the protective wall to the rear of the emplacement was casemated with rooms—both to increase its strength and to provide convenient storage. These were carried through to the similar space behind Butterfield. Also, on the left rear of the battery was located a large power plant to serve all of the batteries at this end of the reservation. Work was done in 1898-1900. Transfer was made on March 7, 1902 for a cost of $110,000. It was named on General Orders No. 30 of March 19, 1902 for Major General Francis Barlow of Civil War service. The battery was armed with two 10-inch Model 1895 Watervliet guns on Model 1896 disappearing carriages (#5/#54 and #10/#66). This armament had been received on June 3 and July 15, 1899. Changes to loading platforms, hoists, and the battery commander's station were made over the next ten years. In October 1914 the original gun tubes were removed, being replaced with Model 1888 Bethlehem tubes #47 and #48. Then in turn in 1934 these tubes were replaced with Model 1888 Bethlehem tubes #41 and #45 (previously mounted at Battery Walbach at Fort Wetherill). The battery served until discontinued right before World War II, sometime in 1938-1940. The emplacement still exists but is partly filled in and damaged due to use as waste transfer area. The battery site is open to the public.

- **CLINTON**: A two-pit mortar battery built on Fishers Island. The site selected was to the east of the main gun sequence. Plans were submitted on June 28, 1898 for an emplacement for eight 12-inch Model 1890M1 mortars. It generally followed type plans, and did include space for a relocator room. Also, it had a large power plant incorporated at this site behind the main traverse. It had wide pit spacing (67-feet wide) and magazines in the flanks and central traverse. Work was done from November 1900 to mid-1902 for transfer on October 24, 1902 at a construction cost of $114,895. The original armament consisted of eight 12-inch Model 1890M1 Watervliet mortars on Model 1896 carriages (#131/#263, #84/#264, #82/#267, #108/#269, #74/#261, #130/#262, #136/#265, and #115/#266). The battery was named on General Orders No. 30 of March 19, 1902 for Brigadier General James Clinton of the Continental Army. Telephone booths were added in 1904. In about 1908 carriages #261 and #262 were replaced with #291 and #242 respectively (the removed carriages eventually being mounted at Battery Harlow in Hawaii). Four mortars and carriages were removed in 1917-1918. The final four mortars served until removed under authority granted in May 1943. The emplacement was abandoned postwar and for years was used as a community trash dump. Due to environmental concerns the dump was remediated and hazardous material placed in one pit and buried. The other pit is still used as a work area. The battery is closed to the public.

- **Dynamite Battery**: Despite the conceptual failure of the early dynamite batteries emplaced at New York and San Francisco, during the fervor of the Spanish American War Congress approved expenditures for additional batteries. An emplacement for a single 15-inch compressed air gun throwing a dynamite charge was funded and construction assigned for Fort H.G. Wright at Fishers Island. The precise site was selected on September 29, 1898 for a location just inshore from the southwest tip of the island known as Race Point. Work was done in 1898-1899 on the open concrete pad for the gun and the adjacent power and compressor rooms. Concerned over the exposed condition of the delicate equipment, engineers secured another $75,000 in the summer of 1899 to build protective shelters for the machinery. The contractor emplaced the gun in late 1899. It carried one 15-inch Model 1893 gun on Model 1893 carriage (#3/#6). It appears unlikely that the emplacement or the power machinery was ever completely operational. In early 1902 this type of gun and emplacement was abandoned by the service, and the parts at H.G. Wright scrapped. On May 22, 1902 salvageable equipment was transferred to the post engineer department, the sections of the gun and other parts were soon sold at public auction. Remains of the carriage pit and surrounding apron still exist on airport property at Fishers Island. The battery site is open to the public.

- **DUTTON**: An emplacement for three 6-inch disappearing guns emplaced 700-feet to the east of the main batteries, but still generally on an east-west line with them. It fired to the south. Plans were submitted on January 28, 1899. They generally followed type plans, but opportunity was taken to lower the structure and emplace it as a sunken battery, to aid both in concealment and in clearing the field of fire for other batteries. Each of the three separate platforms had thier own magazines on the lower left flank. Hoists were originally installed but removed in 1915. Work was done from December 1898 to mid-1899. It was transferred on March 7, 1901 at a cost of $78,000. It was named on General Orders No. 30 of March 19, 1902 for Colonel Arthur Dutton, mortally wounded at the Bermuda Hundred, VA in 1864. It was armed with three 6-inch Model 1897M1 guns on Model 1898 disappearing carriages (#16/#26, #14/#24, and #18/#25). While being used for practice fire in the summer of 1923 with National Guard units, a serious accident occurred at this battery. On August 1 the No. 3 gun was accidentally discharged with an inert round into the parapet before being raised to firing position. The shell punched through the parapet and plowed a substantial trench for some distance. The gun tube was dislodged and flipped over onto the loading platform. There were no serious personnel injuries, just cuts and bruises. An investigation concluded that the premature firing came from a misaligned firing mechanism. The gun was remounted and the concrete parapet repaired for a cost of $3,500. In the late 1920 all three tubes were removed and replaced (with #25, #26, and #20). These served until the 1940s, the battery being declared for disposal in October 1944. The emplacement still exists, though partly filled in. The battery site is open to the public, but filled in and overgrown.

- **HAMILTON**: A battery for two 6-inch disappearing guns emplaced to the northwest and thus being a sort of prolongation of the main gun sequence, firing to the south, southwest. Preliminary plans were submitted on December 26, 1902 for this site. Detailed emplacement plans were submitted on March 24, 1903. Construction was initiated with funds from the Fortification Act of June 6, 1902. The plan was consistent with the 1903 type mimeograph for these batteries, with two flank emplacements spaced by 128-feet and a shared magazine in the central traverse. Concerns about possible fire from the rear resulted in the construction of a parados in the rear of the battery. Work was done in 1903-1904. Transfer was made on August 29, 1905 at a cost of $52,000. The battery was named on General Orders No. 194 of December 27, 1904 for Colonel John Hamilton of Civil War service. It was armed with two 6-inch Model 1903 Watervliet guns on Model 1903 disappearing carriages (#6/#34 and #51/#33). The guns were removed in 1917 for use on wheeled field carriages, the disappearing carriages were left in place until scrapped in 1920. The battery was not subsequently used for armament. It still exists. The battery site is open to the public as part of a walking tour.

- **MARCY**: A battery for two 6-inch disappearing guns emplaced to the east of Battery Dutton by about 500-feet. It fired to the south. Plans were submitted on May 13, 1903. Generally, in line with Dutton, it was advanced in front by 50-feet to increase the possible field of fire. It generally followed type plans, with two separate flank emplacements separated by 128-feet and a shared traverse magazine. Like Dutton it was considered a sunken plan, with the top of the traverse extending level behind the emplacement, a tunnel giving access between the two-gun positions. Work was done from July 1903 to March 1904 for transfer on April 4, 1906 for a cost of $50,000. It was also named on General Orders No. 194 of December 27, 1904 for Major General Randolph Marcy of Civil War service. It was armed with two 6-inch Model 1903 Watervliet guns on Model 1903 disappearing carriages (#31/#35 and #41/#36). The gun tubes were removed in late 1917 for use on wheeled field carriages, the carriages were left in place until scrapped about 1920. The emplacement was subsequently abandoned for military purposes but still exists. The battery is open to the public.

- **HOFFMAN**: A battery for two 3-inch pedestal guns erected to the north of Hamilton near the north shore of the reservation. It fired to the northwest to cover the channel and mine fields toward Fort Mansfield. It was of mimeograph type plan, with two circular-shaped flank platforms spaced 34-feet apart and with separate magazines on the lower level at each left flank. An observation station was included also on the left flank parapet. Plans were submitted on July 30, 1902. Work was done from September 1902 to February 1904, being transferred on April 4, 1906 at a cost of $13,470.56. It was named on General Orders No. 194 of December 27, 1904 for 2nd Lieutenant Satterlee Hoffman who was killed in action at Churubusco Mexico in 1847. The original armament was two 3-inch Model 1902 Bethlehem guns on Model 1902 pedestal mounts emplaced in 1905 (#53/#53 and #55/#55). In 1923 a new CRF station was added on a rise about 100-feet to the southern rear of the battery. While the battery was scheduled for deletion under the 1932 Review, this was not implemented. It served until authority for removal came on March 15, 1946. The emplacement still exists. The battery is open to the public as part of walking tour.

- **HOPPOCK**: A battery for two 3-inch pedestal erects erected on the eastern end of the H. G. Wright reservation, near the road leading from the mortar battery to the lake near the southern shore. This was assumed to be a good location to cover the mine fields to the south. Plans were submitted on June 12, 1903. It was of general 1903 type plan, with two platforms separated by 62-feet and a central traverse holding two magazines and a storeroom. Work was done from August 1903 to July 1905. It was transferred on August 29, 1905 for a cost of $17,250. It was named on General Orders No. 194 of December 27, 1904 for Captain John L. Hoppock who was killed in action in 1813 during the War of 1812. The original armament was two 3-inch Model 1902 guns and pedestal carriages (#54/#54 and #56/#56). In 1913 this armament was exchanged with guns from Battery Greble, Fort Terry. It then carried two 3-inch Model 1903 guns and pedestal carriage mounted by July 28, 1913 (#61/#87 and #62/#88). In 1914 a new CRF station was built just to the west of the battery. Later these guns were also removed and replaced with the Model 1903 guns from Battery Hagner, Fort Terry, when that battery was abandoned in the mid-1930s. The final armament was Model 1903 guns and carriage #59/#55 and #88/#87, which served until removal authority of March 11, 1946. The emplacement still exists. The battery is open to the public, but overgrown.

- **Battery #215**: A 1940 Program dual 6-inch barbette battery emplaced at the western tip of Fishers Island, near Race Point, and firing to the southwest. In 1940 it was assigned the relatively high national priority of #8, and funding began under the FY-1942 Budget. It was of standard 200-series plan, trunnion heights of the site were listed at 17.3-feet. Work was done from January 10 to November 1, 1942. It was transferred on December 13, 1943 at a cost of $228,769. The battery was never named, just known during construction and service as Battery Construction No. 215. It was armed with two 6-inch guns Model 1903A2 on Model M1 barbette carriages (#16/#19 and #28/#20). By late 1944 it was fully operational and serving in the local Long Island Sound defenses as Tactical Battery No. 8. The battery was deleted and disarmed postwar in 1946. The emplacement still exists on local airport property. The battery is open to the public.

- **AMTB #913**: A 1943 Program AMTB battery for two 90mm fixed and two 90mm mobile guns emplaced atop old Battery Butterfield in mid-1943. It was built from May 19 to August 19, 1943 for transfer on December 20, 1943 for a cost of $7,458. It had a trunnion height of 27.7-feet and consisted of concrete foundation blocks for the two fixed mounts and a concrete BC station. The mobile guns were sited outside the middle two fixed mounts. The battery served until being dismounted in 1946. The BC station still exists, but the blocks are heavily overgrown and not easy to find, if they still exist. The battery site is open to the public.

North Hill Military Reservation (1917-1950) is located at the northern end of Fishers Island at North Hill separate and distinct from the H.G. Wright reservation. Located at this small reservation were New Battery Hackleman (1944-1946) and a planned unnamed 5-inch gun battery (1917–1919).

North Hill M. R. Gun Batteries

- A World War I temporary battery for four 5-inch guns was installed at North Hill, on Fishers Island. The guns were emplaced by December 27, 1917, and probably were simple concrete gun blocks. Confirmation is lacking, the guns were probably two Model 1900 guns on Model 1903 pedestal) mounts and two Model 1896 guns on Model 1897 balanced pillar carriages coming from Fort Mansfield's batteries Connell and Crawford. In any event they served only a short while before being disarmed at war's end. No trace of the blocks or battery has ever been located.

- **HACKLEMAN**: A battery for two 3-inch Model 1903 pedestal guns erected at the Fishers Island North Hill reservation. The guns were relocated here from Battery Hackleman at Fort Constitution, Portsmouth New Hampshire. Work was done from June 5 to September 15, 1942 for transfer on January 8, 1944 for a cost of $11,801. It consisted of simple concrete gun blocks, separated by some 100-feet, and a concrete BC station between them. The battery was not formally named, continuing to carry the name Battery Hackleman. It was armed with two 3-inch Model 1903 guns and pedestal mounts (#88/#63 and #89/#54). The battery was probably disarmed shortly after the end of the war, under authority to abandon dated March 11, 1946. The blocks and station still exist. The battery site is closed to the public.

Mount Prospect/Wilderness Point Military Reservation (1908/1943-1949/1960s) includes Battery 111 (1944, never armed) on Mt. Prospect, Battery 214 (1944, never armed) at Wilderness Point, and the underground Harbor Defense Command Post / Harbor Entrance Control Post on Mt. Prospect. A two-gun AA battery (1918, armed in 1920, third gun added 1930s) was also located on Mt. Prospect. Four fire-control stations are still located on Mt. Prospect. The U.S. Navy took over this parcel in 1949 for the Naval Underwater Sound Laboratory. The U.S. Navy still retains this reservation and access is restricted.

Mount Prospect/Wilderness Point M. R. Gun Batteries

- **Battery #111**: A 1940 Program dual 16-inch barbette battery built on Fishers Island in a new reservation known as Mount Prospect/Wilderness Point. One of the original heavy battery projects for the Sound, it was given a lower priority that the matching batteries at Camp Hero. It was assigned national priority #27 of the September 11, 1940 list and #28 on that of August 11, 1941. Work was done from February 15, 1943 to May 26, 1944. The work was transferred on July 14, 1944 for a total cost of $1,782,001. It was never named, just known as Battery Construction No. 111 during work. The design was conventional in all respects, the trunnion elevation being at 45.67-feet. A separate, single-entry PSR building was also included in the project. It had fifteen base-end stations assigned, the greatest number for any individual American gun battery. The guns assigned were two 16-inch MkIIM1 tubes (#47 and #105) and two barbette carriages M1919M4 (#58 and #36). RThe armament was never actually mounted but stored in the battery when work was suspended on November 20, 1943. However, all structural concrete work was completed. The incomplete battery was deleted postwar, but the emplacement still exists. The battery is closed to the public.

- **Battery #214**: A 1940 Project dual 6-inch barbette battery emplaced to the east of Battery #111 at the Wilderness Point Military Reservation, firing to the southeast. It was given a low priority, not being funded until the FY-1943 Budget. The plan was strictly consistent with standard 200-series batteries; it had a trunnion elevation of 42.3-feet. The construction was done from April 2, 1943 to May 23, 1944. It was transferred on August 7, 1944 for a cost of $236,354. It was to have been armed with two 6-inch M1(T2) guns, but these were never delivered with the suspension of the project in 1944. The M4 barbette carriages #49 and #50 were delivered, but it is not clear if they were actually mounted. Although assigned local Tactical Battery No, 13, without armament it never entered service. Also, it was never named, just known as Battery Construction No. 214 during work. It was abandoned postwar, being sold as excess property in 1947. Today the site is privately owned, with a residence built on top of the battery. The battery is closed to the public.

East Point Military Reservation (1943-1946) is located East Point on Fishers Island. AMTB Battery 916 (1943-1946) was planned for this location but was probably not built. Four observation stations on nearby Hill 90.

East Point M. R. Gun Battery

- **AMTB #916** A 1943 Program AMTB battery for two fixed and two mobile 90mm guns, to be emplaced at the eastern tip of Fishers Island, firing to the east. Records are not definitive, but it appears that no work was done for this emplacement, or that if blocks were built any armament was ever stationed here.

Fort Michie (1896-1948) is located on Great Gull Island at the eastern end of Long Island Sound between Plum Island and Fishers Island. This 17-acre island was converted into a heavily fortified island during the Endicott Program. The primary armament was 12-inch and 10-inch disappearing batteries. Secondary armament was two 6-inch batteries and a 3-inch battery. The fort also had a controlled submarine mine casemate. It was named on General Orders 138 of 1899 for 1st Lt. Denis M. Michie who was KIA at San Juan, Santiago, Cuba on July 3, 1898. Due to the exposed location the island received a complete "rip-rap" seawall and facilities to self-supporting for extended periods. During the Interwar Period, a most unusual structure was construction on Fort Michie. The 10-inch disappearing battery was removed and one-of-kind battery for a single 16-inch disappearing gun was built. Known as Battery M.K. Davis, this massive concrete structure housed the 16-inch gun in huge pit on an all-round disappearing carriage. To the rear of gun pit were magazines, power plant, plotting rooms, battery commander station, and mining casemate. During World War II, a 90mm AMTB battery was installed in front of Battery Davis. Today, Great Gull Island is a bird sanctuary controlled by the American Museum of Natural History and permission to visit is controlled during the nesting season.

Fort Michie Gun Batteries

- **PALMER:** A 12-inch disappearing gun battery located across the center of Great Gull Island. After some confusion as to whether the carriages to be employed were to be of Model 1896 or 1897 (the latter was finally selected), work was done in 1897-1899. It was of generally typical design for this type of emplacement, though a large mine casemate was located immediately behind the central traverse slightly after completion in 1908. The battery was transferred on May 12, 1900 for a cost of $129,930. It was initially armed with two 12-inch Model 1895 Watervliet guns on disappearing carriages Model 1897 (#26/#2 and #27/#1). These were mounted by October 1899. The battery was named on General Orders No. 30 of March 19, 1902 for Colonel Innis N. Palmer of Mexican and Civil War service. Over the next ten years the battery had modifications to its chain hoists, platforms, and had a new battery commander station added. In 1924 the entire battery was re-wired. Also, two 3-inch anti-aircraft blocks were placed just outside of each 12-inch gun pits in the early 1920s. At some point gun tube #27 was removed and replaced with Model 1895 gun #2. The battery was retained in the 1932 Review but scheduled for deletion in the 1940 Program. That was somewhat delayed pending the completion of the new 16-inch battery at Wilderness Point, authority for disarming not coming until late in 1945. The emplacement still exists on the island. The battery is open, but the island is closed to the public.

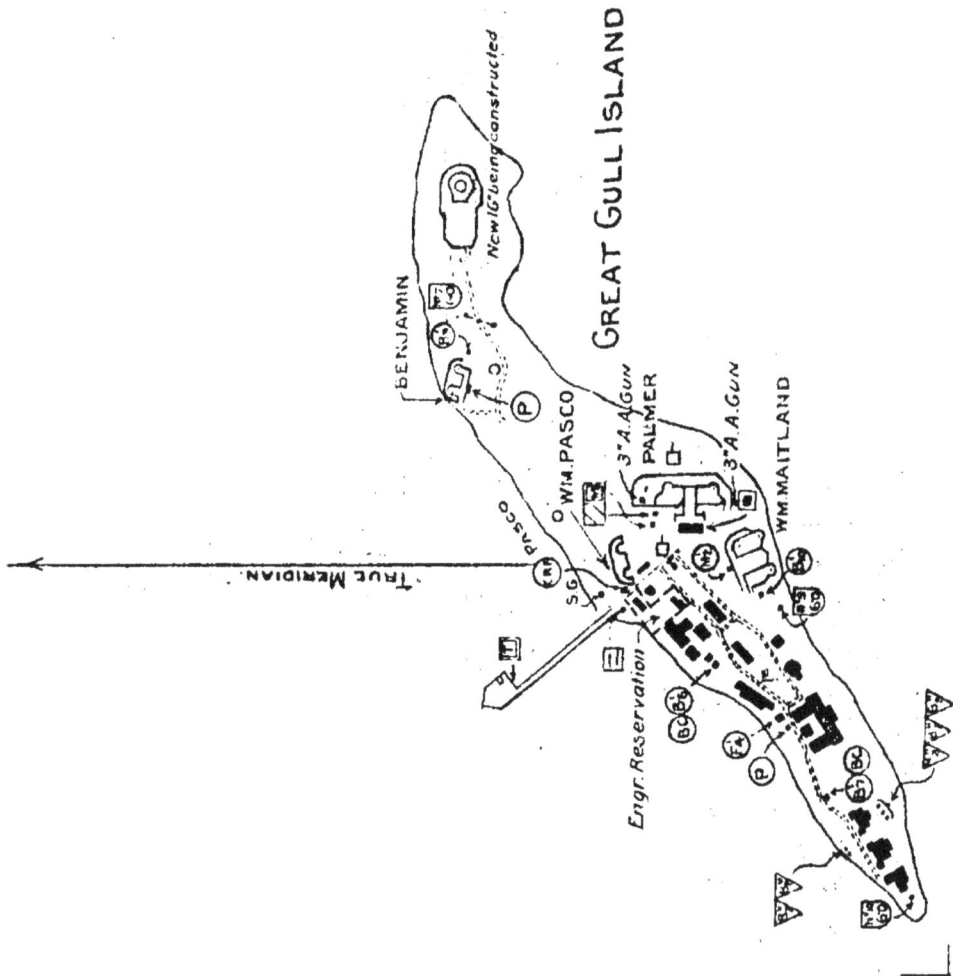

LONG ISLAND SOUND
FORT MICHIE
GENERAL MAP

LITTLE GULL ISLAND

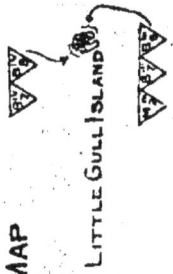

SERIAL NUMBER 124

EDITION OF MAR. 3, 1919.
REVISIONS: FEB. 13, 1920.
MAR. 28, 1921.

TRUE MERIDIAN

GREAT GULL ISLAND

New IG being constructed

BENJAMIN

WM. PASCO

S.G.

PASCO

3" A.A. Gun
PALMER
3" A.A. Gun

WM. MAITLAND

Engr. Reservation

SCALE OF FEET
0 100 200 300 400 500 600   1200

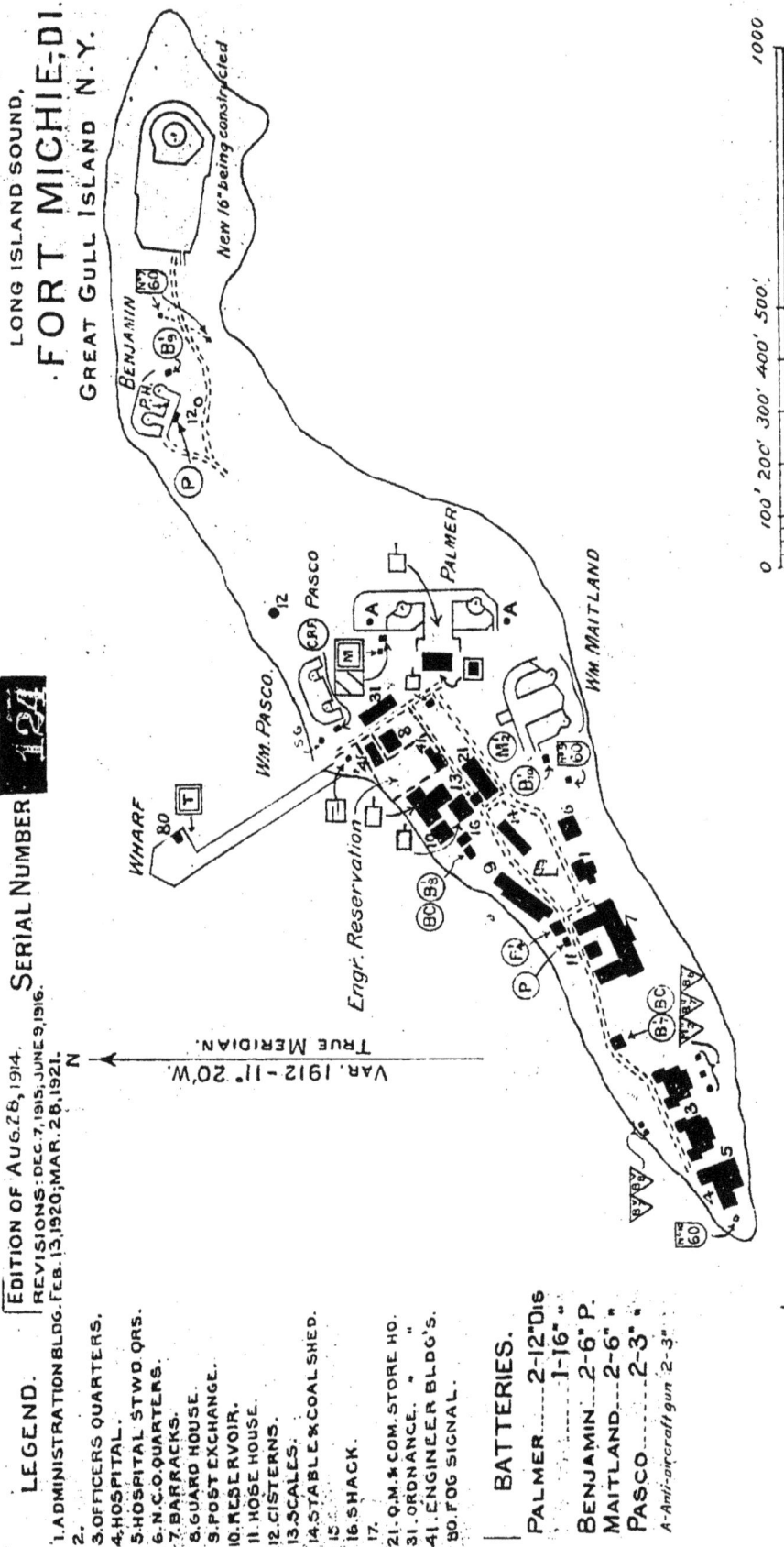

LONG ISLAND SOUND,
FORT MICHIE-D1.
GREAT GULL ISLAND N.Y.

SERIAL NUMBER **124**

EDITION OF AUG. 28, 1914.
REVISIONS: DEC. 7, 1915; JUNE 9, 1916.
BLDG. FEB. 13, 1920; MAR. 28, 1921.

TRUE MERIDIAN.
VAR. 1912 - 11° 20' W.

New 16" being constructed

Engr. Reservation

WHARF

WM. PASCO.

WM. MAITLAND.

PALMER

BENJAMIN

PASCO

LEGEND.

1. ADMINISTRATION BLDG.
2.
3. OFFICERS QUARTERS.
4. HOSPITAL.
5. HOSPITAL STWD. QRS.
6. N.C.O. QUARTERS.
7. BARRACKS.
8. GUARD HOUSE.
9. POST EXCHANGE.
10. RESERVOIR.
11. HOSE HOUSE.
12. CISTERNS.
13. SCALES.
14. STABLE & COAL SHED.
15.
16. SHACK.
17.
21. Q.M. & COM. STORE HO.
31. ORDNANCE.   "
41. ENGINEER BLDG'S.
80. FOG SIGNAL.

BATTERIES.

PALMER.....2-12" Dis.
      "   ..... 1-16" "
BENJAMIN..2-6" P.
MAITLAND..2-6" "
PASCO.....2-3" "
A-Anti-aircraft gun 2-3"

0 100' 200' 300' 400' 500'   1000'

LONG ISLAND SOUND

FORT MICHIE-D2.

LITTLE GULL ISLAND, N.Y.

Scale of Feet

SERIAL NUMBER

124

EDITION OF MAR. 3,1919.
REVISIONS: FEB. 13,1920.

LEGEND

80. DWELLING.
81. BOAT HOUSE.
82. SHED.
83. COAL BIN.
84. FOG SIGNAL HOUSE.
85. DRAIN INLET.
86. WELL.

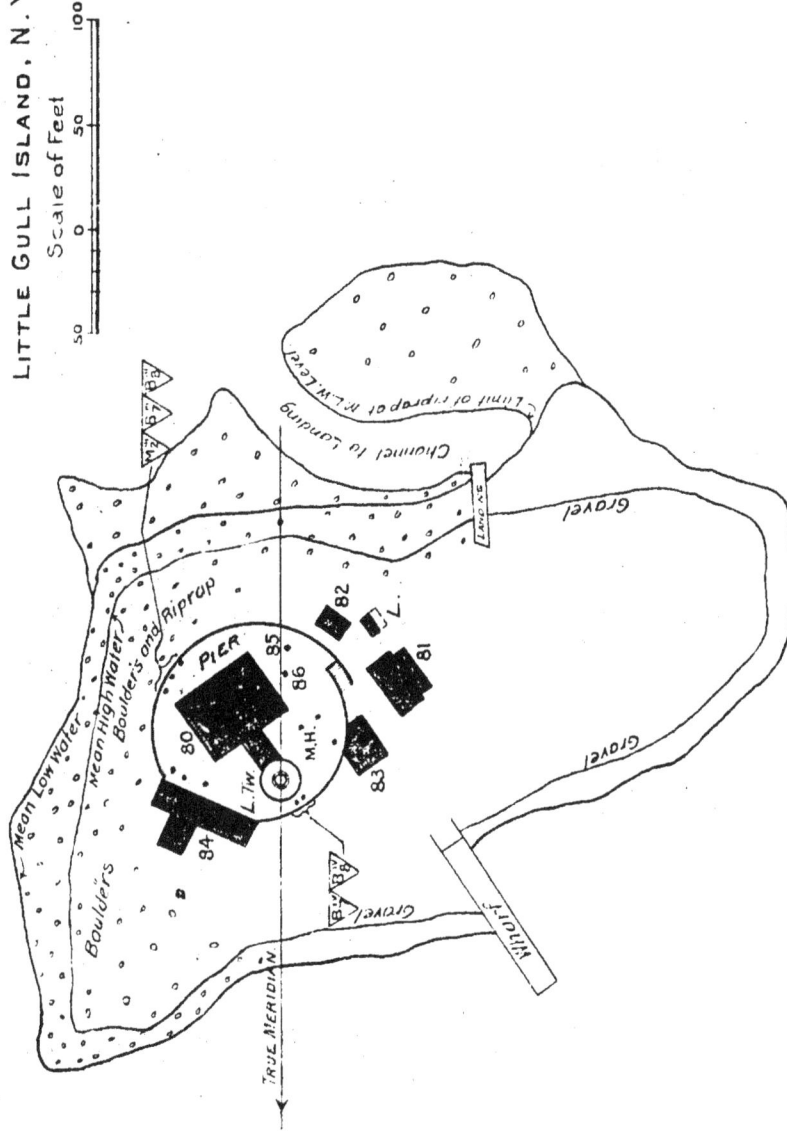

Mean Low Water

Mean High Water

Boulders and Riprap

Boulders

PIER

L.Tw.

M.H.

TRUE MERIDIAN

Gravel

Gravel

Gravel

Channel to Landing

Limit of riprap at M.L.W. Level

80

81

82

83

84

85

86

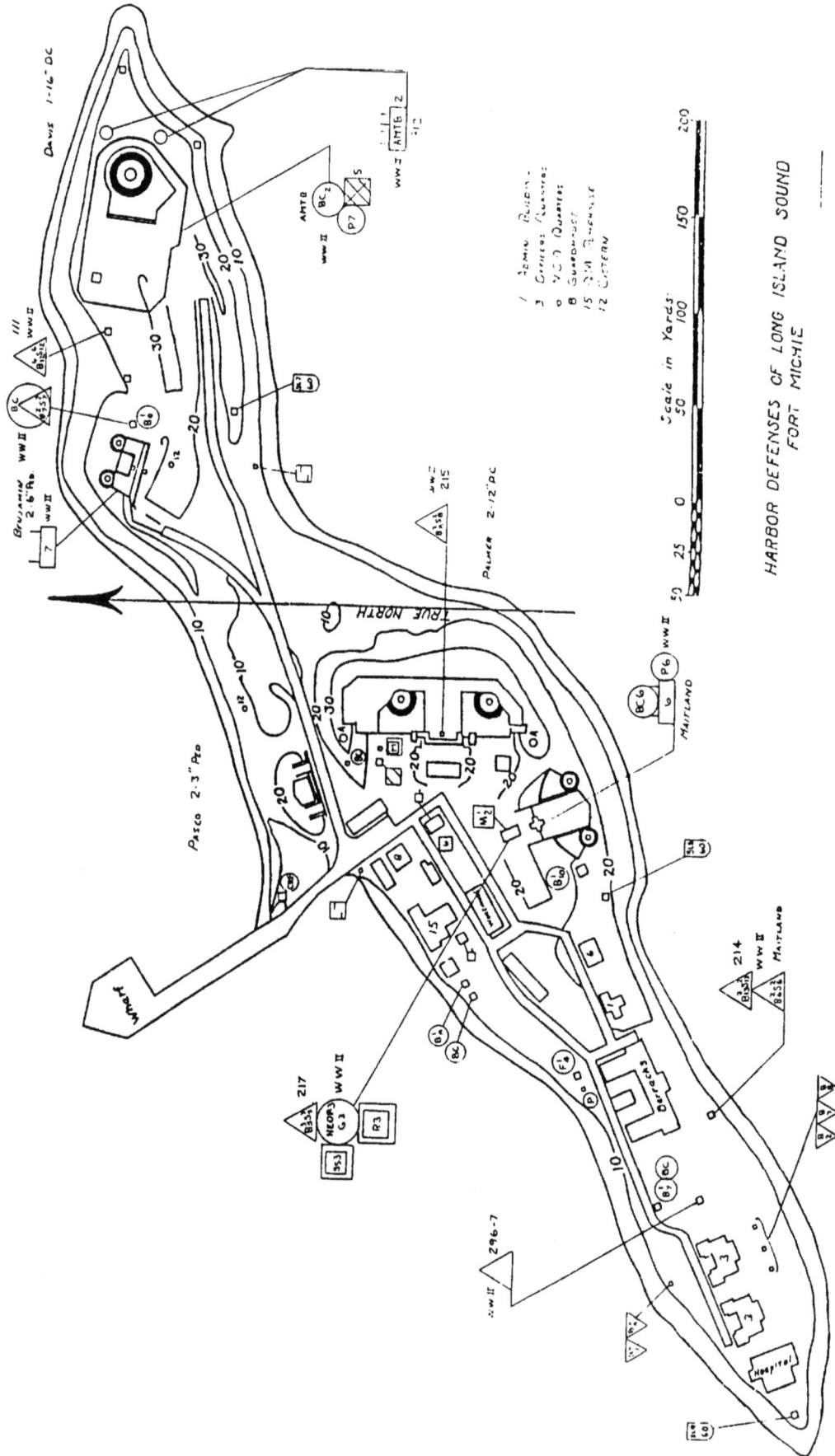

HARBOR DEFENSES OF LONG ISLAND SOUND
FORT MICHIE

Scale in Yards

1 Admin. Bldg.
3 Officer Quarters
6 N.C.O. Quarters
8 Guardhouse
15 P.X. Exchange
12 Cistern

Davis 1-16" DC

Brosius 2-6" Bb. WW II

Pasco 2-3" AA

Palmer 2-12" DC

Wharf

TRUE NORTH

Hospital

Barracks

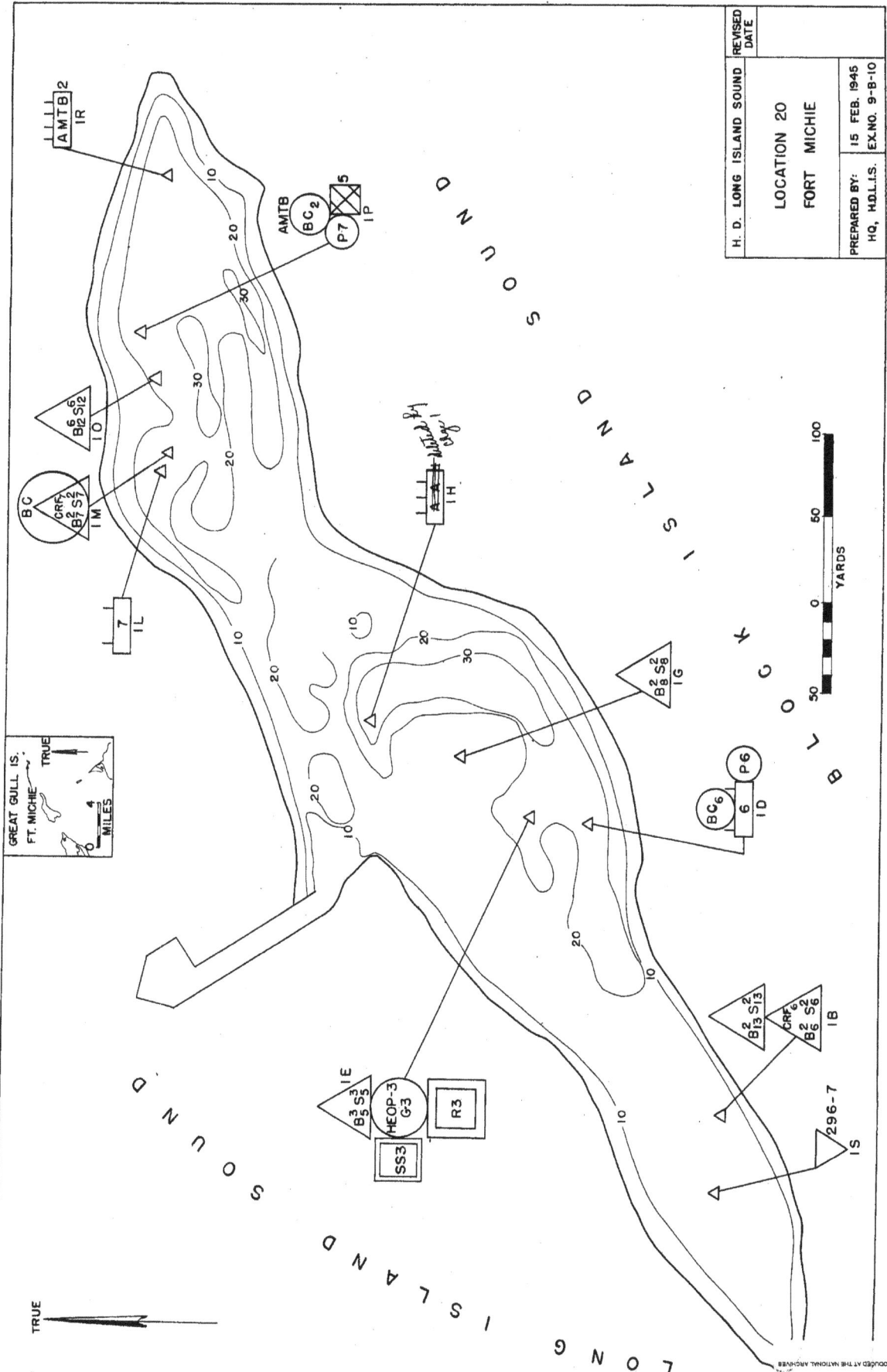

Fort Michie1920s (NARA)

Fort Michie 1924 (NARA)

Battery Palmer and Battery Maitland Fort Michie, Great Gull Island (Terry McGovern)

Battery JMK Davis Fort Michie, Great Gull Island (Terry McGovern)

- **NORTH:** A 10-inch disappearing gun battery located at the eastern tip of Great Gull Island. The battery plan was submitted on March 22, 1896. The narrow point of the island only had room for a single emplacement, so the two guns of this battery were placed so that the rear emplacement could fire over the lower, front emplacement. Also, they had a somewhat different orientation, with the rearward gun firing more to the northeast. Work was done in 1897-1898. It was transferred on May 12, 1900 for a cost of $100,000. The armament was received on August 10, 1898. That consisted of two 10-inch Model 1888MII Bethlehem guns on Model 1896 disappearing carriages (#18/#37 and #22/#47). The battery was named on General Orders No. 30 of March 19, 1902 for Brigadier General William North, Adjutant General during the Revolutionary War. Changes were made in the following years to the platforms and new chain hoists were installed. The armament was removed in 1917. The emplacement was completely destroyed for the construction of Battery J. M. K. Davis in this location in 1919-1920. No trace remains of it today.

- **J.M.K. DAVIS:** A 1915 Board of Review Program battery for one 16-inch disappearing gun built at the eastern tip of Fort Michie on Great Gull Island. One of the most significant projects for the 1915 Board was for new, heavy armament at the eastern end of Long Island Sound to prevent access to the approaches to northern New York. A variety of plans for 16-inch guns were suggested for the islands. The original 1915 proposal was for forts H.G. Wright and Terry to each get two 16-inch barbette guns and a battery of 16-inch mortars; low-lying Fort Michie was to get two 16-inch in a turret. Realities of postwar budgets finally reduced the project to a single battery with the very powerful new 16-inch Model 1919. On October 20, 1917 tentative approval was given for a 16-inch battery to be built at Wright's Mt. Prospect. By 1919 the project evolved to an emplacement for a single 16-inch disappearing gun to be built at the eastern tip of Great Gull Island, replacing Battery North. The battery's role was to prevent a "run-by" of enemy ships through the Race between Fishers and Great Gull Island. As this called for only a desired gun range of 5000-yards, it was determined that the location could make use of the single 16-inch disappearing carriage authorized rather than a much more expensive (and never-to-be-developed) turret mount. Fort Michie received the only 1915 Program battery for the Long Island Sound, and the only Model 1919 disappearing carriage ever built. The battery plan was unique. The large pit with complete 360-degree traverse was on the eastern tip of the emplacement, behind directly to the west were the same-level magazines. Shells were carried by overhead triplex blocks to shell tables where they were loaded on carts and taken to the gun. Battery North was entirely destroyed for this construction. Work took place between October 1919 and June 1922. It was transferred on May 17, 1923 for $728,286.49. The armament was one 16-inch Model 1919 Watervliet gun on disappearing carriage Model 1919 (#1/#1). It was named by General Orders No. 13 of March 27, 1922 for Brigadier General John M. K. Davis, army officer who died on May 20, 1920. Unfortunately, the disappearing carriage never really proved successful. The 16-inch gun was too powerful, and even the supposedly rugged design of the carriage did not prove adequate. The gun was almost never fired and even exercises with sub-caliber equipment occurred only rarely. It was kept in inventory, but mostly as a reserve to the defenses, until being deleted with completion of the 1940 Program new batteries. Authority for removal finally came on February 2, 1944 and it was removed shortly thereafter. The emplacement still exists. The battery is open, but the island is closed to the public.

- **BENJAMIN**: A battery for two 6-inch pedestal guns erected to the west and behind Battery North on the northern end of Great Gull Island. The sites for the two 6-inch, rapid-fire gun batteries destined for Fort Michie were approved in January 1899. Benjamin was located just to the rear of Battery North (later J.M.K. Davis), firing to the northeast. The plan was submitted on May 28, 1901. Initially it was designed with the two guns per platform mimeograph, but that was changed during early construction to a plan with two separate platforms. The constricted topography meant that the platforms were separated by only 60-feet (vs. a recommended 94-feet). Magazines had to be placed below the platforms, with shells lifted by hand-operated hoists. Work was done from February 1902 to early 1903. Guns were mounted in July 1905. Transfer was not made until November 25, 1908 for a cost of $54,742. It was named on General Orders No. 30 of March 19, 1902 for 1st Lieutenant Calvin Benjamin killed during the assault on Mexico City in 1847. It carried two 6-inch Model 1900 guns and Model 1900 carriages (#37/#9 and #38/#10). The battery served successfully with this armament for many years. It was eventually disarmed in 1946. The emplacement still exists. The battery is open, but the island is closed to the public.

- **MAITLAND**: A battery for two 6-inch pedestal guns built just to the southwest of Palmer, firing to the southeast. It was constructed between June 1903 and March 1904, receiving its armament in July of 1906. The battery was located on the south shore with an opposite field of fire to the southeast across the island from sister Battery Benjamin. Plan was submitted on May 9, 1903. Room was a little more spacious in this site, the battery was designed and built from the start with two platforms, a full 94-feet between gun centers and adjacent-level, central traverse magazines that required no hoists. Work was done in 1903-1904. Armament was received and mounted in July 1906. The battery was transferred on April 29, 1908 for a cost of $47,500. It was armed with two 6-inch Model 1900 guns and pedestals (#18/#8 and #17/#7). The battery was named on General Orders No. 194 of December 27, 1904 for Brevet Captain William S. Maitland, who died on August 19, 1837 during the Seminole War. Like Battery Benjamin, the battery had a long service life through the 1920s and 1930s. It was finally deleted postwar in 1946. The emplacement still exists. The battery is open, but the island is closed to the public.

- **PASCO**: A 3-inch, rapid-fire emplacement sited on the left flank of Battery Palmer, firing general to the north aligned with the direction of the island's dock. The plan was submitted on October 8, 1902. It generally followed plans for early pedestal batteries, with both platforms curve-shaped as flank emplacements and having 64-feet gun centers. The exposed position prompted the use of thicker concrete protection than usually proscribed. Work was done from 1902-1903. Transfer was made on July 30, 1908 for a cost of $22,800. It was armed with two 3-inch Model 1903 guns and pedestal mounts (#7/#35 and #41/#36). It was named on General Orders No. 194 of December 27, 1904 for 2nd Lieutenant William D. Pasco, 8th U.S. Infantry who was killed during the Philippine Insurrection. It was abandoned in 1934 due to heavy storm damage to the emplacement. In damaged condition, substantial remains still exist. The battery is open, but the island is closed to the public.

- **AMTB-912:** A 1943 AMTB Program for two 90mm fixed and two 90mm mobile guns for Fort Michie. It was located on the eastern end of the island, actually on the riprap, below the berm and retaining wall of 16-inch Battery J. M. K. Davis. Work was done from April 23, 1943 until June 18th and turned over on December 20, 1943 at a cost of $8584. Accompanying it was AMTB-312 with its two 40mm guns. The fixed 90mm gun blocks still exist. The battery is open, but the island is closed to the public.

Fort Terry (1898-1948) is located on Plum Island off the tip of Orient Point which the north end of Long Island. This fort was built on the 840-acre island during the Endicott Program. Constructed were one mortar battery, one 10-inch disappearing battery, three 6-inch disappearing batteries, one 5-inch battery and five 3-inch batteries. The location had an extensive garrison area and a controlled mine facility. It was named on General Orders 134 of 1899 for Maj. Gen Alfred H. Terry, U.S. Army of Civil War and Indian War service. The western end of the island faces along Plum Gut where the controlled mine complex was located, including a large torpedo storehouse and mine casemate. Also located on the western end of the island near Pine Point were two 3-inch batteries and a 6-inch battery. The center part of Plum Island is the cantonment area. Moving toward East Point and the main concentration of batteries is an interesting overlaid 10-inch disappearing gun battery. In a small area of East Point, where seven Endicott Program batteries were located. The defenses were upgraded during the 1940 Program with a 155mm battery which was later replaced by a shielded 6-inch battery and a 90mm AMTB battery. Following World War II, the island was transferred to the US Dept. of Agriculture which used the island to study infectious diseases in animals. Eventually a new Animal Disease Lab was built on the island, and the old military structures were abandoned. Current plans are to close the lab and move its operations to Kansas, and the island is still under Department of Homeland Security and access is restricted to lab personnel.

Fort Terry Gun Batteries

- **STEELE:** A battery for two 10-inch guns erected on the hill on the eastern end of Plum Island. It was the first new battery for this reservation, plans being submitted on May 12, 1897. The constricted space available and required field of fire led to a design highly adapted to the site. The two-gun emplacements were staggered. The No. 1 emplacement faced mostly east and was at a trunnion height of 60-feet. The emplacement to the rear (No. 2) was at 80-feet and could fire over part of No. 1. In addition, its field of fire was a very wide 210 degrees to allow it to fire on either side of the island. The arrangement of the magazines and ammunition lifts were also modified from standard plans to fit. Work was done in 1897-1899. It was transferred on March 31, 1900 for a cost of $102,593. The battery was armed with two 10-inch Model 1888MII Bethlehem Steel guns on Model 1896 disappearing carriages (#4/#45 and #5/#44). It was named in General Orders No. 30 of March 19, 1902 for Brevet Major General Frederick Steele of Mexican War and Civil War service. The original guns were removed in 1917 for use on railway carriages. However, they were replaced by new tubes removed from Battery Church at Fort Monroe (10-inch Model 1888MII Bethlehem tubes #25 and #31). The battery was declared obsolete and slated for abandonment in the 1932 Review. Actual date of disarmament is unclear, notes of June 3, 1936 state that the battery is no longer required, but another authorization of October 10, 1942 declares it ready for salvage. In 1940 a battery for four 155mm GPF guns on Panama Mounts was built across the apron of No. 1 gun— clearly suggesting that no serviceable armament was left in at least that emplacement. The emplacement still exists, but in rather poor condition. The battery is open, but the island is closed to the public.

SERIAL NUMBER 124

LONG ISLAND SOUND

FORT TERRY,
GENERAL MAP
PLUM ISLAND, N.Y.

EDITION OF MAR. 3, 1919.
REVISIONS: FEB. 13, 1920.
MAR. 28, 1921.

N
True Meridian.
Var. 1912-11°20'W.

BATTERIES.

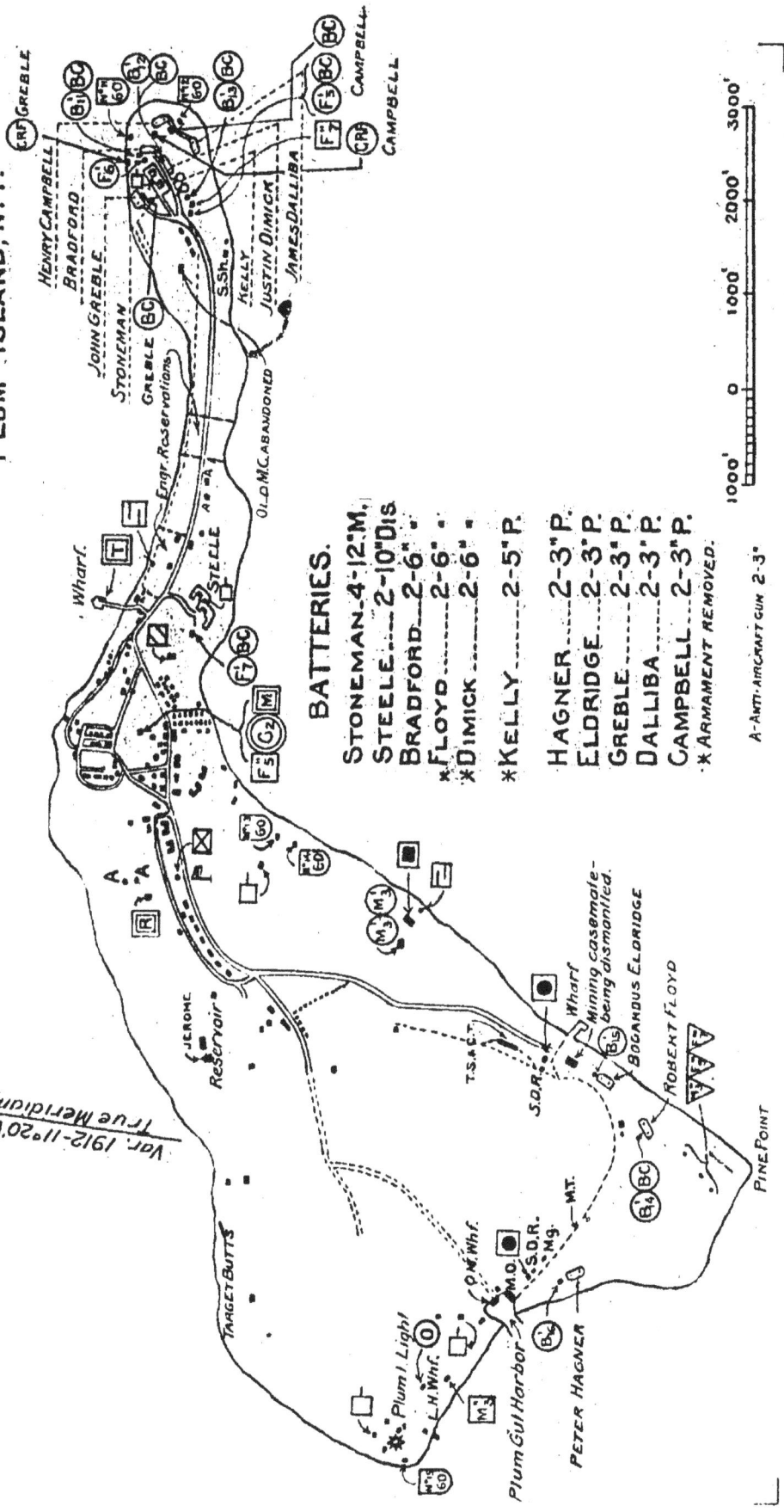

STONEMAN—4-12" M.
STEELE—2-10" Dis
BRADFORD—2-6" "
*FLOYD—2-6" "
*DIMICK—2-6" "

*KELLY—2-5" P.

HAGNER—2-3" P.
ELDRIDGE—2-3" P.
GREBLE—2-3" P.
DALLIBA—2-3" P.
CAMPBELL—2-3" P.
*ARMAMENT REMOVED.

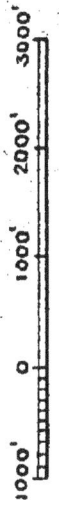

A–ANTI-AIRCRAFT GUN 2-3"

1000'   0   1000'   2000'   3000'

HENRY CAMPBELL
BRADFORD
JOHN GREBLE
STONEMAN
GREBLE
S. STEELE
KELLY
JUSTIN DIMICK
JAMES DALLIBA
CAMPBELL

ERF GREBLE
CAMPBELL

Engr. Reservations
OLD M.C. ABANDONED

Wharf.
STEELE

JEROME
Reservoir

TARGET BUTTS

Plum I. Light
L.H. Whf.

Plum Gut Harbor
Q.M. Whf.
M.O.
S.D.R.
Mg.
M.T.

PETER HAGNER

T.S.A.C.T.
S.D.R.
Wharf
Mining casemate—
being dismantled.
BOGARDUS ELDRIDGE
ROBERT FLOYD

PINE POINT

SERIAL NUMBER **124**

LONG ISLAND SOUND
FORT TERRY-DI.
PLUM ISLAND, N.Y.

EDITION OF MAR. 3, 1919.
REVISIONS: FEB. 13, 1920; MAR. 28, 1921.

96
205 Q.M.BARRACKS.

BATTERIES.
STONEMAN----4-12" M.
STEELE------2-10'Dis
BRADFORD----2-6 "
*DIMICK-----2-6 "
*KELLY------2-5'P.
GREBLE------2-3'P.
DALLIBA-----2-3'P.
CAMPBELL----2-3'P.
*ARMAMENT REMOVED.

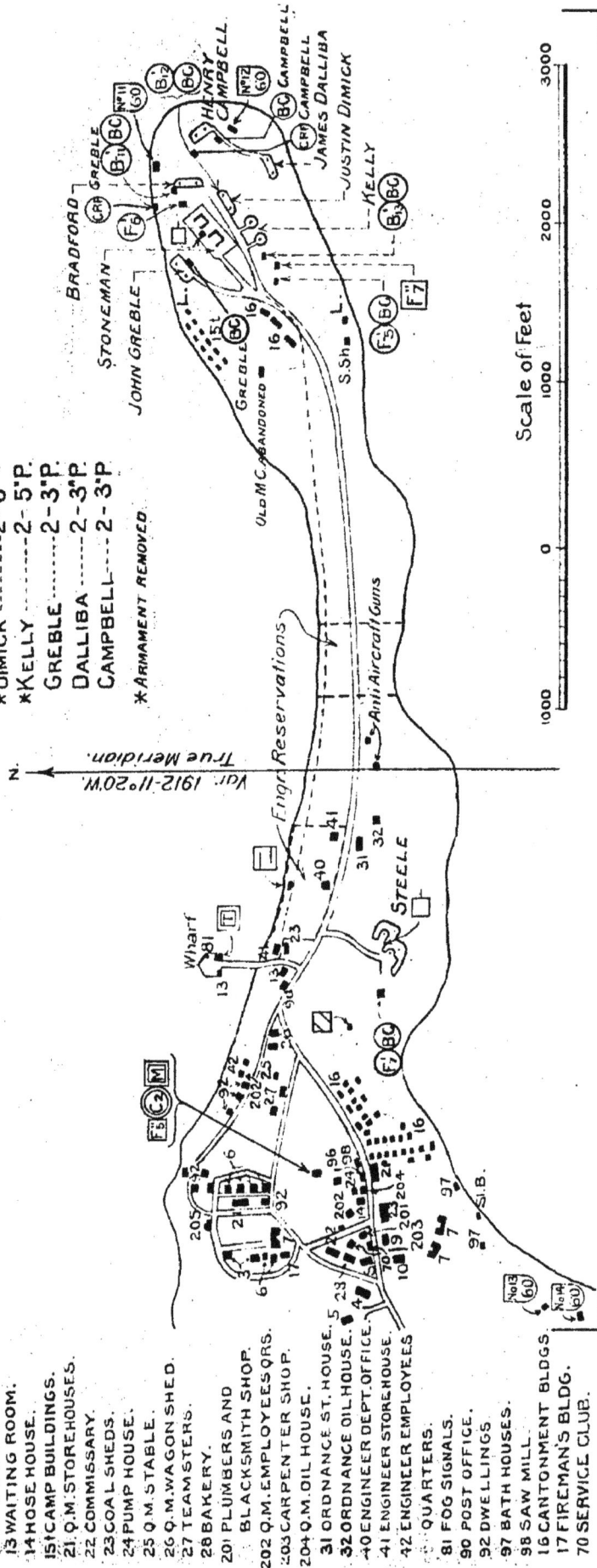

LEGEND.
3 OFFICERS QRS.
4 HOSPITAL.
5 HOSPITAL STD. QRS.
6 NON-COMMISSIONED
  OFFICERS QUARTERS.
7 BARRACKS.
8 GUARD HOUSE.
9 POST EXCHANGE.
10 GYMNASIUM.
13 WAITING ROOM.
14 HOSE HOUSE.
15 †CAMP BUILDINGS.
21 Q.M. STOREHOUSES.
22 COMMISSARY.
23 COAL SHEDS.
24 PUMP HOUSE.
25 Q.M. STABLE.
26 Q.M. WAGON SHED.
27 TEAMSTERS.
28 BAKERY.
201 PLUMBERS AND
    BLACKSMITH SHOP.
202 Q.M. EMPLOYEES QRS.
203 CARPENTER SHOP.
204 Q.M. OIL HOUSE.
31 ORDNANCE ST. HOUSE.
32 ORDNANCE OIL HOUSE.
41 ENGINEER STOREHOUSE.
42 ENGINEER EMPLOYEES
   QUARTERS.
81 FOG SIGNALS.
90 POST OFFICE.
92 DWELLINGS.
97 BATH HOUSES.
98 SAW MILL.
16 CANTONMENT BLDGS.
17 FIREMAN'S BLDG.
70 SERVICE CLUB.

N ← True Meridian.
Var: 1912-11°20'W.

Eng. Reservations

Anti Aircraft Guns

Scale of Feet
1000   0   1000   2000   3000

EDITION OF MAR.3,1919.
REVISIONS: FEB.13,1920; MAR.28,1921.

SERIAL NUMBER **124**

LONG ISLAND SOUND
FORT TERRY-D2.
PLUM ISLAND, N.Y.

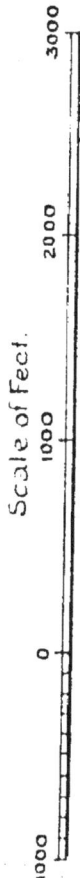

Scale of Feet.
1000   0   1000   2000   3000

True Meridian
Var. 1912-11°20'W.

N

3"Anti-Aircraft Guns

Reservoir
JEROME
Target Butts

LEGEND.
1 ADMINISTRATION BLDG.
2 COMMANDING OFFICERS QRS.
3 OFFICERS QUARTERS.
4 HOSPITAL.
5 HOSPITAL STW'D. QRS.
7 BARRACKS.
24 PUMP HOUSE.
29 CREMATORY.
91 LAUNDRY.
93 HENCOOPS.
94 GREENHOUSE.
95 BARNS.
96 GARAGE.
97 BATH HOUSE.
71 POST CHAPEL.

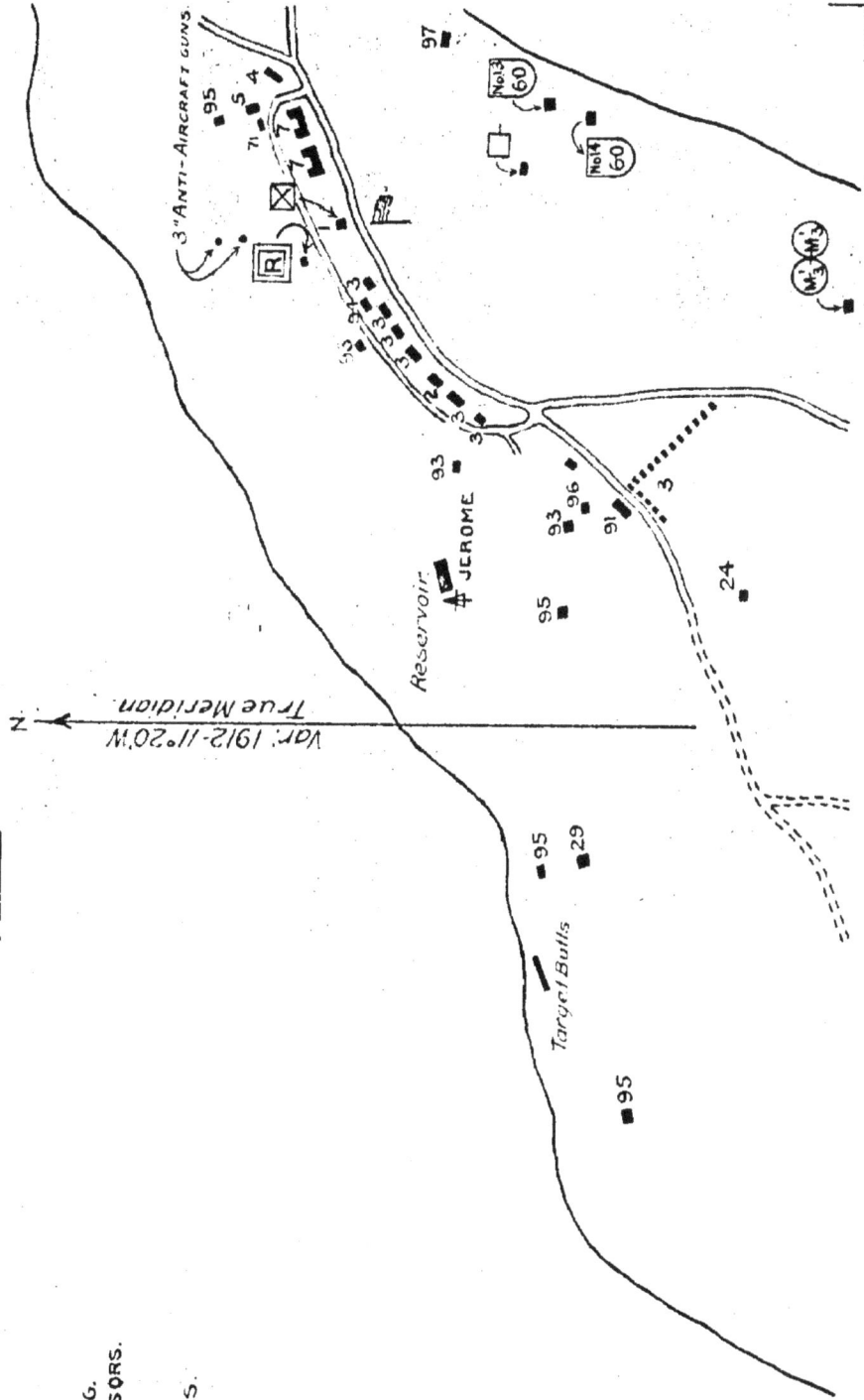

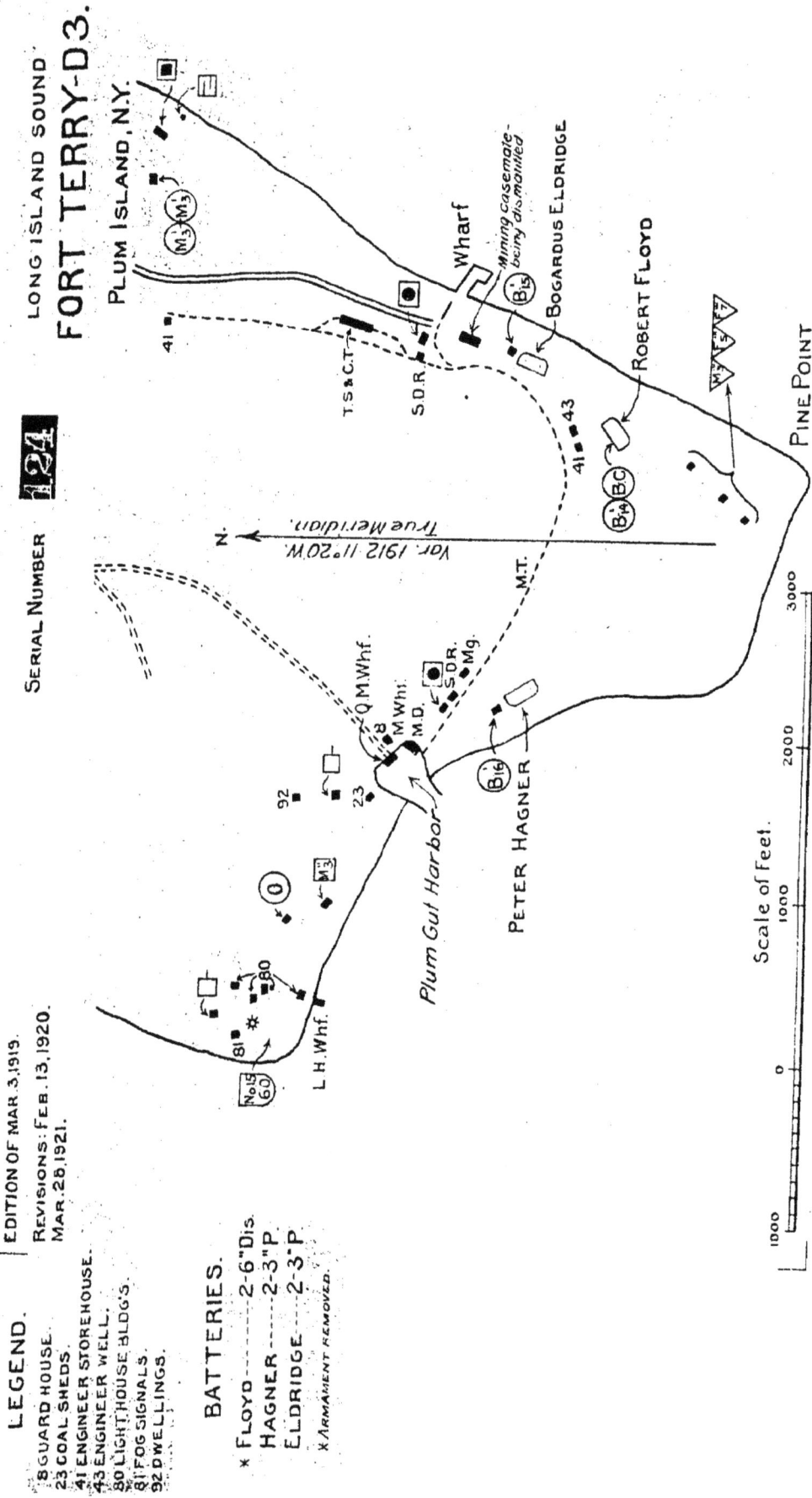

FORT TERRY-D3.

LONG ISLAND SOUND

PLUM ISLAND, N.Y.

SERIAL NUMBER 124

LEGEND.
8 GUARD HOUSE.
23 COAL SHEDS.
41 ENGINEER STOREHOUSE.
43 ENGINEER WELL.
80 LIGHT HOUSE BLDG'S.
81 FOG SIGNALS.
92 DWELLINGS.

BATTERIES.
* FLOYD------2-6"DIS.
HAGNER-----2-3"P.
ELDRIDGE---2-3"P.
* ARMAMENT REMOVED.

EDITION OF MAR. 3, 1919.
REVISIONS: FEB. 13, 1920.
MAR. 28, 1921.

Scale of Feet.

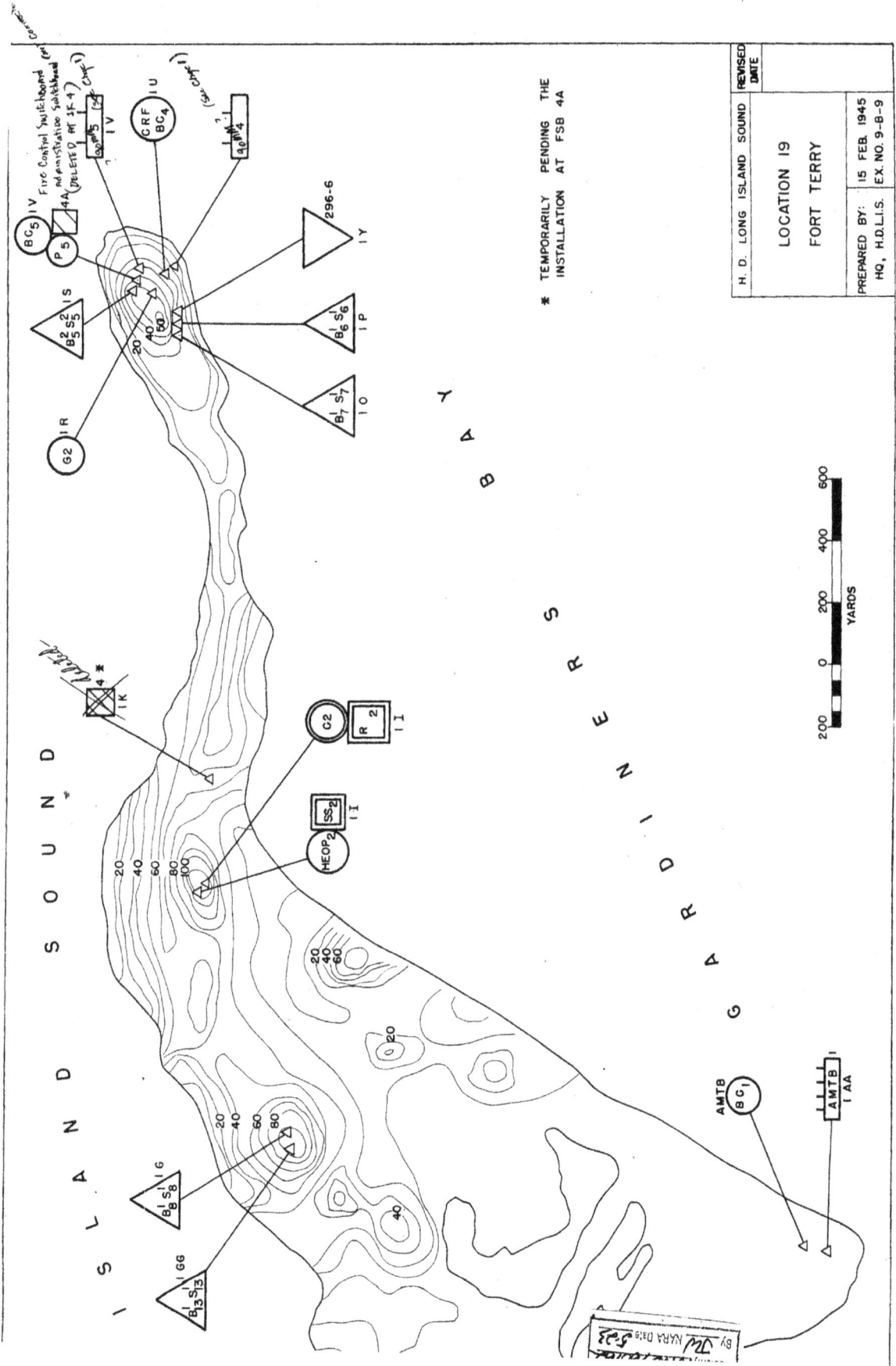

H. D. LONG ISLAND SOUND

LOCATION 19
FORT TERRY

PREPARED BY: 15 FEB. 1945
HQ, H.D.L.I.S. EX. NO. 9-B-9

REVISED DATE

* TEMPORARILY PENDING THE
INSTALLATION AT FSB 4A

Fort Terry 1924 (NARA)

Fort Terry 1924 (NARA)

Plum Island (Fort Terry) (Terry McGovern)

Ferry Landing and USDA Research Lab, Plum Island (Terry McGovern)

Plum Island batteries (Terry McGovern)

Battery Steele Plum Island (Terry McGovern)

- **STONEMAN**: The mortar battery planned as part of the Long Island Sound defenses located on Plum Island. The selected battery site was just to the rear (west) of Battery Steele, on a small plateau between low ridges. With funding from the Act of May 7, 1898 a submission plan was submitted on June 7, 1898. It was of conventional, wide-pit design with adjacent shell and powder magazines on either flank as well as in the central traverse between the two pits. The front wall was further advanced than in the type plan to ease loading all-round in case it was necessary to fire to the rear. An almost complete parados protected the rear of the emplacement, as it would have been possible to have fire approach from ships that passed the island. Work was done in 1898-1900. It was transferred on March 4, 1901 for a cost of $112,000. The battery was armed with eight Model 1890M1 mortars on Model 1896 carriages (Niles tube #13/#162 and the rest of the tubes were Bethlehem Steel #39/#159, #42/#161, #43/#181, #49/#192, #50/#163, #51/#158 and #52/#160). These had been received for mounting on July 18, 1899. In 1908 mortar tube #52 (slightly damaged) was replaced with Watervliet tube #171 originally from Battery Habersham, Fort Screven. The battery was named on General Orders No. 30 of March 19, 1902 for cavalry Major General George Stoneman of Civil War service. In 1905 telautograph booths were added by each pit. Four mortars and carriages were removed in 1918 (forward mortars, No. 2 and 4 from each pit) #42/#161, #49/#192, #50/#163 and #51/#158. The remaining four mortars remained throughout the 1920s and 1930s. The battery was scheduled for deletion pending completion of new defenses under the 1940 Program, and authority was granted on October 5, 1942 for removal. The guns and mortars were probably taken out in May 1943. The emplacement still exists in fair condition on Plum Island. The battery is open, but the island is closed to the public.

- **BRADFORD**: A battery for 6-inch disappearing guns emplaced at the Plum Island reservation of Fort Terry. It was located on the eastern end of the island, just to the east and in front of the mortar battery, with a clear field of fire directly to the west. Plans were submitted on March 29, 1899, funding was included in the fortification Act of March 3, 1899. It was of standard type plan, though the exposed location forced it to be fully sunken and thus flush with the surrounding ground. Lower-level magazines supplied the two separate loading platforms for the guns (though the battery's hoists were removed in 1915, and service thereafter was by hand carrying). Work was done from July 1898 to October 1900 for transfer on March 4, 1901 for a cost of $50,000. It was armed with two 6-inch Model 1897M1 guns on disappearing carriages Model 1898 (#23/#14 and #29/#13). It was named on General Orders No. 30 on March 19, 1902 for Captain James Bradford, U.S. Artillery, killed in action during St. Clair's Defeat in Ohio in 1791. A new large BC station and plot was added in 1914. It had a long service life, being retained well into World War II. Authority for deletion was given in October 1944 and accomplished soon thereafter. The emplacement still exists on Plum Island. The battery is open, but the island is closed to the public.

- **DIMICK**: A later battery for two 6-inch disappearing guns emplaced on the eastern end of the island, following the primary road and just to the right flank (southeast) of Battery Bradford. It fired to the southeast. Location was submitted on January 27, 1903, detailed construction plans on April 20, 1903. It followed mimeograph plans; it had two flank emplacements of standard dimensions, sharing a single central traverse magazine with hand-delivery ammunition service. Work was done in 1903-1904. Transfer came on August 31, 1905 for $57,500. It was named for 1st Lieutenant Justin Dimick killed at Chancellorsville in 1863 by General Orders No. 194 of December 27, 1904. It received two 6-inch Model 1903 Watervliet guns on Model 1903 disappearing carriages (#35/#37 and #54/#38). The battery served for only a short while, the tubes being removed in 1918 and the carriages scrapped in place in 1920. The emplacement still exists on Plum Island. The battery is open, but the island is closed to the public.

- **FLOYD:** A battery for two 6-inch disappearing guns emplaced to the southeast on Plum Island, near and protecting the mine wharf on that shore. Plans were submitted on May 14, 1903. The site was to the south of the wharf, soon 3-inch Battery Eldridge would be built between this battery and the wharf. It fired to the southeast. It was of standard mimeograph design, with two flank emplacements, platforms separated by 127-feet, and conventional hand-delivery ammunition service from a central traverse shared magazine. Work was done from 1903-1904, for transfer on November 22, 1906 for a cost of $51,626. It was named on General Orders No. 194 of December 27, 1904 for 2nd Lieutenant Robert Floyd, U.S. Artillery, killed in action at Chickamauga, GA in 1863. It was armed with two 6-inch Model 1903 Watervliet gun tubes on Model 1903 disappearing carriages (#36/#42 and #46/#41). The gun tubes were removed in 1917 for use on wheeled field artillery mounts, and the carriages were scrapped in place in 1920. The emplacement still exists on Plum Island. The battery is open, but the island is closed to the public.

- **KELLY:** A battery for two rapid-fire guns built during the Spanish American War at Fort Terry. In April 1898 authority was granted to emplace a single, purchased Armstrong 4.72-inch gun on an emplacement at Plum Island. Following the plans for a single 5-inch gun, a platform and adjacent protected magazine was built along the curving road that ran south, southwest of Battery Bradford (due south of the mortar battery). It fired to the southeast. Work was done in 1898. It was transferred on March 31, 1900. It carried one 4.7-inch Armstrong gun and pedestal carriage (#11007/#11003). Then on September 5, 1899 a plan was submitted and approved for a separate, but virtually identical additional emplacement but this time for a 5-inch balanced pillar gun. The two platforms were 550-feet apart but were similar in design and both fired to the southeast. Work was done in 1899 on the second emplacement. It was transferred in turn on March 4, 1901. The cost of both emplacements together was $14,500. It carried one 5-inch gun Model 1900 on Model 1903 pedestal carriage (#14/#7) and one 4.7 Armstrong gun. Originally the 5-inch emplacement was built for the Brown wire-wound type of gun that was never procured, so the replacement gun was allocated, but due to the length of time needed to design and produce, many years went by with the emplacement empty. Both emplacements were named on General Orders No. 20 of May 25, 1903 for Captain Patrick Kelly killed in action at Petersburg, VA in 1864. Sometime between 1904 and 1909 the Armstrong 4.7-inch gun was removed and replaced with another 5-inch Model 1900 gun on Model 1903 pedestal (#21/#8). The two 5-inch guns served until removed in 1917, after which the emplacement was abandoned. While heavily overgrown, the two emplacements still exist at Plum Island. The battery is open, but the island is closed to the public.

- **DALLIBA:** One of five two 3-inch gun pedestal emplacements constructed at Fort Terry. It was emplaced on the heavily fortified eastern end of the island, on the southern shore of the eastern tip, south of Battery Campbell. Work was done in 1903-1904. It was of conventional mimeograph design, with two platforms separated by 62-feet, and three rooms (two magazines and a storeroom) under the central traverse. It was transferred on August 28, 1905 for a cost of $17,500. It was near the beach (platform height of just 34-feet) and fired to the southeast. It was named on General Orders No. 194 of December 27, 1904 for Major James Dalliba of War of 1812 service. It was armed with two 3-inch Model 1903 guns and pedestals (#53/#85 and #65/#86). Later these two tubes were removed to rearm Battery Eldridge but replaced in turn with Model 1903 guns #56 and #57 sometime between 1936 and 1940. Dalliba served with this armament as local Tactical Battery No. 4 until deleted on March 11, 1946. The emplacement still exists at Plum Island but is being eroded and breaking apart.

- **ELDRIDGE:** A battery for two 3-inch pedestal guns. Plans were submitted on August 20, 1902. The battery was to be emplaced on the south shore road on the southeast end of the island protecting the mine casemate and mine wharf on this shore, on the left flank of Battery Floyd. It generally followed standard plans with two circular platforms separated by 62-feet, and a central traverse covering two magazine and one storerooms. However, due to the exposed location here, the thickness of protection was increased over the type plan. It fired to the southeast. Work was done in 1903, for transfer on November 22, 1906 for a cost of $15,924. It was named on General Orders No. 194 of December 27, 1904 for Captain Bogardus Eldridge who was killed during the Philippine Insurrection in 1899. It was armed with two 3-inch Model 1903 guns on pedestal carriages (#56/#53 and #57/#54), mounted on July 23, 1909. A new large CRF station was built on the emplacement flank in 1921. The battery was retained in the 1932 Review but was severely damaged by the hurricane of September 21, 1938. Repairs were not possible to the structure, so the guns were removed to go into local storage and the emplacement abandoned by May 1939. However, it was found possible to restore the battery to service about 1940, the battery received guns from Battery Dalliba to place on its original carriages (#65/#54 and #53/#53). These were then carried until disarmament in 1943 and ultimate abandonment of the site in March 1946. Eroded badly on the current shoreline, the emplacement still exists at Plum Island. The battery is open, but the island is closed to the public.

- **HAGNER:** A battery for two 3-inch pedestal guns emplaced on the southern end of Plum Island to protect Plum Gut Harbor. Far from all the other batteries on the island, this battery was by itself southeast of the little harbor on the western end of the island, firing to the southwest. It was of standard mimeograph design, with two platforms separated by 62-feet and magazines below the central traverse. For some unknown reason the steps and blocks for ammunition transfer to the platform of the No. 1 gun are unusual, being replaced by a stair stretching across the entire width of the platform. The battery was built in 1903-1904. It was transferred on November 22, 1906 for a cost of $15,924. It was named on General Orders No. 194 of December 27, 1904 for Brigadier General Peter Hagner of Mexican and Civil War service. It was armed with two 3-inch Model 1903 guns and pedestal carriages (#59/#90 and #55/#89). By the 1930s the shoreline on this coast was shifting due to periodic storm erosion, so the battery was abandoned. The two guns (the carriages were scrapped) were moved to arm Battery Hoppock at Fort H.G. Wright about 1932. After its abandonment the battery did not serve military purposes. It is now entirely gone.

- **GREBLE:** Another battery for two 3-inch pedestal guns, emplaced at the northern extremity of the island, firing on the reverse side of the major batteries to the northwest. The plans for the battery were submitted on May 9, 1903. They were of standard mimeograph type, with two platforms separated by 62-feet, and a central traverse covering two magazines and a storeroom. Work was done in 1903-1904, it was reported ready for armament on June 30, 1905. It was transferred on August 28, 1905 at a cost of $16,686. It was named with the other 3-inch batteries with General Orders No. 194 of December 27, 1904 for Brevet Lt. Colonel John Greble who was killed in action at Big Bethel, VA in 1861. The battery was armed with two 3-inch Model 1903 guns and pedestals (#61/#87, and #62/#88). It carried this armament from 1909 until exchanging the M1903 guns for the Model 1902 guns and carriages (#54/#54 and #56/#56) from Battery Hoppock at Fort H. G. Wright in 1913. These guns in turn were subsequently removed in the early 1930s when the battery was abandoned. The emplacement still exists at Fort Terry. The battery is open, but the island is closed to the public.

- **CAMPBELL**: The final battery for two 3-inch pedestal guns, emplaced at the very eastern tip of Plum Island, firing to the northeast. It was approximately between batteries Bradford and Dalliba. Work was done in 1903-1904, and it was reported completed in June 1905. The work was transferred on August 28, 1905 for a cost of $16,686.50. It was of standard mimeograph plan, two circular-shaped platforms with low parapets separated by 62-feet, and a central traverse covering two magazine and one storeroom. It was named on General Orders No. 194 of December 27, 1904 for Brevet Captain Henry Campbell of War of 1812 service. The battery was armed with two 3-inch Model 1903 guns and pedestal mounts (#58/#84 and #28/#83) mounted in July 1908. This armament was carried into the 1930s, when the battery was declared surplus in the 1932 Review. The armament was removed in 1934 and sent to the Philippines. The emplacement subsequently suffered heavily from shore erosion. Almost entirely broken up, just large pieces of the emplacement remain on the shore at Plum Island.

- *Battery #114* (planned): A planned dual 16-inch battery for Fort Terry from the 1940 Program. It was listed in both the July and September 1940 Program projections. Shortly after conception, its tentative location was switched to Watch Hill on the mainland. Soon after that it was cancelled in its entirely without any construction ever taking place.

- **Battery #217**: A 1940 Program dual 6-inch barbette battery built among the other Endicott batteries at the eastern end of Plum Island. The battery was of mostly conventional design, but with a peculiar curved rear entrance. It was not funded until FY-1943 and given a relatively low national priority compared to other 6-inch batteries. Work was done from May 18 to October 31, 1943. It was transferred to the Coast Artillery on July 1, 1944 for a cost of $249,478 It was to have mounted two 6-inch Model M1(T2) guns on Model M4 barbette carriages. Carriages #23 and #51 were eventually received for the battery, but the tubes were never received. The battery was never named, simply being known during construction as Battery Construction No. 217. The completed concrete emplacement still exists at Plum Island. The battery is open, but the island is closed to the public.

- **AMTB #911**: A 1943 Program AMTB battery for two 90mm fixed and two 90mm mobile guns. It was built at the southern tip of Plum Island, near old Battery Floyd (and it used the BC station and magazines of that battery for the new emplacement). It consisted mostly of just two concrete gun blocks for the two M3 fixed mounts. Work was done from May 18th to June 21, 1943. It was transferred on December 20, 1943 at a cost of $12,051. It was armed by May 1944 and disarmed once again in mid-1945. The two-gun blocks still exist at Plum Island. The battery site is open, but the island is closed to the public.

Fort Tyler (1898-1920) is located on a detached sand spit off Gardiner's Point which is part of Gardiner's Island. A unique Endicott Program construction which has two batteries located back-to-back to form a square. Designed for two 8-inch guns and two 5-inch guns, the fort was never armed and occupied only for a short time. It was named on General Orders 194 of 1904 for Brig. Gen. Daniel Tyler, USV of Civil War service. After the U.S. Army abandoned the fort, the U.S. Navy used it as a bombing range until 1975. It has suffered heavily from storms and bomb hits, so little remains today. Locally, it is known as "The Ruins". Access is by boat only.

Fort Tyler Gun Battery

- **SMITH**: Fort Tyler and Battery Smith was a unique Endicott work built on a small sand spit, sometimes connected to the mainland at Gardiner's Island in Long Island Sound. While anticipated as the site for defense works, nothing was undertaken until March 1898 with the start of the Span-

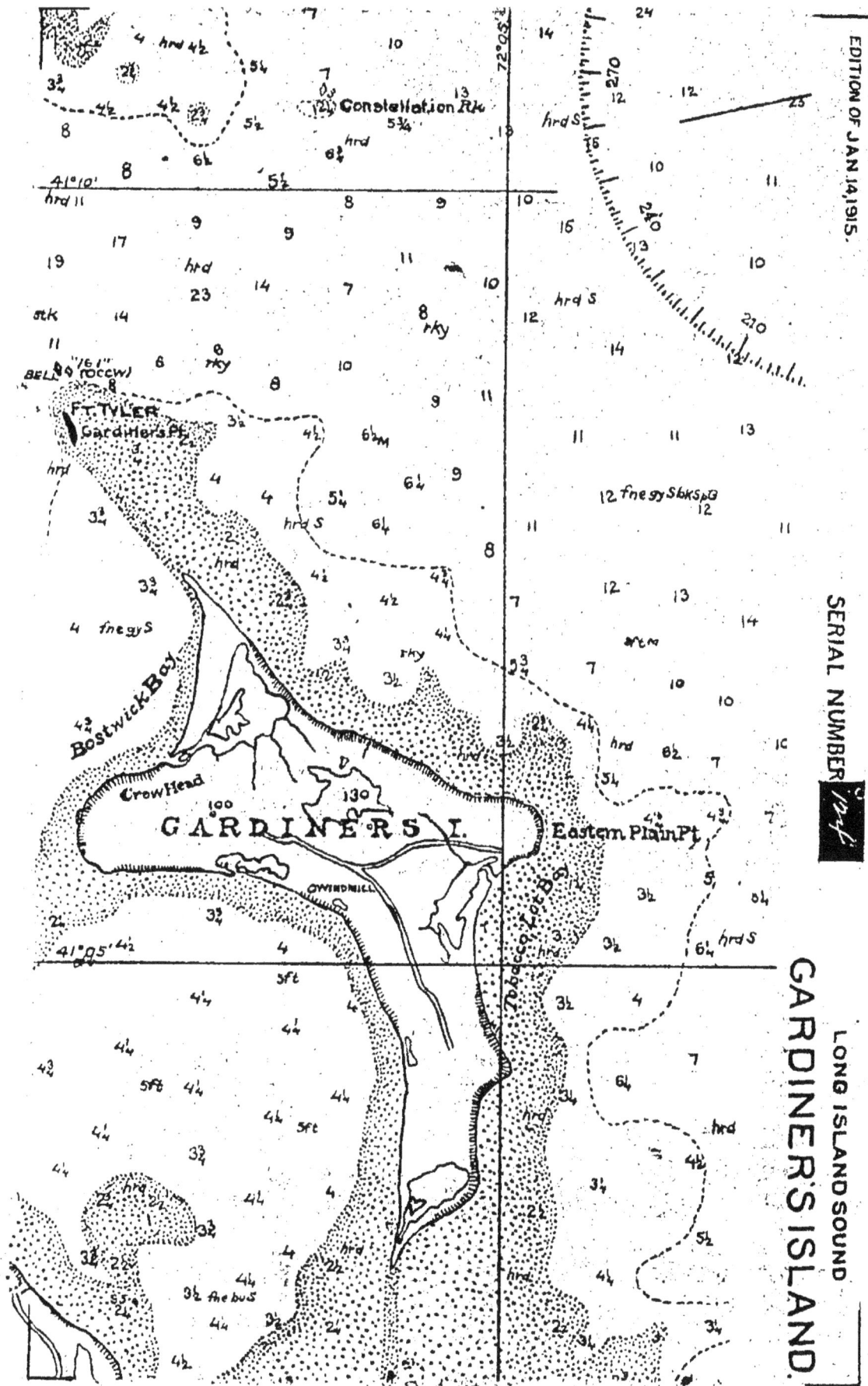

EDITION OF JAN.14,1915.

SERIAL NUMBER

LONG ISLAND SOUND

GARDINER'S ISLAND.

Constellation Rk.

BELL

FT. TYLER
Gardiners Pt.

Bostwick Bay

Crow Head

100

130

GARDINERS I.

WINDMILL

Eastern Plain Pt.

Tobacco Lot Bay

41°10'

41°05'

72°05'

SERIAL NUMBER **124**

LONG ISLAND SOUND,

FORT TYLER.

GARDINERS PT.

POST ABANDONED, ARMAMENT REMOVED.

BATTERIES.

SMITH ---- {2-8" DIS.
{2-5" P.

EDITION OF MARCH 4.1914.
REVISIONS: DEC.7.1915, FEB.13.1920.

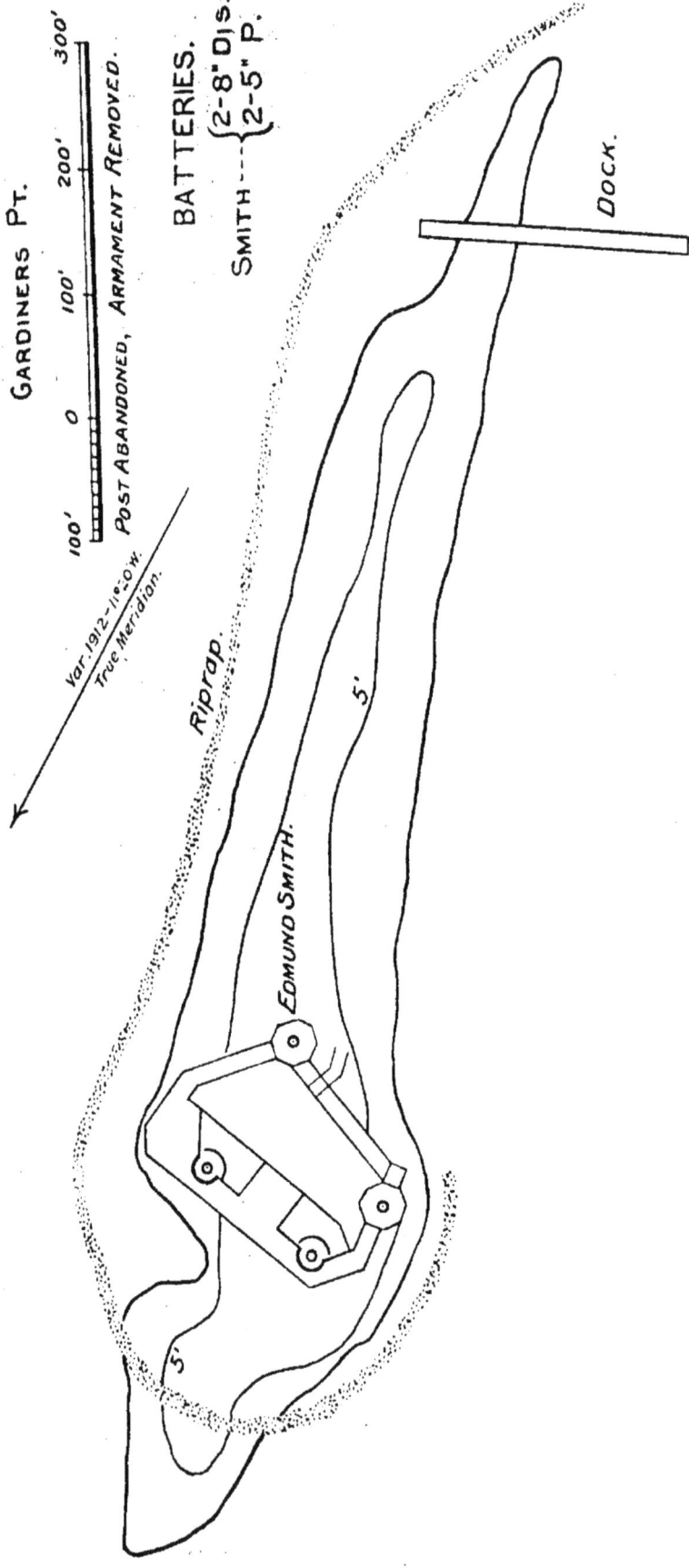

Var. 1912 – 11°.0'W.
True Meridian.

Riprap.

Edmund Smith.

5'

5'

Dock.

The "Ruins" Fort Tyler (Terry McGovern)

Montauk Point State Park Battery #112 and Nike radar facility (Terry McGovern)

Montauk Point State Park Battery #216 (Terry McGovern)

Montauk Point State Park Battery Dunn(Terry McGovern)

ish American War. After securing the land (by transfer from the Light House Service), a plan was submitted. Because of the low site and vulnerability all-round, designed was a polygonal "box" of concrete, with two 8-inch gun emplacements on the main channel side, and two towers at the opposite corners for 5-inch pedestal guns capable a firing all-round, even over the 8-inch battery. The 8-inch gun battery was intended for disappearing guns, but with the urgency of the war on May 14, 1898 it was reconfigured to hold pintles and racers for Rodman carriages to carry two 8-inch Model 1888 breechloading guns. However, the pits were to be filled with concrete of a weakened strength so they could be more easily restored to disappearing carriages later. The work resembled a strong citadel, complete with an entry portal. The 8-inch battery was somewhat reduced in length. Work was well underway by late summer of 1898; in October it was found necessary to move the sallyport to a location more convenient to the proposed wharf. The gun tubes were delivered to the site on August 30, 1898 (Watervliet Model 1888 #9 and #11), but the two reinforced Rodman carriages only made it as far as Fort H. G. Wright where they went into storage. The 5-inch tower emplacements were designed for the Brown wire-wound gun, which was never perfected or procured. Authority was given to change the positions for 5-inch Model 1903 pedestal mounts, but the armament was never received. The battery (all four emplacements) was named on General Orders No, 194 of December 27, 1904 for Captain Edmund Smith, mortally wounded in action during the Philippine Insurrection in 1900. In 1901 substantial changes were made in waterproofing the magazines. Except for a short period in 1902, and even then, without any installed armament, the fort was never garrisoned and mostly remained in a minimal caretaker status. In 1907 new power facilities were recommended, but in general by 1910 it was decided to abandon the site. In the 1920s it was frequented by rum-running pirates, and even later was used as a bombing target. In very poor condition, the emplacement still exists but with very limited boat only access and on private land.

Camp Hero (1940-1957) is located on Montauk Point off State Highway 27 on the southern side of the Point. Camp Hero was established as a quarantine station for troops returning from the Spanish-American War. The 1940 Program built two casemated 16-inch batteries (Battery Dunn and #112) and one shielded 6-inch battery (#216). It was named in General Orders 58 of 1942 for Maj. Gen. Andrew J. Hero, Jr., U.S. Army. After the war, the site was used for anti-aircraft training by the U.S. Army. The 278-acre reservation was turn over to the Air Force in 1957 for use as an air defense radar installation, including a large AN/FPS-35 antenna. In 1980, the radar operation was ended and in 1984 Camp Hero became part of the Montauk Point State Park. The sites are accessible to the public, but the entrances to the batteries have all been sealed. The Montauk Point lighthouse (and fire control tower) is also in the park.

Camp Hero Gun Batteries

- **DUNN:** One of two modern 16-inch casemated barbette batteries built at the Montauk Point reservation of Camp Hero. Part of the 1940 Program it was emplaced in the central area of the Camp Hero reservation, about 2000-feet west from sister Battery #112 with a field of fire to the southeast. Originally assigned national priority #15 on August 11, 1940, this was increased to #5 on September 11, 1941 (switching priority with Battery Construction No. 112 to allow the earlier completion of Dunn with its more favorable field of fire). The original plan for the Long Island Sound project had called for three 16-inch batteries, one each at Fort H.G. Wright, one at Ft. Terry, and one at Montauk. The Ft. Terry unit was cancelled and replaced with this second battery at Montauk Point. Appropriations first came under the FY-1942 Budget of $625,000. Concrete work was done between March 23, 1942 and a year later in March of 1943. It was fully complete by December 1943 and transferred on January 12, 1944 at a cost of $1,369,528. During construction it was

TRUE

P-1   PSB FSB 2   296-3   P-2   296-4

1B   1H   1D   1I

582

HEOP-1

B5 S5 / 12 12

2C

B3 S3 / 3 3

MONTAUK POINT LIGHTHOUSE RESERVATION

ARMY CONTROLLED

B3 S3 / 2 2

2B

BC3 1F

P-3   FSB 3

1G

I   C-1 G-1   R1   2   M1 SS1   B5 S5 / T T   3

1A   1E   1FF   1J   1G

BC1   BC2

1C

TRUE BLOCK IS. SOUND

LONG ISLAND   LOC. 16 ATLANTIC OCEAN

1 0 1 2 3 4 5
HUNDREDS OF YARDS

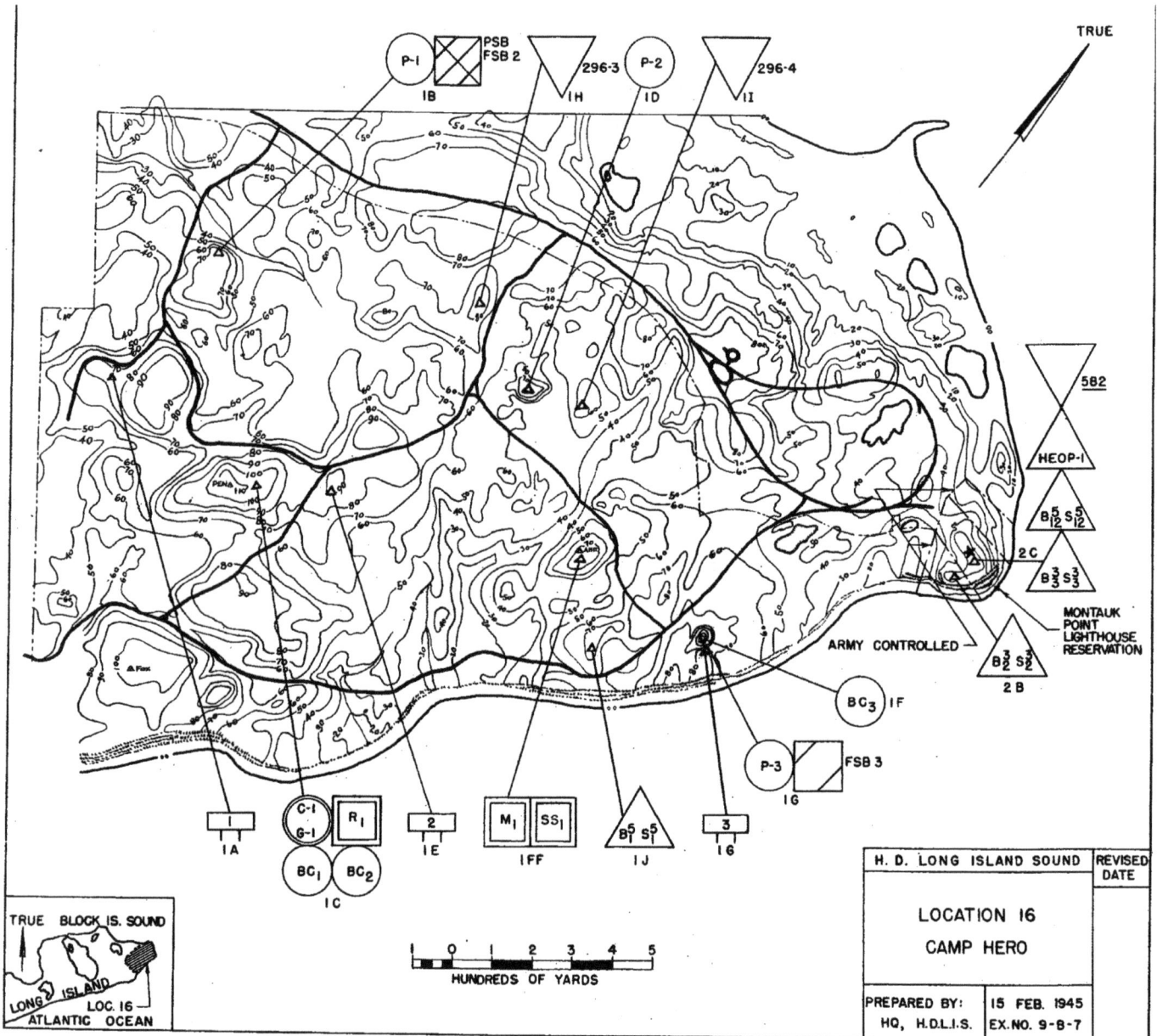

H. D. LONG ISLAND SOUND	REVISED DATE
LOCATION 16 CAMP HERO	
PREPARED BY: HQ, H.D.L.I.S.	15 FEB. 1945 EX. NO. 9-B-7

known as Battery Construction No. 113. It was armed with two 16-inch guns Model MkIIM1 on barbette carriages Model M1919M4 (#45/#27 and #48/#28). It was of standard 100-series design, with a separate PSR station to the north and six assigned base-end stations. It was named on General Orders No. 45 of September 17, 1942 for Colonel John M. Dunn, U.S. Army. The work was disarmed and the battery deactivated in 1947. The emplacement still exists at the Montauk Point State Park. The battery site is open, but the interior is closed to the public.

- **Battery #112**: The second modern 16-inch barbette battery built at Camp Hero. It was one of three 16-inch batteries approved in the 1940 Project for the defenses of Long Island Sound. It was placed to fire south on the western side of the reservation, some 2000-feet from sister Battery #113 (Battery Dunn). Originally given a high national priority #8 on August 11, 1940, its ranking was switched with Battery #113 and given a lower priority of #12 on September 11, 1941. Appropriations of $30,000 in FY-1941 and $595,000 in FY-1942 were used to initiate construction. The design was entirely conventional and consistent with standard 100-series type emplacements. The battery had a separate PSR room to the north and five base-end stations were assigned for fire control. Concrete work was done from March 23, 1942 to April 1943. It was armed with two 16-inch guns MkIIM1 on Model 1919M4 barbette carriages (#73/#43 and #92/#42). It was never named and just known at Battery Construction No. 112. By November 1944 it was considered complete and in active service as Tactical Battery No. 1. It was retained in the 1946 Review, but then the work was disarmed and the battery deactivated in 1947. The emplacement still exists at the Montauk Point State Park. The battery site is open, but the interior is closed to the public.

- **Battery #216:** A 1940 Program dual 6-inch barbette battery planned and built at Camp Hero. It was emplaced to the east of Battery Dunn at a considerable distance and fired almost due east. It was given an initial high national priority and funded under authority of the FY-1942 Budget. It was of standard 200-series design. Construction took place between May 26, 1942 and April of 1943. It was transferred on January 12, 1944 at a cost of $198,008.51. The battery was never named, just known during construction and service at Battery Construction No. 216. It was armed with two 6-inch Model 1903A2 guns on barbette carriages Model M1 (#31/#97 and #56/#96). It served locally as Tactical Battery No. 3. The work was disarmed and the battery deactivated in 1947. The emplacement still exists at the Montauk Point State Park. The battery site is open, but the interior is closed to the public.

Battery JMK Davis in 2005 on Great Gull Island (Mark Berhow)

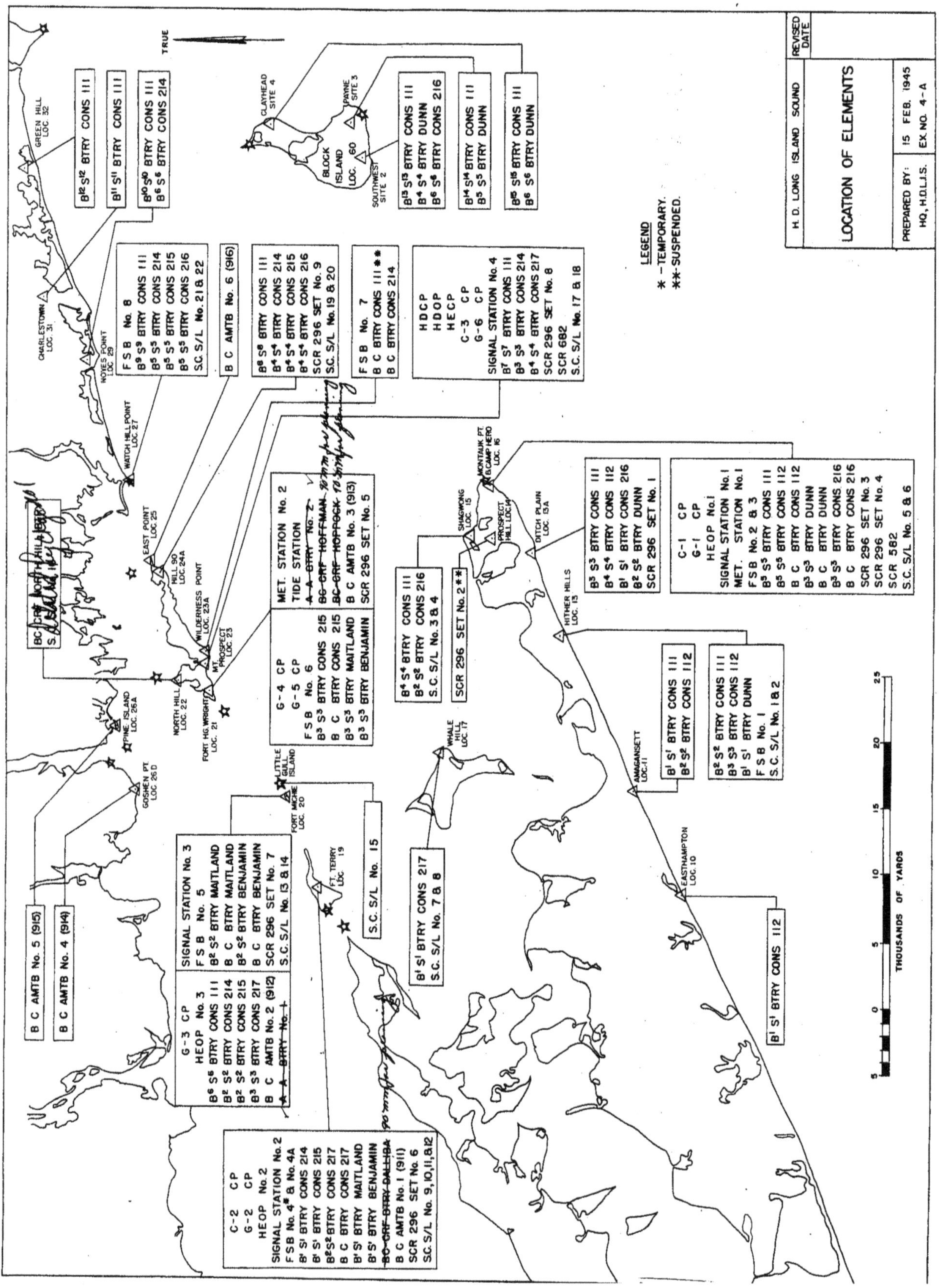
TRUE
H. D. LONG ISLAND SOUND
LOCATION OF ELEMENTS
PREPARED BY: HQ, H.D.L.I.S.
15 FEB. 1945
EX. NO. 4-A
REVISED DATE
LEGEND
* -TEMPORARY.
** -SUSPENDED.
THOUSANDS OF YARDS
GREEN HILL LOC. 32
CHARLESTOWN LOC. 31
NOYES POINT LOC. 29
WATCH HILL POINT LOC. 27
EAST POINT LOC. 25
HILL 90 LOC. 24A
WILDERNESS POINT LOC. 23A
MT. PROSPECT LOC. 23
NORTH HILL LOC. 22
FORT HG WRIGHT LOC. 21
PINE ISLAND LOC. 26A
GOSHEN PT. LOC. 26 D
LITTLE GULL ISLAND
FORT MICHIE LOC. 20
FT. TERRY LOC. 19
WHALE HILL LOC. 17
CLAYHEAD SITE 4
PAYNE SITE 3
SOUTHWEST SITE 2
BLOCK ISLAND LOC. 60
MONTAUK PT. & CAMP HERO LOC. 16
SHAGWONG LOC. 15
PROSPECT HILL LOC. 14
DITCH PLAIN LOC. 5A
HITHER HILLS LOC. 13
AMAGANSETT LOC. 11
EASTHAMPTON LOC. 10
B12 S12 BTRY CONS 111
B11 S11 BTRY CONS 111
B10 S10 BTRY CONS 111
B6 S6 BTRY CONS 214
FSB No. 8
B9 S9 BTRY CONS 111
B5 S5 BTRY CONS 214
B5 S5 BTRY CONS 215
B5 S5 BTRY CONS 216
S.C. S/L No. 21 & 22
B C AMTB No. 6 (916)
B6 S6 BTRY CONS 111
B4 S4 BTRY CONS 214
B4 S4 BTRY CONS 215
B4 S4 BTRY CONS 216
SCR 296 SET No. 9
S.C. S/L No. 19 & 20
FSB No. 7
B C BTRY CONS 111 **
B C BTRY CONS 214
HDCP
HDOP
HECP
C-3 CP
G-6 CP
SIGNAL STATION No. 4
B7 S7 BTRY CONS 111
B3 S3 BTRY CONS 214
B4 S4 BTRY CONS 217
SCR 296 SET No. 8
SCR 682
S.C. S/L No. 17 & 18
B13 S13 BTRY CONS 111
B4 S4 BTRY DUNN
B6 S6 BTRY CONS 216
B14 S14 BTRY CONS 111
B5 S5 BTRY DUNN
B15 S15 BTRY CONS 111
B6 S6 BTRY DUNN
MET. STATION No. 2
TIDE STATION
A-A BTRY No. 2
BC GRF HOFFMAN
BC GRF HOPPOCK
B C AMTB No. 3 (913)
SCR 296 SET No. 5
G-4 CP
G-5 CP
FSB No. 6
B3 S3 BTRY CONS 215
B C BTRY CONS 215
B3 S3 BTRY MAITLAND
B3 S3 BTRY BENJAMIN
B4 S4 BTRY CONS 111
B2 S2 BTRY CONS 216
S.C. S/L No. 3 & 4
SCR 296 SET No. 2 **
B3 S3 BTRY CONS 111
B4 S4 BTRY CONS 112
B1 S1 BTRY CONS 216
B2 S2 BTRY DUNN
SCR 296 SET No. 1
C-1 CP
G-1 CP
HEOP No. 1
SIGNAL STATION No. 1
MET. STATION No. 1
FSB No. 2 & 3
B5 S5 BTRY CONS 111
B5 S5 BTRY CONS 112
B C BTRY CONS 112
B3 S3 BTRY DUNN
B C BTRY DUNN
B3 S3 BTRY CONS 216
B C BTRY CONS 216
SCR 296 SET No. 3
SCR 296 SET No. 4
SCR 582
S.C. S/L No. 5 & 6
B1 S1 BTRY CONS 111
B2 S2 BTRY CONS 112
B2 S2 BTRY CONS 111
B3 S3 BTRY CONS 112
B1 S1 BTRY DUNN
FSB No. 1
S.C. S/L No. 1 & 2
B1 S1 BTRY CONS 112
B1 S1 BTRY CONS 217
S.C. S/L No. 7 & 8
S.C. S/L No. 15
B C AMTB No. 5 (915)
B C AMTB No. 4 (914)
G-3 CP
HEOP No. 3
B6 S6 BTRY CONS 111
B2 S2 BTRY CONS 214
B2 S2 BTRY CONS 215
B3 S3 BTRY CONS 217
B C AMTB No. 2 (912)
A-A BTRY No. 1
SIGNAL STATION No. 3
FSB No. 5
B2 S2 BTRY MAITLAND
B C BTRY MAITLAND
B2 S2 BTRY BENJAMIN
B C BTRY BENJAMIN
SCR 296 SET No. 7
S.C. S/L No. 13 & 14
C-2 CP
G-2 CP
HEOP No. 2
SIGNAL STATION No. 2
FSB No. 4 & No. 4A
B1 S1 BTRY CONS 214
B1 S1 BTRY CONS 215
B2 S2 BTRY CONS 217
B C BTRY CONS 217
B1 S1 BTRY MAITLAND
B1 S1 BTRY BENJAMIN
BC GRF BTRY DALLIBA
B C AMTB No. 1 (911)
SCR 296 SET No. 6
S.C. S/L No. 9, 10, 11, & 12

Long Island Sound World War II-era Site Locations. Stations housed in a single structure are connected by dashes (-)

location	Loc#	Purpose
East Hampton	10	BS1/112
Amagansett	11	BS1/111-BS2/112
Hither Hills	13	BS3/112-BS2/111, BS1/Dunn, SBR
Ditch Plain	13A	BS3/111-BS4/112, BS1/216, BS2/Dunn, SL 1,2
Prospect Hill	14	SCR296-1
Shagwong	15	BS4/111, BS2/216, SL 3,4
Camp Hero	16	Batt. Tact. #1 BCN 112, Batt. Tact. #2 Dunn, Batt. Tact. #3, BCN216, BC-BS3/216, BC-BS3/Dunn, BC-BS5/112, SCR296-3, SCR296-4, SCR582, C1, G1, Met, HDOP1
Montauk Point	16	B5/111
Gardiners Island/ Orient Point	17	B1/217, SL 1,2 (Whale Hill) hut
Fort Terry	19	Batt. Tact. #4 Daliba, Batt. Tact. #5 BCN 217, C2, G2, HDOP2, SBR, AMTB 1, SCR296-6, BS1/214, BS1/215, BC-B2/217, BS1/Maitland, BS1/Benjamin, BC/Daliba, SL 9,10,11,12
Fort Michie	20	Batt. Tact. #6 Maitland, Batt. Tact. #7 Benjamin, G3, HDOP3, SCR296-7, BS6/111, BS2/214, BS2/215, BS3/217, BC-BS2/Maitland, BC-BS2/Benjamin, AMTB2 SL 13,14,15
Fort H.G. Wright	21	Batt. Tact. #8 BCN 215, Batt. Tact. #9 Hoffman, Batt. Tact. #10 Hoppack, G4, G5, SBR, Met, T, BC-BS3/215, BS3/Maitland, BS3/Benjamin, BC/Hoffman, BC/Hoppack, AMTB3, SCR296-5
North Hill	22	Batt. Tact. #11, BC-3-inch
Mount Prospect	23	HDCP-HECP, HDOP, C3, G6, BS7/11, BS3/214, BS4/217, SCR296-8, SCR682, SL 17,18
Wilderness Point	23A	Batt. Tact. #12 BCN 111, Batt. Tact. #13 BCN 214, BC/111, BC/214, SBR
Hill 90	24A	BS8/111, BS4/214, BS4/215, BS4/216, SCR296-9
East Point	25	AMTB6
Pine Island	26A	AMTB5
Goshen Point	26B	AMTB4
Watch Hill	27	SBR, BS9/111, BS5/214, BS5/215, BS5/216, SL 21,22
Noyes Point	29	BS10/111, BS6/214
Charleston	31	BS11/111
Green Hill	32	BS12/111
Block Island sites	2	BS13/111, BS4/Dunn, BS6/216
Block Island sites	3	BS14/111, BS5/Dunn
Block Island sites	4	BS15/111, BS6/Dunn

Bramson, Ruth A, Fleming, Geoffrey K., and Folk, Amy K. *A World Unto Itself: The Remarkable History of Plum Island, New York.* Southhold Historical Society, Southhold, NY, 2014

Rafferty, Pierce, and John Wilton. *Guardian of the Sound, A Pictorial History of Fort H.G. Wright,* Fishers Island, NY. Mount Mercer Press, NY, 1998.

THE HARBOR DEFENSES OF THE EASTERN ENTRANCE TO NEW YORK

The East River connects Long Island Sound to New York's inner harbor separating Long Island from The Bronx and Manhattan Island. The eastern end of the East River is bounded by Throngs Neck on the mainland and Willets Point on Long Island before opening up to Long Island Sound. Based on wartime experiences during the Revolutionary War and War of 1812, the importance of defending this "backdoor" to New York's harbor was recognized. During the Third System, a major work was built on Throngs Neck, and another work was started at Willets Point to close off this approach to New York. The defenses were upgraded with 14 new concrete batteries at three locations during the Endicott Program. The defenses were inactivated by 1928, but portions retained by the U.S. Army as the eastern end of the sound was not technically closed until the World War II defenses there were completed in 1944-45.

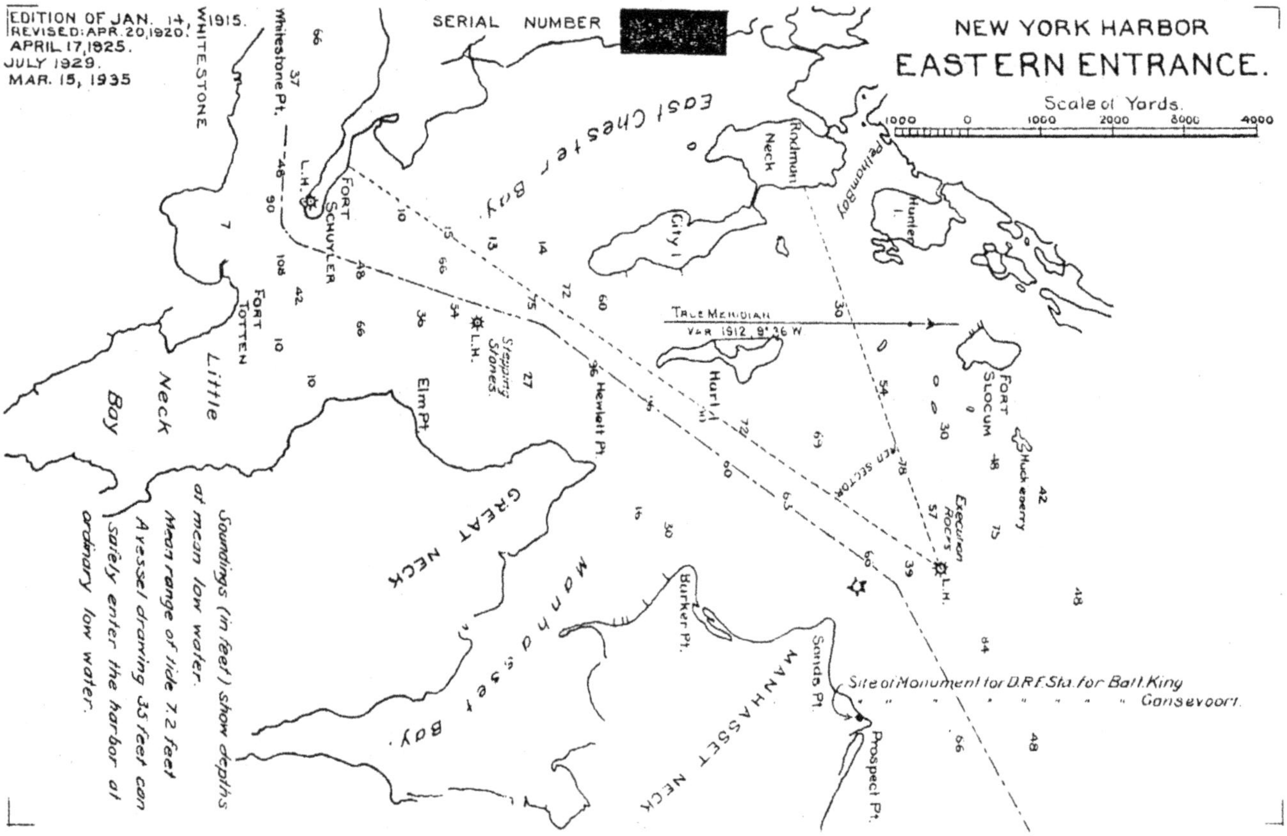

Fort Slocum (1878-1928) is located on David's Island in Long Island Sound, just offshore from the City of New Rochelle, New York. The 80-acre fort was established in 1861 as a military hospital with a coast artillery role being added in 1878. Starting in 1897, four Endicott Program batteries were constructed. It was named on General Orders 28 of 1896 for Maj. Gen. Henry W. Slocum, USV of Civil War service. These batteries were disarmed around 1919. Fort Slocum then became in succession a U.S. Army recruiting center, a training school, an Air Force base, another training school, a Nike missile base, and finally a training school again until it closed in 1965. Owned by the City of New Rochelle, the island has been an abandoned condition for the last sixty years awaiting a successful development scheme. During this period the fort's structures were being destroyed by neglect as vandalism and fires took their toll on many of Fort Slocum's buildings, so in 2008 the U.S. Army Corps of Engineers demolished all remaining structures. The mortar battery (with one pit destroyed), along with a Spanish-American era battery, are today's only remaining batteries. Currently it's an undeveloped park and permission from the city is needed to visit the island.

BATTERIES.

OVERTON ....
HASKIN ....
KINNEY ....
FRASER ....

NEW YORK HARBOR
FORT SLOCUM.
DAVIDS ISLAND.

True Meridian
Var. 1913, 9°36'W.

Kinney.
Fraser.
Haskin.
Overton.
Mg.
S. Sh.
W. Tr.

LEGEND.

1 ADMINISTRATION B'LDG.
2 COMMANDING OFFICERS QRS.
3 OFFICERS QRS.
4 HOSPITAL.
4a RECEPTION HOSPITAL.
4b ISOLATION
4c RECEIVING VAULT.
5 HOSPITAL STEWARDS QRS.
6 NON-COMMISSIONED.
    OFFICERS QUARTERS.
7 BARRACKS.
8 GUARDHOUSE.
9 POST EXCHANGE AND
    GYMNASIUM.
10 OFFICERS CLUB.
11 DRILL HALL.
12 BOATHOUSE.
13 STABLE.
14 SHED.
15 CREMATORY.
16
17 HD. QRS. BUILDINGS.
20 Q.M. ST. HO. AND OFFICE.
21 Q.M. STOREHOUSE.
22 COMMISSARY ST. HOUSE.
23 COAL SHED.
24 Q.M. SHOP.
25 OIL HOUSE.
30 ORDNANCE STOREHOUSE.
40 ENGINEER STOREHOUSE.
70 CHAPEL.
71 Y.M.C.A.
50 TRANSFORMER STATION.
26 WAITING ROOM.

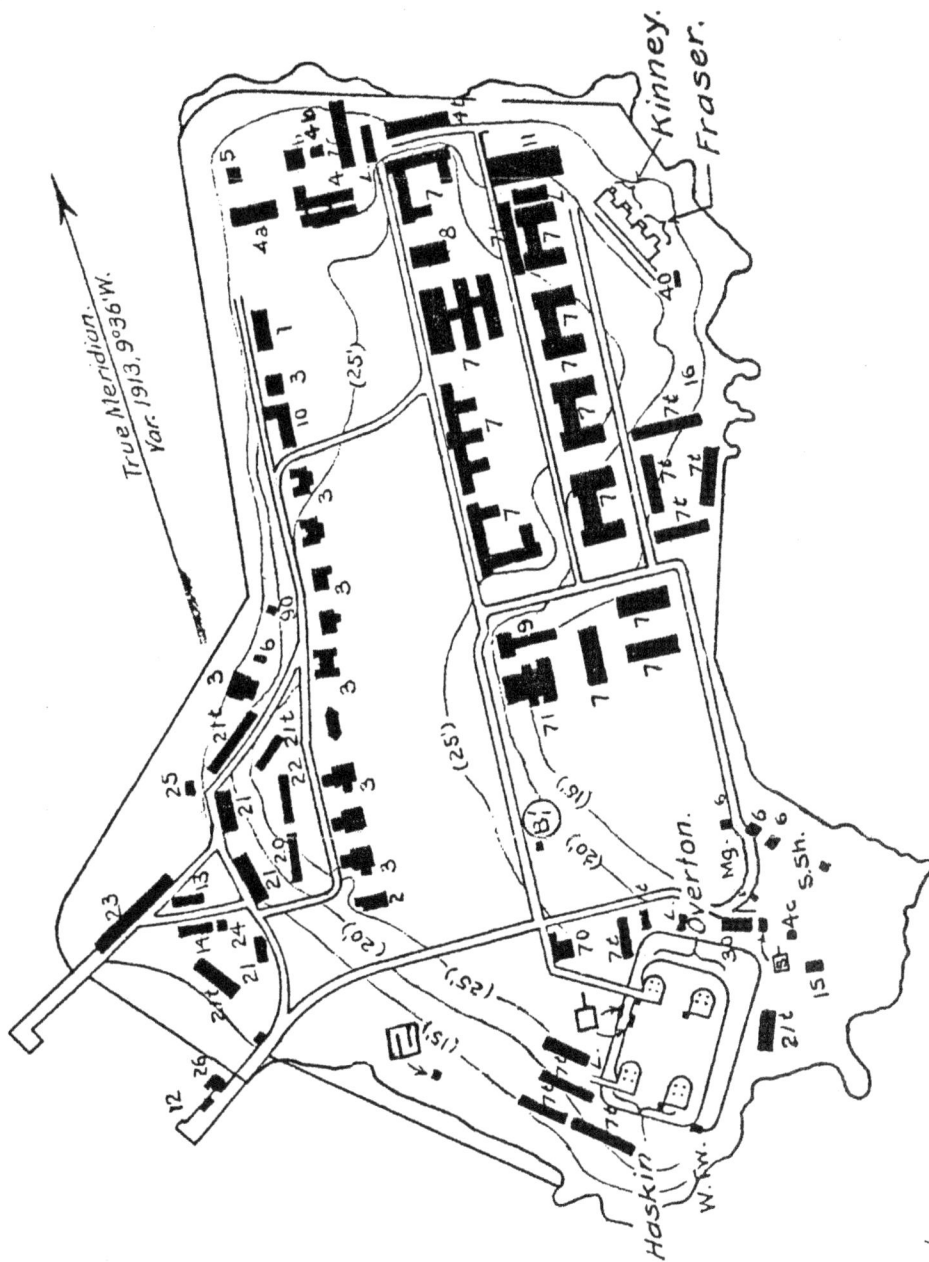

EDITION OF JAN.14, 1915.
REVISIONS: DEC.7, 1915, APR.10, 1916, FEB.9, 1970.
APR.20, 1920, MAY 23, 1921, AUG.24, 1921.
APR.17, 1925

Fort Slocum 1924 (NARA)

Fort Schuyler 1920s (NARA)

Fort Slocum Gun Batteries

- **HASKIN – OVERTON**: The mortar battery for Fort Slocum on David's Island. It was one of the first such Endicott batteries, proposed in November 1890 to be built along with that for Sandy Hook, the Presidio in San Francisco and at Grover's Cliff in Boston. The site selected for the northern entrance to New York was influenced by the fact that Slocum was already a military reservation and additional land purchase would not be necessary. Plans were finalized in mid-1891 and work began soon after. The plan adopted was to the dimensions of other early quadrangular mortar emplacements. However the thickness of protection on the unthreatened northwest and southwest fronts was reduced, and no need for a ditch or counterscarp land defense position was perceived. Work was done in 1891-1893, then suspended until 1895 when the carriage type was decided on and foundation rings could be placed. Transfer was finally made on October 13, 1897 for a construction cost of $199.018.92. It was originally named on General Orders No. 80 of March 19, 1902 for Brevet Brigadier General Joseph Haskin of Mexican War and Civil War service. The battery was armed with sixteen 12-inch Model 1886 cast-iron, steel-hooped mortars on Model 1891 carriages (#1/#39, #3/#40, #9/#41, #10/#42, #11/#43, #12/#50, #31/#52, #34/#54, #55/#56, #56/#57, #59/#59, #60/#63, #68/#64, #69/#66, #71/#5, and #73/#24). In January 1906, like all other sixteen-gun mortar batteries, this unit was split tactically to form two dual-pit batteries. The southern two its remained Battery Haskin, and the two northern pits were given the name Battery Overton. This was done in General Orders No. 6 of January 25, 1906 with the battery named for Captain Clough Overton, U.S. Cavalry officer killed in action during the Philippine Insurrection in 1903. The batteries continued to serve until dismounted sometime between 1917 and 1920 (the mortars were certainly gone by April 15, 1920). In subsequent years the eastern pit of Haskin was destroyed, probably during World War II to facilitate using the area as a firing range. The emplacement still exists, though heavily overgrown and in poor condition. The island is city property and closed to the public.

- **KINNEY**: An emplacement for two 6-inch rapid-fire guns approved for a site on the left flank and closely adjacent Battery Fraser. Plans were submitted on May 31, 1901. Work was done in 1901-1902, and it was reported ready for armament on May 19, 1902. It was built to the first design plan for the 6-inch Model 1900 type, with just a single wide platform holding two guns immediately adjacent and just 20-feet apart. In July 1902, before the guns were mounted, orders were issued to change the emplacement in favor of a more conventional design with each gun on its own platform, separated from others by a protective traverse. A new gun platform (No. 2) was erected on the left flank and the gun shifted to that position. There was now 66-feet between guns. Transfer was made on November 28, 1904 for a cost of $28,000. It was named on General Orders No. 78 of May 25, 1903 for Captain Joseph Kinney, killed in action at Lundy's Lane in 1814. It was armed with two 6-inch Model 1900 guns on pedestal mounts (#26/#18 and #44/#19). The battery served only a short time, being disarmed in early 1917. The two 6-inch guns and pedestals were sent to Fort Tilden for re-emplacement in a new battery on February 17, 1917. Demolition of the now empty emplacement was recommended in 1930, and that was accomplished within a few years. Nothing remains of the battery today.

- **FRASER**: A battery for two 5-inch rapid-fire pedestal guns erected at Fort Slocum on the northeastern point of David's Island, a couple thousand feet away from the mortars on the other end of the island. Plans were submitted on July 2, 1899 for construction of the battery, to standard plans with two platforms and flank traverse magazines. Guns were separated by 45-feet. Work was done in 1899-1900. Transfer was made on February 26, 1901 for a construction cost of $18,392.73. As it was built for the proposed Brown wire-wound 5-inch gunthat were never produced, there was a considerable delay before the armament was received and mounted. Eventually it received two 5-inch Model 1900 guns on Model 1903 pedestal mounts (#8/#1 and #9/#10). It was named on General Orders No. 78 of May 25, 1903 for Captain Upton S. Fraser killed in action against the Seminole Indians in 1835. The armament was removed in 1917 and sent to a temporary battery at Fisherman's Island in the Chesapeake Bay. The empty emplacement was destroyed following a recommendation of 1930. Nothing remains of the battery today.

- A temporary emplacement for a single 8-inch Model 1888 gun on reinforced 15-inch Rodman carriage emplaced at Fort Slocum. This was one of 21 such gun combinations emplaced for the emergency of the Spanish American War in 1898. These were excess gun tubes without carriages that were placed as an expediency during the emergency. The spot selected for David's Island was at an old Rodman practice battery on the southeastern corner of the island, directly to the east of the mortar emplacement. Local engineers were ordered to prepare the emplacement on May 12, 1898. Mostly just involving repairs to the old site and cleaning out the adjacent magazine, work was quickly done. Mounted was Model 1888 gun (#5 by West Point Foundry) which had been shipped here on August 28, 1898. Next to it also at the practice battery was an older Rodman 8-inch converted rifle. Apparently, the gun served for just a short time, it was gone at least by 1903. Some partial remains of the emplacement and its breast-high wall still exist in poor condition.

Fort Schuyler (1833-1928) is located on Throg's Neck in the Bronx at the western end of Long Island Sound. Fort Schuyler, a granite Third System fort, was constructed between 1833 and 1856 to guard the eastern entrance to the New York's harbor via the East River. It was named on General Orders 32 of 1833 for Maj. Gen Philip Schuyler of the Continental Army. Construction of the four Endicott Program batteries was undertaken on the grounds of the old fort from 1898 to 1900. A mine casemate was built in 1875 in the southwest corner of the fort, reused for the Endicott Program. The fort became inactive in the 1920s and the property was transferred to State of New York in 1934 for the State University of New York Maritime College. Today, the Third System fort is used as an administration building by the maritime college as well as the Maritime Industry Museum, which includes exhibits on the history of fort. Its open to the public on weekdays. Only part of one Endicott battery remains visible, as the rest have been destroyed or buried by the construction of the Throg's Neck Bridge or by new college buildings.

Fort Schuyler Gun Batteries

- **GANSEVOORT**: A battery for two separate, single 12-inch guns emplacements at the Throg's Neck reservation of Fort Schuyler. It was built under two separate authorizations. One emplacement was submitted on April 17, 1897 for a single gun on a Model 1896 disappearing carriage for a position in the north bastion of the old fort's counterface hornwork. It had most of the standard features, though somewhat adjusted to fit in the constricted space of the bastion. Work was done in 1897-1898. It was reported completed on February 28, 1898. It was transferred on August 17, 1899. The second emplacement was submitted the next year on July 12, 1898. This was located on open ground not far to the northwest of the previous emplacement. Work was done in 1898-1899, transfer being made on August 15, 1900. Construction cost of both emplacements was calculated at $93,442.22. Emplacement No. 1 (in the hornwork) carried one 12-inch Model 1888M1 Watervliet gun on Model 1896 disappearing carriage (#44/#6). Emplacement No. 2 had one 12-inch Model 1895 Watervliet gun on a Model 1897 disappearing carriage (#9/#5). It was named on General Orders No. 78 of May 25, 1903 for Brigadier General Peter Gansevoort, Continental Army. In

NEW YORK HARBOR
FORT SCHUYLER

EDITION OF JAN.14,1915.
REVISIONS: DEC.7,1915; APR.10,1916, FEB.17,1920.
JAN.3,1922, APR.17,1925, JULY 1929.
MAR. 15, 1935.

BATTERIES.

* GANSEVOORT. 1-12"DIS
   HAZZARD---- 2-10" "
† BELL -------- 5 "
† BEECHER ------ 3 ".
* Empl. No.2 vacant
   'A-Anti-aircraft gun 2-3"

† Guns dismounted

TRUE MERIDIAN.
Var. 1912, 9°26'W.

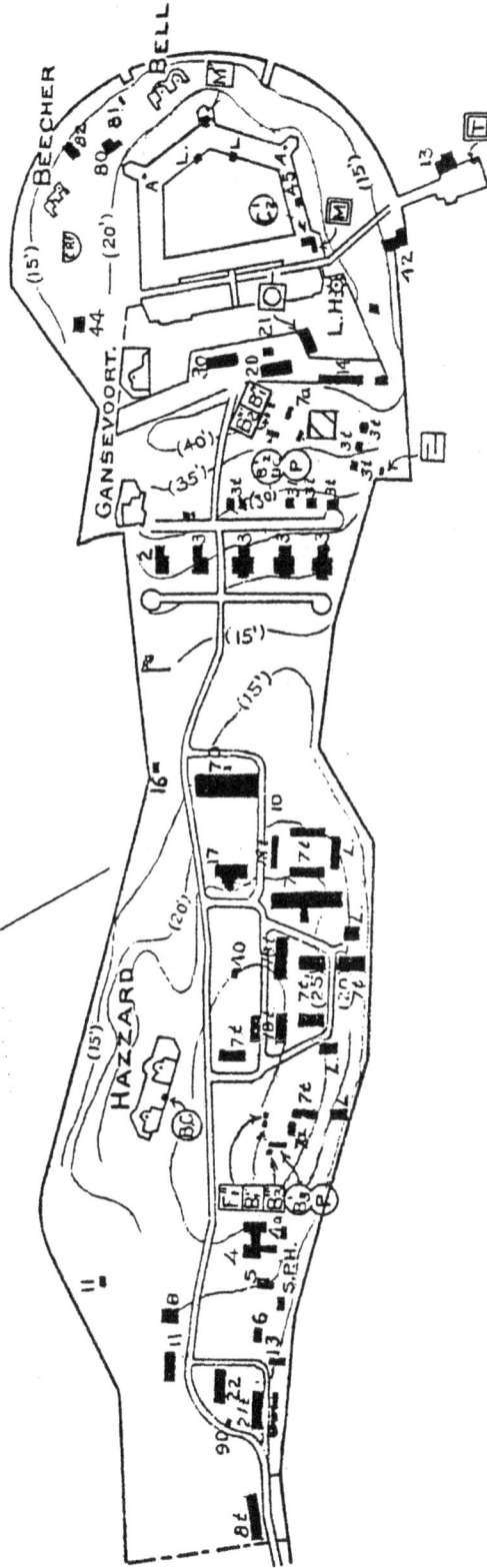

LEGEND.

1
2 COMMANDING OFF.QRS.
3 OFFICER'S QUARTERS.
4 HOSPITAL.
4a DEAD HOUSE.
5 HOSPITAL STWD.QRS.
6 N.C.OFFICERS QRS
7 BARRACKS.
7a DORMITORY.
8 GUARDHOUSE.
9 POST EXCHANGE.
10 HOSE HOUSE.
11 STABLE.
12
13 BOAT HOUSE.
14 COAL SHED.
15 BAKERY.
16 BATH HOUSE.
17 MESS HALL.
20 Q.M.OFFICE & ST. HO.
21 Q.M.STORE HOUSE.
22 COMMISSARY ST. HO.
30 ORDNANCE ST. HO.
40 ENGINEER OFFICE.
41
42 ENGINEER QUARTERS.
43
70 Y.M.C.A.
80 LIGHT KEEPER'S QRS.
81 FOG BELL.
90 WATER METER.
44 ENGR. STORE HOUSE.
45 CARPENTER SHOP.
52 LIGHT KEEPER'S STABLE.
18 MESS HALL & KITCHEN.

the years after completion, it was given a new BC station and adjustment made to its loading platforms and hoists. The gun carriages were modified for increased elevation in 1916-1917. The gun in emplacement No. 2 was removed in 1918 when it was taken for railway carriage use. The carriage was scrapped about 1920. Emplacement No. 1 remained armed into the 1930s. The gun was still mounted in 1928 but listed as decommissioned. It remained as a district spare in 1935. Its final removal date is not known. Postwar the No.2 emplacement was entirely destroyed, but the hornwork position has partially survived, though mostly filled it. It is on the grounds of the SUNY Maritime College. The battery site is closed to the public.

- **HAZZARD:** This emplacement for two 10-inch disappearing guns was the first new work for Fort Schuyler. Plans were submitted on July 11, 1896 for a two-gun emplacement located a distance to the northwest of the old fort, with a field of fire beyond to the northeast. Work was done in 1896-1898, for transfer on May 5, 1898 at a construction cost of $74,432.61. It was armed with two 10-inch Model 1888MII Watervliet guns on Model LF 1896 disappearing carriages (#59/#1 and #63/#10). The battery was named on General Orders No. 78 of May 25, 1903 for Captain George Hazzard who was mortally wounded at White Oak Swamp, VA in 1862. The usual emplacement modifications were made from 1905-1912, including a battery commander's station, enlarged loading platforms and new chain hoists. The guns were left in place and not removed during World War I. However, they were dismounted sometime between 1928 and 1932. In 1936 the hoists were removed and used elsewhere. The emplacement itself was eventually destroyed completely, probably to open room for the maritime college in the early 1950s. Nothing remains today.

- **BELL:** A battery for two 5-inch balanced pillar guns submitted on August 24, 1898. The approved location was close to the old work, just to the east of the central bastion on the seaward curtain wall. Concrete work was done in 1899-1900 for transfer on July 7, 1900 for a cost of $15,222.89. It was armed with two 5-inch Model 1897 Bethlehem guns on Model 1896 balanced pillar carriages (#14/#5 and #13/#6). The battery was named on General Orders No. 78 of May 25, 1903 for Captain Jacob Bell killed in action at Stones River, TN in 1862. The guns were taken out in 1917, and the carriages destroyed in 1920. The battery emplacement itself was destroyed later, probably in the 1934-1938 time frame. Nothing remains today.

- **BEECHER:** A small battery for two 3-inch Model 1898 masking parapet guns close to old Fort Schuyler on its northeast side. Plans were submitted on January 21, 1899. Funding came from the Act of July 7, 1898 and most concrete work was done in 1899-1900. It was of standard design with gun centers of 29-feet and composed of one flank and one square-shaped internal platform. It was reported as complete and ready for transfer on December 8, 1900. Transfer was made on December 22, 1900 for a cost of $8011.96. It was armed with two 15-pounder, 3-inch Model 1898 guns on masking parapet mounts (#26/#26 and #27/#27). It was named on General Orders No. 78 of May 25, 1903 for 1st Lieutenant Frederick Beecher who was killed in action with hostile Indians in Kansas in 1868. The carriages were converted to M1898M1 fixed pedestals in 1917. In 1920 the guns and carriages were removed when this type of 3-inch mount was declared obsolete. The emplacement was subsequently destroyed, noting remains at the site.

Fort Totten (1862-1946) is located on Willets Point in the borough of Queens at the western end of Long Island Sound. The "Fort at Willets Point", a granite Third System fort, was started in 1862 but was never completed to guard the eastern entrance to the New York's harbor via the East River. A 270-foot-long tunnel was built in 1871 from the bluff batteries to the water battery. An underground Grand Magazine was built behind the water battery in 1871. Construction of the seven Endicott Program batteries, mounting a total of 22 guns and mortars, was undertaken on the military reservation from 1897 to 1904. It was named on General Orders 106 of 1898 for Maj. General Joseph G. Totten, Chief of Engineers. The fort served for a time as the Engineers School and a school for Mine Defense. During World War II, Fort Totten provided the only active batteries for the Harbor Defenses of Eastern New York and was the headquarters for the anti-aircraft defense of New York. After the war, Fort Totten became the regional headquarters for U.S. Army Air Defense Command, which was in charge of New York's Nike defenses and other anti-aircraft defenses. While the post closed in 1974 (when the Nike defenses were phased-out), the post continued to serve as an U.S. Army reserve center. Eventually the reserve center was downsized, and the remainder of the reservation was transferred to the City of New York, which uses the complex as a police lab, a fire academy, and a public park. Many of the post garrison buildings remain and are used by city, while others are abandoned. The mine complex is being used by the NY City police department. The mortar battery has been buried. The fort's ordnance storehouse is being used as visitor center for the stone fort. The focus of the historical interpretation is the unfinished third system fort, the main gun line is behind a fence barrier and neglected.

Fort Totten Gun Batteries

- **KING**: A 12-inch mortar battery for the Willets Point reservation of Fort Totten. A prototype mortar battery with a new type of layout had been designed by Corps of Engineer officer Henry Abbot was built here in the early 1870s. It featured four individual pits (each with four mortars), surrounded by high parapets and connected to each other and magazines by tunnels through the parapet. At the time it was armed with old Civil War-era 13-inch smoothbore mortars. It was never really completed or armed, but the experience with the design was favorable, and an updated, enlarged version armed with new rifled breechloaders became the first standardized type of the Endicott generation. When it came time in the 1890s to build the mortar battery assigned to Fort Totten, an effort was made to try and re-use this work from twenty years earlier. Plans were submitted on May 12, 1897 to recondition two of the old pits, each to hold four mortars. New carriage platforms would be required, along with a new drainage system. While the tunnels were very narrow (just six feet wide) and the magazine space inadequate (shot and shell had to be stored in the rear of the gun pits themselves), they were retained to conserve on expense. Work was done in 1897-1898, for transfer on August 6, 1900 for a cost of only $19,329. It was armed with eight 12-inch Model 1890M1 mortars on Model 1896 carriages (Builders Iron Foundry tubes #11/#47, #13/#55, #14/#51, and Watervliet tubes #16/#50, #17/#52, #18/#48, #27/#54 and #31/#53). It was named on General Orders No. 43 on April 4, 1900 for Lieutenant Colonel William R, King, former Chief of Engineers. The use of the old emplacement proved a mistake. It was simply too cramped for efficient operation. The pits were found too small to allow reloading of the mortars without returning them to a particular orientation. A report in 1903 recommended a complete reconstruction, if not that, then the removal of the exterior parapets in order to free up room. Rebuilding was not authorized. The completion of the Long Island Sound defenses had provided enough additional coverage of the eastern entrance to New York to reduce the need of the defense at Fort Totten. Henceforth the battery, not yet disarmed, was essentially abandoned. The eight mortars and carriages were not physically removed until 1935-1936. In April 1938 the pits were filled in. Eventually the entire battery was buried and covered over, with the intent that the interior magazine might still be used

NEW YORK HARBOR
FORT TOTTEN
WILLETS POINT.

TRUE MERIDIAN.
VAR. 1912, 9°26'W.

MOTOR PARK
(for A.A.G.)

Gun
Shed

OF TOG

POND

S.PH.

KING

SUMNER
GRAHAM
BAKER
MAHAN
STUART
BURNES

EDITION OF JAN.14,1915.
REVISIONS: DEC.7,1915; APR.10,1916; FEB.27,1917; FEB.17,1920.
JAN.3,1922; APRIL 17,1925; JULY 1929.
MAR.15,1935

LEGEND.
1. ADMINISTRATION BLDG.
2. COMDG. OFFICERS QRS.
3. OFFICER'S QRS.
4. HOSPITAL.
5. HOSPITAL STW'D. QRS.
6. N.C.O.QUARTERS.
7. BARRACKS.
8. GUARD HOUSE.
9. POST EXCH.& Y.M.C.A.
10. GYMNASIUM
11. TORPEDO WORK SHOP.
12. LABORATORY.
13. BAND QUARTERS.
14. OFFICERS MESS.
15. BAKERY
16. KITCHENS & MESSHALL
17. STEAM LAUNDRY.
18. AUTO. SHED.
19. BAND STAND.
100. HOSE REEL HOUSE
101. STOREHOUSE
102.
103. BOAT HOUSE.
104. LAUNCH HOUSE.
105. OIL HOUSE.
106. ICE HOUSE
107.
108. BLACKSMITH SHOP
109. GREENHOUSE.
110. STABLE.
111. CREMATORY.
112. SOLDIER'S QUARTERS.
113.
114.
115. PAINT SHED.
20. Q.M. STOREHOUSE.
21. COMMISSARY ST. HQ.
22. Q.M. SHOP.
23. Q.M. WAGON SHED.
24. Q.M. COAL SHED.
30. ORDNANCE ST. HO.
40. PLUMBER'S QRS.
41. ENGINEER SHOP ST.HO
70. CHURCH.
90. WATER-METER HOUSE
91. TRANSFORMER STA.
25. Q.M. SHED.
116.
117. AUTO REPAIR SHOP.
118. OFFICER'S GARAGE.
119.
120. ANTI-AIRCRAFT GUN STHD
71.
72. Y.M.C.A.
121. GARAGE.

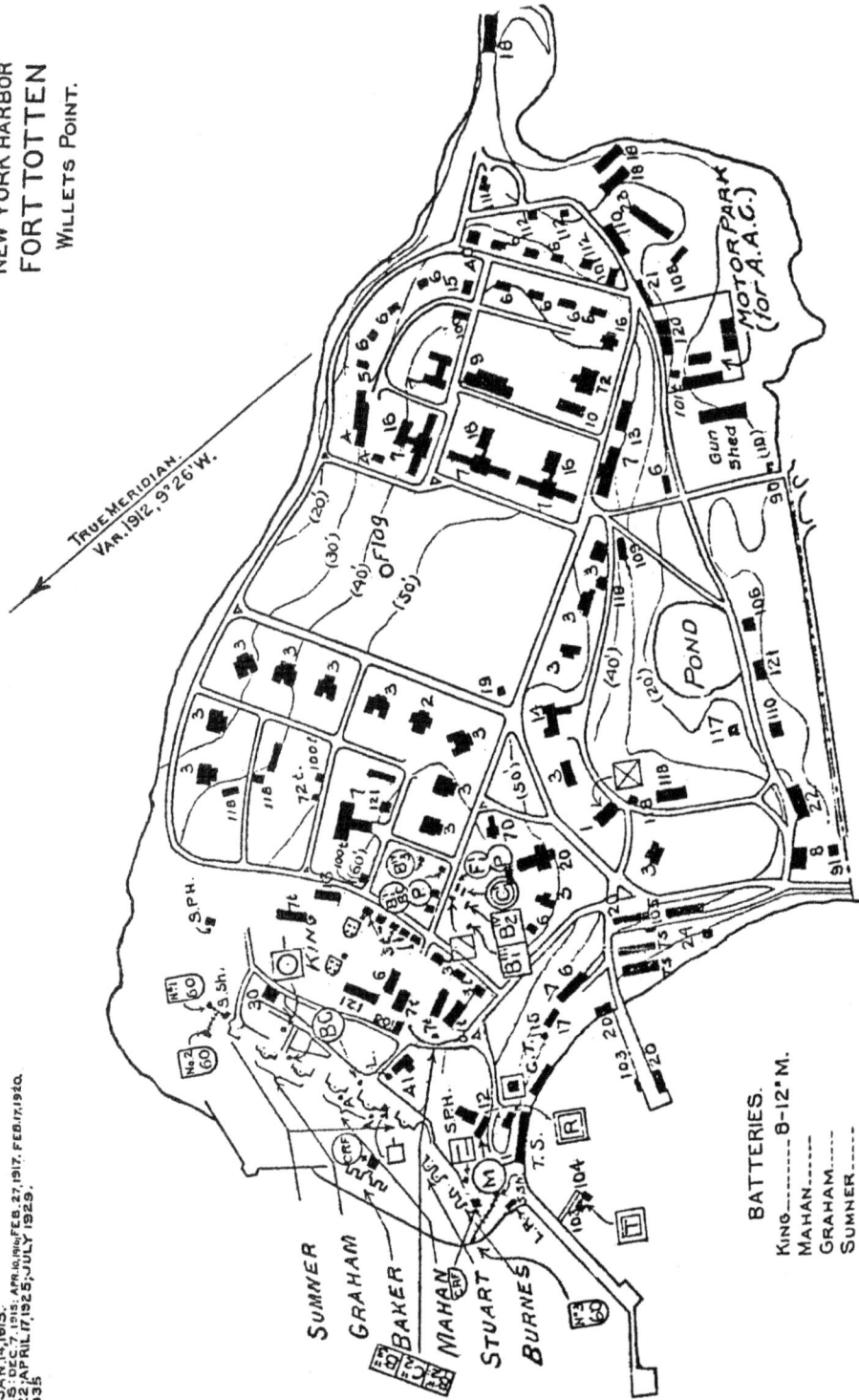

BATTERIES.
KING -------- 8-12" M.
MAHAN --------
GRAHAM --------
SUMNER --------
STUART -------- } 2-3" P.

∠ *BAKER ---- 2-3" P.
∠ · BURNES ---- 2-3" P.

A-Anti-aircraft gun 2-3" *Empls. 3, 4" Vacant.
(being dismounted
by Ordnance)

Fort Totten 1924 (NARA)

Fort Totten 1936 (NARA)

as a harbor defense command post in the early 1940s. That was not subsequently done. It is not clear, but it may be that some of this emplacement still exists under the earth at Fort Totten Park.

- **SUMNER**: A battery for two 8-inch guns on disappearing carriages built at Fort Totten. The two emplacements are on opposite ends of a gun line and were built at different times. Originally in the early 1890s approval was granted to construct three emplacements on the eastern end of an 1870s battery on the bluff overlooking uncompleted old Fort Totten. Submission was made for the three emplacements on May 4, 1891. At the time the size and type of carriage was yet unknown, and they were thought capable of eventually carrying either 8 or 10-inch guns. They followed early plans for protected platforms with magazines at a lower level and ammunition service only with cranes. Work was done on the three emplacements in 1891-1893. In 1895 the separated No. 1 emplacement was ordered to be completed for an 8-inch Model 1894 disappearing carriage. While the two western, immediately adjacent emplacements became 10-inch Battery Graham. All three were transferred together on December 27, 1897 for $218,036. It carried one 8-inch gun Model 1888M1 Watervliet gun on Model 1894 disappearing carriage (#30/#12). Then on April 19, 1897 submission was made for another single 8-inch emplacement on the western end of the gun line, beyond and separated slightly from Batteries Graham and Mahan. This became No. 2 emplacement of Battery Sumner. Work was done in 1897-1898 for transfer on March 4, 1899 for a cost of $45,000. It also carried a single 8-inch Model 1888MII Watervliet gun, but this time on a slightly more modern Model 1896 disappearing carriage (#51/#5). Both emplacements were named on General Orders No. 78 of May 25, 1903 for Brigadier General Jethro Sumner of the Continental Army. In its early years the emplacements had their loading platforms modified, were given a new BC station, and the No. 1 emplacement was equipped with modern ammunition chain hoists. In common with many other 8-inch batteries, Sumner was disarmed in mid-1918 to provide tubes for railway guns. It was not subsequently used or rearmed. At times heavily overgrown, it still exists at the Fort Totten Park. The battery is closed to the public.

- **GRAHAM**: A battery for two 10-inch guns built with the No. 1 emplacement of Battery Sumner immediately on its eastern flank. Original plans for the three emplacements were submitted on May 4, 1891. Work of the western two, adjoining emplacements (which would become Battery Graham) was done from 1891-1894. In June 1895 the decision was made to have these two emplacements hold 10-inch guns on Model 1894 disappearing carriages, and they were completed as such. They were of early design, with small loading platforms, lack of communications between emplacements, and just hand hoists. These features were modernized through the 1910s. The battery was completed for transfer on December 27, 1897 for a cost (combined with 8-inch Battery Sumner No. 1) of $219,036. It was armed with two 10-inch Model 1888 guns on Model 1894 LF disappearing carriages (Watervliet tubes #8/#2 and #16/#9). It was named on General Orders No. 78 of May 25, 1903 for Lt. Colonel William M. Graham, 11th U.S. Infantry killed in action at Moleno del Rey Mexico in 1847. The battery was disarmed in 1917-1918 with the decreased importance of the Eastern New York defenses and the need for gun tubes for railway artillery during World War I. The emplacement was not subsequently used. It does still exist on Fort Totten Park grounds. The battery is closed to the public.

- **MAHAN**: A battery for two 12-inch disappearing guns built as emplacements No. 4 and No. 5 of the main gun line at Fort Totten (Battery Graham on the right flank, and the final emplacement of Battery Sumner of the left flank). Plans were submitted on May 19, 1898 for two guns to go on Model 1897 disappearing carriages. It was funded with $100,000 from the Fortification Act of May 7, 1898. Work was done on 1898-1900. It was of fairly consistent mimeograph type plan design, with flank traverse magazines and equipped with ammunition hoists. A large battery commander's station was added in 1907. The battery was transferred on October 12, 1900 for a construction cost of $88,536. Naming was conferred by General Orders No. 43 of April 4, 1900 for Dennis H. Mahan, professor of engineering at the military academy from 1832-1871. It was armed with two 12-inch Model 1895 Watervliet guns on Model 1897 disappearing carriages (#21/#9 and #23/#8). This armament was removed in 1918 for use of the gun tubes on railway carriages. The guns were not returned, and the carriages left in place were scrapped about 1921. The emplacement still exists on Fort Totten Park grounds. The battery is closed to the public.

- **STUART**: A battery for two 5-inch rapid fire guns for Fort Totten. It was located west of the gun line among the old 1870s earthen battery located there. It was placed directly between two former traverse magazines, which provided crew shelters and equipment rooms though the battery did have a new, dedicated magazine. Submission for the emplacement was made on November 27, 1897 for an estimated $9000 cost. Work was done in 1898-1899, it being reported ready for transfer on November 12, 1900. Transfer was made on December 11, 1900 for a cost of only $9300. The battery was of relatively simple design, following mimeograph guidelines. It was armed with two 5-inch Model 1897 Watervliet guns on Model 1896 balanced pillar mounts (#4/#28 and #5/#29). It was named on General Orders No. 43 of April 4, 1900 for Captain Sidney E. Stuart of the army's ordnance department. These guns were dismounted in 1917 for use on wheeled field mounts and never returned. The carriages were scrapped in 1920. It still exists on Fort Totten Park grounds. The battery is closed to the public.

- **BURNES**: A battery for two 3-inch Model 1902 guns and pedestals built as a dual emplacement on the left flank of Battery Stuart in the old 1870s earthen battery. It was of standard design, with one rounded flank emplacement (on the left) and one square internal emplacement. Each block had its own flank magazine. Work was done in 1902-1903, for transfer made on May 5, 1904 for a cost of $10,000. It was armed with two 3-inch Model 1902 guns and pedestal mounts (#57/#57 and #58/#58). The battery was named on General Orders No. 194 of December 27, 1904 for 2nd Lieutenant Thomas Burnes, mortally wounded at Hatcher's Run, VA in 1864. Some later documents and maps mistakenly misspell the name as "Burns." In 1920 it received a coincidence range finder station. It was one of the few batteries at Fort Totten to remain armed after World War II. It was retained in the 1932 Review and as part of the 1940 Program. Locally during World War II, it served as Tactical Battery No. 27. It was disarmed in 1946. The emplacement still exists on Fort Totten Park grounds. The battery is closed to the public.

- **BAKER**: A battery for four 3-inch guns, constructed in two sections. An earlier emplacement for a pair of 3-inch, 15-pounder guns on masking parapet mounts was later joined by a pair of 3-inch pedestal guns. The battery was located on a shelf just to the west of old Fort Totten near the shoreline. The first two guns were submitted on January 31, 1899 for an estimated construction cost of $8000. Work was done in 1899-1900. The battery was of fairly standard mimeograph design. It was transferred on August 6, 1900 for a cost of $8150. It was armed with two 3-inch M1898 guns and masking parapet mounts (Driggs-Seabury #28 and #29). It was named on General Orders No. 78 on May 25, 1903 for 2nd Lieutenant William L. Baker killed in action at Antietam, MD in 1862. Then on January 30, 1903 two additional Model 1902 3-inch guns on pedestal mounts were submitted to be placed on the right flank of the existing guns. Work was done in 1903-1904 and transferred on May 5, 1904 for a cost of $10,000. It was armed with two 3-inch Model 1902 guns and pedestals (#60 and #59). The two masking parapet guns were modified in 1916 to M1898M1 fixed types. While a CRF for this section was added in 1919, the two guns were removed and scrapped in 1920. The two M1902 pedestal guns were retained through World War II. There is evidence of gadding to the front of the pits, perhaps indicating the use of new, larger box shields in the war. The pedestal guns were not removed until probably 1946. The emplacement still exists at the Fort Totten Park grounds. The battery is closed to the public.

U.S. Military Academy (1790-Present) is located on strategic high ground overlooking the Hudson River 50 miles north of New York City. The Federal military reservation was created in 1790. The first engineer cadet class enrolled in 1802. Coastal artillery training batteries once located on campus included Battery Schofield (disappearing carriage), Seacoast Battery (barbette), and Battery Byrne (Mortar Battery until 1911) (mortars). Live-fire training for all gun types was conducted at Fort Hancock, Sandy Hook, New Jersey. While the U.S. Military Academy still operates, all seacoast training batteries have been destroyed or buried.

Modern Batteries at West Point

- There is photographic evidence of a West Point interim mortar battery of a 12-inch mortar Model 1890M1 on an older Model 1886 carriage. Ordnance records indicate mortar #2 Watervliet was received at West Point on April 26, 1895 and carriage Model 1891 West Point Foundry #73 received August 3, 1894. Just where this was emplaced and when removed is not currently known.

- **BYRNE:** The original cadet practice battery was emplaced at the edge of the plain and armed in the 1890s with three 8-inch and three 10-inch smoothbore mortars. It was rebuilt in 1906-1907 for two 12-inch mortars Model 1890M1 on Model 1896 carriages. These were transferred here from Battery Kearny at Fort Preble. Renamed Battery Byrne at some point. The emplacement was buried in the 1920s.

- **SCHOFIELD:** The primary modern practice battery for West Point Military Academy. At first it was armed with one 8-inch gun Model 1888 on barbette carriage Model 1892 (Watervliet #10/#9) shipped here on May 19, 1894. In 1907-08 a modern style dual 6-inch disappearing battery was also built. It was armed with two 6-inch guns and carriage transferred here from Battery Livingston at Fort Hamilton. It was armed with two 6-inch guns Model 1905 on Model 1905 disappearing carriages (#30/#1 and #9/#2). These guns were emplaced at least as late as 1956, though obviously not still in use as training weapons. They were recovered in the 1960s by the National Park service and one each put on display at Fort Winfield Scott and Fort Pickens. The emplacement itself was later destroyed.

8-inch gun on barbette mount at the Seacoast Battery, West Point (NARA)

THE HARBOR DEFENSES OF THE SOUTHERN ENTRANCE TO NEW YORK — NEW YORK AND NEW JERSEY

The great natural harbor of New York was settled by the Dutch in 1624, taken over by the British in 1667 and 1674, and the harbor played a key role during the American Revolution. New York harbor was controlled by the British throughout most of the war and served as their pollical and military base. Fort Amsterdam was the first several defenses constructed for the southern entrance to New York harbor during colonial times. These defenses were expanded by both the State of New York and federal Government during the First, Second and Third Systems with works around the inner harbor and at the Narrows between Staten Island and Brooklyn. The defenses were modernized with an extensive set on concrete batteries at the Narrows and on Sandy Hook, New Jersey, upgraded after World War I, and again during World War II. The military reservations were retained through the Cold War years for use as training centers, housing, and Nike missile sites. With the base closures of 1989, much of the property was transferred to the National Park Service becoming part of the new Gateway National Recreation Area in 1972.

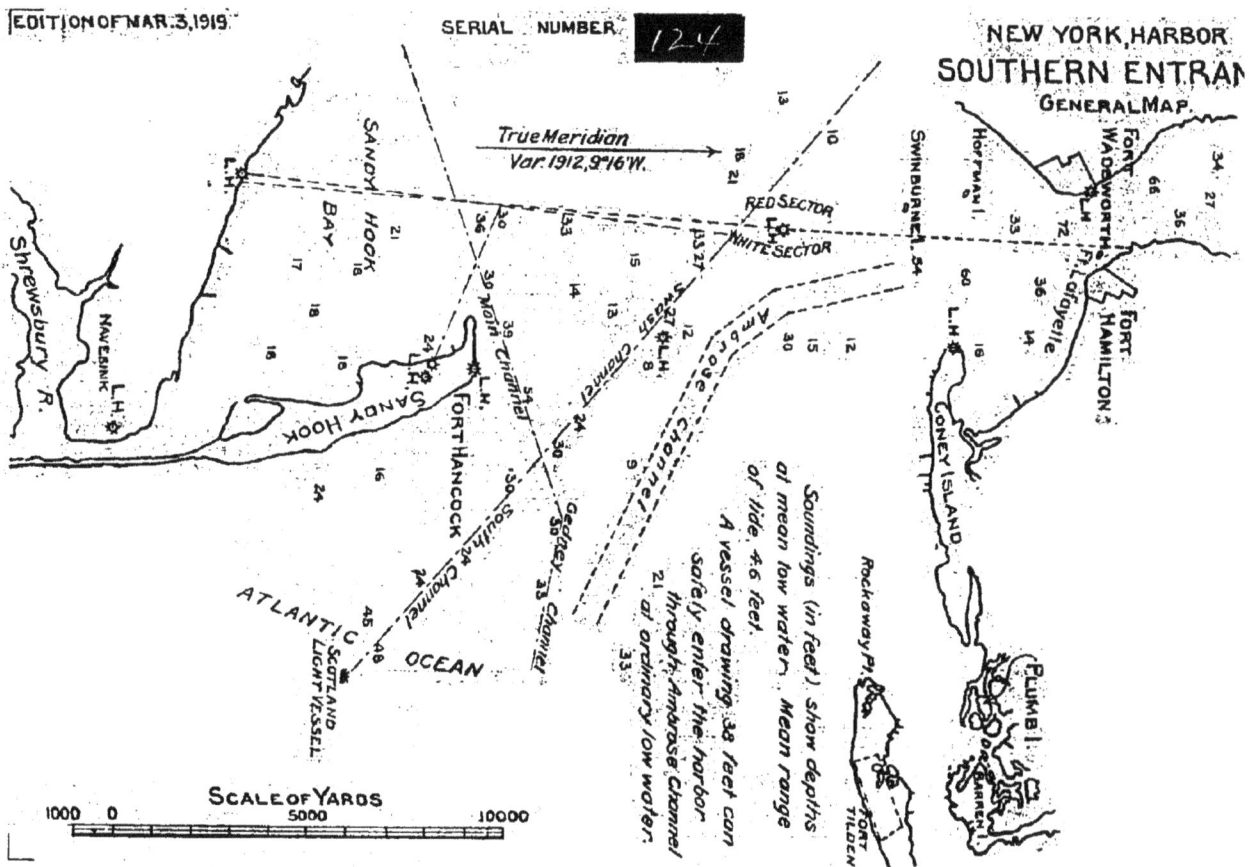

Jamaica Bay Military Reservation (1941-1942) was located on Jamaica Bay (also known as Grassy Bay) at the southern portion of the western tip of Long Island, New York. The estuary is partially man-made, and partially natural. The bay connects with Lower New York Bay to the west, through Rockaway Inlet, and is the western most of the coastal lagoons on the south shore of Long Island. The proposed site for a 1940 Program 16-inch battery was an area of marshland on Jamaica Bay, which included the Idlewild Golf Course as well as a summer hotel and a landing strip called the Jamaica Sea-Airport. Instead of coast defense battery, the title to the land was conveyed to the city at the end of December 1941 for a large airport. Construction began in 1943 of Idlewild Airport (later JFK Airport). Battery #117 was to be built on a point of land at nearby JFK Airport, but it was considered to be too close to the airport. That site is now under one of the airport's taxiways.

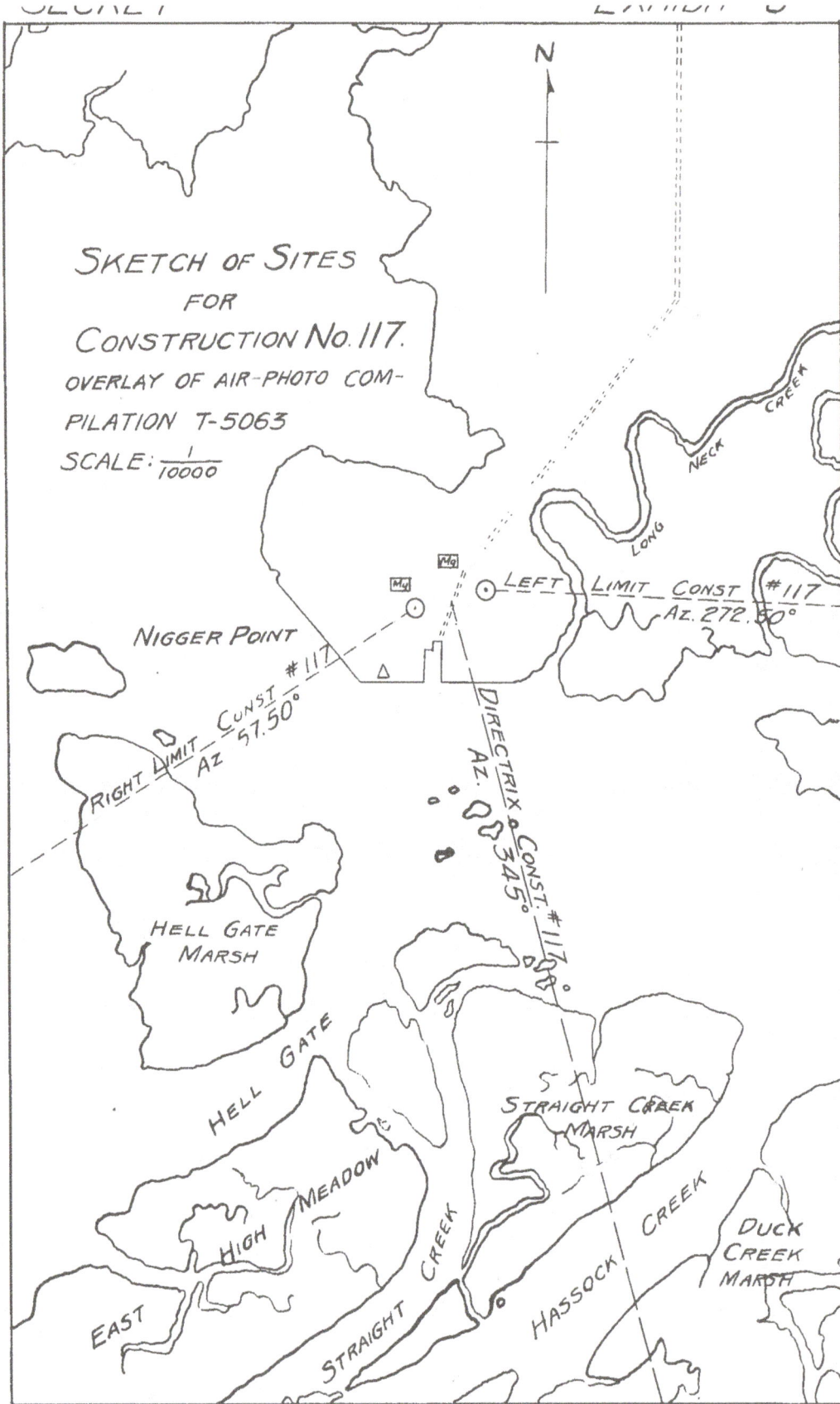

SKETCH OF SITES
FOR
CONSTRUCTION No. 117.
OVERLAY OF AIR-PHOTO COM-
PILATION T-5063
SCALE: $\frac{1}{10000}$

N

LONG NECK CREEK

LEFT LIMIT CONST #117
Az. 272.50°

NIGGER POINT

RIGHT LIMIT CONST #117
Az. 57.50°

DIRECTRIX CONST #117
Az. 345°

HELL GATE MARSH

HELL GATE

STRAIGHT CREEK MARSH

HIGH MEADOW

STRAIGHT CREEK

HASSOCK CREEK

DUCK CREEK MARSH

EAST

Jamaica Bay M.R. Battery

- *Battery #117* (planned): A standard 1940 Program dual 16-inch casemated battery planned for the New York defenses. Originally the plans called for the addition of another gun to each of Battery Harris' single emplacements. This was soon changed instead to a new standard emplacement at a site know at the time as Nigger Point in the Jamaica Bay area near Idlewild Airport in southern Long Island. This location was approved by a board meeting of October 23, 1941. Both because of its low priority (#21) and the poor location near a major airport, the project was never started and eventually it was cancelled in its entirety on November 13, 1942.

Fort Tilden (1917-1975) is located on Rockway Beach on Long Island in the borough of Queens. Fort Tilden was built during the Interwar Period as the site of two 16-inch Model 1919MII guns on barbette carriages Model 1919. It was named in General Orders 100 of 1917 for New York politician Samual Tilden. Only three batteries of these powerful coast artillery weapons were ever constructed. Originally these guns were in open emplacements that allowed for 360-degree firing to a maximum range of 50,000 yard. During World War II, both guns were given overhead protection in the form of separate reinforced concrete casemates. Fort Tilden was also the site of an anti-aircraft battery, a balloon hanger, and two temporary batteries of 6-inch Model 1900 guns. Located near Fort Tilden was a temporary World War I emplacement for four 12-inch seacoast mortar. The 1940 Program resulted in a #200 Series battery for 6-inch guns constructed on the fort's grounds, along with a mine casemate. Nearby, two AMTB batteries were installed at Norton Point and at Rockway Point. After World War II, the fort's guns were removed, and Fort Tilden was used first as anti-aircraft gun base and then as a Nike missile base until 1970. It then served as a U.S. Army Reserve post until 1978. The reservation is now the Fort Tilden Historic District, which is part of the Gateway National Recreation Area, and most emplacements remain in abandoned condition. The site open to the public.

Fort Tilden Gun Batteries

- **HARRIS:** This was a long-planned, heavy battery for 16-inch guns to be emplaced near Rockaway Beach and provide crossfire to the heavy guns at Sandy Hook. The first preliminary plans were cast in mid-1915. On August 6, 1917 a detailed proposal for two 16-inch emplacements was submitted. The final approved plan had to await the adoption of the Model 1919 16-inch gun and its barbette. The first money allocated was $20,000 from an appropriation of December 1920. Built on the south, center of a new reservation, the battery faced and fired to the south. Following the preferred layout of the time, it was just two massive, but simple gun blocks with racers to hold the carriages. The two blocks were 1000-feet apart. They were connected to light, dispersed magazines by a rail system. There were four magazines and two power rooms. Basic construction was done from March 1921 to September 1923. Transfer was made on December 26, 1924 for a cost of $114,460. It was named on General Orders No. 13 in March 1922 for Colonel Henry L. Harris. It was armed with two 16-inch Model 1919MII Watervliet guns on Model 1919 barbettes (#4/#2 and #5/#3). Continuous modifications began after construction. A telephone switchboard and electric lighting was completed in 1924-1926. And in 1931 a new central magazine with splinter protection was added. During the Interwar Period it was the single most important battery of New York's defenses. Just before World War II a new covered plotting room was built, and the gun positions were provided with heavy overhead, casemate protection from October 14, 1941 to January 1943 at a cost of $695,266. Each position was casemated, but they were left separated. Gun shields were fabricated and delivered in June of 1944, but as they would not fit and needed modification, they

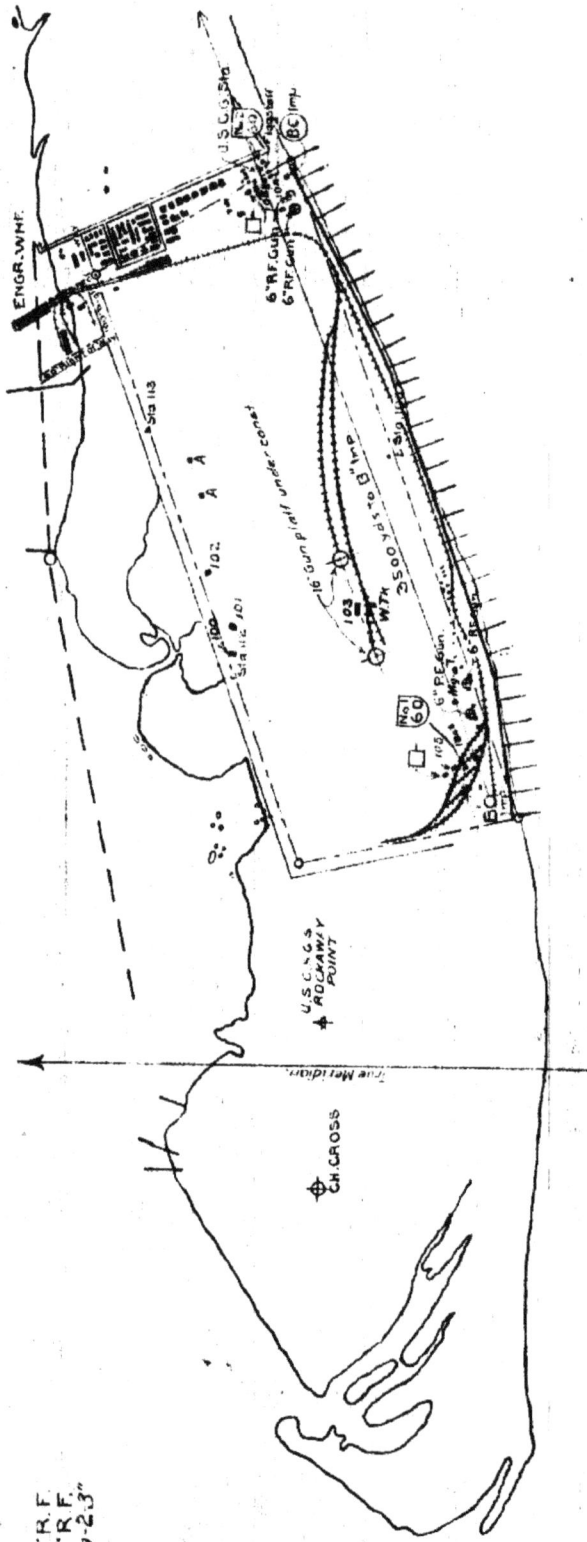

SERIAL NUMBER *103*

NEW YORK, HARBOR

FORT TILDEN

Rockaway Beach, N.Y.

GENERAL MAP

LEGEND

EDITION OF MAR. 3, 1919.
REVISIONS. FEB. 20, 1920. APR. 20, 1920.
MAY 23, 1921; AUG. 27, 1921.
NOV. 28, 1921.

100. GAS GEN. HO.
101. HANGAR (BALLOON.)
102. INCINERATOR.
103. CEMENT HOUSE.
104. WATCH TOWER.
105. OLD 25 KW. SHELTER.
7. BARRACKS.

BATTERIES

* EAST   2-6" R.F.
* WEST   2-6" R.F.
* A Anti-aircraft gun-2-3"

*Temporary.

Scale of Feet

1000 500   0   1000   2000   3000   4000

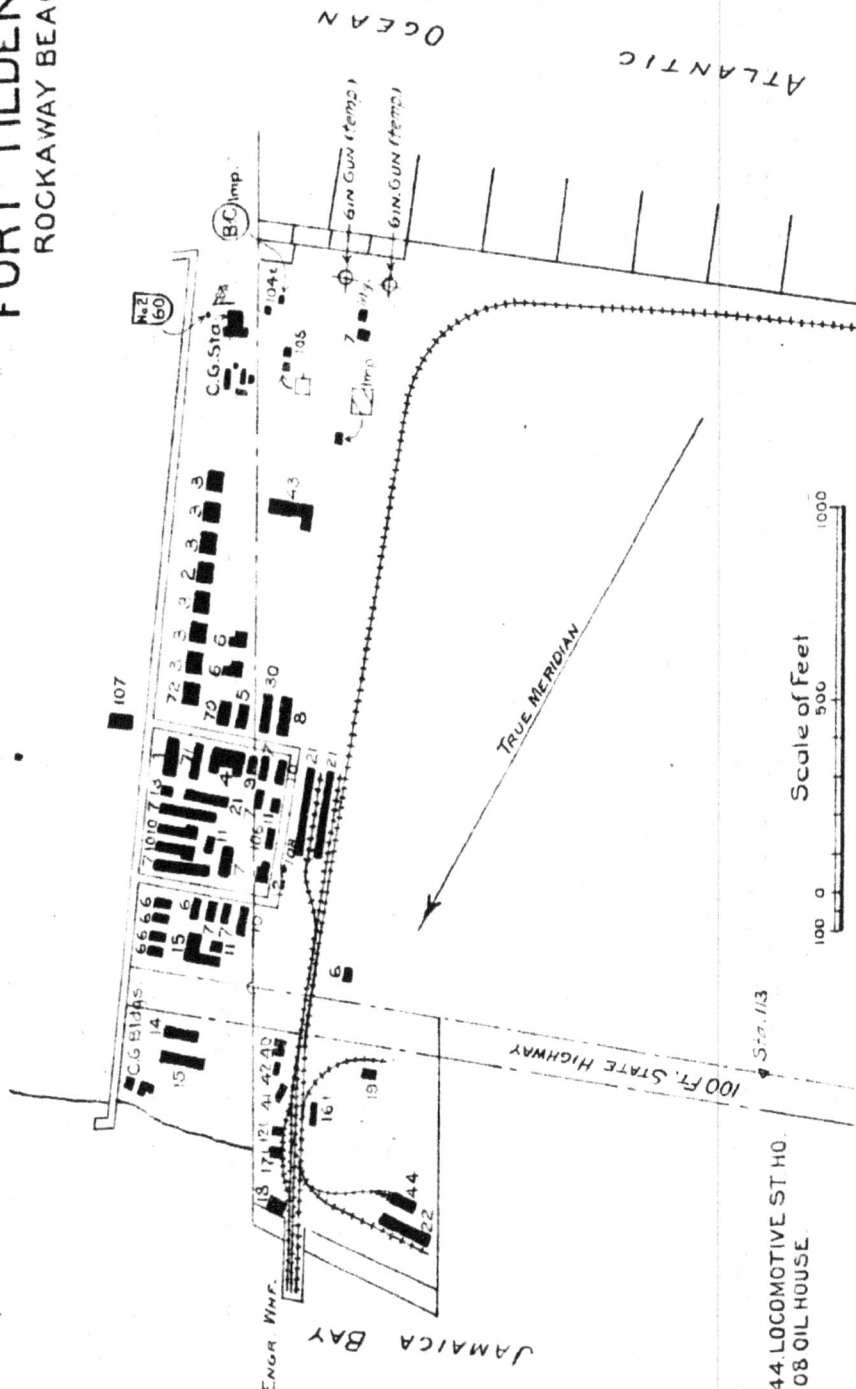

NEW YORK HARBOR
FORT TILDEN D-1.
ROCKAWAY BEACH

SERIAL NUMBER 103

EDITION OF MAY 23, 1921.
REVISIONS: AUG. 27, 1921.
NOV. 28, 1921.

ATLANTIC OCEAN

JAMAICA BAY

TRUE MERIDIAN

Scale of Feet
100 0   500   1000

100 FT. STATE HIGHWAY

LEGEND
1 ADMINISTRATION BLDG.
2 COMMDG. OFF. QRS.
3 OFFICERS' QRS.
4 HOSPITAL.
5 HOSPITAL STEWD. QRS.
6 N.C.O. QUARTERS.
7 BARRACKS.
8 GUARD HOUSE.
9 POST EXCHANGE.
10 MESS HALLS.
11 BATH HOUSE.
12 GARAGE.
13 POST OFFICE.
14 WAGON SHED.
15 STABLES.
16 CARPENTER AND
   BLACKSMITH SHOP.
17 LOCOMOTIVE SHED.
18 200 TON COAL BIN.
19 SHACK (WOOD)
104 WATCH TOWER.
105 OLD 25 KW SHELTER.
106 CARPENTER PAINT &
    PLUMBING SHOPS.
107 FIRE ENGINE HOUSE.
21 Q.M. STOREHOUSE.
22 Q.M. COAL BIN.
30 ORDNANCE OFFICE &
   STOREROOM.
40 ENGR OFFICE
41 ENGR. CARPENTER SHOP.
42 WOODEN SHACK (ENGR.)
43 ENGR. QUARTERS.
70 SERVICE CLUB.
71 Y.M.C.A.
72 Y.M.C.A. SECY. OFFICE.
44 LOCOMOTIVE ST. HO.
108 OIL HOUSE.

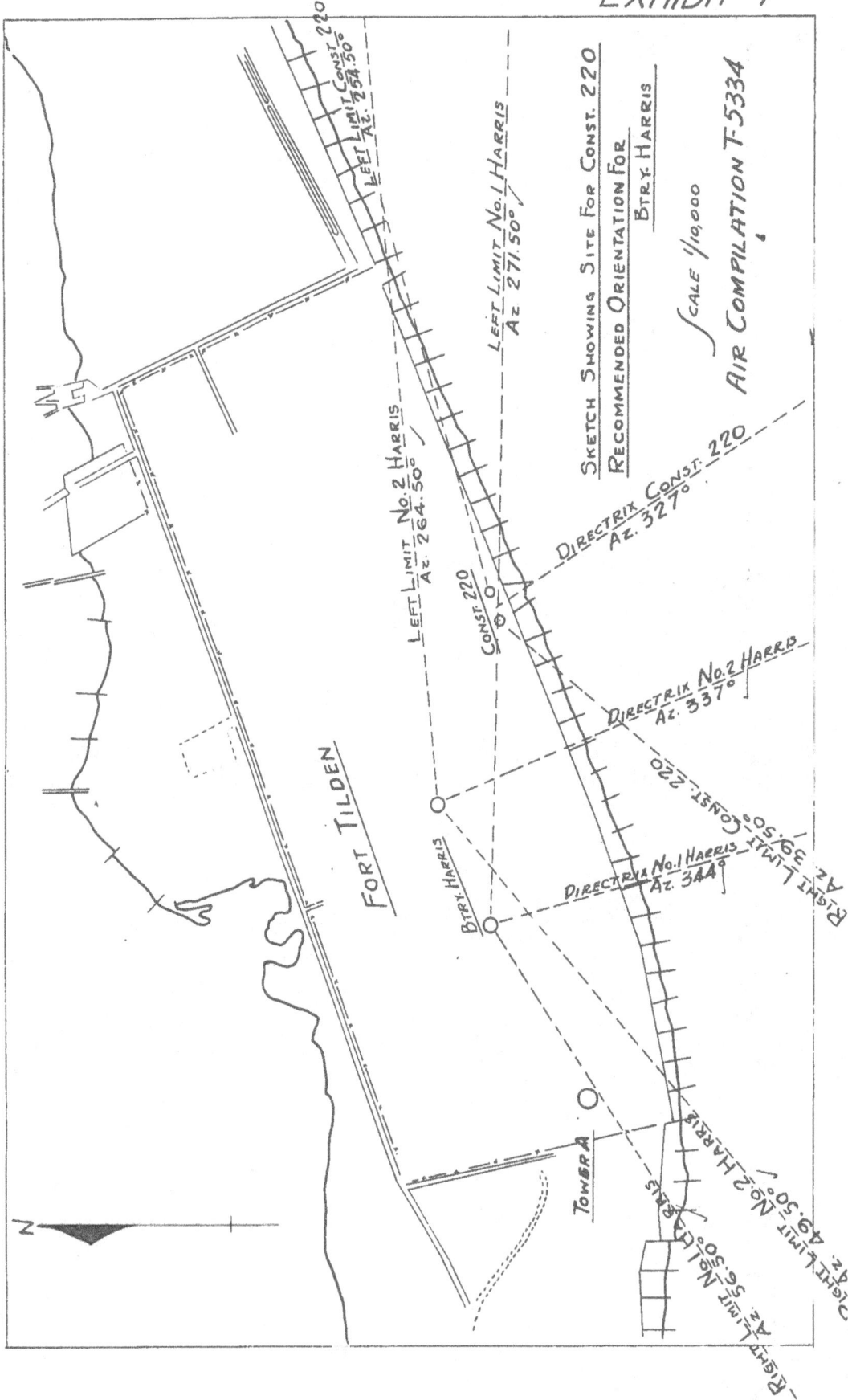

EXHIBIT 7

SKETCH SHOWING SITE FOR CONST. 220
RECOMMENDED ORIENTATION FOR
BTRY. HARRIS

SCALE 1/10,000

AIR COMPILATION T-5334

LEFT LIMIT CONST. 220
AZ. 254.50°

LEFT LIMIT NO.1 HARRIS
AZ. 271.50°

LEFT LIMIT NO.2 HARRIS
AZ. 264.50°

CONST. 220

DIRECTRIX CONST. 220
AZ. 327°

DIRECTRIX NO.2 HARRIS
AZ. 337°

RIGHT LIMIT CONST. 220
AZ. 39.50°

DIRECTRIX NO.1 HARRIS
AZ. 344°

RIGHT LIMIT NO.2 HARRIS
AZ. 49.50°

RIGHT LIMIT NO.1 HARRIS
AZ. 56.00°

FORT TILDEN

BTRY. HARRIS

TOWER A

N

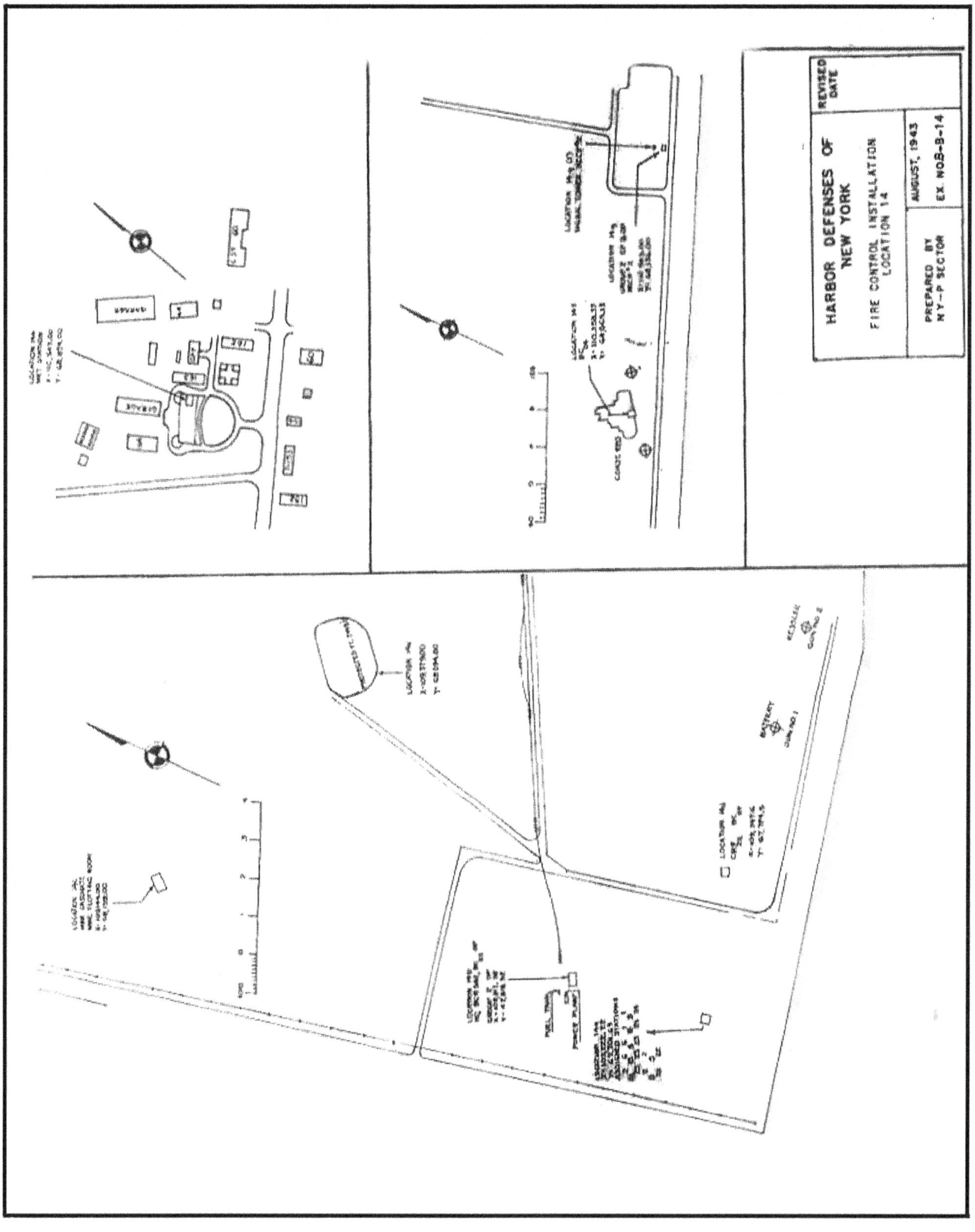
REVISED DATE
HARBOR DEFENSES OF NEW YORK
FIRE CONTROL INSTALLATION LOCATION 14
AUGUST, 1943
EX. NO.B-S-14
PREPARED BY N Y-P SECTOR

were not installed. The battery served as Tactical Battery #2 for New York. It was not finally deleted until 1948. The emplacements and much or the magazine and support buildings at the Fort Tilden section of the Gateway National Recreation Area. The battery site is open to the public.

- In February of 1917 it was recommended that four mortars be emplaced at a new reservation near or at Rockaway Beach. As there were not enough spare mortars, it was assumed that these would be relocated from an existing mortar battery somewhere. Soon the proposal was to move four weapons from Battery Piper at Fort Hamilton here. Acquiring property from New York City parks, construction work was authorized on March 27, 1917 for an expected cost of $15,000. Just concrete platforms would be built, magazines and other support structures would be temporary wood buildings with galvanized iron roofs. The first block was ready on May 25, 1917. Transferred were four Model 1890M1 mortars on Model 1896M1 carriages (Watervliet tubes #93/#202, #101/#224, #102/#225 and #103/#203). They served just a short time, being removed in 1920. No trace remains of the blocks in what is today the Jacob Riis Park.

- **FERGUSON**: In February 1917 a project was approved to relocate four 6-inch Model 1900 rapid-fire guns from existing Endicott batteries to new locations near Rockaway Beach. The two guns removed from Battery Kinney at Fort Slocum eventually became Battery Ferguson. At first the new emplacements were simple gun platforms with wooden magazines and battery commander's station. Work was authorized on February 18, 1917 with a $50,000 appropriation. It was completed for proof firing on May 14, 1917. It was located about 300-feet behind the beach, near the Coast Guard station on the eastern side of the new reservation. Initially it was known just as "Battery East." It carried two 6-inch Model 1900 guns and shielded pedestal mounts (#44/#19 and #26/#18). The guns had 330-foot distance between them. In 1922 wooden shelters were built around each platform and its gun. The battery was finally named on General Orders No. 185 of December 1, 1939 for Brigadier General Frank K. Ferguson, Coast Artillery Corps. On account of its relatively poor position, the battery was not recommended for retention under the 1940 Program. However, with the delays in completing nearby Battery #220, Ferguson was not abandoned or scrapped until after the war in 1946. The few visible parts of the emplacement were soon grown over, and while some evidence may still exist, it has not been located in recent years. The battery site is open to the public.

- **KESSLER**: A second set of 6-inch guns relocated to the Rockaway Beach reservation in 1917. The two guns removed from Battery Burke at Fort Hamilton eventually became Battery Kessler. At first the new emplacements were simple gun platforms with wooden magazines and battery commander's station. Work was authorized on February 18, 1917 with a $50,000 appropriation. It was completed for proof firing on May 14, 1917. It was also about 300-feet from the shoreline, and about 800-feet west of Ferguson. Similarly, it was known for years as "West Battery," then it was finally named on General Orders No. 185 of December 1, 1939 for Colonel Percy M Kessler, a coast artillery officer and past commander of the Harbor Defenses of Sandy Hook. It was armed with two 6-inch Model 1900 guns and pedestal mountss (#24/#20 and #25/#21). It was in an important location and recommended just before World War II for complete reconstruction. A new concrete traverse magazine was constructed from October 25, 1941 to January 1943. The plan of this magazine/plot/power room was similar to the prototype 200-series plan used at Battery #224 at Fort Story. The guns were not replaced, and in fact continued in service during the construction of the new magazine structure. The battery served as New York Tactical No. 22 throughout most of the war. The emplacement still exists on the Fort Tilden unit of the Gateway National Recreation Area. The battery site is open to the public.

- **Battery #220**: A standard 1940 Program dual 6-inch battery emplaced at Fort Tilden. It was not funded until FY-1943 and held a relatively low priority for construction. It was sited to the south of Harris, close to the shore and between the older pair of temporary 6-inch batteries. The intention was that it would replace Battery Ferguson upon completion. Work was done from October 5, 1942 until suspended in mid-1943. While never fully completed, the concrete gun pads and traverse magazine were structurally finished. No guns were ever mounted, though two M4 carriages were delivered in 1944 (#41 and #42). By 1946 it was still incomplete without gun tubes. While there was an official recommendation that the battery should be completed, that was never done, and it was abandoned in 1948. The emplacement still exists on the Fort Tilden unit of the Gateway National Recreation Area. The battery site is open to the public.

Fort Tilden 1920s (NARA)

Norton Point Military Reservation (1942-1946) is located in Sea Gate, on the west end of Coney Island, Brooklyn, in New York City, east of New York Harbor's main channel. Located here were New Battery Catlin (AMTB 18) (1942-1946), and AMTB Battery 19 (1943-1946). Other World War II military support structures that were once here no longer exist. Seagate is a private gated community; public access is restricted.

Norton Point M.R. Gun Batteries

- **CATLIN**: Four 3-inch guns on pedestal were relocated to Norton Point from Battery Catlin at Fort Wadsworth. The four guns were erected on simple gun blocks pursuant to authority granted on April 9, 1942. They were 3-inch Model 1902 guns and pedestals #32/#32, #33/#33, #34/#34 and #35/#35. It served until shortly after the end of the war as Tactical Battery #18. These gun blocks are buried. The battery site is closed to the public.

- **AMTB #19:** A 1943 Program AMTB battery for two 90mm fixed and two 90mm mobile guns emplaced at Norton Point pursuant to authority of November 20, 1942. Two fixed gun blocks and a battery commander's station were constructed in early 1943. The guns served as New York Tactical Battery #19, probably between 1943 and 1946. Two mounts remain exposed in surf, located along Atlantic Ave. between Beach 40th and Beach 42nd Streets.

Rockway Point Military Reservation (1942-1946), also known as Breezy Point, is located in the New York City borough of Queens, located on the western end of the Rockaway peninsula, between Rockaway Inlet and Jamaica Bay to the north and the Atlantic Ocean to the west and south. Rockaway Point had AMTB 20 with two 3-inch M1902 guns and AMTB 21 with four 90 mm guns.

Rockaway Point M.R. Gun Batteries

- **AMTB #20:** A pair of 3-inch guns on pedestals were moved to Rockaway Beach when Battery Catlin at Fort Wadsworth was dismounted in 1942. Two Model 1902 guns and pedestals (#30/#30 and #31/#31) were moved pursuant to authority granted February 21, 1941 to serve in the anti-motor torpedo boat role pending the completion of the 90mm batteries. The guns remained on their simple concrete blocks until removed shortly after the end of the war. For many years the blocks survived on the beach at Rockaway Point.

- **AMTB #21:** A 1943 Program AMTB battery of two 90mm fixed and two 90mm mobile guns. Fixed gun blocks were erected at Rockaway Point under authority of November 20, 1942. The guns served until removed shortly after the war.

Swinburne Island Military Reservation (1943-1946) is a 4-acre artificial island in Lower New York Bay, east of Staten Island in New York City. It was used for quarantine of immigrants. Swinburne Island is the smaller of two nearby islands, the other being Hoffman Island to the north. The AMTB battery at Swinburne Island consisted of two 90mm gun mounted on 90mm M3 gun mounts and two mobile 90mm guns. Concrete pads for both these guns were laid at south end of the island. Both Swinburne and Hoffman Islands are now part of Gateway National Recreation Area with no public access to protect the wildlife.

Swinburne Island M.R. Gun Battery

- **AMTB #12:** A 1943 Program AMTB battery of two 90mm fixed and two 90mm mobile guns. Two fixed platforms for the fixed 90mm were emplaced on this island in late 1943. It served locally at Tactical Battery #12 for New York. The guns were probably removed in 1946. Two-gun mounts remain. The battery site is closed to the public.

Swinburne Island post war (NARA)

Battery Harris, Fort Tilden Unit, Gateway National Recreation Area (Terry McGovern)

Fort Hamilton (Terry McGovern)

Fort Hamilton 1831-1948) is located on the eastern side of the entrance (also known as the "Narrows") to Upper New York Bay on Long Island in borough of Brooklyn. The Third System granite fort, completed in 1831, was paired with Fort Lafayette (located on Hendricks Reef) as the defense of east side of the Narrows. Fort Hamilton was designed to provide land defenses for Fort Lafayette (located on Hendricks Reef and destroyed in the 1960s to make room for the eastern tower of the Verrazzano Narrows Bridge). During the Endicott Program a two-pit mortar battery and an impressive gun line of 5 major caliber batteries and 5 minor caliber batteries were built along the shoreline to cover the Lower New York Bay. Several of these batteries were built into the seaward side of the Third System fort. It was officially named on General Orders 2 of 1938 for Alexander Hamilton, the first Secretary of the Treasury. At the high point of the Endicott Program defense, the fort mounted 40 guns and mortars ranging from 3-inch to 12-inch in size. By the end of World War II, only two batteries of rapid-fire guns remained active, and fort's primary purpose was as a mobilization center. During the late 1950s the U.S. Army destroyed of all the Endicott Program batteries and remaining seaward side of the 3rd System fort, for the construction of Army housing. The landward side of the 3rd System fort is now a service club. In the 1960s, the U.S. Army's Chaplain School was located at the fort. The U.S. Army maintains a small museum on harbor defense of New York in the 3rd System's detached caponier. In the park area next to the post entrance is a display of a number of artillery pieces, and in the near-by city park is one of two existing 20-inch Rodman smoothbore cannons. Fort Hamilton remains an active U.S. Army installation as the headquarters for the New York Command, which supports several U.S. Army Reserve and New York Army National Guard units in the New York City area. Access to the fort is controlled and visitors require passes.

Fort Hamilton Gun Batteries

- **SPEAR:** A battery for ultimately three 10-inch guns on disappearing carriages built east of the old fort following the bluff line. It was first started as a single emplacement (though also including the magazine for an adjacent gun) as a project on October 18, 1892. Then on August 25, 1896 plans were submitted for the adjacent two guns. These together became emplacement No. 5, 6, and 7 for the seven-gun series of 10-inch disappearing mounts. There were a number of changes in the plans for magazine of these emplacements. The guns were emplaced along the line of old 1870s Battery 2, for 15-inch Rodmans with intervening traverse magazines. In the new construction these old magazines were retained, but with the intent to convert them to casemated batteries of 5-inch RF guns. A total of eight such emplacements were to be used, and access tunnels were lengthened to use them. While still unarmed in 1898 it was determined that the plan of these casemates resulted in them being potential funnels for incoming shells that could not only knock out these guns but lead to dangerous penetration of the magazines for the 10-inch guns behind. They were ordered eliminated, filled with concrete and earth and the front parapet slope smoothed over. Concrete work for the three 10-inch emplacements was spread out over 1893-97. Transfer came on March 12, 1898. Total cost for all seven emplacements was calculated at $288,985.30. It was named on General Orders No. 20 of January 25, 1906 for Brevet Brigadier General Samuel Spear of Civil War service. It was armed with three 10-inch Model 1888 Watervliet guns on Model 1894 disappearing carriages (#45/#23, #14/#20, and 17/#21). The usual changes were made in 1908-1912 to loading platforms, battery commander's stations and ammunition lifts. The battery was disarmed in 1917 to release guns for railway artillery usage, and not subsequently rearmed. The emplacement was destroyed in the 1960s.

NEW YORK HARBOR

FORT HAMILTON

NORTHERN PART

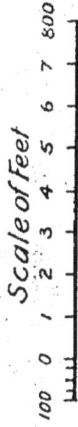

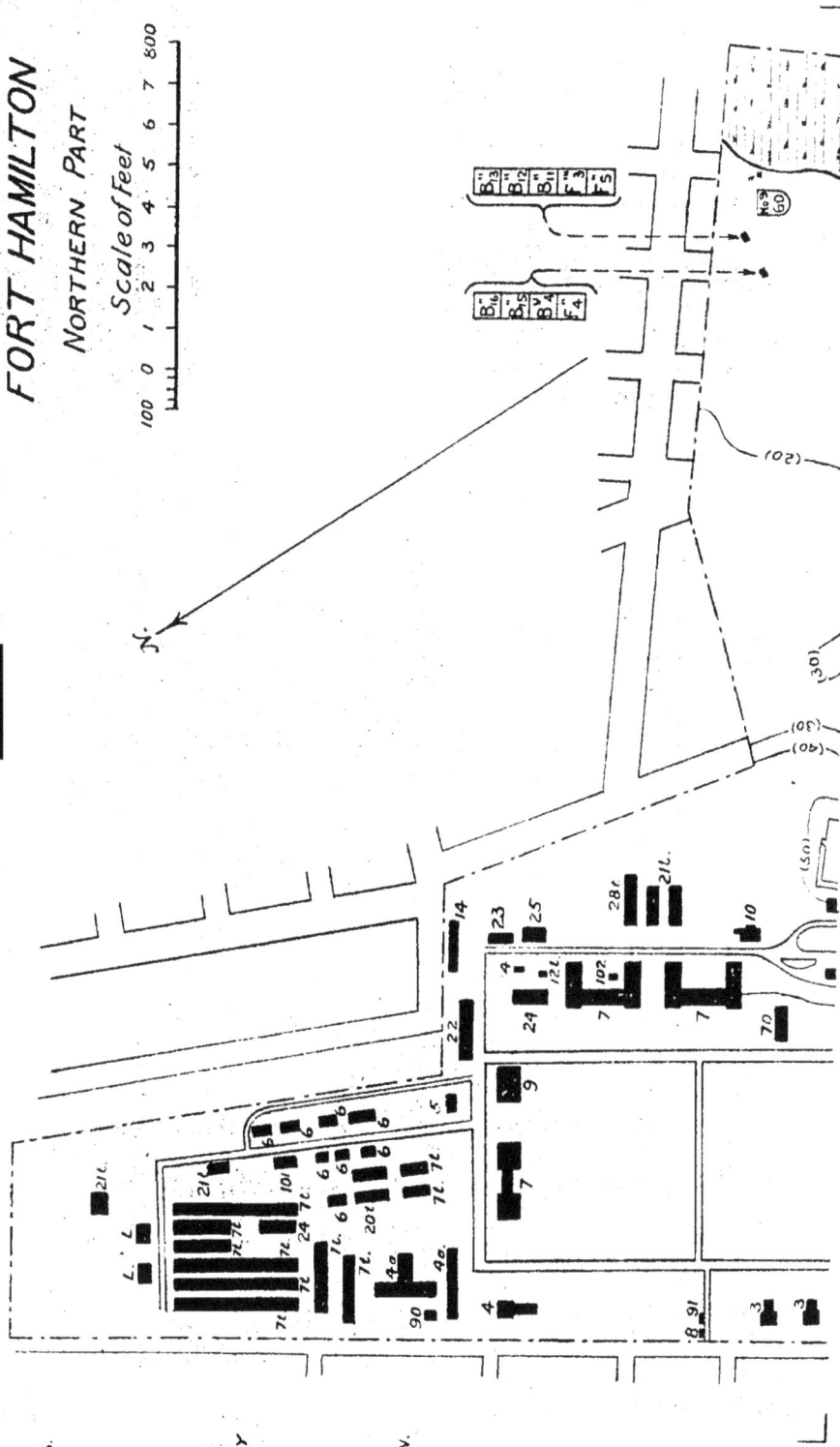

Scale of Feet

100 0 1 2 3 4 5 6 7 800

SERIAL NUMBER 124

EDITION OF NOV. 28, 1921.

LEGEND

3. OFFICERS QRS.
4. HOSPITAL.
4a HOSPITAL PAVILION.
5. HOSPITAL STWDS. QRS.
6. N.C. OFFICERS QRS.
7. BARRACKS.
8. GUARD HOUSE.
9. POST EXCHANGE.
10. BAKERY.
12. DORMITORY.
14. WAGON SHED.
101. GARAGE.
102. HOSE REEL HOUSE.
20. Q.M. AND COMMISSARY
OFFICE & STORE HO.
21. Q.M. STOREHOUSE.
22. FORAGE STOREHOUSE.
23. Q.M. WORK SHOP.
24. Q.M. STABLE.
25. HAY SHED. (Q.M.)
28. Q.M. SALVAGE WARE HO.
70. Y.M.C.A.
90. TRANSFORMER STATION.
91. GATE HOUSE.

NEW YORK HARBOR
FORT HAMILTON
Southern Part

SERIAL NUMBER 103

EDITION OF NOV. 28, 1921.

BATTERIES
PIPER   4-12" M.
DOUBLEDAY 2 12" Dis
BROWN   2-12 "
NEARY   2-12" N.Dis.
GILLMORE 4-10" Dis.
SPEAR
BURKE
LIVINGSTON 2-6" P.
MENDENHALL
JOHNSTON 2-6 "
GRIFFIN   2-3" P.

Griffin Empls. 1 2 vacant.
Livingston Empls 3 4 vacant
A Anti-aircraft gun - 2.3"

LEGEND
1. ADMINIS. BLDG.
2. COMDG. OFFICERS QRS.
3. OFFICERS QRS.
6. N.C OFFICERS QRS.
7. BARRACKS
8 GUARD HOUSE.
11 BAND STAND
12 DORMITORY
13. GREEN HOUSE.
14 WAGON SHED.
15 WOOD SHED.
16 MESS HALL.
17 MANEUVER CAMP SHED
18 FIRE ENGINE HOUSE
19 COAL SHED
100 CREMATORY.
101 GARAGE.
102. HOSE REEL HOUSE.
103. BOAT HOUSE.
20 Q.M. AND COMMISSARY
    OFFICE & STORE HO
21 Q.M. STOREHOUSE.
26. Q.M. SCALES
27. Q.M. SHED
30. ORDNANCE OFFICE
    AND STORE HOUSE.
40. ENGINEER OFFICE.
41 ENGINEER ST. HO.
42 ENGINEER BLACKSMITH
    AND CARPENTER SHOP.
70 Y.M.C.A
71. K OF C

Scale of Feet

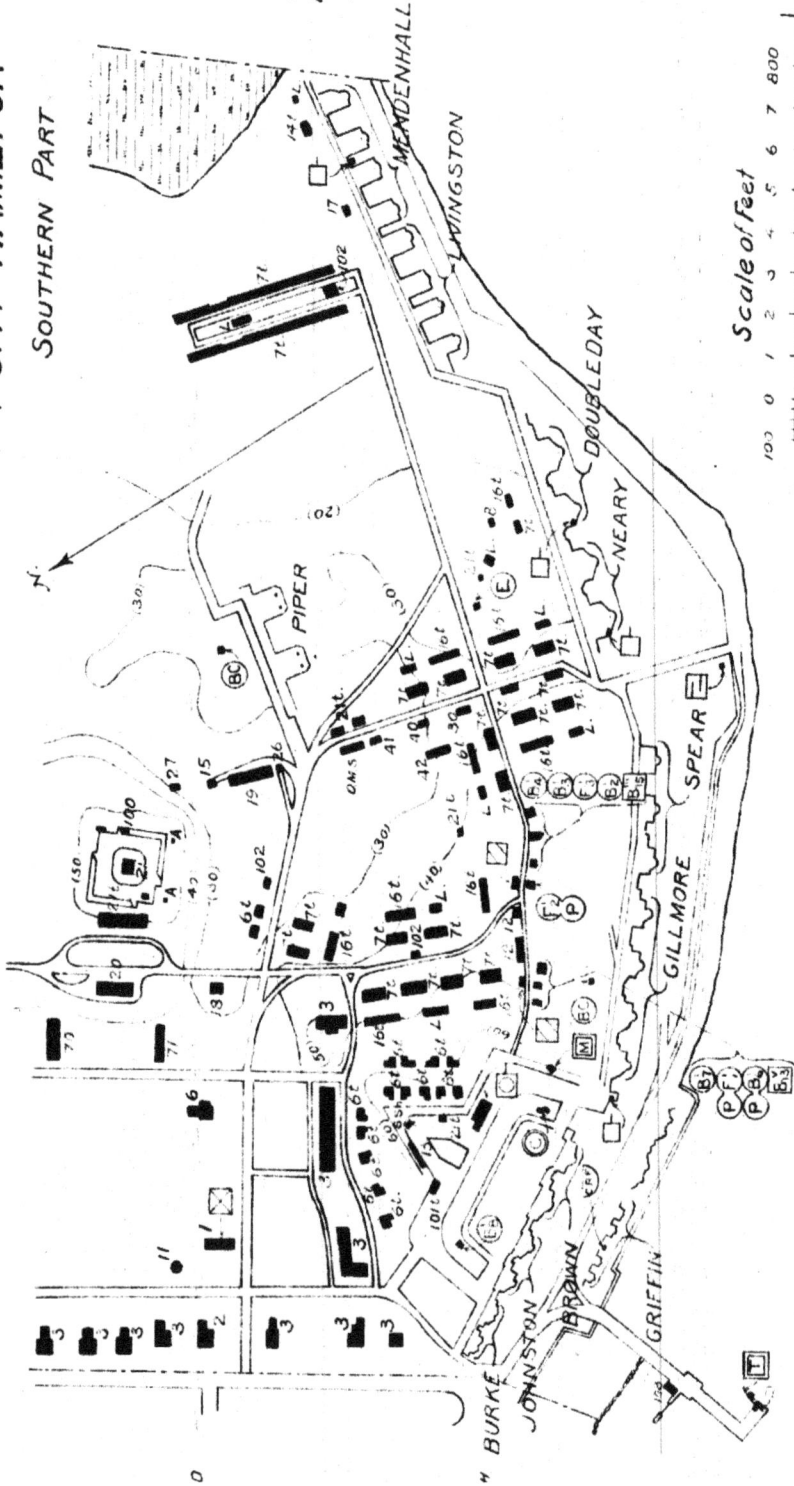

Fort Hamilton 1940 (NARA)

Fort Hamilton 1924 (NARA)

- **GILLMORE:** A battery for four 10-inch guns on disappearing carriages built between the old third system fort and the three guns of Battery Spear. It resulted in a continuous series of seven 10-inch guns built just behind the trace of old 1870s Battery No. 2. One gun plan was submitted in May of 1897 and the last three on March 18, 1898. Work was done in 1897-1899. The battery was of typical mimeograph plans; it was two-story and by this date supplied with ammunition hoists. In 1899 it received a new gallery to connect all the platforms. Also, these were designed for the newer Model 1896 LF carriage, vs. the Model 1894s used in Spear. Transfer was made for these four units on September 26, 1899. Total cost for all seven emplacements of Spear and Gillmore was calculated at $288,985.30. It was armed with two 10-inch Model 1895 guns on Model 1896 disappearing carriages (emplacements No.1 and No. 2 Watervliet tube #6/#58 and Watervliet tube #4/#62) and two 10-inch Model 1888 guns on Model 1896 disappearing carriages (emplacements No. 3 and No. 4 Bethlehem tube #8/#31 and Watervliet tube #38/#14). The battery received modifications to its communications, hoists, and BC station before World War I. The armament served until authorized for removal on November 12, 1942—though the guns weren't actually dismounted until 1944. The emplacement was destroyed in the 1960s.

- **PIPER:** The mortar battery built at Fort Hamilton. Plans were submitted for a typical 8-gun battery of two pits on June 7, 1898. Funds for the battery were obtained from the Fortification Act of May 7, 1898. Construction was done in 1898-1899 for transfer on March 4, 1901 at a cost of $100,000. It was armed with eight Model 1890M1 steel mortars on Model 1896M1 carriages (Watervliet mortar tubes #65, #89, #93, #100, #101, #102, #103 and Builders mortar tube #46 with carriages #223, #224, #225, #226, #227, #201, #202, and #203). The battery was named on General Orders No. 30 of March 19, 1902 for Colonel Alexander Piper of Civil War service. In 1917 four mortars and carriages were removed from the battery and transferred to a new temporary emplacement at Fort Tilden. The final armament of four mortars was authorized for removal in 1942. The emplacement itself was destroyed in the 1960s for development at the post.

- **BROWN:** A battery for two 12-inch disappearing guns built along the southwestern portion of old Fort Hamilton's seafront curtain wall. Engineering construction plans were submitted on May 20, 1899. Work commenced almost immediately. The plans closely follow proscribed mimeograph design without any major changes. Work was done in 1899-1900. Transfer was made on December 2, 1902 for a construction cost of $94,000. It was named on General Orders No. 30 of March 19, 1902 for Brevet Major General Harvey Brown of Mexican and Civil War service. The battery carried two 12-inch Model 1895M1 guns on Model LF 1897 disappearing carriages (Bethlehem tube #2/on carriage #29 and Watervliet tube #32/on carriage #28). Like Doubleday this armament was removed and replaced with older Model 1888 guns in 1917-18 (receiving Watervliet tubes M1888 1½ #36 and #34 on remaining carriages #29 and #28). These guns were transferred here from Baltimore's Battery Stricker at Fort Howard. Authorization for final removal came on October 13, 1942. The emplacement was destroyed in the 1960s to make room for more military housing.

- **DOUBLEDAY:** A battery for two 12-inch disappearing guns built 1400-feet to the east of old Fort Hamilton, still following the coastal bluff line. It was built contiguous to Battery Neary of two 12-inch barbette guns to its immediate right (western) flank. The construction plan was submitted on June 23, 1898 with an estimated cost of $100,000. Work began quickly and was reported complete on November 2, 1900. The design plan closely followed current engineer guidelines. It was transferred on November 20, 1900 for a cost of $91,500. It was named on General Orders No. 43 on April 4, 1900 for Major General Abner Doubleday of Civil War service. It was armed with two 12-inch Model 1895 on disappearing carriages Model 1897 (Watervliet gun tubes #17/

on carriage #12 and tube #1/on carriage #16). The battery received changes in its platforms and battery commander stations around 1910. In 1916-18 the gun carriages were modified in place to allow higher elevation and thus range. These guns were removed in late 1917 or early 1918 and soon replaced with Watervliet tubes Model 1888 (#37 and #38 previously at Battery Read, Fort Dupont). This new armament was retained under the 1932 Review and not finally authorized for removal until October 23, 1943. The emplacement was destroyed in the 1960s when Fort Hamilton expanded its military housing.

- **NEARY**: A battery for two 12-inch barbette guns built 1200-feet to the east of old Fort Hamilton, still following the coastal bluff line. It was built contiguous to Battery Doubleday of two 12-inch disappearing guns to its immediate left (eastern) flank. The battery plan was submitted on May 23, 1898. Although for barbette guns, it closely followed the design characteristics then current with the engineers for 12-inch disappearing gun emplacements. With the urgency of the Spanish American War, the platform was reported available for armament mounting on July 27, 1898. The emplacement was transferred on November 20, 1900 for a cost of $87,635.37. It was armed with two 12-inch Model 1888MII Watervliet guns on Model 1892 barbette carriages (#32/#19 and #46/#20). It was named on General Orders No. 20 of January 25, 1906 for 1st Lieutenant William C. Neary. This name had been used previously for a couple of years for a single gun 8-inch battery at Fort Colombia, Washington. Changes were made in the early 20th century to hoists, battery commander station, and commander's walk. The battery was retained for some time but then recommended for deletion in the 1932 Review. That was accomplished and the armament scrapped in 1936-1938. The emplacement was destroyed in the 1960s with the expansion of the Fort Hamilton military housing.

- **JOHNSTON**: A battery for two 6-inch disappearing guns built next to Battery Brown at the old seacoast curtain wall of Fort Hamilton. The battery design was submitted by local engineers on March 8, 1899. Work was done in 1899-1900. Transfer was made on December 2, 1902 for a cost of $49,500. The battery was named on General Orders No. 20 on January 25, 1906 for Captain Abraham Johnston of Mexican War service in 1846. It was armed with two 6-inch Model 1897M1 guns on Model 1898 disappearing carriages (#1/#27 and #9/#1). The battery was retained through World War I. It was finally deleted and the guns scrapped under authority of October 23, 1943. Like all emplacements at Fort Hamilton, it was destroyed in the 1960s.

- **BURKE**: A battery for four 6-inch rapid-fire pedestal guns built to the west and adjacent to Battery Johnston along the line of the western curtain wall of old Fort Hamilton. The original battery plan, the first such designed and built for the Model 1900 pedestal gun, was submitted on April 3, 1901. It had two wide platforms each to be armed with two, side-by-side 6-inch guns. Magazines were below in the central traverse between the platforms. On May 20, 1902 the platforms were ready for their guns. However slightly before arming, it was decided that the two guns would potentially interfere too much with each other, and plans changed to requiring just a single gun per platform. It was too late to modify the battery's construction, so the emplacement only received a single gun to put onto each platform. As it turned out the "extra" guns and pedestals already assigned to Fort Hamilton were diverted and used to arm part of Battery Livingston. Transfer was made on September 14, 1903 for a construction cost of $38,965.85. It was named for Lt. Colonel Martin Burke of Civil War service on General Orders No. 107 of October 24, 1902. Its final armament was two 6-inch Model 1900 guns and pedestal mountss (#24/#20 and #25/#21). In 1917 this armament was removed and sent to arm a temporary emplacement on the beach at Fort Tilden. The empty emplacement did not serve for armament purposes again and was destroyed in the 1960s.

- **MENDENHALL:** A battery for four 6-inch Model 1903 disappearing guns built near the eastern limit of the Fort Hamilton reservation. It was immediately adjacent to Battery Livingston (on its western flank) and had a field of fire directly to the south. The plan for eight 6-inch guns (what became batteries Mendenhall and Livingston) was submitted on May 4, 1903. The first to be started would become Battery Mendenhall. Work was undertaken in 1903-1905 and reported completed on November 16, 1905. It was transferred along with Livingston on November 23, 1905 for a combined cost of $135,000. The battery was named on General Orders No. 194 of December 27, 1904 for Colonel John Mendenhall of Civil War service. It was armed in 1906 with four 6-inch Model 1903 guns and disappearing carriages (#43/#3, #44/#4, #49/#5 and #55/#6). The gun tubes were removed in 1917-18 for use on wheeled field mounts for European service. The carriages were left in place until scrapped in 1921. The emplacement itself was destroyed for new military housing in the early 1960s.

- **LIVINGSTON:** A battery for four 6-inch guns (two disappearing and two pedestal) built on the eastern portion of the Fort Hamilton reservation. Battery Mendenhall was on its immediate eastern (left flank) and with Battery Doubleday not far along the hill line to the west. The plan for eight 6-inch guns (what became batteries Mendenhall and Livingston) was submitted on May 4, 1903. The second battery, which would become Livingston, was allocated two 6-inch Model 1903 guns and disappearing carriages, but the guns in emplacements No. 1 and 2 would be the 6-inch Model 1900 guns and pedestals recently released from Battery Burke. Work was undertaken in 1903-1905 and reported completed on November 16, 1905. It was transferred along with Livingston on November 23, 1905 for a combined cost of $135,000. It was named on General Orders No. 194 of December 27, 1904 for Captain Manning Livingston of Civil War service. It was armed with two 6-inch Model 1900 guns and pedestals (#22/#22 and #23/#23) and two 6-inch guns Model 1903 on Model 1903 disappearing carriages (#21/#1 and #33/#2). The two disappearing guns were removed after a very short time. They were taken out in November 1907 and sent to West Point to be installed in an instructional battery. However, the pedestal guns had a long service life, not being removed until 1948 and sent to Battery Peck at Fort Hancock. The emplacement itself was destroyed in the 1960s.

- **GRIFFIN:** The only battery for light, rapid-fire guns at Fort Hamilton. It had a mixed armament of both Armstrong 4.7-inch and 3-inch guns. It was located on the old 1870s Rodman battery part way down the slope in front of the old third system work. Work was begun in 1898 with National Defense Act funds to emplace two 4.7-inch rapid-fire guns on temporary platforms. They were installed on two simple concrete circular platforms of 17-foot diameter. In 1899 authority was granted to add two 3-inch Model 1898 masking parapet guns, and then finally on July 16, 1902 a plan was submitted for two 3-inch pedestal guns. When completed the two 4.7-inch Armstrong guns and pedestals (#10999/#9083 and #11000/#9031) were transferred on December 1, 1899 for $2290.45. The two 3-inch M1898 masking parapet guns and mounts (#87/#87 and #88/#88) were transferred on September 10, 1902 for $5956.17. And then finally the two 3-inch Model 1903 guns and pedestal mounts (#26/#43 and #32/#44) were transferred on February 4, 1903 for $6630.27. As arranged the first two (right flank) were the M1898s, then the M1903s, then on the left flanks the two 4.7s. The battery of six guns was named on General Orders No. 107 on October 24, 1902 for Colonel Charles Griffin of Civil War service. The battery was disarmed gradually. The two 4.72-inch guns were removed in 1913 and sent for the land defense program in Oahu. The two 3-inch masking parapet guns and pillars were removed when this type of gun was declared obsolete in 1920. The two pedestal 3-inch guns served through World War II, serving as Tac-16 in the 1940s. They weren't removed until 1946. Most of the battery was destroyed in the 1960s, but unlike most of the other emplacements at Fort Hamilton, there are a few physical remains at the battery site. The battery site is open to the public, but the fort has restricted access.

Fort Wadsworth (1847-1948) is located on the western side of the entrance (also known as the "Narrows") to Upper New York Bay on Staten Island. Fortified since the Colonial Period, the heights overlooking the Narrows was one of the most important points in the defense of New York Harbor. The State of New York undertook the building replacement forts for the works that were located at this site during Colonial and Revolutionary War periods. These defenses were known as Fort Richmond, Fort Tompkins, Fort Morton, and Fort Hudson allowing up to 164 cannons to be installed. By 1835, these fortifications were in ruins, so the federal government undertook total reconstruction in 1847. Two Third System granite forts, Fort Tompkins and Fort Richmond, were constructed. Fort Tompkins commanded the top of heights and provide a land defense, while Fort Richmond (later named Battery Weed) was at the water's edge and was design solely for defense against naval attacks. Several 15-inch Rodman batteries were built during the 1870s. Construction of twelve Endicott Program batteries from 1899 to 1904 saw the entire military reservation be named Fort Wadsworth, with 40 breech loading guns ranging from 12-inch to 3-inch. The new construction included new garrison buildings and a mine facility. The post was formally named on General Orders 16 of 1902 after Brig. Gen. James S. Wadsworth, who was KIA in the Battle of the Wilderness on May 6, 1864. Fort Wadsworth served as a coast artillery and an infantry post through the end of World War II. A new 6-inch battery (#218) was built during World War II while a planned 16-inch casemated battery was cancelled. AMTB batteries were installed nearby at Miller Field and Swinburne Island. After World War II, Fort Wadsworth served various U.S. Army requirements (air defense during the 1950s) until it was transferred to U.S. Navy as the Naval Station New York in 1979. As part of this plan much of the reservation received new housing and many of the Endicott Program batteries were partially buried. Parts of the fort was also affected by the construction of the Verrazano Narrows bridge in the 1960s. In 1995 large sections of Fort Wadsworth were transferred to the National Park Service as part of the Gateway National Recreation Area. Other sections of the reservation are still used for military housing, an U.S. Army Reserve center, and the US Coast Guard. The NPS has a large visitors center and provides a number of guided tours of the Third System works. A number of the Endicott batteries are partially buried and fenced off, or heavily overgrown.

Fort Wadsworth Gun Batteries

- **DUANE**: The first modern Endicott battery for Fort Wadsworth was this battery ultimately having emplacements for five 8-inch disappearing guns. Plans for this battery just south of old Fort Tompkins were cast as early as 1890. At this time the fort reservation was quite small, private property to the west and south had not yet been acquired. Work was done intermittently between 1891 and 1894. On February 8, 1893 four positions were reported as completed (except for gun base rings as a type had not yet been adopted), the fifth and final unit was being planned. The fifth emplacement was designed somewhat differently, with its magazines moved back as far as possible. Submission of this final emplacement was made on April 18, 1894. By this time the first Model 1894 disappearing carriage was adopted, and the battery's emplacements were finished for this carriage. Transfer was made on June 16, 1896 for a cost of $267,513. It was named on General Orders No. 43 of April 4, 1900 for Brigadier General James C. Duane, former Chief of Engineers, U.S. Army. The battery was armed with five 8-inch Model 1888 Watervliet guns on disappearing carriages Model 1894 (#5/#2, #7/#3, #17/#1, #14/#4, and #24/#53). Perhaps an indication of the early use of natural Rosendale cement, the battery was plagued with structural defects and water problems. Parts began to crumble almost as soon as constructed. In 1903 it was recommended to rebuild the

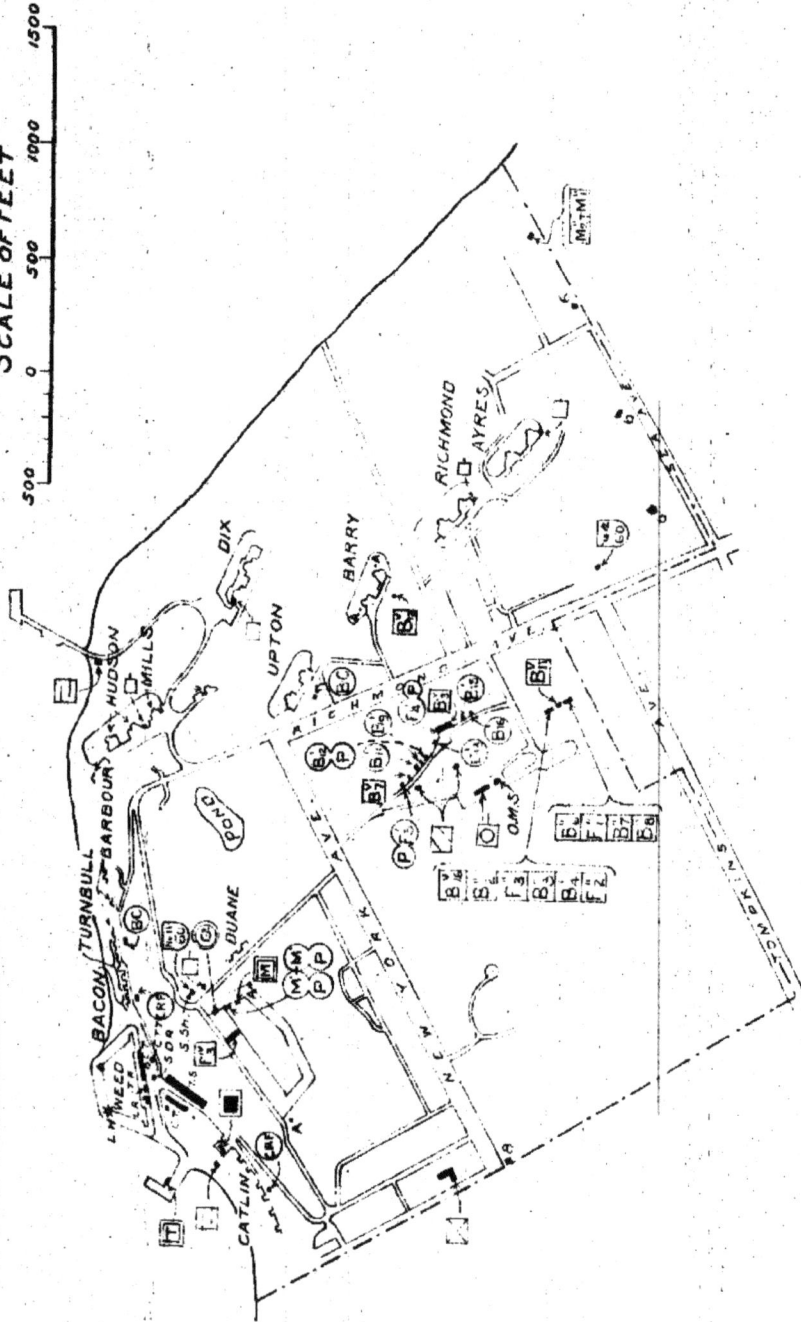

NEW YORK HARBOR

FORT WADSWORTH

GENERAL MAP

SCALE OF FEET

SERIAL NUMBER 103

TRUE MERIDIAN.

LEGEND

6 N.C.O.QUARTERS.

8 GUARD HOUSE.

EDITION OF AUG. 27, 1921.
REVISIONS: NOV. 26, 1921.

BATTERIES

RICHMOND ... 2-12" Dis.
AYRES .... 2-12"
DIX .... 2-12"
HUDSON ...... 2-12"
UPTON .... 2-10"
BARRY ....
DUANE ....
* MILLS .... 1-6" Dis.

† BARBOUR ... 1-4.7" P.

TURNBULL ... 6-3" P.
CATLIN ...... 6-3" P.
BACON ....

A-Anti-aircraft gun 4-3"

* Emplacement No 2 vacant.
† Emplacements 2-4 vacant.

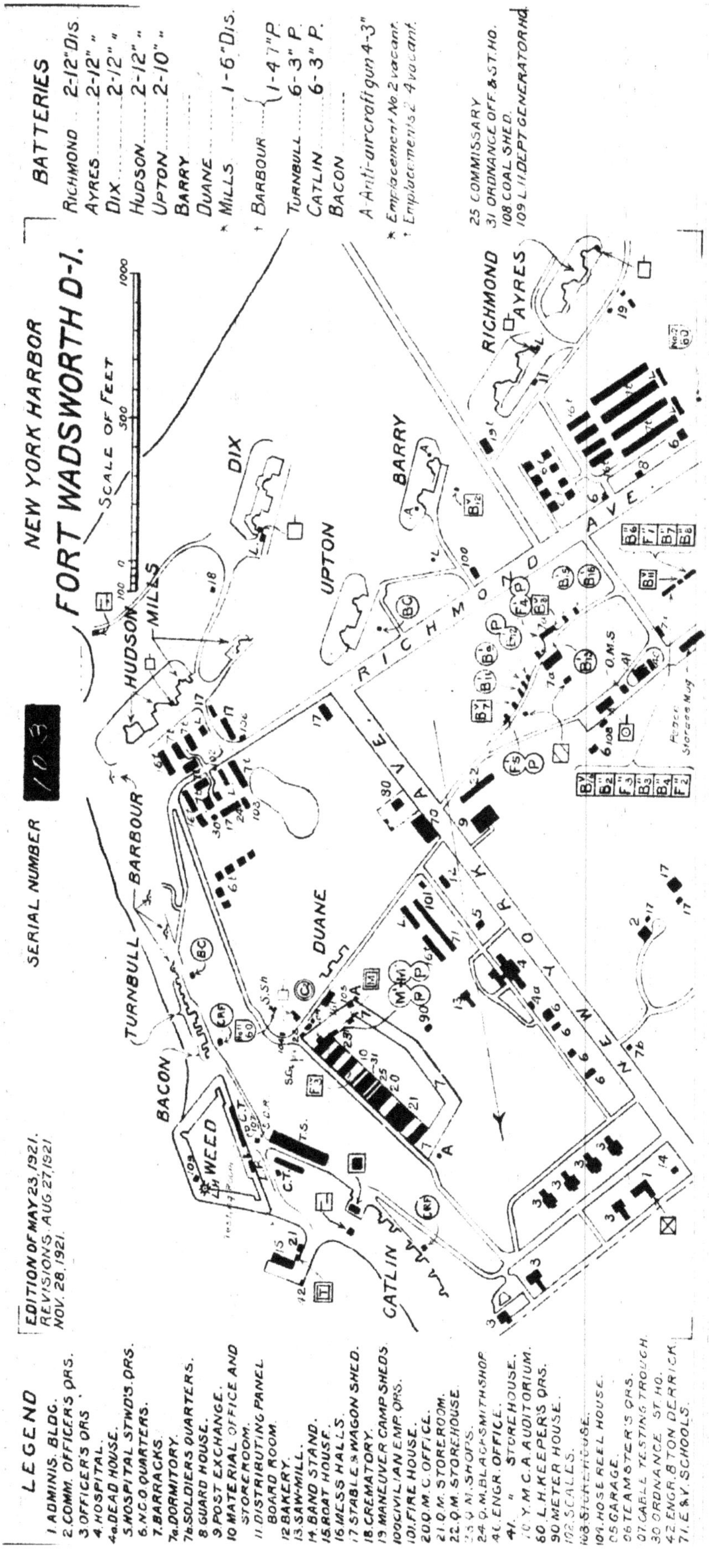

SERIAL NUMBER 103
NEW YORK HARBOR
FORT WADSWORTH D-1.
SCALE OF FEET
100 0 500 1000
EDITION OF MAY 23, 1921.
REVISIONS. AUG 27, 1921
NOV. 28, 1921.
BATTERIES
RICHMOND 2-12" DIS.
AYRES 2-12" "
DIX 2-12" "
HUDSON 2-12" "
UPTON 2-10" "
BARRY
DUANE
* MILLS 1-6" DIS.
† BARBOUR 1-4.7" P.
TURNBULL 6-3" P.
CATLIN 6-3" P.
BACON
A-Anti-aircraft gun 4-3"
* Emplacement No 2 vacant.
† Emplacements 2 4 vacant.
25 COMMISSARY
31 ORDNANCE OFF. & ST. HO.
108 COAL SHED.
109 L.I. DEPT GENERATOR HO.
LEGEND
1 ADMINIS. BLDG.
2 COMM. OFFICERS QRS.
3 OFFICER'S QRS.
4 HOSPITAL.
4a DEAD HOUSE.
5 HOSPITAL STWD'S. QRS.
6 N.C.O. QUARTERS.
7 BARRACKS.
7a DORMITORY.
7b SOLDIERS QUARTERS.
8 GUARD HOUSE.
9 POST EXCHANGE.
10 MATERIAL OFFICE AND STOREROOM.
11 DISTRIBUTING PANEL BOARD ROOM.
12 BAKERY.
13 SAW-MILL.
14 BAND STAND.
15 BOAT HOUSE.
16 MESS HALLS.
17 STABLE & WAGON SHED.
18 CREMATORY.
19 MANEUVER CAMP SHEDS.
100 CIVILIAN EMP. QRS.
101 FIRE HOUSE.
20 Q.M.C. OFFICE.
21 Q.M. STOREROOM.
22 Q.M. STOREHOUSE.
23 Q.M. SHOPS.
24 Q.M. BLACKSMITH SHOP
40 ENGR. OFFICE.
41 " STOREHOUSE.
70 Y.M.C.A. AUDITORIUM.
60 L.H. KEEPER'S QRS.
90 METER HOUSE.
102 SCALES.
103 STOREHOUSE.
104 HOSE REEL HOUSE.
105 GARAGE.
106 TEAMSTER'S QRS.
107 CABLE TESTING TROUGH.
30 ORDNANCE ST. HO.
42 ENGR. 8 TON DERRICK.
71 E.&Y. SCHOOLS.
RICHMOND AVE.
NEW YORK AVE.
WEED
CATLIN
BACON
TURNBULL
BARBOUR
HUDSON
MILLS
DIX
UPTON
BARRY
DUANE
RICHMOND
AYRES
O.M.S.

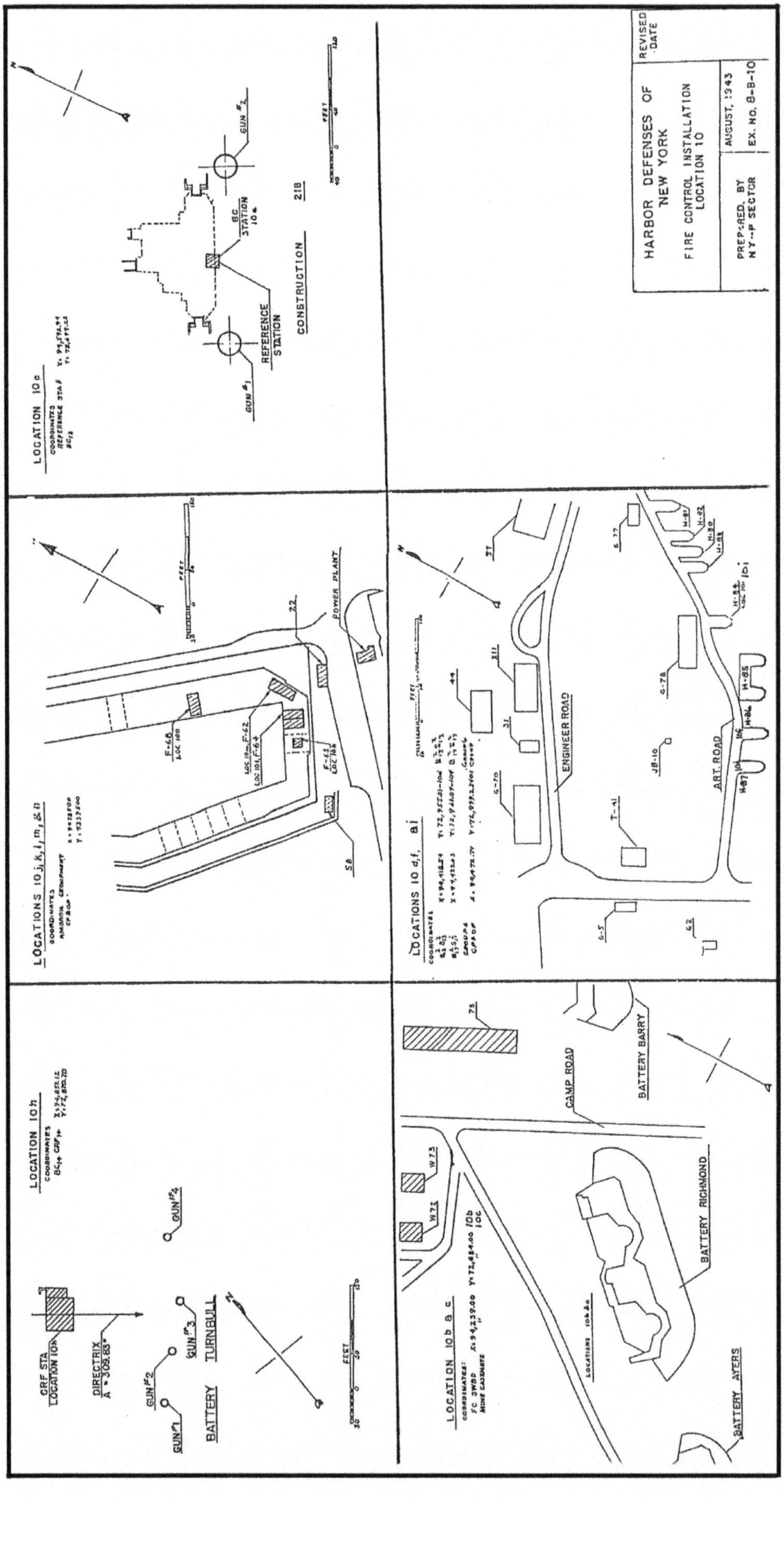
HARBOR DEFENSES OF NEW YORK
FIRE CONTROL INSTALLATION LOCATION 10
PREPARED BY NY-P SECTOR
AUGUST, 1943
EX. NO. 8-B-10
REVISED DATE
LOCATION 10a
REFERENCE STATION
BC STATION 10a
CONSTRUCTION
218
GUN #1
GUN #2
LOCATIONS 10 j, k, l, m, & n
POWER PLANT
LOCATIONS 10 d, f, & i
ENGINEER ROAD
ART. ROAD
LOCATION 10h
GRF STA LOCATION 10h
DIRECTRIX
BATTERY TURNBULL
GUN #3
GUN #4
LOCATION 10b & c
CAMP ROAD
BATTERY BARRY
BATTERY RICHMOND
BATTERY AYERS

FT. WADSWORTH

(0104-870N-14)(6-6-27-11A)(20-2100) Fort Wadsworth, Staten Id., N.Y.

Fort Wadsworth 1927 (NARA)

Fort Wadsworth Unit, Gateway National Recreation Area (Terry McGovern)

Fort Wadsworth Unit, Gateway National Recreation Area (Terry McGovern)

battery; but by then the larger, better-built disappearing guns were installed at the post and Duane's importance had diminished. Three guns and carriages were removed in 1908 and sent for mounting in new Battery Arrowsmith at Fort Hancock. The final two guns (in the latest, best-condition emplacements) served until removed in November 1915. Badly damaged, with significant portions destroyed, the battery remains still exist on the grounds of the Gateway National Recreation Area's Fort Wadsworth unit. The battery site is open to the public.

- **BARRY**: The first of two dual gun 10-inch batteries placed in close proximity to the west of Fort Tompkins on what was then called the Fort Newton reservation (later combined with the rest of Fort Wadsworth). Submission of the plan was made on July 6, 1896. The battery was located on newly purchased land just south of Richmond Avenue. It followed current suggested mimeograph plans closely for this type. Work was done in 1896-1897, Transfer was made on December 18, 1897 for a cost of $61,498.67. The battery was named on General Orders No. 16 of February 14, 1902 for Colonel William F. Barry of the 2nd U.S. Artillery, Army of the Potomac during the Civil War. The battery was armed with two 10-inch Model 1888 Watervliet guns on Model LF 1896 disappearing carriages (#23/#21 and #24/#22). These guns were removed temporarily in 1909 to allow modifications to the platforms and hoists at the battery. A new BC station was also added before World War I. The armament was removed under orders of 1917 in May of 1918 for use on railway carriages. The battery was not subsequently used for armament. The emplacement still exists, heavily overgrown, at the Gateway National Recreation Area's Fort Wadsworth unit. The battery is closed to the public.

- **UPTON**: The second of the two similar dual gun 10-inch batteries placed in close proximity to the west of Fort Tompkins on what was then called the Fort Newton reservation (later combined with the rest of Fort Wadsworth). Submission of the plan was made on July 6, 1896 along with Battery Barry. The battery was located on newly purchased land just south of Richmond Avenue. It followed current suggested mimeograph plans closely for this type. Work was quickly undertaken and as soon as April 1897 the platforms were reported ready for their carriages. It was transferred on December 18, 1897 at a cost of $61,499. It was named on General Orders No. 16 of February 14, 1902 for Colonel Emory Upton, U.S. Artillery officer. The battery was armed with two Model 1888 10-inch Watervliet guns on Model 1896 disappearing carriages (#11/#23 and #19/#9). Extensive repairs and modifications were made in the ten years after completion; including widened loading platforms, a battery commander station, and removal of powder hoists. In 1918 one-gun tube (#19) was removed and replaced with Watervliet 10-inch tube #45 moved here from Battery Spear at Fort Hamilton. The battery continued to serve throughout the 1930s, though at times at a reduced status or even "out of service". Authority to abandon and scrap the armament came ultimately in 1943 through authority of November 12, 1942. The emplacement still exists, though at times heavily overgrown at the Gateway National Recreation Area's Fort Wadsworth unit. The battery is closed to the public.

- **RICHMOND**: This was the first 12-inch gun battery emplaced at Fort Wadsworth. Plans were submitted on March 22, 1898 for a location to the west of old Fort Tompkins on the recently acquired King and Martin properties, to the southwest of recently built Battery Upton and Barry. Work was done in 1898-1899. Transfer was made on September 5, 1899 at a cost of $96,992. The battery was named on General Orders No. 43 on April 4, 1900 for old Fort Richmond located nearby. The battery was of standard design layout. It was armed with two 12-inch guns Model 1888MII on Model 1896 disappearing carriages (Bethlehem Steel tube #5/carriage #26 and Watervliet tube #28/carriage #10). The battery received the usual modifications to loading platforms, hoists, battery commander station and walk in its early years. The battery had a long service life, and was not disarmed during World War I. In 1916-17 modifications were made with the carriages to permit 5-degrees of increased elevation. Authority for removal was given on November 12, 1942, and the armament probably removed in 1943 or even early 1944. Mostly now buried and covered over, the emplacement still exists at the Gateway National Recreation Area's Fort Wadsworth unit. The battery is closed to the public.

- **HUDSON**: The second dual 12-inch disappearing battery emplaced at Fort Wadsworth. Emplacement plans for a battery using two Model 1896 disappearing carriages west of Battery Tompkins were submitted on June 10, 1898. The immediately available plant and contract labor allowed the work to start right away. The No. 2 emplacement of 6-inch Battery Mills was placed adjacent on the emplacement's right flank. The battery was transferred on August 18, 1900 for a cost of $99,463. It was named on General Orders No. 43 on April 4, 1900 for the earthen battery at the site preceding the Endicott work. Original armament was two 12-inch guns Model 1888 on Model 1896 disappearing carriages (Bethlehem tube #3/carriage #11 and Watervliet tube #20/carriage #21). Gun tube #3 was soon shifted to a new battery in San Francisco, being replaced with 12-inch Model 1895 Watervliet gun #41. Then even this tube was removed in 1918 and replaced with 12-inch Model 1888 Watervliet gun #21. The battery received the usual modifications to loading platforms, hoists, battery commander station and walk in its early years. In 1916-17 modifications were made with the carriages to permit 5-degrees of increased elevation. The final armament served until deleted under authority of November 12, 1942. Scrapping was accomplished in 1944. Partially buried (the lower magazine levels), the battery still exists at the Gateway National Recreation Area's Fort Wadsworth unit. The battery is closed to the public.

- **AYRES**: This was the third two-gun disappearing guns for 12-inch guns emplaced on the recently acquired land to the west and south of old Fort Tompkins. Plans were submitted on June 28, 1900 using funds from the Fortification Act of May 25, 1900. The plans followed standardized mimeographs, but with the addition of ammunition truck recesses and also ready ammunition niches to hold 12 rounds near the loading platform. Work was done in 1900-1901, for transfer on July 30, 1902 for $87,000. The battery was initially armed with two 12-inch guns Model 1895 on carriages Model 1897 (Bethlehem tube #1/carriage #24 and Watervliet tube #19/carriage #25). It was named on General Orders No. 16 of February 14, 1902 for Colonel Romeyn Ayres, U.S. Volunteers. In 1916-17 modifications were made with the carriages to permit 5-degrees of increased elevation. The two Model 1895 gun tubes were removed in 1918 for use on railway carriages but soon replaced with two 12-inch Model 1888 tubes formerly from Battery Towson in Baltimore on July 15, 1918 (#26 and #45). Ayres retained this armament until authority for removal on November 12, 1942. Now partially buried the battery still exists at the Gateway National Recreation Area's Fort Wadsworth unit. The battery is closed to the public.

- **DIX**: The final large disappearing gun battery intended for Fort Wadsworth. Emplacement plans were submitted on August 3, 1901, but were changed on April 18th, 1902 to use the newest Model 1901 12-inch disappearing carriage. Work was soon undertaken, and the battery was reported completed on June 6, 1904. It was transferred on June 17, 1904 for a construction cost of $100,000. The battery was named on General Orders No. 78 of May 25, 1903 for Major General John Dix of Civil War service. The battery was known while building as Battery Hudson No. 2. When completed it was armed with two 12-inch Model 1900 Watervliet guns on Model 1901 disappearing carriages (#6/#5 and #9/#6). Unlike many other emplacements with this type of gun, they were retained and never replaced. In 1916-17 modifications were made with the carriages to permit an additional 5-degrees of increased elevation. The battery, as the most modern heavy unit in Wadsworth's defenses, served actively between the wars. It was authorized for deletion on October 25, 1943 and the armament removed in 1944. The battery still exists in relatively good condition at the Gateway National Recreation Area's Fort Wadsworth unit. The battery is closed to the public.

- **BARBOUR:** A battery for a mixed armament of British 4.7-inch and 6-inch rapid-fire guns emplaced as part of the emergency defenses during the Spanish American War. Local engineers were told to select locations for purchased British RF guns on March 19, 1898. Utilizing platforms and magazines from the numerous 15-inch Rodman positions built in the 1870s, two emplacements for 4.7-inch guns were built in April, and two for 6-inch guns in September. These were basically spread among the old emplacements of the South Cliff Battery. From north to south were: oen 6-inch at old emplacement No. 8 (Vickers #12135/#11160), one 4.7-inch at South Cliff emplacement No. 5 (Armstrong #9721/#10840), one 6-inch at South Cliff emplacement No. 1 (Vickers #12136/#11161) and one 4.7-inch at the last South Cliff emplacement next to Battery Hudson (Armstrong #9720/#10837). All four emplacements were transferred on December 20, 1899 for a cost of $3,802 (for the two 6-inch) and $2,681.49 (for each of the two 4.7-inch). It was named on General Orders No. 78 of May 25, 1903 for Captain Philip Barbour killed in action at Monterrey, Mexico in 1846. In 1909 two guns were dismounted and sent to Sandy Hook for tests involving new ammunition types. All of these types of guns used were declared obsolete in 1919, and the remaining guns dismounted and abandoned in 1920. While at times difficult to identify, most of the emplacement blocks and magazines still exist at the Fort Wadsworth Unit of the Gateway National Recreation Area. The battery site is open to the public.

- **MILLS:** A battery for two 6-inch disappearing guns built next to 12-inch Battery Hudson at the Fort Wadsworth reservation. Plans were finalized and submitted on February 28, 1899 for 6-inch guns on Model 1898 disappearing carriages. While the preferred location was on the right flank of Battery Hudson, the location of the access road here forced the engineers to build two separate, single-gun emplacements—one on each side of the road. The two emplacements were about 200-feet apart. Work was done in 1899, and transfer was made on October 26, 1900 for a cost of $52,700. It was named on General Orders No. 20 of January 25, 1906 for Brevet Colonel Charles J. Mills killed in action at Hatcher's Run, VA in 1865. The battery was armed with two 6-inch guns Model 1897M1 on disappearing carriages Model 1898 (#7/#28 and #8/29). The gun in the emplacement adjacent to Hudson (#8/#29) was removed in 1917-1918. The second unit continued to serve between the wars and was not finally authorized for removal until October 23, 1943. The emplacement still exists, though parts of it are buried, at the Fort Wadsworth Unit of the Gateway National Recreation Area. The battery is closed to the public.

- **BACON:** A battery for two 3-inch, 15-pounder masking parapet guns located in the old South Cliff Battery. This placed it just south of Battery Weed and just north of what would become Battery Turnbull. Using funds from the Fortification Act of July 7, 1898 the battery plan was submitted on January 12, 1899. It was of very simple, conventional two-gun plan, really just gun blocks with foundation and a single, lower-level magazine. It was transferred on March 12, 1903 for a cost of $4,000. The battery was armed with two 3-inch Model 1898 guns and masking parapet mounts (Driggs-Seabury #91/#91 and #94/#94). It was named on General Orders No, 78 of May 25, 1903 for 1st Lieutenant John Bacon who was mortally wounded at Churubusco Mexico in 1847. These were modified in 1917 to become fixed pedestal mounts M1898M1. This armament was removed in 1920 in common with all of this gun type recently declared obsolete. A CRF station was built on the right rear flank in 1919 right before abandonment. The emplacement still exists on the Fort Wadsworth Unit of the Gateway National Recreation Area. The battery site is open to the public.

- **TURNBULL**: An emplacement for six 3-inch pedestal type guns built in a section of the old 1870s South Cliff Battery. Plans for the battery were submitted on August 23, 1902. The battery followed typical mimeograph-suggested plans. Four emplacements (No. 1 to No. 4 on the right flank) were built together with gun centers of 35-feet distance. Two additional emplacements (No. 5 and 6) were built together to the left after skipping an intervening old 1870s magazine. The latter two had 28-foot gun centers. Work was done 1902-1903, with transfer being made on November 19, 1903 at a construction cost of $30,000. It was named on General Orders No. 194 of December 27, 1904 for Brevet Colonel William Turnbull who served with distinction during the Mexican War. The battery was armed originally with six 3-inch Model 1902 guns and pedestal mounts (#30, #31, #32, #33, #34, and #35). A CRF station was added and placed on the rise behind the battery in 1920. Sometime between 1915 and 1928 these were exchanged with the Model 1903 guns at the fort's Battery Caitlin, ending up with serials #60/#37, #66/#38, #67/#39, #68/#40, #61/#41 and #82/#42. These guns were then retained until the early 1940s. In 1942 the armament was taken to provide guns for various temporary batteries at the Delaware River and here at Fort Wadsworth. The emplacement still exists at the Fort Wadsworth Unit of the Gateway National Recreation Area. The battery site is open to the public.

- **CATLIN**: Another 6-gun, 3-inch battery assigned to Fort Wadsworth. It was located among the platforms of the old North Cliff Battery to the north of old masonry Battery Weed. It was planned in sections, two guns were proposed on June 30, 1903, and another four on July 23rd of the same year, Work was done in 1902-03, all six platforms being reported ready for mounts on December 9, 1903. They were arranged in three two-gun paired emplacements, joined by tunnel access to each other. The battery was transferred on July 22, 1904 for a cost of $37,500. It was named on General Orders No. 194 of December 27, 1904 for Captain Robert Catlin of Civil War service. It was originally armed with six 3-inch Model 1903 guns and pedestal mounts (#60/#37, #66/#38, #67/#39, #68/#40, #81/#41, and #82/#42). These were not mounted for some reason until 1913. Sometime after 1915 these were exchanged for the Model 1902 guns and pedestals at Battery Turnbull also at Fort Wadsworth (#30, #31, #32, #33, #34, and #35). These were retained throughout the 1920s and 1930s. However, all six guns were moved to new locations in 1941-42. Two went to a new emplacement at Rockaway Point, and the other four (along with the battery name) were moved to Norton Point across the harbor entrance. The original emplacement still exists at the Fort Wadsworth Unit of the Gateway National Recreation Area. The battery site is open to the public.

- Two of the 21 temporary 8-inch guns on reinforced Rodman carriages were emplaced at Fort Wadsworth during the emergency of the Spanish American War. Correspondence of April 28, 1898 recommended shipping two 8-inch Model 1888 tubes to Wadsworth for placing on already present 15-inch Rodman carriages (though they would need strengthening modifications). By May 8th this was finished. The site used was one in the old South Cliff Battery and one in Hudson Battery. By 1899 the guns were mounted and in service (Model 1888 8-inch guns West Point Foundry #3 and #4), The guns and carriages were removed around 1903-04. Parts of the platforms used still exist at the Fort Wadsworth National Park unit. The battery site is open to the public.

- *Battery #115* (planned): A 1940 Program planned battery for two 16-inch casemated guns for Fort Wadsworth. It was to be emplaced on the southeastern side of Wadsworth. The project was given a low priority, just #34 of 36 assignments on the September 11, 1940 schedule. By early 1942 the project was dropped, and no work was ever undertaken at the site.

- **Battery #218**: A 1940 Program battery for two 6-inch barbette guns of the "200-series" type planned for Fort Wadsworth. Initially the battery was of low priority, not being funded until the FY-1943 budget. It was to be emplaced southeast of Battery Barry. Work was finally begun on October 22, 1942, but then suspended in 1943. At that point it was structurally complete but without armament or major pieces of equipment. The two 6-inch guns intended were never delivered, but the two Model M4 carriages (#57 and #58) were allocated to the battery and perhaps delivered to Fort Wadsworth. It was to have served as local Tactical No. 23 and had four assigned base-end stations. It was officially deleted and abandoned in 1946. The battery structure still exists at the Fort Wadsworth Unit of the Gateway National Recreation Area. The battery is closed to the public.

- **Tac-14:** Tactical Battery No. 14 was the designation given to the four guns of old Battery Turnbull relocated in mid-1942. In common with other locations, a number of light 3-inch guns were relocated to positions to provide better anti-torpedo boat coverage pending the construction of additional 90mm batteries of the 1943 Program. This battery was located on the southern part of the reservation, in front of constructing Battery #218. Four-gun blocks and a combined BC and CRF station were erected from June 10th to July 15, 1942, for transfer on January 15, 1944 for $10,640.59. The four 3-inch M1903 guns (#60/#37, #66/#38, #67/#39, and #68/#40) were mounted on September 30, 1942. The battery served until dismounted shortly after the end of the war. The emplacement blocks and BC station still exist on Fort Wadsworth Unit of the Gateway National Recreation Area. The battery is open to the public.

Fort Wadsworth New York 1930s (NARA)

Fort Jay/Governer's Island Military Reservation in the inner harbor of New York was used as headquarters, recruiting station and a supply warehouse facility during 20th century. Currently it is managed by three public agencies and is open to the public.

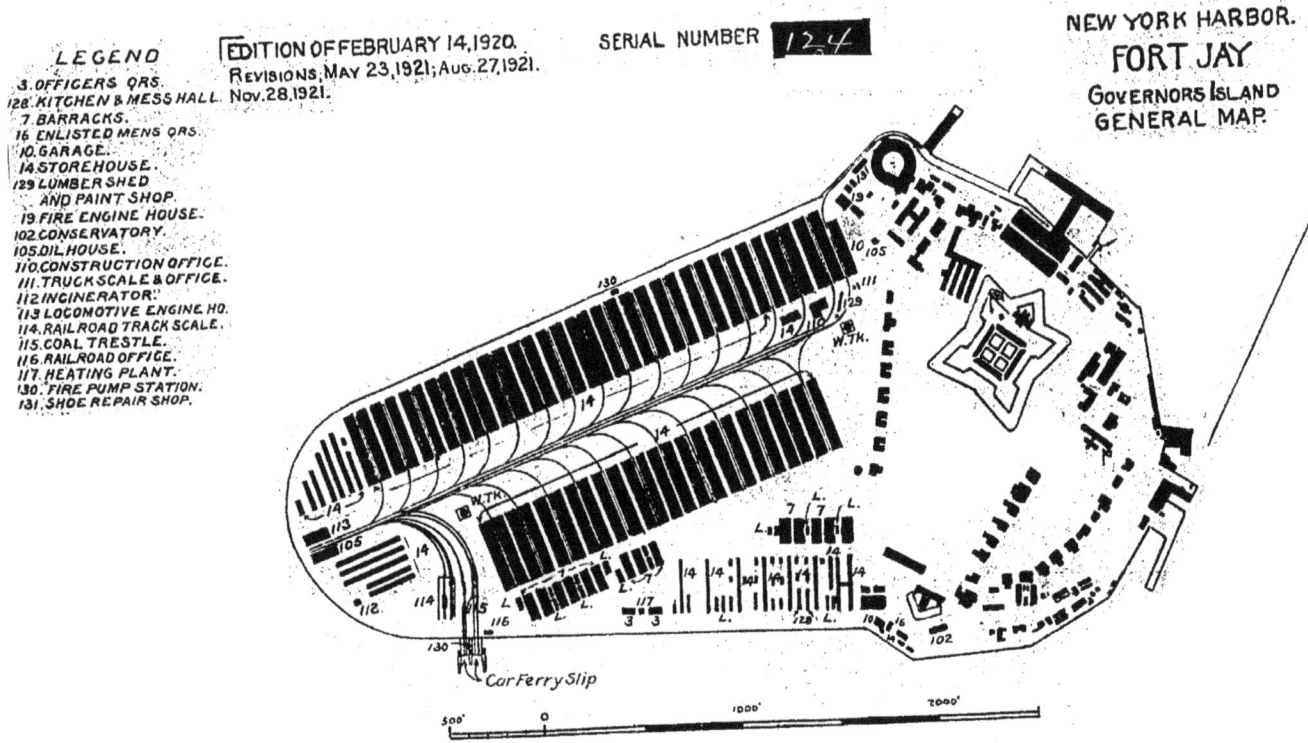

LEGEND

3. OFFICERS QRS.
128. KITCHEN & MESS HALL.
7. BARRACKS.
16 ENLISTED MENS QRS.
10. GARAGE.
14. STOREHOUSE.
129 LUMBER SHED AND PAINT SHOP.
19. FIRE ENGINE HOUSE.
102. CONSERVATORY.
105. OIL HOUSE.
110. CONSTRUCTION OFFICE.
111. TRUCK SCALE & OFFICE.
112. INCINERATOR.
113 LOCOMOTIVE ENGINE HO.
114. RAILROAD TRACK SCALE.
115. COAL TRESTLE.
116. RAILROAD OFFICE.
117. HEATING PLANT.
130. FIRE PUMP STATION.
131. SHOE REPAIR SHOP.

EDITION OF FEBRUARY 14, 1920.
REVISIONS, MAY 23, 1921; AUG. 27, 1921.
Nov. 28, 1921.

SERIAL NUMBER 124

NEW YORK HARBOR.
FORT JAY
GOVERNORS ISLAND
GENERAL MAP.

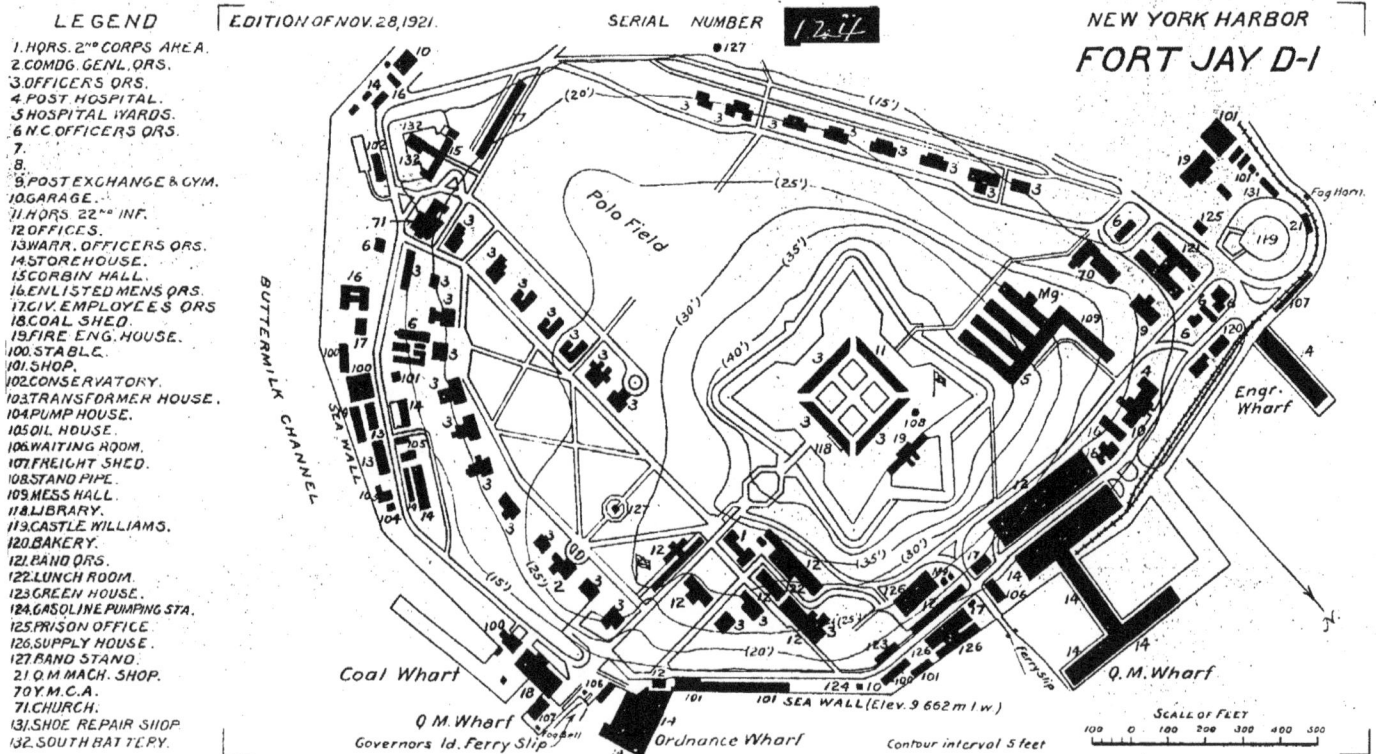

LEGEND

1. HQRS. 2ND CORPS AREA.
2. COMDG. GENL. QRS.
3. OFFICERS QRS.
4. POST HOSPITAL.
5. HOSPITAL WARDS.
6 N.C. OFFICERS QRS.
7.
8.
9. POST EXCHANGE & GYM.
10. GARAGE.
11. HQRS. 22ND INF.
12. OFFICES.
13. WARR. OFFICERS QRS.
14. STOREHOUSE.
15. CORBIN HALL.
16. ENLISTED MENS QRS.
17. CIV. EMPLOYEES QRS.
18. COAL SHED.
19. FIRE ENG. HOUSE.
100. STABLE.
101. SHOP.
102. CONSERVATORY.
103. TRANSFORMER HOUSE.
104. PUMP HOUSE.
105. OIL HOUSE.
106. WAITING ROOM.
107. FREIGHT SHED.
108. STAND PIPE.
109. MESS HALL.
118. LIBRARY.
119. CASTLE WILLIAMS.
120. BAKERY.
121. BAND QRS.
122. LUNCH ROOM.
123. GREEN HOUSE.
124. GASOLINE PUMPING STA.
125. PRISON OFFICE.
126. SUPPLY HOUSE.
127. BAND STAND.
21 Q.M. MACH. SHOP.
70 Y.M.C.A.
71. CHURCH.
131. SHOE REPAIR SHOP.
132. SOUTH BATTERY.

EDITION OF NOV. 28, 1921.

SERIAL NUMBER 124

NEW YORK HARBOR
FORT JAY D-1

Coal Wharf

Q.M. Wharf

Governors Id. Ferry Slip

Ordnance Wharf

Q.M. Wharf

SEA WALL (Elev. 9 662 m l w)

Contour interval 5 feet

SCALE OF FEET
100 0 100 200 300 400 500

(0-55-870B-97TH)(12-9-37-12:00M)(12-2500) GOVERNORS ISLAND, N.Y.

(0194-870N-8)(2-27-33-[00P)(12-2000) FORT JAY, N.Y.

Fort Hancock (1847-1950) is located on Sandy Hook, New Jersey, a five-mile-long sandy peninsula about a half mile in width which is the northern end of New Jersey's barrier islands. This long sand spit has had military importance since the Colonial Era due to the fact that New York Harbor's main shipping channel passed directly offshore. In 1859, construction begun on a large granite Third System fort but due to the Civil War and technical advances this "Fort at Sandy Hook" was never completed and by the turn of century very little of this fort remained. The advancement in military technology caused the U.S. Army to establish a proving ground for heavy artillery at Sandy Hook in 1874 to allow for longer ranges. The Sandy Hook Proving Ground consisted of gun blocks and bombproofs, as well as target butts, as well as the foundations for testing the 14-inch turrets for Fort Drum. These testing function continue at Sandy Hook until 1919 when these operations were transfer to a new proving ground at Aberdeen, Maryland. In 1892, construction of the first of the Endicott Program batteries begun at the northern end of the Hook. These early batteries consisted of Battery Potter, a one-of-kind gun lifts (elevators) for two 12-inch guns on barbette carriages, Battery McCook-Reynolds, a prototype 12-inch mortar battery with built-in land defenses, and Battery Dynamite, a pneumatic system which launched 15-inch high-explosive shells. It was named in General Orders 57 of 1895 for Maj. Gen. Winfield Scott Hancock, U.S. Army of Civil War and Indian War service. By 1900, the fort held a collection of 15 Endicott Program batteries, including what was called the "Nine Gun Battery" (which was a series of connected disappearing gun batteries. The post had an extensive set of fire control stations and a submarine mine complex. The Interwar Period saw two long range 12-inch barbette batteries and several railway artillery units (consisting of 12-inch mortars and 8-inch guns) being added within the fort's boundaries. The World War II period saw few changes in the armament of Fort Hancock except for the addition of casemates to the two long-range 12-inch batteries, a new mine casemate, and addition of Anti-Motor Torpedo Boat and Anti-Aircraft batteries. Sandy Hook ceased being a coast artillery post in 1950, but it continued to serve in an anti-aircraft role until 1974 with Nike radar and missile installations. The former military reservation and state park on Sandy Hook became a major part of the National Park Service's Gateway National Recreation Area in 1975. The Coast Guard uses the area around the old submarine mine complex. NPS is developing tenants to renovate and utilize the remaining buildings. A major tenant is the Marine Academy of Science and Technology. There is limited access to most of the old gun batteries, but old proving grounds are open to the public. Battery Gunnison/New Peck is a restored 6-inch pedestal mount battery with two original guns and carriages, featuring a restored interior plotting room, communications room, shell hoist, and magazines. The Park is open daily, and an admission fee required. There is a Visitor Center, with a wide range of interpretive programs Battery Gunnison and the Nike Radar site are open on select weekends. Primarily oriented to its recreation mission, the park's former military structures are mostly delict and manly closed to the public, but remains of Fort Hancock are still an outstanding site to visit for all students of American Coast Defenses.

Fort Hancock Gun Batteries

- **POTTER:** The Endicott board advocated the development and emplacement of several 12-inch gun lift batteries. The technical concept was for guns mounted on relatively simple barbette carriages that could be lifted on a massive elevator to firing position on the parapet and then lowered for reloading and servicing. This was sort of a first step to the disappearing carriage concept. The prototype gun lift battery (and in fact the only one completed as such) was this unit erected at Fort Hancock. The experimental nature of the equipment made the choice of locating it adjacent to the Sandy Hook Proving Ground obvious. It was located at a central location, to the south of the old masonry fort. The massive structure built was a box-like affair 239 by 162-feet of two stories height. It had 10-foot-thick outer walls, with inner sand fill, and a 40-foot equivalent thick parapet. The lower floor had the magazines and compressed air plant to operate the lifts. The second floor had

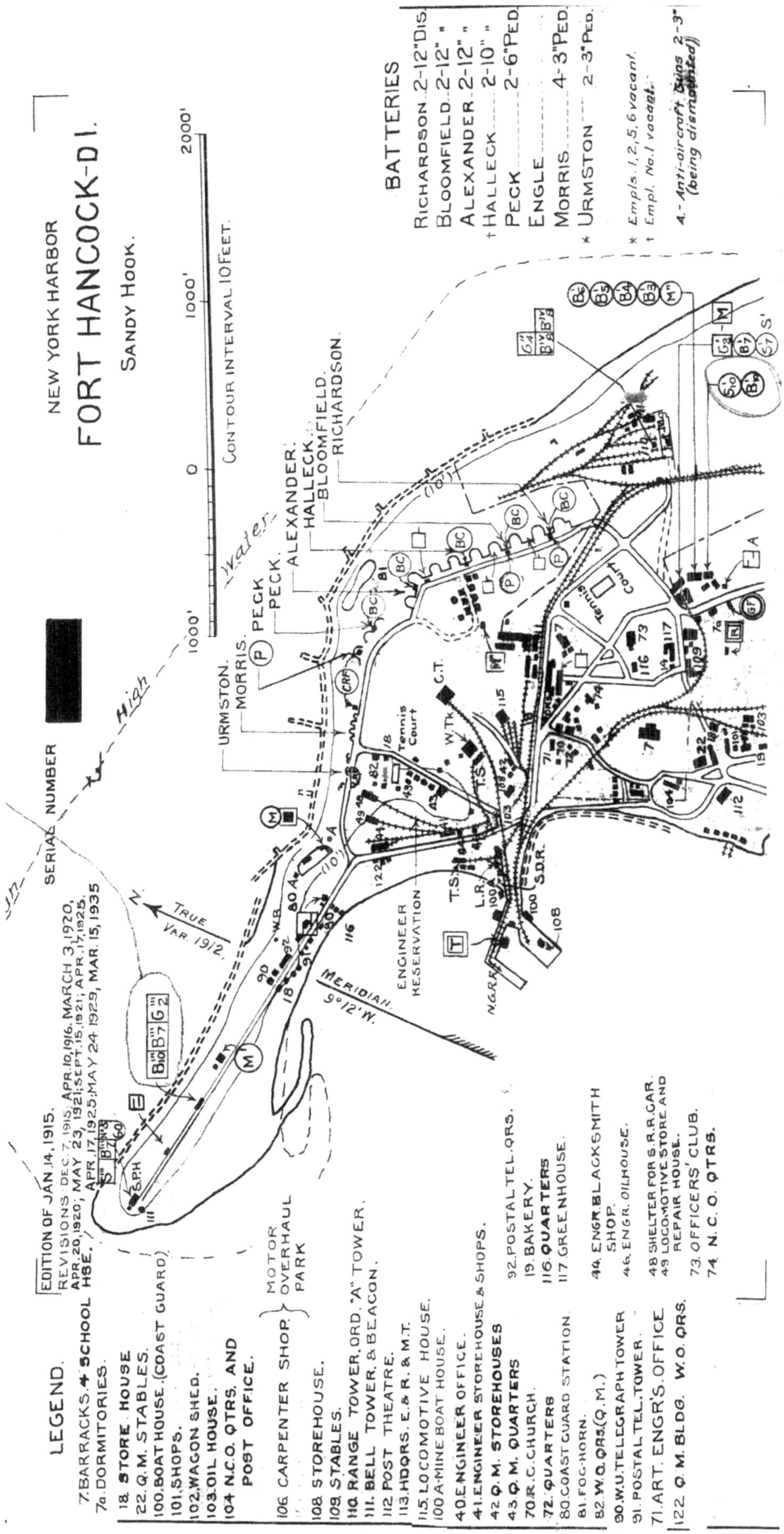

NEW YORK HARBOR
FORT HANCOCK-D.I.

SANDY HOOK.

CONTOUR INTERVAL 10 FEET.

2000' 1000' 0 1000' 2000'

BATTERIES

RICHARDSON 2-12"DIS.
BLOOMFIELD 2-12" "
ALEXANDER 2-12" "
†HALLECK 2-10" "
PECK 2-6"PED.
ENGLE
MORRIS 4-3"PED.
*URMSTON 2-3"PED.

* Empl.s. 1, 2, 5, 6 vacant.
† Empl. No.1 vacant.
4-Anti-aircraft Guns 2-3"
  (being dismantled)

LEGEND.

EDITION OF JAN.14,1915.
REVISIONS-DEC.7,1915; APR.10,1916; MARCH 3,1920;
APR.20,1920; MAY 23, 1921; SEPT.15,1921; APR.17,1925;
APR.17,1925; MAY 24, 1929; MAR. 15, 1935

SERIAL NUMBER

7. BARRACKS. ⊕ SCHOOL HSE.
7a. DORMITORIES.
18. STORE HOUSE.
22. Q.M. STABLES.
100. BOAT HOUSE. (COAST GUARD).
101. SHOPS.
102. WAGON SHED.
103. OIL HOUSE.
104. N.C.O. QTRS. AND POST OFFICE.
106. CARPENTER SHOP. ⎱ MOTOR OVERHAUL PARK.
108. STOREHOUSE.
109. STABLES.
110. RANGE TOWER, ORD. "A" TOWER.
111. BELL TOWER, & BEACON.
112. POST THEATRE.
113. HDQRS. E. & R. & M.T.
115. LOCOMOTIVE HOUSE.
100. A-MINE BOAT HOUSE.
40. ENGINEER OFFICE.
41. ENGINEER STOREHOUSE & SHOPS.
42. Q.M. STOREHOUSES
43. Q.M. QUARTERS
70. R.C. CHURCH.
72. QUARTERS
80. COAST GUARD STATION
81. FOG-HORN.
90. W.U.TELEGRAPH TOWER
90. W.O. QR5.(Q.M.)
91. POSTAL TEL. TOWER.
71. ART. ENGR'S. OFFICE
122. Q.M. BLDG. W.O. QRS.
92. POSTAL TEL. QRS.
19. BAKERY.
116. QUARTERS
117. GREENHOUSE.
44. ENGR. BLACKSMITH SHOP.
46. ENGR. OILHOUSE.
48. SHELTER FOR S.R.R.CAR.
49. LOCOMOTIVE STORE AND REPAIR HOUSE.
73. OFFICERS' CLUB.
74. N.C.O. QTRS.

TRUE
VAR. 19 1/2.

MERIDIAN.
9° 12' W.

ENGINEER RESERVATION

Tennis Court

NEW YORK HARBOR

FORT HANCOCK

SANDY HOOK

GENERAL MAP

SERIAL NUMBER

EDITION OF JAN.14,1915.
REVISIONS: DEC. 7,1915; APR.10,1916,
JUNE 9,1916. FEB.28,1920.
APR.20,1920. MAY 23,1921.
SEPT.15,1921. APR. 17,1925.
MAY 24,1929.
MAR. 15,1935

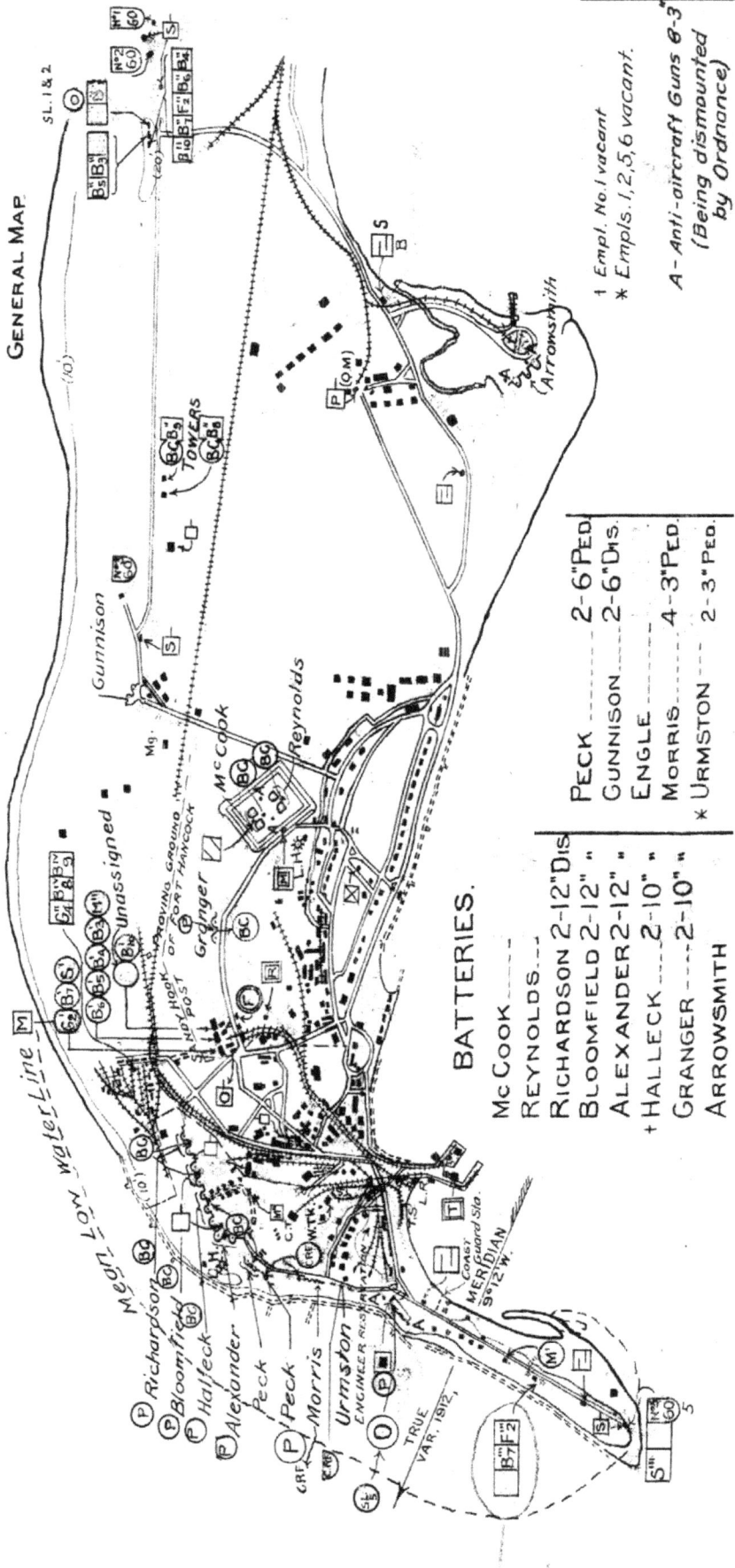

SL.1&2

1 Empl. No.1 vacant.
* Empls.1,2,5,6 vacant.

A - Anti-aircraft Guns 8-3"
(Being dismounted
by Ordnance)

Active Status.

Contour interval 10 Feet.

4000 FT.

3000

2000

1000

0

1000

PECK ---------- 2-6" PED.
GUNNISON ----- 2-6" DIS.
ENGLE ----------
MORRIS -------- 4-3" PED.
* URMSTON ----- 2-3" PED.

BATTERIES.

McCOOK ----------
REYNOLDS ----------
RICHARDSON 2-12" DIS
BLOOMFIELD 2-12" "
ALEXANDER 2-12" "
+HALLECK 2-10" "
GRANGER ----- 2-10" "
ARROWSMITH

Mean Low Water Line

TRUE
VAR. 1912,
9°12'W.

MERIDIAN

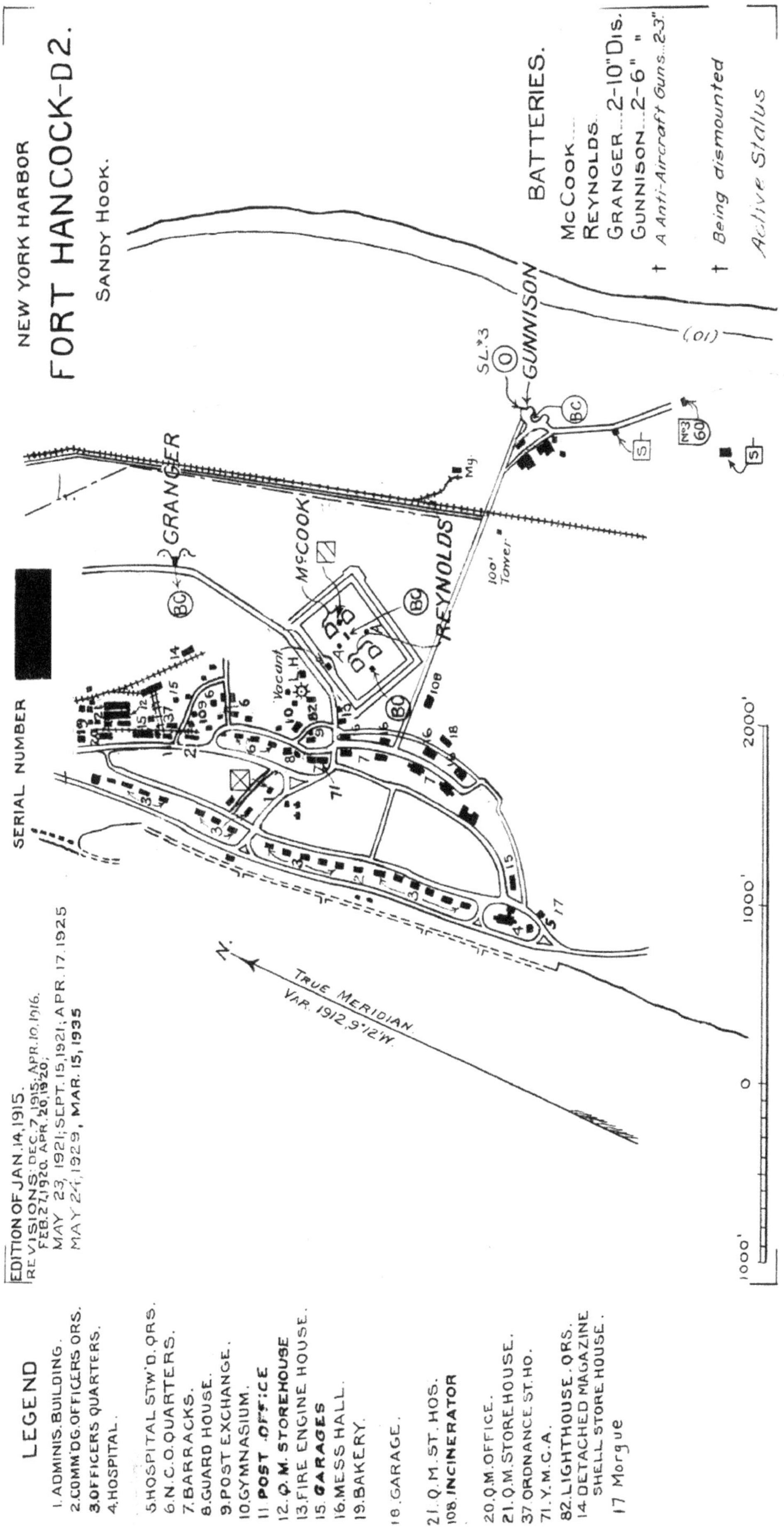

NEW YORK HARBOR
FORT HANCOCK-D2.
SANDY HOOK.

BATTERIES.

McCook ....
REYNOLDS ....
GRANGER...2-10"Dis.
GUNNISON...2-6" "
+ A Anti-Aircraft Guns..2-3"

† *Being dismounted*

Active Status

SERIAL NUMBER

EDITION OF JAN.14,1915.
REVISIONS: DEC.7,1915; APR.10,1916.
FEB.27,1920; APR.20,1920;
MAY 23, 1921; SEPT.15,1921; APR.17,1925
MAY 24,1929, **MAR.15,1935**

N.
TRUE MERIDIAN
VAR. 1912.9°12'W.

1000'   0   1000'   2000'

LEGEND
1. ADMINIS. BUILDING.
2. COMM'DG. OFFICERS QRS.
3. OFFICERS QUARTERS.
4. HOSPITAL.

5. HOSPITAL STW'D. QRS.
6. N.C.O. QUARTERS.
7. BARRACKS.
8. GUARD HOUSE.
9. POST EXCHANGE.
10. GYMNASIUM.
11. **POST OFFICE**
12. Q.M. STOREHOUSE.
13. FIRE ENGINE HOUSE.
15. **GARAGES**
16. MESS HALL.
19. BAKERY.

18. GARAGE.

21. Q.M. ST. HOS.
108. **INCINERATOR**

20. Q.M. OFFICE.
21. Q.M. STOREHOUSE.
37. ORDNANCE ST. HO.
71. Y.M.C.A.

82. LIGHTHOUSE. QRS.
14. DETACHED MAGAZINE
    SHELL STORE HOUSE.
17. Morgue

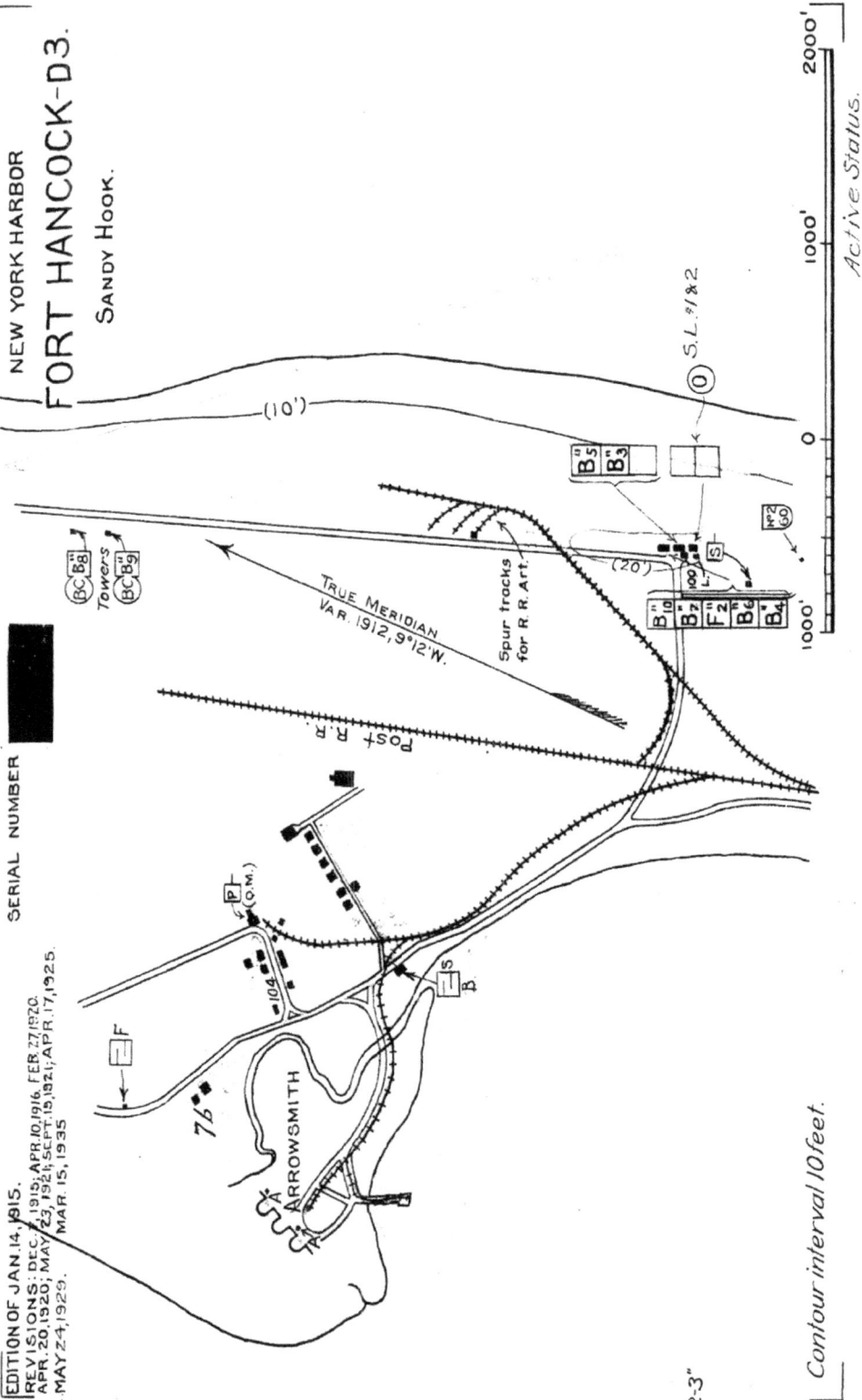

NEW YORK HARBOR
FORT HANCOCK-D3.
SANDY HOOK.

Active Status.

SERIAL NUMBER

EDITION OF JAN.14,1915.
REVISIONS: DEC. 1,1915; APR.10,1916, FEB.27,1920.
APR.20,1920; MAY 23,1921, SEPT.15,1921; APR.17,1925.
MAY 24,1929.   MAR.15,1935.

TRUE MERIDIAN
VAR. 1912, 9°12'W.

Spur tracks for R.R.Art.

Post R.R.

ARROWSMITH

Contour interval 10 feet.

LEGEND.
7b. INCINERATOR
104 N.C.O. QUARTERS
148. BALLOON HANGAR.
100. MESS HALL

BATTERIES.
ARROWSMITH--
† A. Anti-Aircraft Guns. 2-3"
† *Being dismounted*

EDITION OF JULY 1, 1925.
REVISIONS; MAY 24, 1929.
MAR. 15, 1935.

SERIAL NUMBER

NEW YORK HARBOR

FORT HANCOCK—D 4.

SANDY HOOK

BATTERIES

MILLS.....2-12"B.
KINGMAN 2-12"B.
† A-Anti-aircraft-2-3"

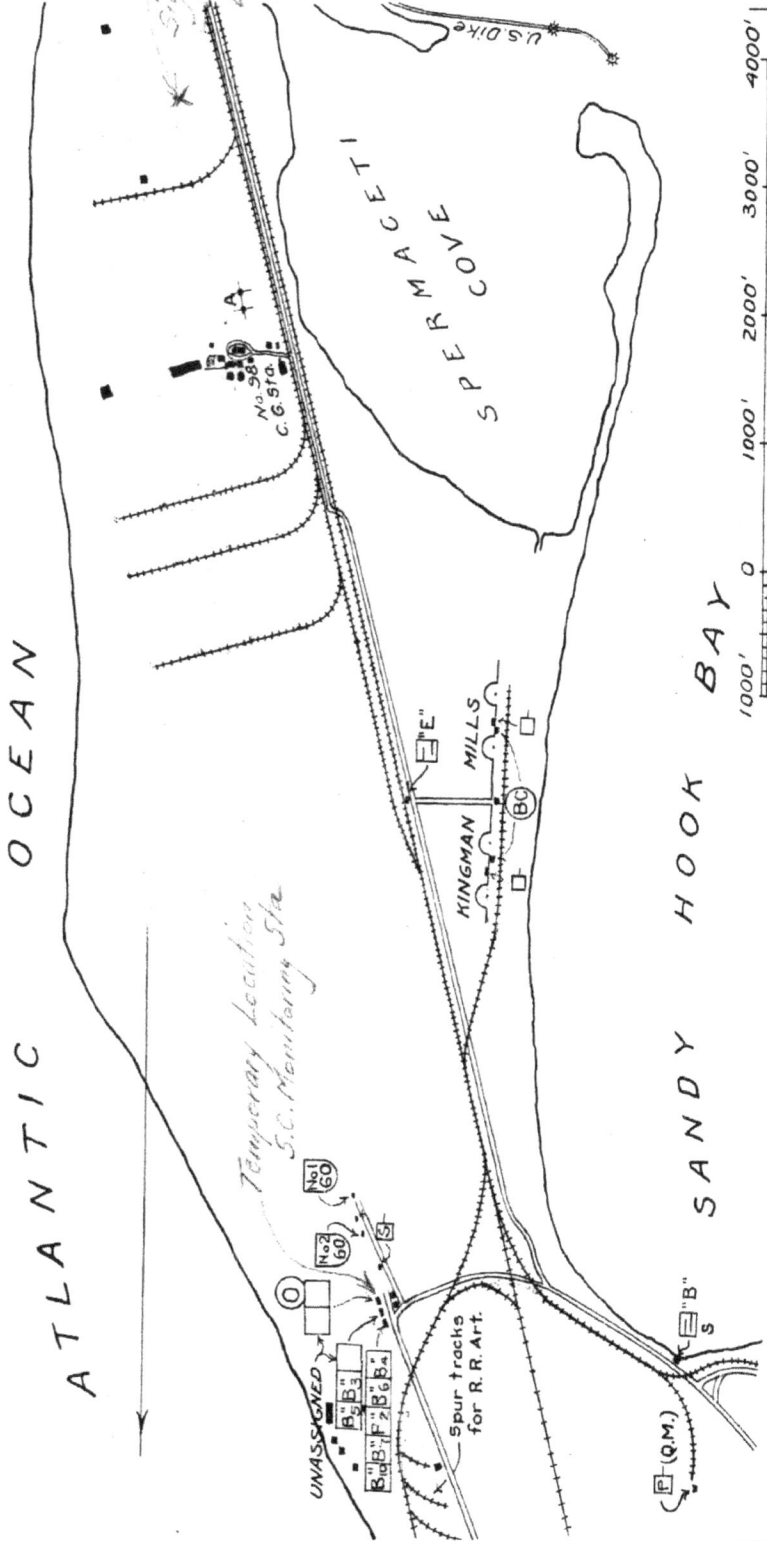

† Being dismounted.

Approx. location
Sig. Corps Monitoring
Station, if funds
become available
676.3 (Sandy Hk.)

U.S. Dike

A T L A N T I C   O C E A N

S P E R M A C E T I   C O V E

No. 98
C.G. sta.

A

Temporary location
S.C. Monitoring Sta.

UNASSIGNED

No.1
60

No.2
60

B"B"
B3

B"B"F
2

B6
B4

Spur tracks
for R.R. Art.

"E"

KINGMAN
MILLS

BC

"B"
S

P"(Q.M.)

S A N D Y   H O O K   B A Y

1000'   0   1000'   2000'   3000'   4000'

Active Status

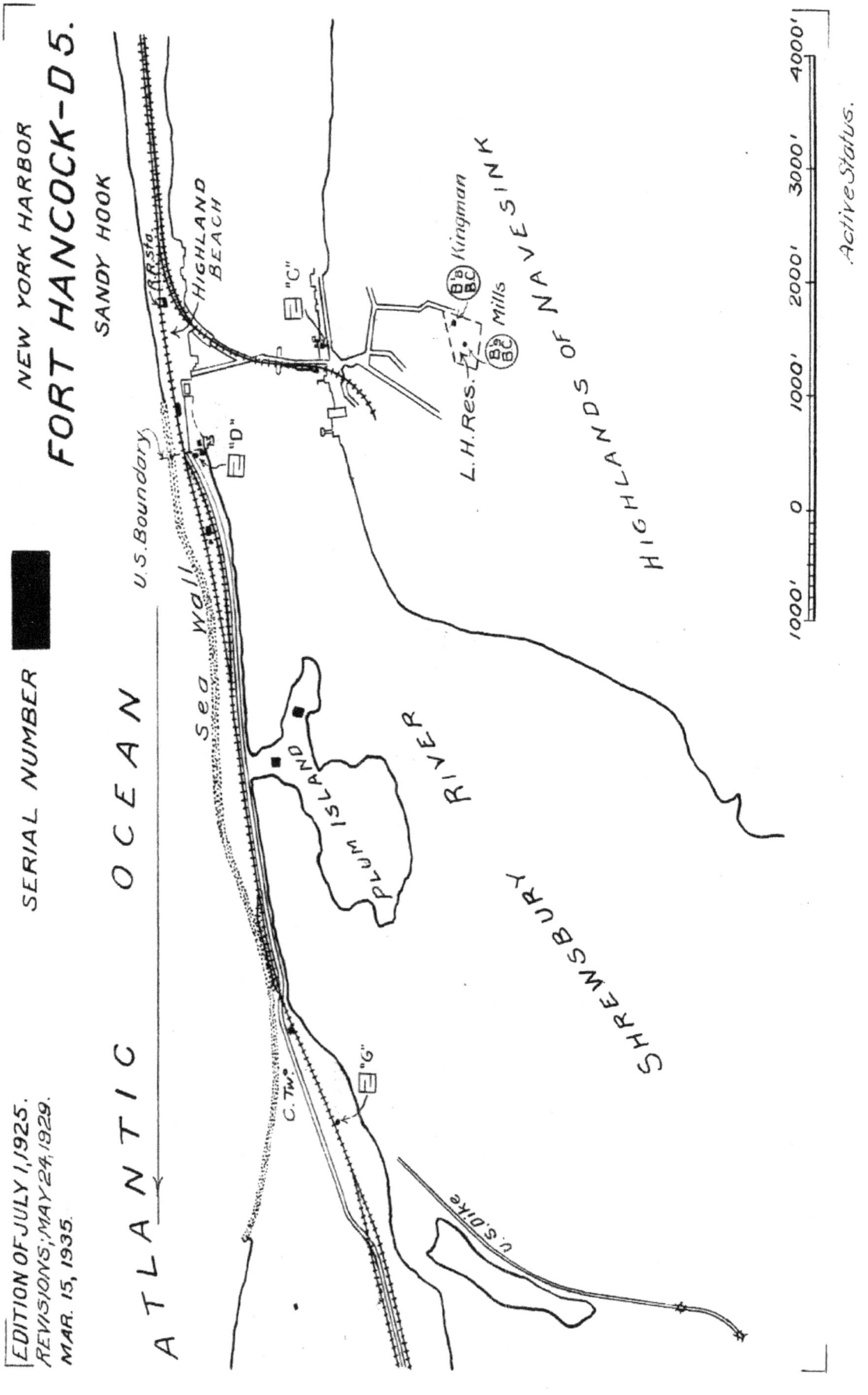

NEW YORK HARBOR
FORT HANCOCK-D5.
SANDY HOOK

SERIAL NUMBER

EDITION OF JULY 1, 1925.
REVISIONS; MAY 24, 1929.
MAR. 15, 1935.

ATLANTIC OCEAN

Sea Wall

U.S. Boundary

R.R. Sta.

HIGHLAND BEACH

"C"

"D"

L.H. Res.

Kingman
B B
B.C.

B
B.C.
Mills

HIGHLANDS OF NAVESINK

PLUM ISLAND

SHREWSBURY RIVER

C. Tw.

"G"

U.S. Pipe

1000'   0   1000'   2000'   3000'   4000'

Active Status.

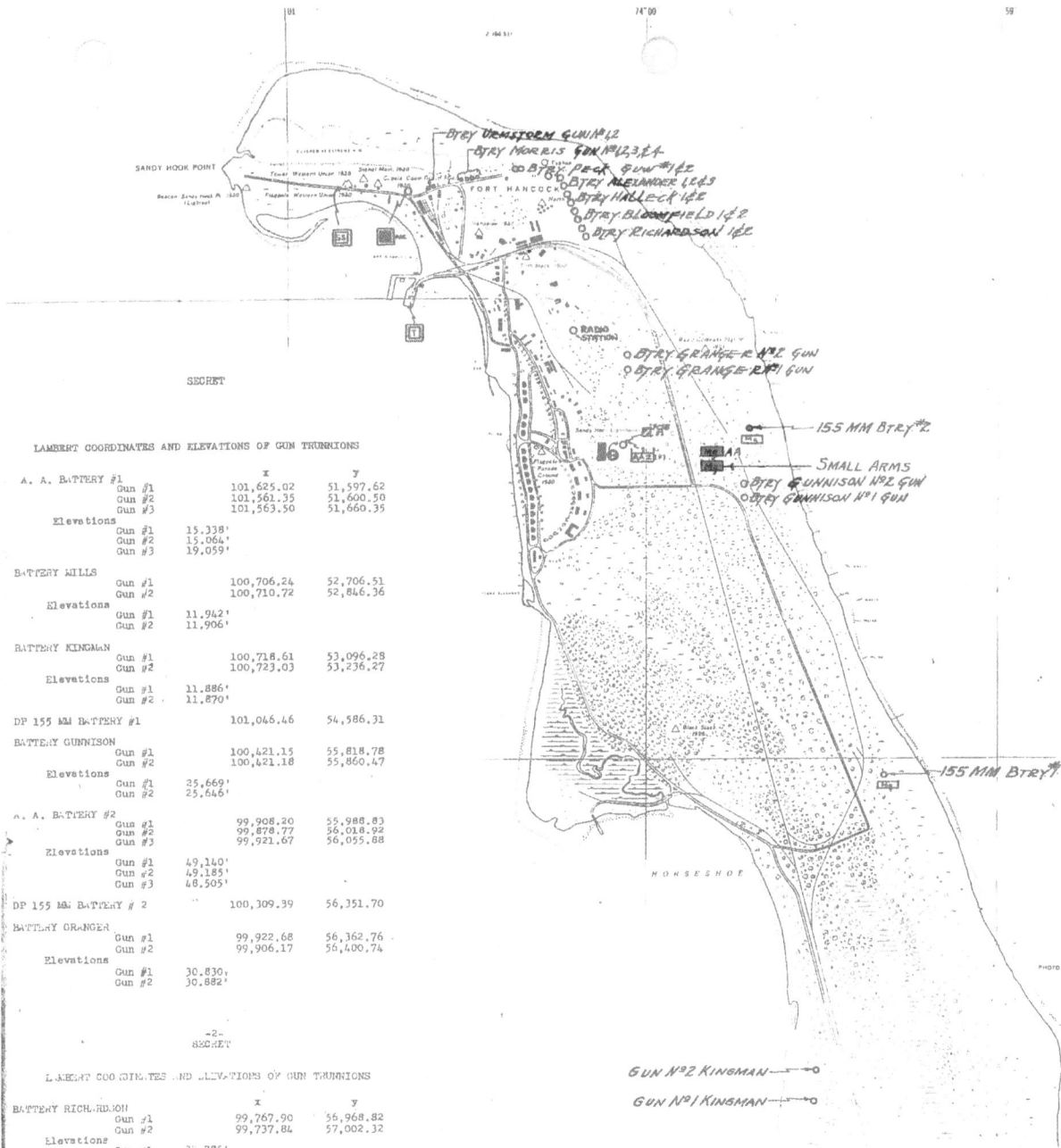

SECRET

LAMBERT COORDINATES AND ELEVATIONS OF GUN TRUNNIONS

		x	y
A. A. BATTERY #1	Gun #1	101,625.02	51,597.62
	Gun #2	101,561.35	51,600.50
	Gun #3	101,563.50	51,660.35
Elevations	Gun #1	15.338'	
	Gun #2	15.064'	
	Gun #3	19.059'	
BATTERY MILLS	Gun #1	100,706.24	52,706.51
	Gun #2	100,710.72	52,846.36
Elevations	Gun #1	11.942'	
	Gun #2	11.906'	
BATTERY KINGMAN	Gun #1	100,718.61	53,096.28
	Gun #2	100,723.03	53,236.27
Elevations	Gun #1	11.886'	
	Gun #2	11.870'	
DP 155 MM BATTERY #1		101,046.46	54,586.31
BATTERY GUNNISON	Gun #1	100,421.15	55,818.78
	Gun #2	100,421.18	55,860.47
Elevations	Gun #1	25.669'	
	Gun #2	25.646'	
A. A. BATTERY #2	Gun #1	99,908.20	55,988.83
	Gun #2	99,878.77	56,018.92
	Gun #3	99,921.67	56,055.88
Elevations	Gun #1	49,140'	
	Gun #2	49.185'	
	Gun #3	48.505'	
DP 155 MM BATTERY #2		100,309.39	56,351.70
BATTERY GRANGER	Gun #1	99,922.68	56,362.76
	Gun #2	99,906.17	56,400.74
Elevations	Gun #1	30.830'	
	Gun #2	30.882'	

-2-
SECRET

LAMBERT COORDINATES AND ELEVATIONS OF GUN TRUNNIONS

		x	y
BATTERY RICHARDSON	Gun #1	99,767.90	56,968.82
	Gun #2	99,737.84	57,002.32
Elevations	Gun #1		

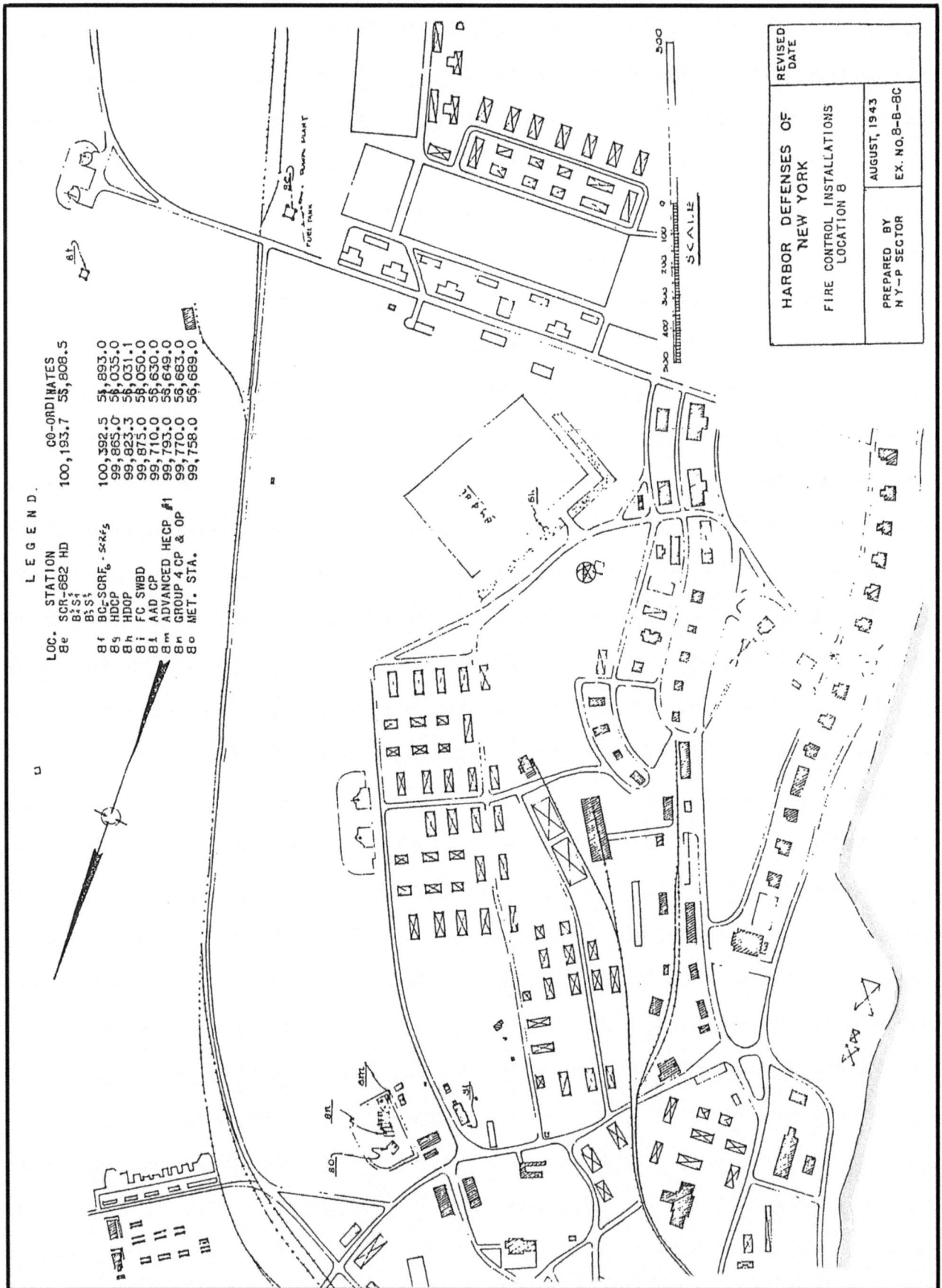

L E G E N D.

Loc.	STATION	CO-ORDINATES	
8e	SCR-682 HD	100,193.7	55,808.5
	B₁S₁		
	B₂S₂		
8f	BC-SCRF₆-SCRf₅	100,392.5	55,893.0
8g	HDCP	99,865.0	56,035.0
8h	HDOP	99,823.3	56,031.1
8i	FC SWBD	99,875.0	56,050.0
8l	AAD CP	99,710.0	55,630.0
8m	ADVANCED HECP #1	99,793.0	55,649.0
8n	GROUP 4 CP & OP	99,770.0	56,683.0
8o	MET. STA.	99,758.0	56,689.0

HARBOR DEFENSES OF
NEW YORK

FIRE CONTROL INSTALLATIONS
LOCATION 8

| PREPARED BY
N Y – P SECTOR | AUGUST, 1943 |
| | EX. NO. 8–B–8C |

| REVISED
DATE |

ATLANTIC OCEAN

Location 8q — 99,288.4   57,209.4

BC₉ , CRF₉

MINE CABLE TANK

WATER TANK

U.S. ENGINEER RESERVATION

LOCATION 8r — 99,009.0   57,167.0

GROUP 3, CP, & OP

COAST GUARD
BOATHOUSE

ENGINEER
WHARF

FLOW AREA

SANDY HOOK BAY

LOCATION 8S — 98,492.0   57,207.0

M²₁₀ , B⁴₅ , S⁴₅

LOCATION 80(1)

S³₅

X - 99,541.15
Y - 57,187.01

N

SANDY HOOK POINT,
LIGHT & BELL

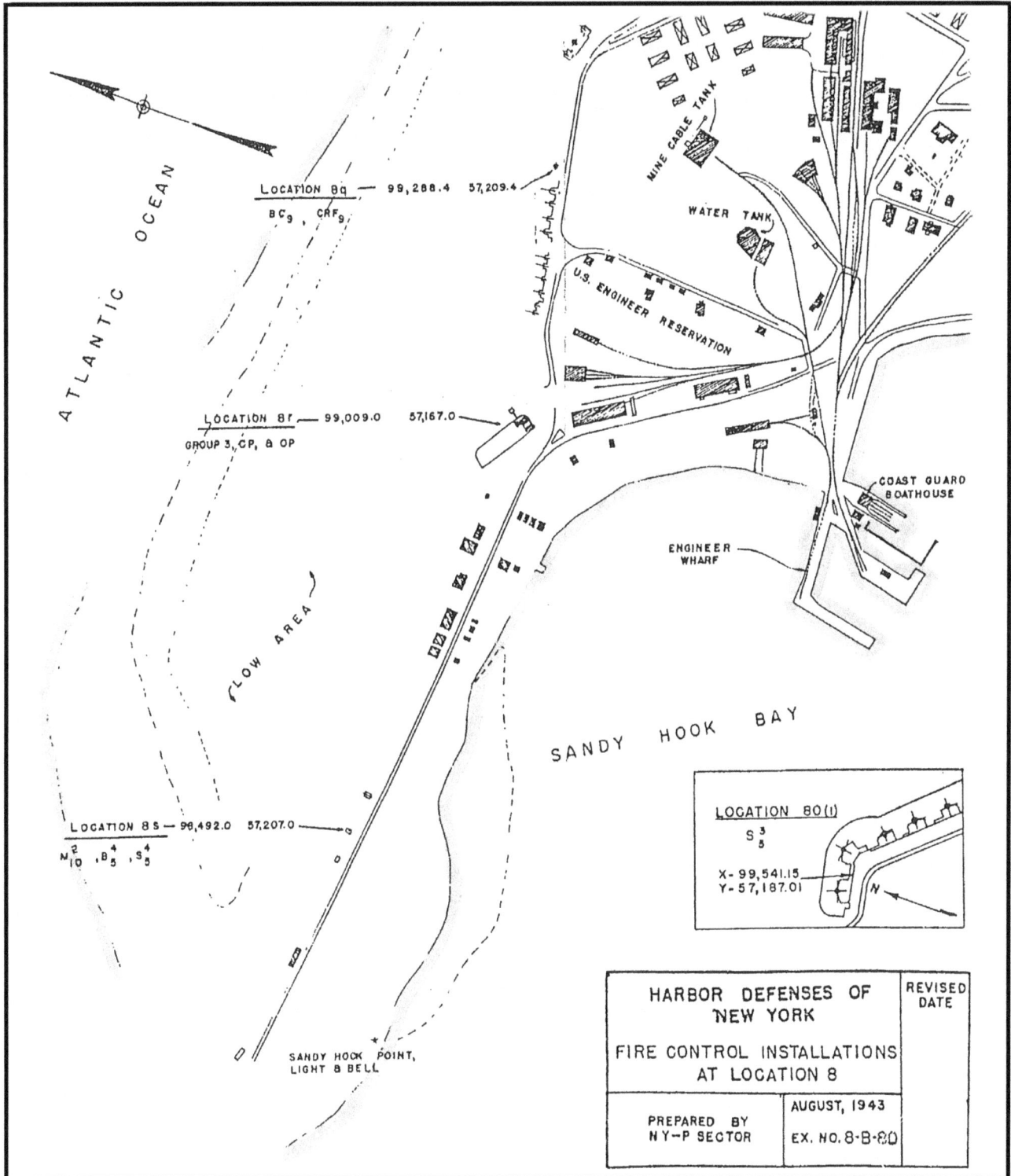

HARBOR DEFENSES OF NEW YORK	REVISED DATE
FIRE CONTROL INSTALLATIONS AT LOCATION 8	

| PREPARED BY N Y–P SECTOR | AUGUST, 1943 |
| | EX. NO. 8-B-80 |

the loading level equipment and crew detachment shelter. The entry into the battery had embrasures for light, defensive guns. Openings for the lifts were 20-by-24 feet. The engineering plan was first submitted on February 8, 1890. Concrete work was mostly accomplished in 1890-92. The steam and compressed air plant was manufactured and installed by the Continental Iron Works from 1891-1894. Understandably tests of the guns, the unique carriages, and the lift machinery took several years. One gun and carriage (the latter imported from France) was installed in 1892, the other of American manufacture was installed in 1896. It was armed with two 12-inch guns Model 1888 guns on Model 1891 gun lift carriages (#2/#1 and #11/#2). The original type 12-inch gun mounted here for trials in 1892, was replaced with Gun No. 1 in 1895. Transfer was eventually made of March 12, 1898 for a cost to date of $357,100. It was not named until General Orders No. 78 of May 25, 1903 for Brigadier General Joseph H. Potter of Mexican and Civil War service. While the concept worked, the expense and resources required for what would be a slow-firing, extremely obvious target was a poor return; no future gun lift batteries were built, and those under construction were cancelled or converted. While armed, the gun was never considered an active part of the defenses. By 1905 it was slated for deletion, the armament was removed the next year. Shortly after the high platform provided by the battery's parapet was used for a number of fire control stations. Some of these served well into World War II. The unique emplacement still exists at the Sandy Hook unit of the Gateway National Recreation Area. The battery is closed to the public.

- **Dynamite Battery**: A battery for pneumatic dynamite guns funded by Congress in the early 1890s. It was emplaced at the western extension of Sandy Hook near the lighthouse. This invention featured the firing of relatively large charges of explosive (usually dynamite but also guncotton) using compressed air designed to detonate on or near enemy warships. While these guns did not have the range of conventional rifled cannon, the damage effect was expected to be much greater. The Army was not enthusiastic about the concept, but Congress was and provided the money to private inventors to develop their weapons, test them, and have selected emplacements built as part of the coast defenses. Then a submission was made for this battery on November 12, 1894 to provide protective wood and canvas covering for the guns and equipment. At this point they were totally exposed to the elements, and much of the connecting piping and tanks susceptible to corrosion. Finally in April 1898, after the start of the Spanish American War, additional funds were requested and granted to build a large, 50-foot-tall sand and concrete parapet completely around the gun blocks and steam plant to protect it from gunfire. The original Model 1886 guns were two of 15-inch bore and one of 8-inch, placed in line on rotating carriages emplaced on concrete foundation pads. It was not formally transferred or ever named. While "finished" they don't appear to have been regularly manned or ever relied upon as part of the active defenses. Tests here and at San Francisco were disappointing concerning the accuracy and most importantly the destructive effects of the charges. By 1902 the guns were deactivated. They were offered for scrap not long after. The emplacement itself was found to be a most convenient location for a protected mine casemate and was used as such subsequently. As modified over the years, some of the emplacement still exists within the US Coast Guard reservation at Sandy Hook. The battery is closed to the public.

- **McCOOK – REYNOLDS**: A mortar battery emplaced on the central area of the Fort Hancock reservation, adjacent to the lighthouse. This was one of the earliest of the new Endicott mortar batteries, using a quadrangular pit design evolved from the one developed in the 1870s. Plans were first submitted on September 28, 1888. It was known as a "half-sunken" type. It had four pits each for four mortars. The pits were protected with an encircling high parapet all-around; each being connected by a covered tunnel to magazines and support structures. It also was to have a complete dry ditch completely around the structure, protected with machine guns or light revolving cannons in counterscarp galleries at opposing corners. However, these land defenses were never really completed or armed. The appropriation of August 18, 1890 included $201,000 specifically for this battery. Work started in late 1890 and was reported completed in February 1894. The mortar platforms were ready in 1893, and by October of that year four mortars and carriages were mounted while they were awaiting delivery of the last carriages. Like many of the initial batteries, transfer was somewhat delayed—it was not made until March 22, 1898 for a construction cost of $276,743.00. It was named on General Orders No. 78 of May 25, 1903 for Brig. Gen. John Reynolds, U.S. Volunteers who was killed at Gettysburg in 1863. Initially this name was for the entire battery, but in 1906 it was administratively split, the two northern pair of pits being named on General Orders No. 25 of January 25, 1906 for Major General Alexander McCook of Civil War service. The battery was armed with sixteen 12-inch Model 1886 cast-iron mortars on Model 1891 spring-return mortar carriages. In Battery Reynolds were mortars/carriages #15/#14, #19/#15, #20/#4, #64/#16, #21/#25, #22/#7, #17/#6, and #70/#1. In Battery McCook were mortars/carriages: #13/#13, #14/#8, #18/#12, #30/#11, #58/#9, #67/#10, #65/#2, and #66/#3. Seven Gatling guns were on hand, but unmounted for ditch defense in 1909, and by then new telautograph booths had also been installed in each pit. In 1917 four mortars and carriages, one from each pit, were removed and installed at the new temporary mortar battery on the Navesink Highlands. The twelve remaining mortars served until declared obsolete and scrapped in 1920. During World War II the interior of the battery was extensively modified and given gas proofing to serve as the advanced Harbor Defense Command Post for New York. The emplacement still exists, though in places damaged, at the Sandy Hook unit of the Gateway National Recreation Area. The battery is open to the public.

- **HALLECK:** A battery for three 10-inch disappearing guns emplaced as the first unit of what would become a nine-gun sequence along the seafront curtain wall of the old, uncompleted masonry fort at Sandy Hook. Original Endicott plans recommended a battery of non-disappearing guns here in iron-armored casemates, but with the successful advent of disappearing carriages, plans changed to the latter. This change was authorized on July 3, 1896, and detailed plans were submitted on October 14, 1896 for three 10-inch guns on Model 1896 disappearing carriages. Funding came from the Act of June 23, 1896. The emplacement for three adjacent emplacements was adapted to the site. The outer granite face of the old curtain wall was retained at the forward face of the internal magazines, but the exterior was covered with a graded slope of earth. In most other features the battery followed mimeo type plans. Work was done in 1896-1897, guns being mounted in 1899. It was transferred on January 6, 1900 for a cost of $73,844. It was named on General Orders No. 43 of April 4, 1900 for Major General Henry W. Halleck, commander-in-chief of the U.S. Army in 1862-1864. The battery was armed with three 10-inch Model 1888 Watervliet guns on Model LF 1896 disappearing carriages (#2/#8, #35/#18 and #37/#43). Hoists were transferred in 1907 and then replaced with a different model in 1911. A new BC station and commander's walk was added in 1907. In 1917 the gun in emplacement No. 1 was removed (Watervliet #2) for use on railway artillery, its carriage was scrapped in place in 1920. The guns in the other two emplacements served until deleted under authority of November 12, 1942. The emplacement still exists at the Sandy Hook unit of the Gateway National Recreation Area. The battery is closed to the public.

- **ALEXANDER:** A battery for two 12-inch guns emplaced on the left (north flank) of Battery Halleck, creating a bend in the sequence, now following the north curtain wall of the old masonry fort. Thus, it fired more to the north and provided heavy gun coverage of the inner entrance to the main channel. Plans were submitted on June 3, 1896. It was somewhat delayed in start of construction.

The battery was funded with the Act of March 3, 1897, work being done in 1897-1898. It was of typical plan, though gun No. 1 was canted outward (towards the east) to provide better integration of the field of fire with neighboring Battery Halleck. Transfer was made on July 15, 1899 for a cost of $94,228. It was armed with two 12-inch guns Model 1888M1 on Model 1896 disappearing carriages (Bethlehem tube #4/carriage #25 and Watervliet tube #7/carriage #8). It was named on General Orders No. 194 of December 27, 1904 for Major General William Alexander of the Continental Army. After construction the usual modifications to hoists, loading platforms, and BC stations were made. As one of the major batteries in the defense, the armament was retained without disruption until World War II. It was authorized for removal on November 12, 1942 and probably had the guns and carriages scrapped by mid-1943. The emplacement was used in 1915 as the test bed for modifying the carriage for an increase of 5-degrees of elevation. The emplacement still exists at the Sandy Hook unit of the Gateway National Recreation Area. The battery is closed to the public.

- **BLOOMFIELD**: A battery for two 12-inch disappearing guns emplaced south of Battery Halleck (and eventually with Battery Richardson on its right flank) along the line of the old seafront curtain wall of the masonry fort at Sandy Hook. It was submitted on June 26, 1898 with funds from the March 1898 National Defense Act. Despite the date of submission, with the urgency of the Spanish American War, work actually began in early April even before formal submission approval. It was transferred on December 20, 1899 for a construction cost of $94,228. It was named on General Orders No. 194 on December 27, 1904 for Brigadier General Joseph Bloomfield of War of 1812 service. Concrete work was finished in mid-1899. It was of typical mimeograph design, though the crest elevation was modified to allow it to fit in the sequence of guns immediately adjacent to the battery. The battery was armed with two 12-inch Model 1888 Watervliet guns on disappearing carriage model 1896 (#42/#19 and #43/#22). The usual changes to the loading platforms, hoists, and BC station and walk were made over the next ten years. The gun carriages were modified to provide an additional 5-degrees of elevation in 1916 or 1917. The guns were retained through World War I. Authority for removal and scrapping came on October 23, 1942, though the guns and carriages weren't actually scrapped until 1944. The emplacement still exists at the Sandy Hook unit of the Gateway National Recreation Area. The battery is closed to the public.

- **RICHARDSON**: A battery for two 12-inch, late model 1901 disappearing carriages in the main sequence of heavy guns at Fort Hancock. It was emplaced as the final unit on the southern flank of Battery Halleck along the curtain wall line of the old masonry work. This site was approved on March 29, 1901, and the engineering plan was submitted on May 4, 1901. The battery was fully in compliance with current mimeograph type plans. Work was relatively quickly accomplished, in 1901-1902. It was transferred on April 23, 1904 for a construction cost of $100,000. The battery was named on General Orders No. 194 of December 27, 1904 for Major General Israel Richardson who mortally wounded at Antietam in 1862. The original armament consisted of two 12-inch Model 1900 guns on Model 1901 disappearing carriages (#1/#4 and #2/#3). With the discovery of rapid bore erosion of this type of gun, it was soon necessary to replace the tubes. Around 1910-12 the tubes were replaced with Model 1895 12-inch (Watervliet tube #64 and Watervliet tube #71, previously unused). In 1915 the emplacement was used as the test bed for modifying the Model 1901 carriage for an increase of elevation from 10 to now 15-degrees. The final armament was carried throughout the 1920s and 30s. In 1940 a proposal was made to destroy the emplacement and replace it with a new 1940 Program standard 6-inch emplacement. However, this was never approved, and the armament of Richardson was authorized for scrapping on October 23, 1943.

The emplacement still exists at the Sandy Hook unit of the Gateway National Recreation Area. The battery is closed to the public.

- **GRANGER:** A battery for two 10-inch disappearing guns emplaced at the Hancock reservation midway between Battery Potter and the mortar battery. The battery plan was submitted on July 21, 1896. It covered the eastern entrance to the Gedney Channel. Concrete work was done in 1896-97. It followed conventional design mimeographs, with powder magazines on the lower, left flank traverse of each platform and ammunition hoists incorporated in the design. It was transferred on March 22, 1898 for a cost of $87,300. The battery was named on General Orders No. 43 of April 4, 1900 for Major General Gordon Granger of Mexican and Civil War service. The battery was armed with two 10-inch Model 1888 Watervliet guns on Model LF 1896 disappearing carriages (#61/#19 and #36/#20). Plans were made to replace these tubes with spare Model 1895 types, but apparently this was never done. The usual changes were made in 1905-1910 to the hoists, loading platforms, and the addition of a new BC station in 1907. The original armament served on until authorized for removal on October 23, 1942. The guns were taken out and the carriages scrapped in 1944. The emplacement still exists at the Sandy Hook unit of the Gateway National Recreation Area. The battery is closed to the public.

- **PECK:** A battery for 6-inch rapid-fire pedestal guns approved for placement just to the west of the end of the main sequence, nine-gun disappearing battery. It was one of a number of light batteries designed to cover the minefield in the channel to the north of Sandy Hook. Plans were submitted on June 10, 1898 for two 6-inch guns in one emplacement, but with room on the right flank for two more if the funding allowed. Concrete work was done in 1901-1902, transfer made on November 20, 1903 for a cost of $33,940. The emplacement was modified during construction, instead of mounting both guns on the same central platform, two new individual gun blocks were built on the flanks, each for a single gun and the center platform left unused except as a command station. It was named on General Orders No. 78 of May 25, 1903 for 1st Lt. Fremont P. Peck, and ordnance officer who was killed in a gun accident at Sandy Hook in 1895. The battery was armed with two 6-inch Model 1900 guns and carriages (#27/#12 and #28/#17). These guns were retained as active armament under the 1932 Review and 1940 Program. However, in 1943 the guns and carriages were removed and transferred to the site of Battery Gunnison, which was modified with new, raised gun platforms. The old Peck emplacement was converted for the emplacment of two 90mm AMTB guns of Tactical Battery No. 8. Thus modified, the emplacement still exists at the Sandy Hook unit of the Gateway National Recreation Area. The battery site is open to the public.

- **GUNNISON:** A battery for two 6-inch guns on disappearing carriages. It was planned to be on the property of the Sandy Hook Proving Grounds to the southeast of most of the other battery locations. Submission was made on August 31, 1903 for two guns. By this point in time the battery plans were becoming highly standardized, and this battery was no exception—it followed the mimeograph recommendations very closely. Work was done from late 1903 through 1905 and was reported completed on November 2, 1905. Transfer was made a month later on December 5, 1905 with a construction cost of $45,000. It was named on General Orders No. 194 of December 27, 1904 for Captain John Gunnison, topographical engineer killed in Utah in 1863. It originally carried two 6-inch Model 1903 guns on Model 1903 disappearing carriages (#5/#52 and #34/#57). The battery served for many years with these guns. Then under authority of February 23, 1943 the disappearing guns and carriages were removed and the gun platforms raised and rebuilt to take Model 1900 pedestal guns. The two guns and mounts from Battery Peck here at Fort Hancock were relocated to these new platforms (#27/#12 and #28/#17). Battery Gunnison also assumed the name Battery Peck (or Battery New Peck). This new armament was retained well through the end of the war, until the guns were removed in 1948 and replaced with a pair moved from Battery Livingston, Fort Hamilton (#22 and #23). The battery was abandoned in 1949, though the armament was left for display purposes. These guns were removed for museum use at a proposed Smithsonian Armed Forces Museum. This museum project never came to be, so the two 6-inch guns were returned to Battery Gunnison. Thus, still armed, the emplacement exists at the Sandy Hook unit of the Gateway National Recreation Area and is being actively restored by a group of volunteers. The battery is open to the public.

- **ENGLE**: A battery for a single 5-inch balanced pillar gun emplaced west of the old fort. Plans for the battery were submitted on August 31, 1897, funds coming from the Act of June 27, 1896 (and at the time estimated to cost just $4000). Somewhat unusually it was intended for just a single mount, but that may have been a result of the need to have an easily accessible type emplacement to better evaluate the first Model 1896 balanced pillar mount. It was built in 1897-1898. It consisted of just a single, sunken circular platform of the pillar and adjacent (right) flank magazine. Transfer was made on July 2, 1898 for a cost of $4700. It was named on General Orders No. 78 of May 25, 1903 for Captain Archibald H. Engle, who died at the Battle of Resaca in 1864. It was armed with one 5-inch Model 1897 gun on balanced pillar mount Model 1896 (Watervliet #1/#2). However, gun #1 was very soon removed and returned to the arsenal in late 1899. The replacement was Watervliet tube #2. This combination served until the tube was removed in 1918 and the carriage authorized for scrapping. Subsequently the emplacement was used as a site for building a coincidence range-finder station for neighboring 6-inch Battery Peck. Thus modified, the emplacement still exists at the Sandy Hook unit of the Gateway National Recreation Area. The battery is open to the public.

- A temporary emplacement for a single 4.72-inch gun and pedestal was erected at Fort Hancock at the beginning of the Spanish American War. Four different imported 4.7-inch guns had been purchased and evaluated at Sandy Hook in the mid-1890s. Several of these suffered significant failures, but the single gun supplied by the French firm Schneider was still on hand at the proofing grounds even though its tests had concluded. With the panic over lack of rapid-fire batteries at American seacoast defenses, it was decided to quickly emplace this gun at nearby Fort Hancock. It was located not far from the site of Battery Engle. Apparently just a simple concrete block with foundation bolts was erected in late April, 1898. The Schneider 4.72-inch RF gun (serial #2599) was emplaced on this block. No transfer was made, any cost was absorbed by local fort accounts, no naming was conferred. Apparently dismounted after the war's end, no trace of the block remains today, but it may just be buried by sand.

- **MORRIS**: A battery for four 3-inch Model 1903 pedestal guns emplaced in a single battery just to the west of 5-inch Battery Engle and close to the Sandy Hook beacon. Plans were submitted on May 5, 1903 along with an estimated cost of $24,000. It had a typical design, with square-shaped internal platforms in the center and curved frontal wall emplacements on either flank. There was 44-feet between emplacements, except a wider 50-foot distance between the two center ones. A year later, on May 29, 1904 it was reported completed. Transfer was made on June 7, 1904 for a cost of $23,000. It was named on General Orders No. 194 of December 27, 1904 for Colonel Lewis Morris, 7th New York Volunteer Artillery who was killed in 1864 at Cold Harbor. The battery was armed with four 3-inch Model 1903 guns and pedestals (#5/#91, #42/#92, #70/#93 and #71/#94). It was retained for a number of years, serving in World War II in an AMTB role as Tactical Battery No. 9. The guns were finally dismounted in 1946. The emplacement still exists at the Sandy Hook unit of the Gateway National Recreation Area. The battery is closed to the public.

- **URMSTON**: A mixed battery of 3-inch pedestal and masking parapet mounts emplaced near the other rapid-fire batteries at the northern end of the Fort Hancock gun line. Two emplacements for a pair of Model 1898 3-inch, 15-pound guns on Model 1898 masking parapet mounts were constructed in 1898. Then on June 5, 1900 another two emplacments were submitted and authorized. The work for these four was done by 1901, and in December they were reported ready to receive armament. The four masking parapet emplacements were transferred together on February 23, 1903 for $13,400. They carried 3-inch M1898 guns and pillars (Driggs-Seabury #89/#89, #90/#90, #101/#101 and #102/#102). On May 5, 1903 a final two 3-inch pedestal guns emplacements were submitted, positioned between the previous paired mounts. They were transferred on June 7, 1904 for a construction cost of $12,000. It carried two 3-inch Model 1903 guns and pedestals (#75/#59 and #76/#60). The battery was named on General Orders No. 194 of December 27, 1904 for Captain Thomas Urmston killed in action at Chapel House, VA in 1864. The four balanced pillar guns were removed and the carriages scrapped in 1920, but the pedestal types remained in service for a considerable time. Authority was granted on April 9, 1942 to move these remaining two guns to new blocks in a site close to Battery Gunnison. The emplacement still exists at the Sandy Hook unit of the Gateway National Recreation Area. The battery is closed to the public.

- **ARROWSMITH**: A battery for three 8-inch disappearing guns relocated from old Battery Duane at Fort Wadsworth to a new emplacement at Fort Hancock. In 1905 it was realized that Fort Hancock was vulnerable to enemy craft running by the Hook and bombarding the fort from the bay to the west. A plan was formulated to build an inexpensive new position to cover these waters. By this date the viability of Battery Duane at Fort Wadsworth was questionable. The early design of the battery and poor quality of cement used made the armament's serviceability doubtful. Also, Fort Wadsworth now had a full complement of 10 and 12-inch guns providing excellent coverage of the adjacent waters. At this same time Hancock's all-round fire Battery Potter was being removed from service. The plan envisioned moving three of the Model 1888 8-inch guns and Model 1894 disappearing carriages from Duane to a new, low-cost emplacement at Camp Low, Sandy Hook. As the emplacement site was well masked from the rear, concrete protection was greatly reduced. Also, the area could be spread out—the emplacement used same-level ammunition service avoiding costs and complexity of having hoists. It could be built for just $50,000. Work was done in 1905-1907. The emplacement was reported completed and ready for transfer on July 17, 1908. It was transferred on May 25, 1909 at a cost of $63,500. Transferred and mounted in the battery were three 8-inch Model 1888M1 Watervliet guns on Model 1894 LF disappearing carriages (#14/#1, #17/#4, and #24/#5). It was named on General Orders No. 101 of June 17, 1908 for Lt. Colonel George Arrowsmith who was killed in action at Gettysburg, PA in 1863. The battery served only for a short while, it was disarmed, the armament being sent away in late 1927. The battery was eventually undermined and damaged by erosion, then partially demolished by the NPS. However, significant parts of the emplacement still exist at the Sandy Hook unit of the Gateway National Recreation Area. The battery is closed to the public.

- **KINGMAN**: A 1915 Board of Review Program battery for two 12-inch guns on long-range barbette carriages. Fort Hancock was assigned a set of four new 12-inch guns. For a period, it was thought to mount these as four separate single-gun emplacements, but eventually built were two dual emplacements, located relatively close to each other on new property to the southwest of the rest of Hancock. Preliminary submission was made on December 15, 1915, with final submission made and approved on February 3, 1917. Kingman was the northern emplacement of the pair. It followed the standard type plan, of the type with an enclosed rear passage. It also had a mechanical range indicator included with the initial design. Work was done from 1917 to 1919. It was transferred on April 2, 1921 at a cost of $297,933.04. It was named on General Orders No. 100 of August 2, 1917 for Brigadier General D. C. Kingman, former Chief of U.S. Army Engineers. It carried two 12-inch Model 1895M1A4 Watervliet guns on Model 1917 barbette carriages (#65/#4 and #53/#5). As part of the 1940 Program the battery was given overhead, casemated protection. This work was done slightly earlier than Battery Mills, from August 12, 1941 to February 1943. The battery was deleted under a 1946 evaluation, with the guns and carriages finally removed for scrapping in 1948. The emplacement still exists at the Sandy Hook unit of the Gateway National Recreation Area. The battery is closed to the public.

- **MILLS:** A 1915 Board of Review Program battery for two 12-inch guns on long-range barbette carriages. Fort Hancock was assigned a set of four new 12-inch guns. For a period, it was thought to mount these as four separate single-gun emplacements, but eventually two dual emplacements, located relatively close to each other were constructed on new property to the southwest of the rest of Hancock. Preliminary submission was made on December 15, 1915, with final submission made and approved on February 3, 1917. Mills was the southern emplacement of the pair. It followed the standard type plan, of the type with an enclosed rear passage. It also had a mechanical range indicator included with the initial design. Work was done from March 1917 to early 1919. It was transferred on April 2, 1921 for a cost of $297,933.04. Naming came on General Orders No. 100 of August 2, 1917 for Major General Albert L. Mills, former Chief of the Militia Bureau. It was armed with two 12-inch Model 1895M1A4 Watervliet guns on Model 1917 barbette carriages (#69/#9 and #46/#8). This armament was mounted in November 1919 and served throughout the battery's service life. It was given overhead casemate protection in World War II, from April 13, 1942 to February 1943. The most southern casemate was canted slightly more to the south to increase the field of fire in that direction. During the war it served as tactical battery No. 3 of the defenses. Battery Mills was disarmed and the guns and carriages scrapped in 1948. The emplacement still exists at the Sandy Hook unit of the Gateway National Recreation Area. The battery is closed to the public.

- **Tac-6:** Two 3-inch Model 1903 guns and pedestals (#75/#59 and #76/#60) were relocated from Battery Urmston to new concrete blocks under authority of April 9, 1942. Two new gun blocks were constructed in front of and near Battery Gunnison. The battery served as Tactical Battery No. 6 until 1946, when they were removed. Though somewhat damaged, the blocks still exist at the Sandy Hook unit of the Gateway National Recreation Area. The battery site is open to the public.

- **Tac-7:** A 1943 Program AMTB battery for two 90mm fixed and two 90mm mobile guns, The blocks for the fixed guns were emplaced according to authority given in a letter of November 20, 1942. Work was done from February 3, 1943 to March 19, 1943, with transfer made on March 22, 1943 for a cost of $8,124.19. It was located on the beach on the eastern shore of Sandy Hook, near Battery Gunnison. The battery served as Tactical Battery No. 7. It was disarmed after the war, but the concrete gun blocks still exist at the Sandy Hook unit of the Gateway National Recreation Area. The battery site is open to the public.

- **Tac-8:** A 1943 Program AMTB battery for two 90mm fixed and two 90mm mobile guns. The blocks for the fixed, M3 carriage 90mm were emplaced on disarmed Battery Peck pursuant to authority granted on November 20, 1942. The battery served until ordered disarmed after the war. The blocks in modified Battery Peck still exist at the Sandy Hook unit of the Gateway National Recreation Area. The battery site is open to the public.

Fort Hancock 1940 (NARA)

Sandy Hook Unit, Gateway National Recreation Area (Terry McGovern)

Garrison area of Fort Hancock, Sandy Hook Unit, Gateway National Recreation Area (Terry McGovern)

Gun batteries, Fort Hancock, Sandy Hook Unit, Gateway National Recreation Area (Terry McGovern)

Battery Mills, Fort Hancock, Sandy Hook Unit, Gateway National Recreation Area (Terry McGovern)

Navesink Military Reservation (1917-1950) is located in Highlands, New Jersey, on the headlands before you descend down to the Sandy Hook peninsula. The 224-acre military reservation is off State Highway 36 to the south before you cross the Shrewsbury River. A temporary 12-inch mortar battery site and fire control stations were located there during World War I. During the 1940 Program, a 16-inch casemated battery (#116) and 6-inch battery (#219) were built here. Following the war the site was used a Missile Master radar and command center for the Nike defenses in the New York Area. With end of the Nike program in 1974, the site was transferred to Monmouth County which has developed the area as Hartshorne Woods County Park. During the 2010s, the county has significantly improved the facilities at the park. The 16-inch casemated Battery Lewis was completely renovated during 2015-2018 with interior display panels and a 16-inch barrel is mounted in one of the battery's casemates. The park is open daily, the Battery Lewis is open to the public while guided tours are offered on the weekends.

Navesink Gun Batteries

- **Unnamed Mortar Battery:** It was planned to relocate a number of mortars to new sites around the southern entrance to New York just before America's entry into World War I. Four mortars from the battery at Fort Hancock were to go to a new site near the Navesink Highlands lighthouse to provide coverage to the southern shores of Sandy Hook. The landowner of the proposed site gave his permission for the battery on March 29, 1917. Work was done rapidly in April, the blocks were ready to receive their rings on May 19, 1917. While these were older obsolete Model 1886 cast-iron mortars on Model 1891 carriages, the limited range needed made them suitable to the temporary nature of the coverage required. One mortar and carriage were removed from each pit of Batteries McCook and Reynolds (#20/#4, #21/#25, #18/#12 and #58/#9). The emplacement was nothing more than the four blocks placed inline in a shallow ravine and an adjacent wood and metal-roofed magazine. The position was reported completed and armed on June 10, 1917. The battery was never transferred or named. The guns served only a short time, being probably removed and scrapped in 1919-1920. Partially buried, the mortar pits still exist on private land. The battery site is open to the public.

- **LEWIS:** The 1940 Program recommended that a modern dual gun 16-inch casemate battery be erected at Fort Hancock. The tentative location was south of Battery Gunnison along the beach line. This 16-inch casemated battery was given the relatively high priority of #9 in the September 1940 schedule and given Battery Construction No. 114. The project site was soon moved to the Navesink Highlands reservation early 1941. No work was ever undertaken at Fort Hancock. Work was physically begun on April 2, 1941. That work was reported completed on June 26, 1943. It was of standard late-type 100-series design, with a separate PSR room and eight dispersed base-end station. Transfer was made on February 16, 1944 for a cost of $1,552,492.63. It was named on General Orders No. 49 on October 1, 1942 for Colonel Isaac Newton Lewis of the Coast Artillery Corps. It was armed with two 16-inch guns MkIIM1 on Model M4 barbette carriages (#70/#29 and #46/#44). The carriages had been received in February 1943, the shields in late 1944, and the armament reported completed in March 1945. The armament was dismounted after the war, in 1948. The emplacement still exists at the Hartshorne Woods Park. The battery wass extensively restored in the 2010s and a 16-inch navy MkVII barrel was installed in one of the emplacments. The battery site is open to the public.

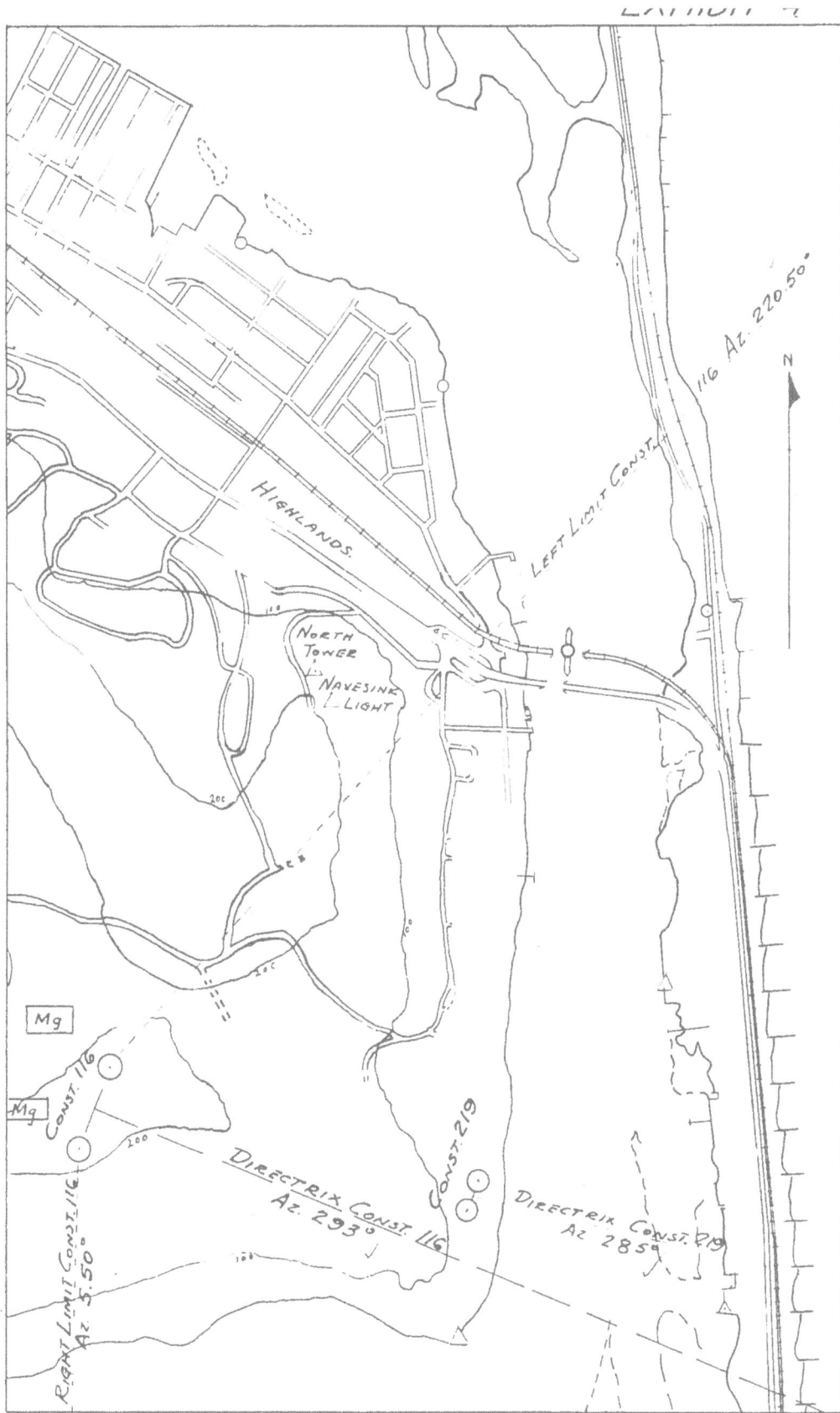

SKETCH OF SITES FOR CONST. 219 AND CONST. 116·—
OVERLAY AIR COMPILATION T·5100

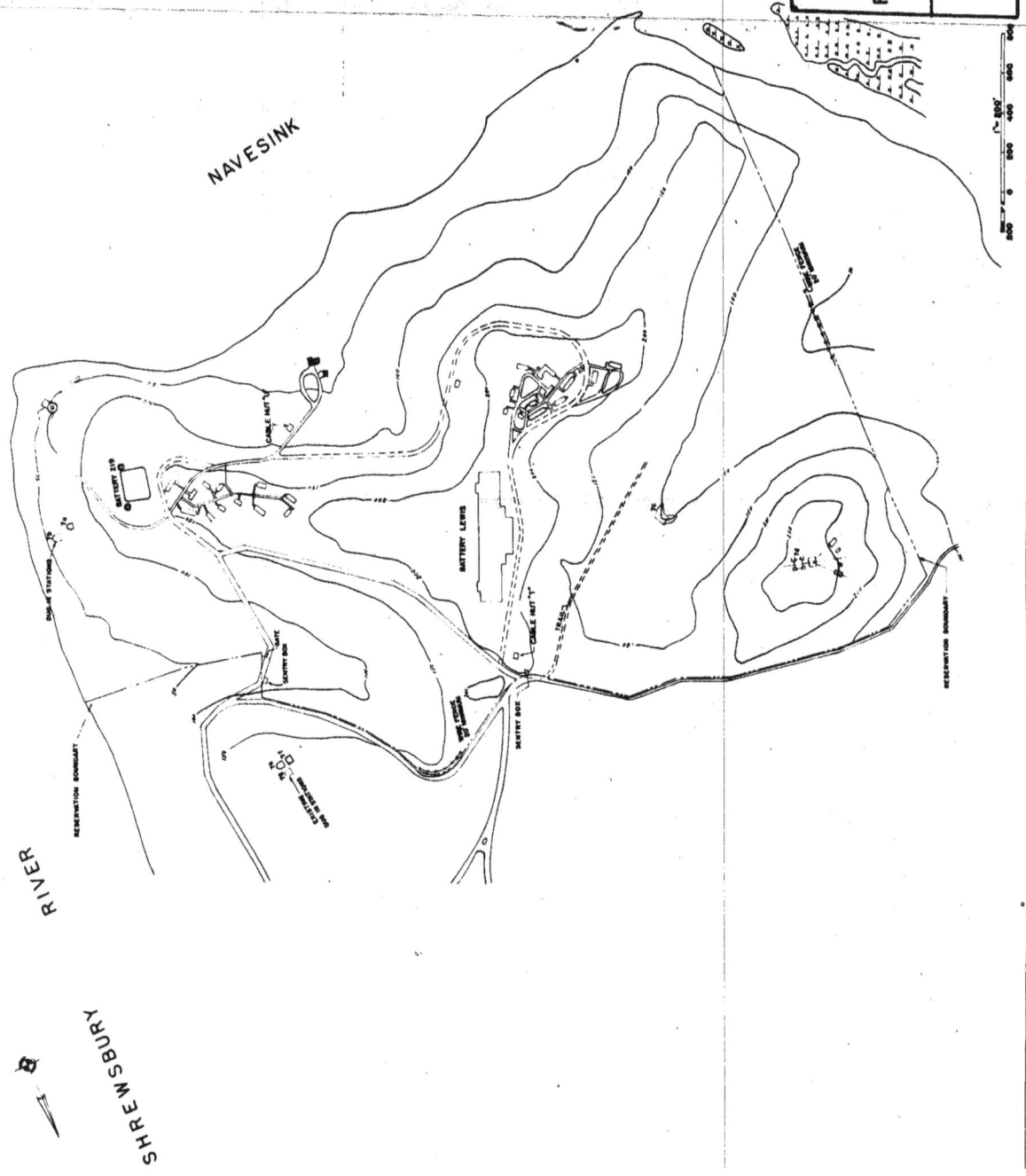

HARBOR DEFENSES OF
NEW YORK

FIRE CONTROL INSTALLATION
AT LOCATION 7

AUGUST, 1943   EX. NO. 8-B-7

PREPARED BY
N Y-P SECTOR

Hartshorne Woods County Park (Navesink Military Reservation) (Terry McGovern)

Battery Lewis, Hartshorne Woods County Park (Terry McGovern)

- **Battery #219**: The 1940 Program also recommended a modern 6-inch battery be located at Fort Hancock. The proposed location was at the site of Battery Richardson, involving the destruction of that emplacement. No work was ever undertaken, the project instead being relocated to the Navesink Highlands reservation. At the Highlands it was positioned east of Battery Lewis along the cliff line. Of high priority (ranked #3 in the U.S. on the schedule of September 11, 1941), work was done from July 8, 1942 to April 24, 1943. It was of standard 6-inch "200-series" design. Transfer was made on February 16, 1944 at a cost of $198,858.67. The battery was never named. It carried two 6-inch guns Model 1903A2 on Model M1 barbette carriages (#53/#17 and #54/#18). The battery was retained for a short period after the war but was disarmed probably in 1948. The emplacement still exists at the Hartshorne Woods Park. The battery site is open to the public.

Gunnery practice at Battery Gunnison, Fort Hancock, Sandy Hook Unit
Gateway National Recreation Area (Mark Berhow)

Dorrance, William H. "History of Fort Slocum (Davids Island Military Reservation)." *CDSG Journal* Vol. 10, No. 4, Nov. 1996, p. 49

Gaines, William C. "Fort Schuyler and the Defenses of the Eastern Approaches to New York Harbor: a Historic Resource Study." *CDSG Journal* Vol. 10, No. 4, Nov. 1996, p. 56

Gaines, William C. "Fort Totten and the Coastal Defenses of Eastern New York." *CDSG Journal* Vol. 11, No. 1, Feb. 1997, p. 41

Gaines, William C. "The Inner Harbor Fortifications of New York." *Coast Defense Journal* Vol. 20, No. 4, Nov. 2006, p. 4

Gaines, William C. "Defending the Narrows: The Harbor Defenses of Southern New York - *Coast Defense Journal* Vol. 22, No. 4, Nov. 2008, p. 4; and Vol. 23, No. 1, Feb. 2009, p. 4

Bearss, Edwin C. *Historic Resource Study, Fort Hancock 1895-1948,* Gateway National Recreation Area, New York/New Jersey. Denver Service Center, Historic Preservation Division, US Dept. of the Interior, NPS, Denver, CO, 1981.

Black, Fredrick R. *Historic Resource Study, A History of Fort Wadsworth, New York Harbor.* Division of Cultural Resources, North Atlantic Regional Office, US Dept. of the Interior, NPS, Boston, MA, 1983.

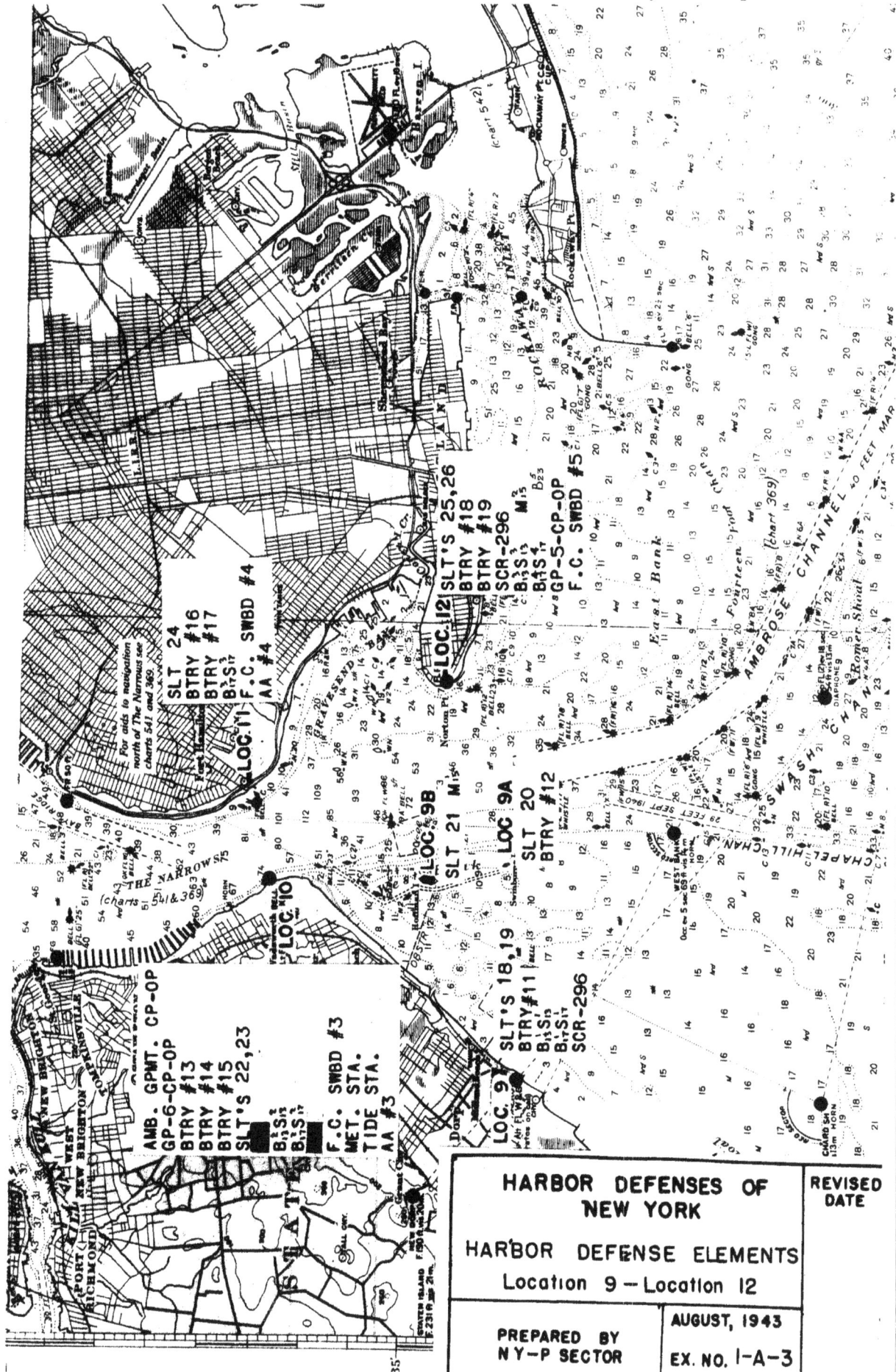

SLT 24
BTRY #16
BTRY #17
B.'S.'3
F.C. SWBD #4
AA #4
LOC. II

For aids to navigation
north of The Narrows see
charts 541 and 369

SLT'S 25,26
BTRY #18
BTRY #19
SCR-296
B.'S.'3
B.'S.'4
GP-5-CP-OP
F.C. SWBD #5
LOC. 12

SLT 21
LOC. 9B
SLT 20
LOC. 9A
BTRY #12

THE NARROWS
(charts 541 & 369)

LOC. 10

AMB. GPMT. CP-OP
GP-6-CP-OP
BTRY #13
BTRY #14
BTRY #15
SLT'S 22,23
B.'S.'3
B.'S.'7
F.C. SWBD #3
MET. STA.
TIDE STA.
AA #3

SLT'S 18,19
BTRY #11
B.'S.'3
B.'S.'7
SCR-296
LOC. 9

HARBOR DEFENSES OF
NEW YORK

HARBOR DEFENSE ELEMENTS
Location 9 – Location 12

PREPARED BY
N Y–P SECTOR

REVISED
DATE

AUGUST, 1943

EX. NO. I-A-3

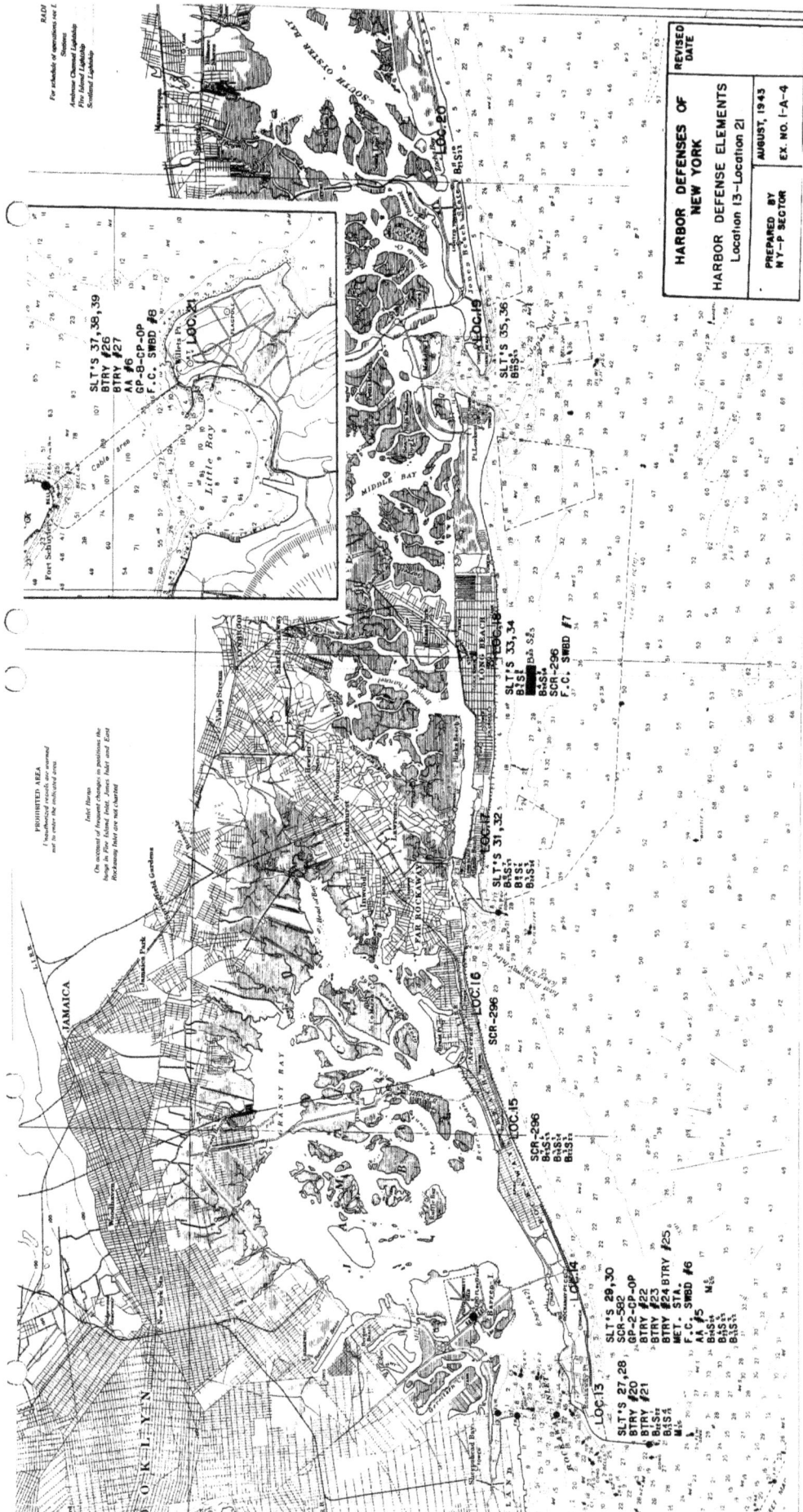

HARBOR DEFENSES OF NEW YORK

HARBOR DEFENSE ELEMENTS
Location 13–Location 21

PREPARED BY
NY–P SECTOR

AUGUST, 1945

EX. NO. I-A-4

New York World War II-era Site Locations. Stations housed in a single structure are connected by dashes (-)

location	Loc#	Purpose
Seagirt/Camp Edison/ Manasquan	1	BS1/Lewis
Shark River	2	BS2/Lewsi-BS1/219
Elberon	3	BS3/Lewis-BS1/Harris-BS1/Mills-BS1/Kingman-BS2/219, SCR296-Mills
North Long Branch	4	BS4/Lewsi-BS2/Harris-BS2/Mills-BS2/Kingman-BS3/219
Monmouth Bearch	5	CR296-Kingman
Seabright	6	SL
Navesink	7	Batt. Tact. #1 BCN 219, Batt. Tact. #2 Lewis, BS1/Peck-BS3/Harris, BS5/Lewis-BS4/219, BC/Lewis, BC/219, PSR/Lewis, BS5/219-BS6/Lewis, BS3/Kingman-BS4/Harris, BS3/Mills-G1, SCR296-Lewis, SCR582
Fort Hancock	8	Batt. Tact. #3 Mills, Batt. Tact. #4 Kingman, Batt. Tact. #5 Peck, Batt. Tact. #6 Urmiston, Batt. Tact. #7 AMTB, Batt. Tact. #8 AMTB, BC Kingman, BC Mills, BS4/Kingman, BS4/Mills, BS7/Lewis, BS2/Peck, SCR296-Peck, BS5/Kingman, BS5/Mills, SCR682, BC Urmiston, HECP-HDCP, SBR
Fort Hancock	8A	Batt. Tact. #9 Morris, Batt. Tact. #10 mine, BS6/219, BS3/Peck, M1, HDOP 1, Bn4, BC/Morris, Bn3, M2
Miller Field	9	Batt. Tact. #11 AMTB, BS1/218-BS1/Livingston, SCR296-Livingston
Swinburne Island	9A	Batt. Tact. #12 AMTB
Hoffman Island	9B	Searchlight
Fort Wadsworth	10	Batt. Tact. #13 BCN 218, Batt. Tact. #14 Turnbull, Batt. Tact. #15 mine, BS2/Livingston, BS2/218, BC/218, BC/Turnbull, Mine, Bn4
Fort Hamilton	11	Batt. Tact. #16 Griffith, Batt. Tact. #17 Livingston, BC/Griffith, BC/Livingston, BS3/Livingston
Norton Point	12	Batt. Tact. #18 Caitlin, Batt. Tact. #19 AMTB, BS5/Harris, Bn5, BS3/218, BS4/Livingston, (M2), BC/CatlinSCR296-218
Rockaway Point	13	Batt. Tact. #21 AMTB, BS1/Kessler, BS4/218,
Fort Tilden	14	Batt. Tact. #22 Kessler, Batt. Tact. #23 Harris, Batt. Tact. #24 BCN 220, Batt. Tact. #25 mine, BC/Harris, BC/Kessler, BC/220, BS1/220, BS2/Kessler, Bn2, BS6/Harris
Seaside	15	BS7/Harris, BS2/220, BS3/Kessler, SCR296-Kessler
Arverne	16	SCR296-220
Atlantic Beach	17	BS8/Harris, BS8/Lewis, BS3/220
Long Beach	18	BS9/Harris, BS9/Lewis, BS4/220, SCR296-Harris
Short Bearch	19	BS10/Harris
Zachs Bay	20	BS11/Harris
Fort Totten	21	Batt. Tact. #26 Baker, Batt. Tact. #27 Burns, Bn8, BC/Baker

Battery Lewis Hartshorne Woods County Park, New Jersey (Mark Berhow)

The restored plotting room at Battery Gunnison, Sandy Hook Unit Gateway NRA (Mark Berhow)

www.ingramcontent.com/pod-product-compliance
Lightning Source LLC
Chambersburg PA
CBHW040259100426
42811CB00011B/1311